Power Microelectronics

Device and Process Technologies

Second Edition

Power Microelectronics

Device and Process Technologies

Second Edition

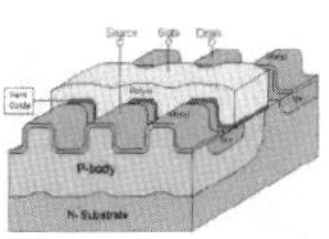
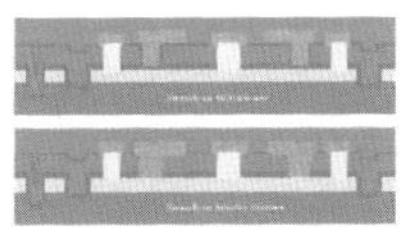
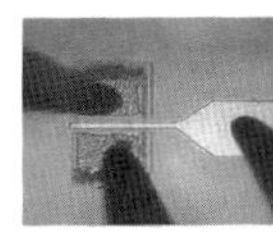
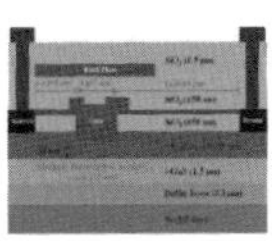
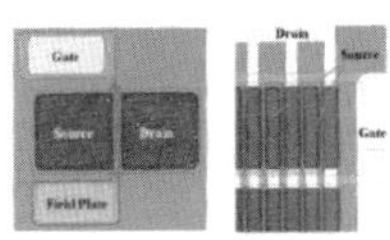
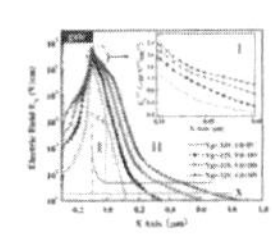

Yung C Liang
National University of Singapore, Singapore

Ganesh S Samudra
National University of Singapore, Singapore

Chih-Fang Huang
National Tsing-Hua University, Taiwan

World Scientific

NEW JERSEY • LONDON • SINGAPORE • BEIJING • SHANGHAI • HONG KONG • TAIPEI • CHENNAI • TOKYO

Published by

World Scientific Publishing Co. Pte. Ltd.
5 Toh Tuck Link, Singapore 596224
USA office: 27 Warren Street, Suite 401-402, Hackensack, NJ 07601
UK office: 57 Shelton Street, Covent Garden, London WC2H 9HE

Library of Congress Cataloging-in-Publication Data
Names: Liang, Yung C., author. | Samudra, Ganesh S., author. | Huang, Chih-Fang, author.
Title: Power microelectronics : device and process technologies / by Yung C Liang (NUS, Singapore), Ganesh S Samudra (NUS, Singapore), Chih-Fang Huang (National Tsing-Hua University, Taiwan).
Description: 2nd edition. | [Hackensack?] New Jersey : World Scientific, [2016] | Includes bibliographical references and index.
Identifiers: LCCN 2016046914 | ISBN 9789813200241 (hc : alk. paper)
Subjects: LCSH: Power semiconductors. | Power electronics. | Microelectronics.
Classification: LCC TK7871.85 .L5425 2016 | DDC 621.3815/2--dc23
LC record available at https://lccn.loc.gov/2016046914

British Library Cataloguing-in-Publication Data
A catalogue record for this book is available from the British Library.

Desk Editor: Herbert Moses

Typeset by Stallion Press
Email: enquiries@stallionpress.com

Printed in Singapore

Acknowledgement

We would also like to thank many people who have helped us with this book, especially Professor Simon Ang at University of Arkansas for providing his lecture notes on bipolar power diodes as a valuable reference for this book; also Mr. Honglin Pan, Mr. Junyang Luo, Dr. Xiaolu Yuan, Mr. Kian Paau Gan, Mr. Yuanzheng Zhu, Mr. Yong Liu, Ms. Changhong Ren, Mr. Andrew JD Lim, Dr. Xin Yang, Ms. Hanmei Zhong, Dr. Yu Chen, Ms. Kavitha Buddharaju, Dr. Huolin Huang, Mr. Jheng-Yi Jiang, Ms. Chia-Hui Cheng, Mr. Tsung-Ying Tsai, Mr. Ruize Sun and Mr. Yun-Hsiang Wang for their technical contribution in design, development, analysis, fabrication, test and measurement of various power semiconductor devices and integrated circuits.

Yung C. Liang
Ganesh S. Samudra
Chih-Fang Huang

About the Authors

Yung C. Liang received the MSc degree from National Tsing Hua University, Taiwan in 1985 and the PhD degree from the University of Sydney, Australia in 1991, both in electrical engineering. He is currently an Associate Professor with the Department of Electrical and Computer Engineering, National University of Singapore, Singapore. He was a visiting academic at the Japan Advanced Institute of Science and Technology, Japan, in 1997 and at the University of California, Berkeley in 2001–2002. His research interests include power microelectronics, silicon sensors and actuators, microfuel cells and ferroelectric devices. He has authored more than 190 journal and conference publications in these areas.

Dr. Liang was the key founding member of the IEEE International Conference on Power Electronics and Drive Systems (IEEE PEDS) in 1995 and the recipient of the IEEE Third Millennium Medal in 2000. He also received the Transactions Paper Prize from the IEEE Power Electronics Society for his work on high-frequency smart power synchronous rectifiers. He currently serves as an Associate Editor for the IEEE Transactions on Power Electronics.

Chih-Fang Huang is currently an Associate Professor in the Department of Electrical and Engineering at National Tsing Hua University (NTHU), Taiwan. He received his PhD from Purdue University, West Lafayette, Indiana, USA in 2003. In 2004, he joined Advanced Power Technology Colorado, Boulder, Colorado, where he engaged in the development of SiC power and RF devices. Since 2005, he has been with NTHU. His current research interests include wide bandgap semiconductors, RF and power devices, and driver ICs for power devices. He is author/co-author of more than 100 journal/conference papers in these fields and is inventor of eight U.S. patents. He is a technical committee member of several international conferences, including *International Conference on Solid State Devices and Materials* (2013–2017), *International Symposium on Power Semiconductor Devices and ICs* (2015–2017), and *International Conference on Silicon Carbide and Related Materials* (2017). He is also an IEEE Senior Member.

Ganesh S. Samudra is currently an Associate Professor in Department of Electrical and Computer Engineering in National University of Singapore. He received his MSc from Indian Institute of Technology, Mumbai in 1976 and MSEE and PhD from Purdue University, USA, in 1985. He was visiting professor at MIT in 2001. His research is in simulation and novel devices. He has published ~300 technical papers in journals and conferences. His recent research papers jointly with Profs. Liang and Huang are on methodologies for the fabrication of normally-off GaN High Electron Mobility Transistors (HEMTs) capable of operating at high temperatures.

He was co-recipient of the IEEE Electron Devices Society's 2008 Paul Rappaport Award and simulation and modeling subcommittee chairman of International Electron Device Meeting (IEDM) in 2010.

Contents

1

Introduction

The global power semiconductor market size, according to the published forecasting data by manufacturers, is likely to attain 30 billion US dollars in the year 2020. Within this segment, for those using next-generation wide bandgap materials such as SiC and GaN, the global market may reach 3 billion US dollars in 2020. The slow growth of the wide bandgap devices is mainly due to issues of cost and reliability. However, the situation is expected to improve.

From the application point of view, in analogy to a human body, the microprocessors or digital signal processors perform "thinking" and calculations to make desirable decisions in electronic systems somehow similar to what the human brain does. Once a decision is made, it frequently needs to be put in work by the peripheral power train, just like human arms and legs. The power train provides high voltage, high current static energy conversion or dynamic motion control involving mechanical actuation to interact with the real world. Power semiconductor devices are the building elements for the power train. They are used to handle the high voltage and high current loads at high temperature.

The boundary to differentiate a power semiconductor device from the semiconductor device group sometimes can be unclear as it can be made from the same CMOS compatible process as well as be seamlessly integrated with other low power semiconductor components. In general, we can look at the main functionality of the device and its power rating in order to identify it to be a power semiconductor device. As indicated in Fig. 1.1, the application of such a device ranges from the linear regulator of a few hundred milliwatts to traction and HVDC of a few hundred megawatts. Clearly, the range of applications is very wide, and these devices vary from low-power CMOS microchips at one end to Megawatt wafer-level large vertical configuration at the other end. Three basic specifications are looked at to determine whether a power device

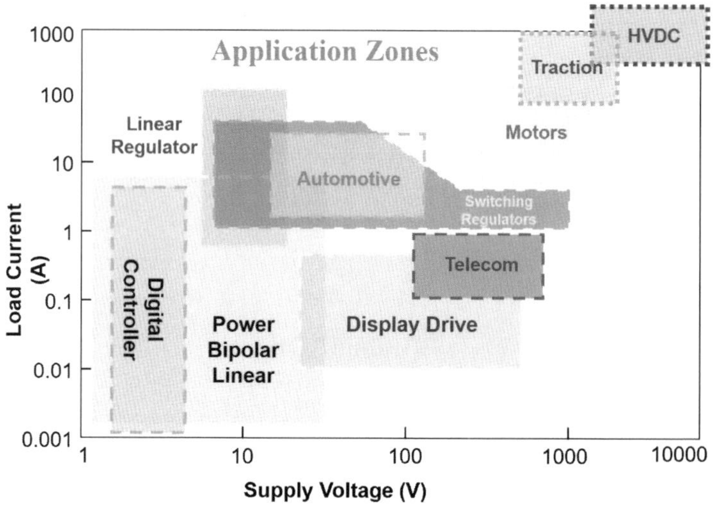

Fig. 1.1. The various application zones of power semiconductor devices in terms of system voltage and load current.

is suitable for a particular application, namely, the voltage rating, the current rating, and the switching frequency (or the switching speed). Or, if the device is to be used for linear application, i.e. non-switching, instead of the switching speed, the small signal gain, the bandwidth, and the power rating will also be looked at.

The desirable features of a power semiconductor device for switching application are to have a high voltage sustaining ability with very low leakage current; a high current rating with very low on-state resistance; a fast switching speed with minimum losses; high noise immunity and to be latch-up free. Similarly, for the linear application, it is to have a high small signal gain with low harmonic distortion; a wide bandwidth and good power rating. In order to achieve all of these, research in several areas, namely, the material, the fabrication processes, the device structures, integration, and packaging were looked into. For the material, although silicon is still the mainstream material for power semiconductor devices, wider bandgap materials such as silicon carbine, gallium nitride, carbon, or strained silicon germanium are getting widely accepted to make devices with better performance. However, the limitation sits on the availability of good quality source wafers and the high-yield fabrication processes. For the fabrication processes, special recipes were constantly developed to meet the requirement of advanced structures. Modern power semiconductor devices can now be categorized into several mature types, namely, the diode, the BJT, the thyristor (and GTO), the MOSFET, the IGBT, and the static induction transistor. These devices, except for MOSFET, are bipolar in nature and suitable for high current conduction. MOSFET, on the other hand, has good switching performance and simple gate drive requirement for medium-power products. Isolation and integration technology for devices of lateral structure are also an important focus for power integrated circuits. Overall, the research and development work is closely linked to fabrication processes which can be rather costly. Industrial companies play an important role in the development of new devices and processes.

Among all power semiconductor devices, MOSFETs and IGBTs are playing major roles in applications and will continue to be the devices in high demand for medium–high voltage applications. The silicon superjunction technology for MOSFET has increased the device voltage rating and reduced its specific on-state resistance, thus making the device to break the unipolar silicon limit. Recently, wide bandgap devices became a popular trend in power semiconductor devices both in research and development, and in industrial commercialization. These involve the growth of suitable substrates, epitaxy buffer layers, new process recipes, and new dielectric materials and device structures. The performance merit of a power device can be rated by the following relationship:

$$\text{Performance Merit} = \left(\frac{V_{\text{Breakdown}}^2}{R_{\text{ON,SP}}} \right), \tag{1.1}$$

where $V_{\text{Breakdown}}$ is the device breakdown voltage in volts and $R_{\text{ON,SP}}$ is the on-state specific resistance in $\Omega \cdot \text{cm}^2$. Typically, for silicon power MOSFET, the performance merit is around 30–50 × 10^6 $\text{V}^2/(\Omega \cdot \text{cm}^2)$. For gallium nitride MISHEMT, it is around 300–500 × 10^6 $\text{V}^2/(\Omega \cdot \text{cm}^2)$. And, for SiC MOSFET, it can be as high as 800–1000 × 10^6 $\text{V}^2/(\Omega \cdot \text{cm}^2)$. Wide bandgap devices are expected to have a sizeable market share.

SiC has always been one of the material candidates for semiconductor devices since the invention of diodes and transistors. However, with the rapid advancement of Si technology, it was ignored most of the time, until the breakthroughs in crystal growth and epitaxy of SiC in 1980s. The availability of electronic grade material greatly facilitates the development of SiC devices for power electronic applications. Today, six-inch SiC wafers with more than 50 μm and less than 1×10^{15} cm^{-3} doped epi-layers are commercially available, serving as the fundamental building block of modern SiC devices. After their introduction into the power device market by Infineon and Cree in 2001, SiC Schottky diodes have demonstrated their capability to significantly improve the system efficiency and power density, and have been widely adopted in some applications. It is fair to say that they have established their reputation and the market share will only grow in the foreseeable future. On the other hand, SiC switches are still facing challenges from the prevailing Si devices, mainly because of the cost issues and unproven reliability. Consensus is that SiC switches will most likely penetrate the market in the high-end and high-voltage sector first where cost is less of an issue. With the help of recent advancements in defect and carrier lifetime control, SiC bipolar devices have started to show excellent performance with much improved reliability for possible ultra-high-voltage applications. Despite the bright future of SiC devices, material costs and peripherals such as packaging and circuitry are the current bottlenecks which will eventually determine the speed and extent of SiC device's success. Nevertheless, the presence of SiC devices pushes the envelope of power electronics limits to higher frequency and power density.

Unlike SiC power devices, the AlGaN/GaN heterojunction induces two-dimensional (2D) electronic polarization charges which form a natural conduction channel. Such a conduction channel makes the AlGaN/GaN device a default normally-on device. The immediate application will be for the RF power amplifier where the operating point is at the normally-on state for large signal swing. Recently, normally-off structures were developed for power electronic applications. To enable normally-off operations, the gate structure is modified from the simple Schottky gate to a p-GaN gate or full trenched gate or fluorinated gate. The trade-off in making normally-off gate is on getting a high positive threshold voltage (above 3.5 V) and a good high temperature threshold stability. Besides, the current collapse phenomenon is another important concern in power electronic application. Multiple field plates and better surface passivation are used to reduce the current collapse. Another option is the usage of two-dimensional hole gas (2DHG) on the top layer to reduce the surface field (RESURF) between the gate and drain in lateral direction, allowing higher voltages to be applied.

The advent of SiC and GaN devices has added newer processing techniques in power device technology. SiC or p-type GaN dopant activation requires higher temperature processing compared to that in silicon technology. Atomic layer deposition (ALD) method and plasma treatment for the introduction of desired impurities have been extensively used in CMOS technology but they are rarely needed in power devices as the junctions are typically deep and dielectric thickness is large compared to what CMOS needs. However, the use of thin dielectric to reduce the gate leakage and low energy F plasma treatment to introduce F with shallow depth to raise threshold voltage has been adapted for GaN devices. Hence, these advanced processing techniques have become part of GaN-based power MISHEMT device processing. Case studies that use these methods are included in this edition.

This book aims to introduce power semiconductor devices and fabrication technology to readers who are interested or involved in this important field of technology. Chapter 2 provides fundamentals for junction electrostatics and breakdown analyses. Chapter 3 focuses on p–n junction rectification aspects. Chapters 4 and 5 are on devices of MOSFET and IGBT. Chapter 6 discusses the new superjunction structure which breaks the unipolar silicon limit in performance merit. Chapter 7 looks into the wide bandgap SiC power devices and Chapter 8 is on the AlGaN/GaN heterojunction polarization and HEMT devices. Chapters 9–11 look into the fabrication process technology and case studies including both the superjunction and wide bandgap devices. Overall, the book is a good reference material for both detailed theory and technology work in the field of power semiconductor devices.

2

Carrier Physics and Junction Electrostatics

2.1. Introduction

To understand the operating characteristics of power semiconductor devices well it requires a good background knowledge in semiconductor physics, in particular on carrier transport and junction electrostatics. For this chapter, some important concepts on carrier transport physics and junction electrostatics with breakdown mechanisms of silicon (Si) semiconductor are reviewed. Materials of gallium arsenide (GaAs) and silicon carbide (SiC) are also briefly mentioned. The aim of this chapter is not to treat the semiconductor physics rigorously in fundamental detail. Rather, it is to highlight those properties that are important and needed in understanding the operational characteristics of power semiconductor devices.

2.2. Crystal Structure and Energy Bands

Silicon belongs to the group IV element and as such it has four valence electrons in its outermost shell. Silicon crystallizes in a diamond structure in which four atoms are located at the corners of opposite diagonals of the cube and they surround one atom in the center of the cube, as shown in Fig. 2.1. The central atom has four covalent shared-pair electron bonds with the surrounding four atoms. The unit cell lattice constant a is 5.43 Å. It has eight corner atoms, six face atoms, and four central atoms for a total of eight atoms in the unit cell. Since there are eight silicon atoms per unit cell and the volume of the unit cell is a^3, it follows that there are 5×10^{22} atoms/cm^3 in the silicon lattice. It is this structure that gives rise to the many unique properties of silicon. For gallium arsenide, another power semiconductor material, it is a compound semiconductor as it consists of gallium and arsenic atoms crystallizing in a structure similar to the diamond structure called the zinc-blende structure as shown in Fig. 2.2. In the zinc-blende structure, the central atoms are occupied

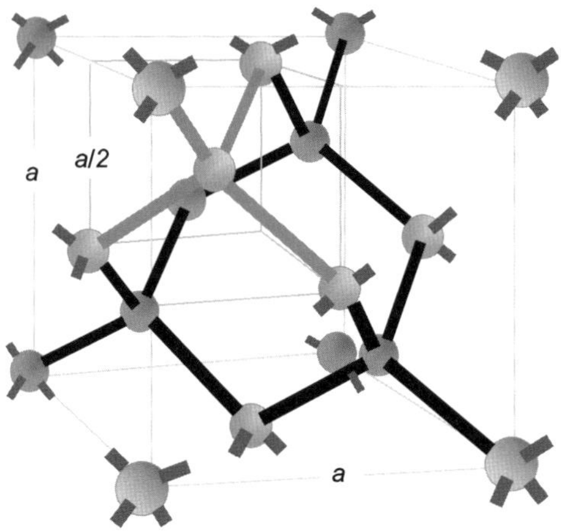

Fig. 2.1. Silicon atomic structure.

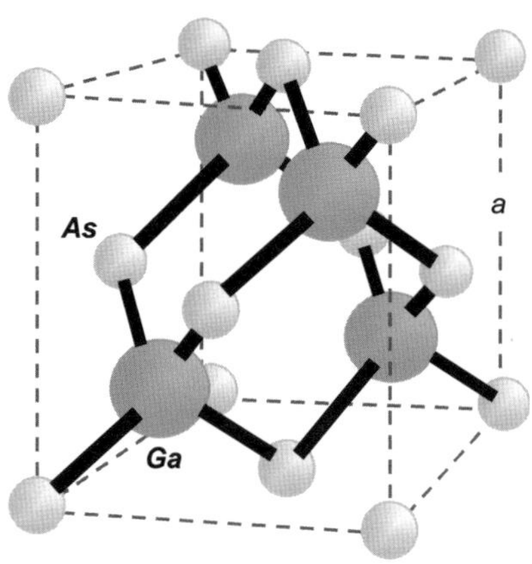

Fig. 2.2. Atomic structure of GaAs.

by the gallium atoms. Thus, the gallium arsenide unit cell has four gallium atoms and four arsenic atoms. Some physical properties of semiconductor materials can be found in Table 2.1.

Electrons in the outermost shell are shared by many atoms, as such, the energy of each can be grouped into various energy bands. In silicon, the valence electrons group together to occupy a band of energy levels, called the valence band. The next higher band of allowed energy levels, called the conduction

Table 2.1. Some physical properties of semiconductor materials.

	Ge	Si	SiO_2	GaAs	$\alpha\beta$-SiC	GaN	AlN	Diamond
Lattice constant (Å)	5.65	5.43	—	5.65	3.09(a) 15.12(c)	3.19(a) 5.19(c)	3.11(a) 4.98(c)	3.57
Density (g/cm^3)	5.32	2.33	2.27	5.65	3.22	—	—	3.52
Energy bandgap (eV)	0.67	1.12	~9.0	1.43	2.4 to 3.26	3.49	6.2	5.45
Dielectric coefficient ($\varepsilon/\varepsilon_o$)	16.3	11.7	3.9	12.8	9.6 to 10	9.0	8.7	5.5
Melting point (°C)	937	1412	1700	1238	2540	—	~2200	~4000
Saturated electron velocity (10^7 cm/s)	0.5	1.0	—	2.0	2.0	2.5	1.8	2.7
Electron mobility ($cm^2/V \cdot s$) @300 K	3900	1350	20	6000	~800	~1500	1100	1900
Hole mobility ($cm^2/V \cdot s$) @300 K	1900	450	~0	330	~120	~300	—	1600
Breakdown field (10^5 V/cm)	0.8	3.0	60	3.5	24–30	~33	117	56
Thermal conductivity ($W/cm \cdot K$)	0.6	1.5	0.014	0.81	4.5	1.5	2.0	20
Direct or indirect	I	I	—	D	I	D	D	I

Note: $\alpha\beta$-SiC denotes either 4H-SiC, 6H-SiC (α), or 3C-SiC(β).

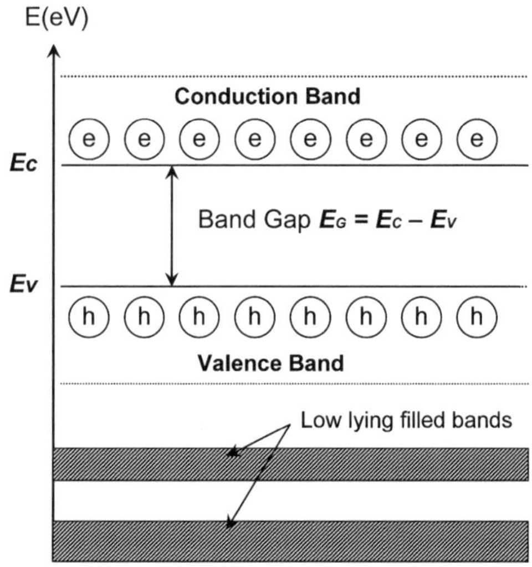

Fig. 2.3. Semiconductor energy band structure.

band, is separated from the valence band by a forbidden energy gap, E_G, as shown in Fig. 2.3. The conduction band is partially occupied by the free electrons. The bandgaps for silicon and gallium arsenide are 1.12 eV (electron-volt) and 1.43 eV, respectively around room temperature ($T = 300\,K$). At room temperature, some electrons in the valence band may acquire enough thermal energy (above 26 meV) to traverse from the valence band to the conduction band. For each successful transition of a valence electron to the conduction band, a hole which is an unoccupied or empty energy level is left behind in the valence band. This process is known as the electron–hole pair generation. Under thermal equilibrium state, the number of electrons in the conduction band and holes in the valence band is equal. Thermal equilibrium is defined as the steady-state condition at a given temperature without any external excitation. These electrons and holes are free to move in the crystal lattice, and they are generally known as free electrons and holes.

One of the models described by the quantum mechanics for the energy bands in solids is called the Kronig–Penney model and an E–k (energy–momentum, where carrier momentum $\propto hk$) dependence diagram can be drawn as in Fig. 2.4 for the energy bands. The interesting concept to be described here is the effective mass of electron. The effective mass is determined by the radius of curvature of the E–k curve at a given energy level, i.e. the effective mass varies with k value. When the E–k curve is concave as the shape of conduction band, the effective electron mass is positive, whereas, when it is convex as

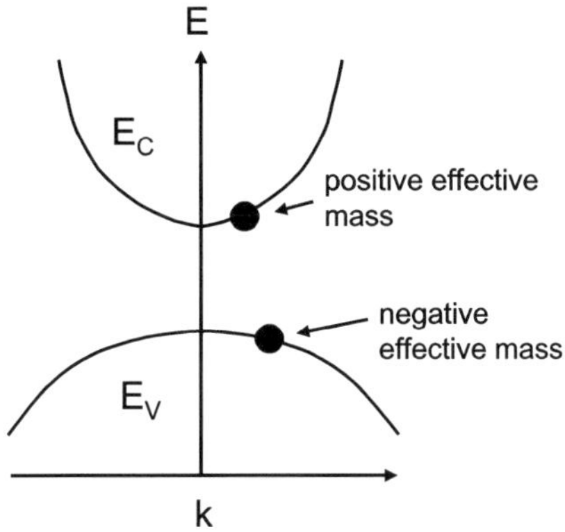

Fig. 2.4. *E–k* diagram with different surface curvatures for conduction band and valence band.

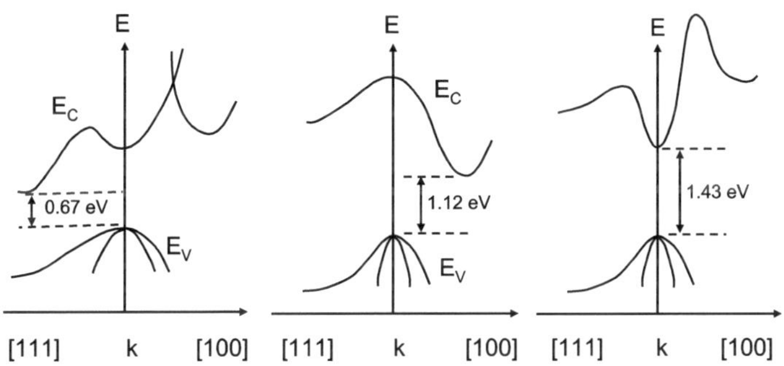

Fig. 2.5. The energy band diagrams in momentum space for germanium, silicon, and gallium arsenide at 300 K (from left to right).

the shape of valence band, the effective electron mass is negative. This means that an electron in the valence band will be accelerated by the field as if it is a positively charged particle and mass. This forms the basic concept of hole carrier.

Depending on the structure of the energy band, a semiconductor can be classified as either a direct-gap or an indirect-gap semiconductor, as shown in the *E–k* diagram of Fig. 2.5 (Bar-Ler, 1984). In a direct-gap semiconductor, the minimum energy of the conduction band coincides with the maximum energy of the valence band in the momentum space, whereas, in an indirect-gap semiconductor, the minimum energy of the conduction band does not align at $k = 0$ but occurs near the zone edge. With this in mind, germanium and silicon materials are indirect-gap semiconductors while gallium arsenide is a direct-gap semiconductor. As such, they have distinctly different carrier recombination processes. In silicon, electrons make transitions to the valence band by a change of both the momentum and energy simultaneously. This means that other quantum particles, such as phonons which are the vibrational modes of the lattice, must participate to remove or contribute the necessary excess momentum. A phonon has therefore a relatively high momentum but low energy. In GaAs, since the minimum of the conduction band aligns with the maximum of the valence band, electrons in the conduction band recombine with holes in the valence band by making direct transition from the conduction band to the valence band without a change in their momentum. The energy given up by the electron will be emitted as a photon, the quantum of light. Thus, gallium arsenide is the material of choice in photonic devices.

Another material of choice for power semiconductor devices is the silicon carbide, SiC for its larger bandgap and higher thermal conductivity. A unique crystal property of the SiC is on the polytypism. If we designate a SiC atom pair in an A-plane in close packing as Aa, in the B-plane as Bb, and in the C-plane as Cc, then we can generate a series of SiC structures by the variation of stacking along the principal crystal axis (Choyke and Pensl, 1997). For example, by having AaBbAaCcAaBbAaCc ... structure, we generate the 4H-SiC polytype. Or, for AaBbCcAaCcBb ... structure, we can generate the 6H-SiC polytype. They are hexagonally symmetric. Both 4H-SiC and 6H-SiC polytypes are available in bulk wafer form and are useful for power semiconductor device applications (Casady *et al.*, 1998; Ramungul *et al.*, 1996). The 4H-SiC has two equivalent sites with 8 atoms per unit cell while the 6H-SiC has three equivalent sites with 12 atoms per unit cell. Therefore, 4H-SiC has the possibility of two donors or two acceptors for a particular substitutional impurity, while 6H-SiC may have three donors or three acceptors. Both structures have a hexagonal crystal structure. The 6H-SiC has a bandgap energy of 3.03 eV and lattice parameters of 3.081 Å and 15.117 Å while the 4H-SiC has a bandgap of 3.26 eV and lattice parameters of 3.073 Å and 10.053 Å. Besides hexagonal (H) polytypes, there are also cubic (C) and rhombohedral (R) polytypes. More than 200 polytypes of SiC have been discovered; among them the common types are 3C, 2H, 4H, 6H, 8H, 9R, 10H, 14H, 15R, 20H, 21H, and 24R. 3C-SiC possesses the smallest bandgap energy of about 2.4 eV and it has the highest electron mobility of 800 $cm^2 \cdot V^{-1} \cdot s^{-1}$ among all polytypes.

Recently, SiC power MOSFETs have achieved a blocking voltage of 6.1 kV and performance figure-of-merit 70 times higher than the ideal silicon counterpart. However, still the current problem faced in the SiC power MOSFET device development is the low carrier mobility at the oxide–SiC interface. This

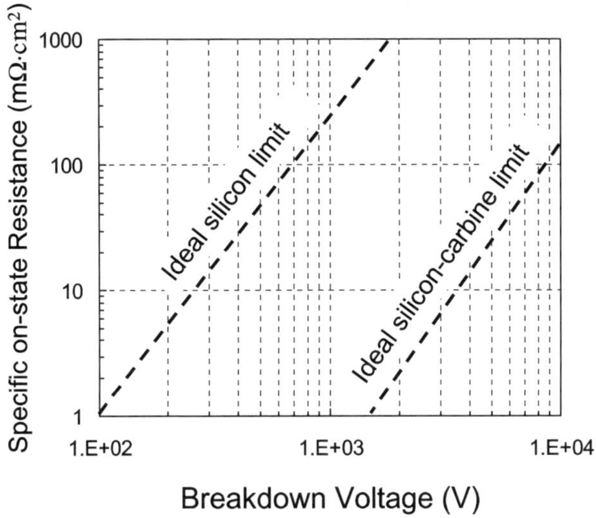

Fig. 2.6. Lines of ideal silicon and silicon carbine limits for majority carrier devices.

requires a better understanding of the interface material properties with an incorporation of proper processes for post-oxidation anneal (Cooper *et al.*, 2002). Another important focus on SiC device development is that the SiC bulk breakdown field of 5 MV/cm is close to that of the SiO_2 breakdown field. Special protective layers need to be placed to avoid gate oxide breakdown, especially at the corner of a U-MOS gate oxide. Figure 2.6 shows the ideal silicon limit and SiC limit for majority carrier devices, such as the Schottky diodes and MOSFETs. As it can be seen, SiC material has a much better value of merit (the breakdown voltage divided by the on-state resistance) than that of the silicon material.

The (aluminum) gallium nitride power device with the best properties is at the infancy development stage. The progress is slow due to the lack of GaN substrate wafer and due to the high cost in epi-growth. The advantage of having the GaN transistor is its capability of sustaining power density above 10 W/mm of gate width (Eastman and Mishra, 2002), while amplifying signals at 10 GHz. For a brief comparison, the silicon-based power transistors can efficiently handle signals up to 3 GHz but at a lower power density. The silicon carbide devices achieve a power density at 7 W/mm but at a lower frequency of 3.5 GHz operations. Gallium arsenide and silicon germanium devices can handle high-frequency operations over 10 GHz, but they cannot withstand high power.

Silicon is expected to continue to be the dominant material for most of the power semiconductor devices for the next 10 years. However, materials such as SiGe for its higher carrier mobility and tuneable bandgap, GaN for its high piezoelectric constant and good carrier transport properties, etc. are also used to produce power semiconductor devices, e.g. HEMT (High Electron Mobility Transistor) for high performance applications. The major concern on SiGe material is on the critical layer thickness which cannot exceed the critical thickness for a given Ge mole fraction. In the recent times, the SiGe transistor research is driven by the wireless applications where low operating voltage and high power-added efficiency are needed for a frequency range between 2 GHz and 5 GHz. Hybrid-material devices, such as having SiC, SiGeC, and GaN materials can also play another important role in device development when both high thermal conductivity and good current carrying capability are needed in application. The material technology brings forth the possibility of making new SiGe BiCMOS large-scale integrated circuits and high-frequency system-on-chip solutions in future.

2.3. Carrier Concentration and Fermi Level

The electronic properties of semiconductor depend on the number of free electrons and holes available for current conduction. It should be noted that electron–hole pair generation at room temperature only contributes to a relatively small number of free carriers for current conduction. The electron density in the conduction band can be obtained if the density-of-state function $N_e(E)$ and its distribution function $f_e(E)$ are known. The density-of-state function describes the number of states that could be occupied by electrons and is given by

$$N_e(E) = \frac{4\pi}{h^3}(2m_e)^{\frac{3}{2}}(E - E_C)^{\frac{1}{2}}, \tag{2.1}$$

where h is the Planck's constant ($h = 6.626 \times 10^{-34}$ J·s), m_e is the effective mass of electron which is equal to $1.18 \times m_{eo}$ for silicon, m_{eo} is the electron mass which is equal to 9.11×10^{-31} kg, and E_C is the conduction band energy level. The probability that an energy level E is occupied by an electron is described by the Fermi–Dirac distribution function

$$f_e(E) = \frac{1}{e^{\frac{E-E_f}{kT}} + 1}, \tag{2.2}$$

where k is the Boltzmann's constant ($k = 1.3806 \times 10^{-23}\,\mathrm{J \cdot K^{-1}}$), E_f is the Fermi level which is the energy level whereby the probability of finding an electron is 50% at any temperature. This is also known as the electron occupancy probability. For energy levels higher than $3\,kT$ above E_f, the Fermi–Dirac distribution function can be approximated by the Maxwell–Boltzmann distribution function

$$f_e(E) \approx \frac{1}{e^{\frac{E-E_f}{kT}}} \tag{2.3}$$

as the exponential term is very much larger than unity if $(E–E_f)$ is greater than $3\,kT$. The electron density in the conduction band can be found by integrating the product of the density of states function and the occupancy probability

$$n(T) = \int_{E_C}^{\infty} f_e(E) \cdot N_e(E)dE. \tag{2.4}$$

This is shown pictorially in Fig. 2.7. Performing the integration, the electron density in the conduction band is

$$n(T) = 2 \times \frac{(2\pi m_e kT)^{\frac{3}{2}}}{h^3} \times e^{-\frac{(E_C-E_f)}{kT}} = N_C e^{-\frac{(E_C-E_f)}{kT}}, \tag{2.5}$$

where

$$N_C = 2 \times \frac{(2\pi m_e kT)^{\frac{3}{2}}}{h^3} = 4.83 \times 10^{15} \times T^{\frac{3}{2}} \tag{2.6}$$

is known as the effective density of states in the conduction band (in units of cm^{-3}) for silicon. At 300 K, it is 2.86×10^{19} cm^{-3}. By analogy, the density of allowed energy states in the valence band is given by

$$N_h(E) = \frac{4\pi}{h^3}(2m_h)^{\frac{3}{2}}(E_V - E)^{\frac{1}{2}}, \tag{2.7}$$

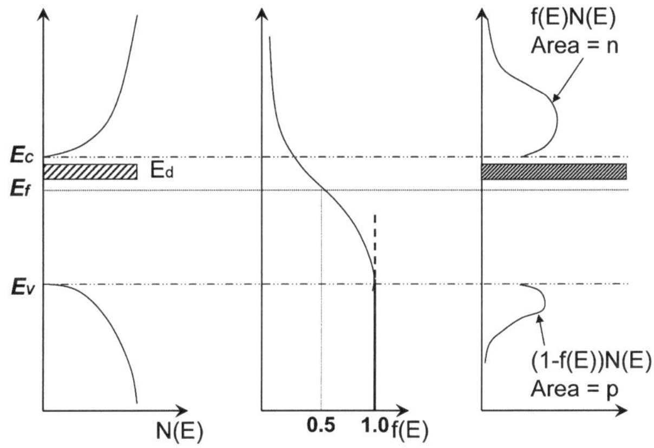

Fig. 2.7. Density of states and the occupancy probability functions on energy state diagram.

where m_h is the effective mass of hole which is equal to $0.5 \times m_{eo}$ for silicon. The hole occupancy probability is simply the probability that a level is not occupied by an electron, and is given by

$$f_h(E) = 1 - f_e(E) = \frac{1}{e^{\frac{(E_f - E)}{kT}} + 1}. \tag{2.8}$$

Thus, the density of holes in the valence band can be obtained by integrating the product of the density of states for holes and its occupancy probability

$$p(T) = \int_{-\infty}^{E_V} f_h(E) \cdot N_h(E) dE. \tag{2.9}$$

The hole density in the valence band is

$$p(T) = 2 \times \frac{(2\pi m_h kT)^{\frac{3}{2}}}{h^3} \times e^{-\frac{(E_f - E_V)}{kT}} = N_V e^{-\frac{(E_f - E_V)}{kT}}, \tag{2.10}$$

where

$$N_V = 2 \times \frac{(2\pi m_h kT)^{\frac{3}{2}}}{h^3} = 1.71 \times 10^{15} \times T^{\frac{3}{2}} \tag{2.11}$$

is known as the effective density of states in the valence band for silicon. At 300 K, it is 1.04×10^{19} cm^{-3}. The product of the electron and hole densities is expressed as

$$n(T)p(T) = N_C N_V e^{-\frac{(E_C - E_V)}{kT}} = N_C N_V e^{-\frac{E_G}{KT}}, \tag{2.12}$$

where $(E_C - E_V)$ is the bandgap energy E_G. It should be noted that the bandgap energy is a function of temperature and for silicon it can be expressed by $E_G = 1.1785 - 9.025 \times 10^{-5} \times T - 3.05 \times 10^{-7} \times T^2$. At room temperature of 293 K, the bandgap energy is 1.126 eV. Under a constant temperature, the $n(T)p(T)$ product in Eq. (2.12) is a constant in equilibrium and it is independent of the Fermi-level position. In an intrinsic semiconductor, the electron density is exactly equal to the hole density due to electron–hole pair generation. Thus,

$$n(T)p(T) = np = n_i^2. \tag{2.13}$$

This is known as the mass-action law and is valid for both intrinsic (or called pure) and extrinsic (or called doped) semiconductors under thermal equilibrium. Extrinsic semiconductors are those semiconductors that are doped with impurities (called dopants) to increase their free carrier concentrations.

Table 2.2. Effective densities of states and intrinsic carrier concentrations.

	Germanium (cm^{-3})	Silicon (cm^{-3})	Gallium arsenide (cm^{-3})
N_C	1.04×10^{19}	2.86×10^{19}	4.7×10^{17}
N_V	6.0×10^{18}	1.04×10^{19}	7.0×10^{18}
n_i	2.4×10^{13}	1.45×10^{10}	9×10^{6}

Table 2.2 shows the typical values of effective densities of states and the intrinsic carrier concentrations in germanium, silicon, and gallium arsenide.

2.3.1. Intrinsic Semiconductor

With Eqs. (2.12) and (2.13), the intrinsic carrier concentration n_i can be expressed as

$$n_i(T) = \sqrt{N_C N_V} e^{-\frac{E_G}{2kT}}. \tag{2.14}$$

As can be seen, the intrinsic carrier concentration is exponentially dependent on its temperature. It is an important parameter in semiconductor device as it determines not only the temperature performance, but also its leakage current. The intrinsic carrier concentrations in silicon and gallium arsenide in are $1.0 \times 10^{10}\,cm^{-3}$ and $1.8 \times 10^{6}\,cm^{-3}$, respectively at 300 K. From Eq. (2.14), it can be seen that the larger the bandgap energy is, the smaller the intrinsic carrier concentration will be. This is due to the fact that the electron in the valence band of a larger bandgap semiconductor must gain a higher energy to make the transition to the conduction band which is more difficult to occur. Diamond, for example, with a bandgap energy of 5.6 eV is used as a semiconductor material for high-temperature diode for its very low intrinsic carrier concentration.

The Fermi level in an intrinsic semiconductor E_{fi} can be found by equating Eqs. (2.5) and (2.10) and by setting $E_f = E_{fi}$:

$$E_{fi} = \frac{1}{2}(E_C - E_V) + \frac{3}{4} kT \times \ln \frac{m_h}{m_e}. \tag{2.15}$$

In silicon, the intrinsic Fermi level lies at 0.0073 eV below the midgap as $m_h = 0.69 \times m_e$. This small deviation from the midgap is typically neglected in most applications. The intrinsic carrier concentration can be found with Eqs. (2.5) and (2.10) to yield

$$n_i(T) = N_C e^{-\frac{(E_C - E_{fi})}{kT}} = N_V e^{-\frac{(E_{fi} - E_V)}{kT}}. \tag{2.16}$$

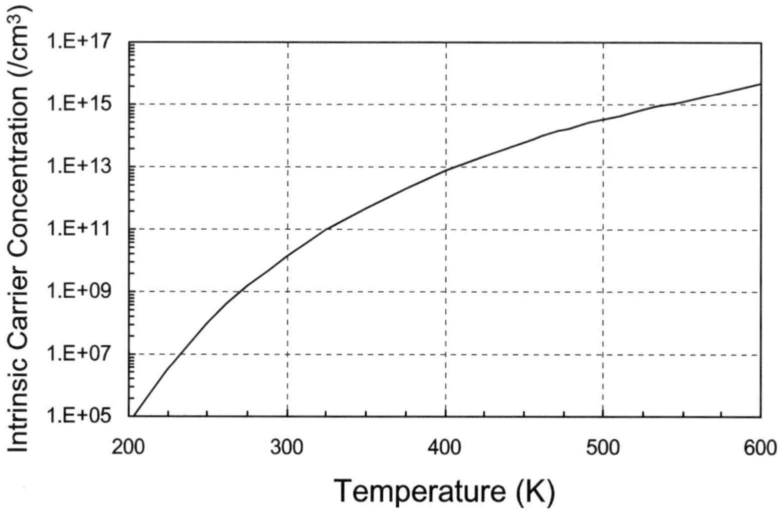

Fig. 2.8. Intrinsic carrier concentration at various temperatures in lightly doped silicon.

Figure 2.8 shows the variation of carrier intrinsic density as a function of temperature in lightly doped silicon. Thus, the electron and hole densities can be rewritten as

$$n(T) = n_i(T)e^{\frac{(E_f - E_{fi})}{kT}}, \tag{2.17}$$

$$p(T) = n_i(T)e^{\frac{(E_{fi} - E_f)}{kT}}. \tag{2.18}$$

These two equations are valid for both the intrinsic and extrinsic semiconductors. As it can be seen, the further the extrinsic Fermi level moves away from its intrinsic position, the higher the concentration of electrons or holes in the semiconductor it will have.

2.3.2. Extrinsic Semiconductor

The intrinsic carrier concentration in a semiconductor is generally too small for any useful on-state current conduction. The concentration of electrons in the conduction band can be increased from its intrinsic concentration to a much higher one by impurity doping. In silicon, the introduction of group V elements such as phosphorus, arsenic, and antimony introduces donors in the energy levels between the intrinsic Fermi level and conduction band energy. As a result, the Fermi level shifts upward from the intrinsic position. Assuming that all the donors are ionized or activated by the thermal energy, the condition of space-charge neutrality in a uniformly doped silicon requires that

$$p - n + N_d^+ = 0, \tag{2.19a}$$

$$n = p + N_d. \tag{2.19b}$$

Substituting Eq. (2.19) into Eq. (2.13),

$$n_n(n_n - N_d) = n_i^2 \tag{2.20a}$$

or

$$n_n^2 - n_n N_d - n_i^2 = 0, \tag{2.20b}$$

where n_n is the electron concentration in an n-type semiconductor. Using quadratic formula, the electron concentration is

$$n_n = \frac{\sqrt{N_d^2 + 4n_i^2} + N_d}{2}. \tag{2.21}$$

Similarly, the hole concentration in the n-type semiconductor, p_n, is

$$p_n = \frac{\sqrt{N_d^2 + 4n_i^2} - N_d}{2}. \tag{2.22}$$

Since the donor concentration is much larger than the intrinsic concentration, i.e. $n_i/N_d \ll 1$, Eqs. (2.21) and (2.22) can be approximated as

$$n_n \approx N_d + \frac{n_i^2}{N_d} \approx N_d \tag{2.23}$$

$$p_n \approx \frac{n_i^2}{N_d}. \tag{2.24}$$

The location of the Fermi level, E_{fn}, can be found by substituting Eq. (2.23) into Eq. (2.5) to yield

$$E_{fn} = E_C - kT \times \ln \frac{N_C}{N_d} \tag{2.25a}$$

for n-silicon. Similarly,

$$E_{fp} = E_V + kT \times \ln \frac{N_V}{N_a} \tag{2.25b}$$

for p-silicon to be mentioned below. Because the concentration of the electrons is much larger than that of the holes in an n-type semiconductor, the electron

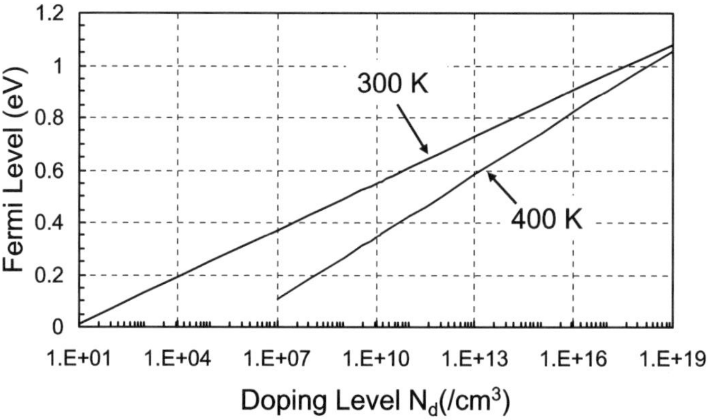

Fig. 2.9. Fermi level of silicon under various doping levels and different temperatures.

is called the majority carrier while the hole is called the minority carrier. The Fermi level of an n-silicon at different doping levels and temperatures can be found in Fig. 2.9. As temperature goes up, the electron effective density of states increases. And, this makes the Fermi level to be lower compared to that at low temperature.

In silicon, the introduction of group III element such as boron or aluminum introduces acceptors in the energy level between the valence band and the intrinsic Fermi level. The equilibrium hole concentration is given by

$$p_{\mathrm{p}} = N_{\mathrm{a}}, \tag{2.26}$$

where N_{a} is the acceptor concentration and the equilibrium electron concentration is

$$n_{\mathrm{p}} = \frac{n_{\mathrm{i}}^2}{N_{\mathrm{a}}}. \tag{2.27}$$

The semiconductor is of p-type. Holes are the majority carriers and electrons are the minority carriers. In practical semiconductor devices, the doping concentration required for contact formation is generally very high. As such, the Fermi level is located very close to either the conduction or valence band. Since $|E_{\mathrm{C,V}}-E_{\mathrm{f}}| < 3kT$, the exact Fermi–Dirac distribution must be used to obtain the correct carrier concentration. The semiconductor is said to be degenerated. A list of extrinsic dopant materials are listed in Table 2.3 together with their elevation energy values, i.e. $E_{\mathrm{C}} - E$ or $E - E_{\mathrm{V}}$ (Bar-Lev, 1984).

Under thermal equilibrium, the Fermi level inside a semiconductor maintains its energy state along the material. The energy constancy of the Fermi

Table 2.3. List of dopant materials and the elevation energy values.

	Dopant	Si (eV)	Ge (eV)	GaAs (eV)
p-type	B	0.045	0.0104	
	Al	0.057	0.0102	
	Ga	0.065	0.0108	
	In	0.16	0.0112	
	Zn			0.0307
	Cd			0.0347
	Si			0.0345
n-type	P	0.044	0.012	
	As	0.049	0.0127	
	Sb	0.039	0.0096	
	S			0.0061
	Se			0.0059
	Te			0.0058

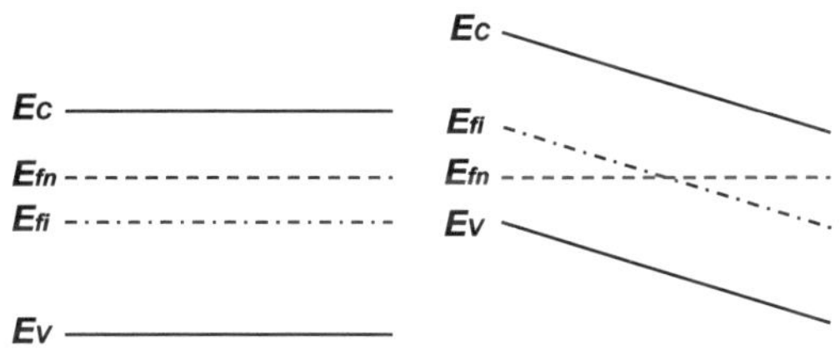

Fig. 2.10. Fermi levels within uniformly doped and nonuniformly doped semiconductors.

level means that E_f is to remain as a horizontal line on the equilibrium energy band diagram. This is true for both uniformly doped and nonuniformly doped semiconductors, as shown in Fig. 2.10.

2.4. Carrier Transport

Under thermal equilibrium, the free carriers in silicon are in random thermal motion. From statistical mechanics, the average velocity of thermal motion for electrons in silicon at room temperature is approximately 10^7 cm/s. The silicon atoms are in constant thermal vibrations which can be treated quantum-mechanically as phonons. The mobile electrons make frequent collisions with the vibrating silicon atoms. The collision of the electrons with the phonon is known as lattice or phonon scattering. This is the predominant collision or scattering process at or above room temperature for lightly doped silicon. Lattice scattering increases with increasing temperature due to increased lattice vibration. It should be noted that under thermal equilibrium, the random motion of electrons leads to zero current in any direction.

2.4.1. Carrier Drift

The movement of electrons and holes are fundamental to the operation of the semiconductor devices. When an external electric field is applied, the electrons in the conduction band, which are negatively charged, gain kinetic energy and move in the opposite direction of the applied electric field at an average velocity called the drift velocity. From Newton's second law of motion, the average drift velocity between two consecutive collisions is

$$v_{\mathrm{dn}} = -\frac{q\tau_{\mathrm{m}}}{m_{\mathrm{e}}}\hat{E} = -\mu_{\mathrm{n}}\hat{E}, \tag{2.28}$$

where $\hat{E}$ is the magnitude of the applied electric field, q is the magnitude of electron charge, τ_{m} is the mean free time between collision (about 1 ps in silicon), m_{e} is the effective mass of electron, and μ_{n} is called the electron mobility and is defined as

$$\mu_{\mathrm{n}} = \frac{q\tau_{\mathrm{m}}}{m_{\mathrm{e}}}. \tag{2.29}$$

Holes in the valence band, which are positively charged, also gain kinetic energy and moves in the same direction as the applied field. The average drift velocity for the hole is

$$v_{\mathrm{dp}} = \mu_{\mathrm{p}}\hat{E}, \tag{2.30}$$

where μ_{p} is the hole mobility and is defined as

$$\mu_{\mathrm{p}} = \frac{q\tau_{\mathrm{m}}}{m_{\mathrm{h}}}. \tag{2.31}$$

The carrier mobility describes the ease of movement of the carrier and is a measure of the average velocity of free carriers in the presence of an applied electric field. A higher carrier mobility means that the carrier can move better at ease in the lattice. Thus, it is an important parameter in power semiconductor devices. Due to the differences in the shapes of the conduction and valence band minima which gives rise to different effective masses of electrons and holes, the mobilities of the electrons and holes are different in silicon and gallium arsenide. Table 2.4 gives the electron and hole mobilities in silicon and gallium arsenide. In lightly doped silicon, carrier mobility decreases with increasing temperatures. This is due to an increase in the lattice or phonon scattering as temperature increases. Also, in silicon, carrier mobility decreases with increasing doping concentrations above $10^{16}\ \mathrm{cm}^{-3}$. This is due to an increased

Table 2.4. Mobilities at 300 K for lightly doped Si and GaAs semiconductors.

	μ_n (cm^2/V · s)	μ_p (cm^2/V · s)
Si	1350	450
GaAs	6000	330

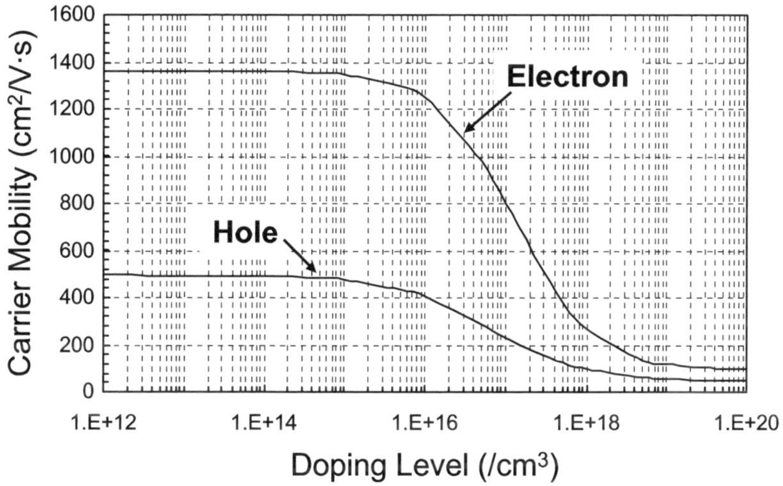

Fig. 2.11. Carrier mobilities in silicon at various doping concentrations at 300 K.

ionized impurity atom scattering. At a doping concentration of 10^{19} cm^{-3}, the electron mobility decreases to a value of about 90 cm^2/V · s and the mobility of hole decreases to a value of about 50 cm^2/V · s. Figure 2.11 shows the carrier mobilities of electron and hole as a function of doping level at 300 K.

Carrier mobility is also dependent on the magnitude of the applied electric field. Above an electric field of 10^5 V/cm, the average carrier mobility decreases inversely with electric field strength due to saturation of the free carrier velocity. At an electric field of 10^6 V/cm, the average electron and hole mobilities decrease to about 10 cm^2/V · s. This is an important concern in the design of high-field power semiconductor devices. Figure 2.12 shows the change of carrier mobilities as a function of electric field. The carrier mobility is also dependent on the carrier injection level. At high carrier injection level, the electron and hole mobilities in silicon decrease to about 100 cm^2/V · s at an injection level of 10^{19} cm^{-3} due to enhanced carrier–carrier scattering. This

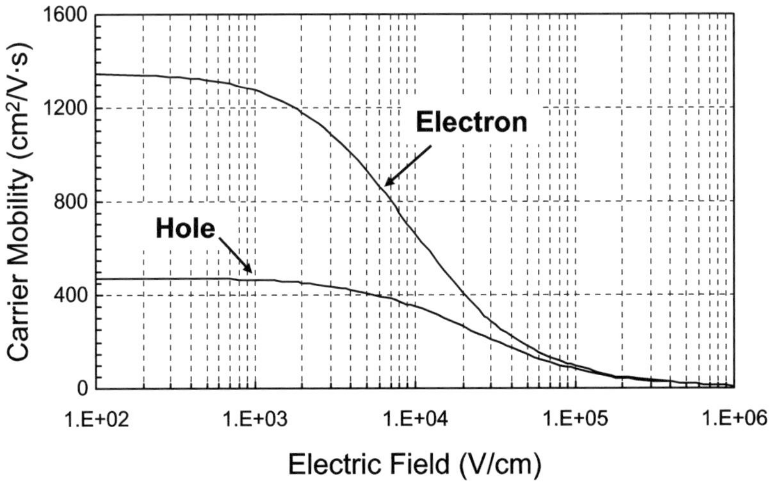

Fig. 2.12. Carrier mobilities at various electric field levels in silicon at 300 K.

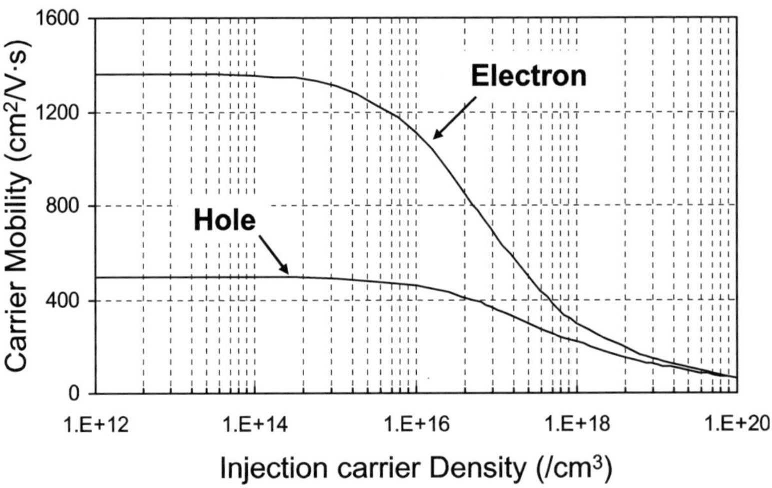

Fig. 2.13. Carrier mobilities at various injection levels in silicon at 300 K.

scattering mechanism occurs frequently in lightly doped base region of power bipolar transistor during forward conduction, where the density of electrons becomes approximately equal to the density of holes to preserve space-charge neutrality. Figure 2.13 describes the variation of carrier mobilities as a function of excess carrier injection level.

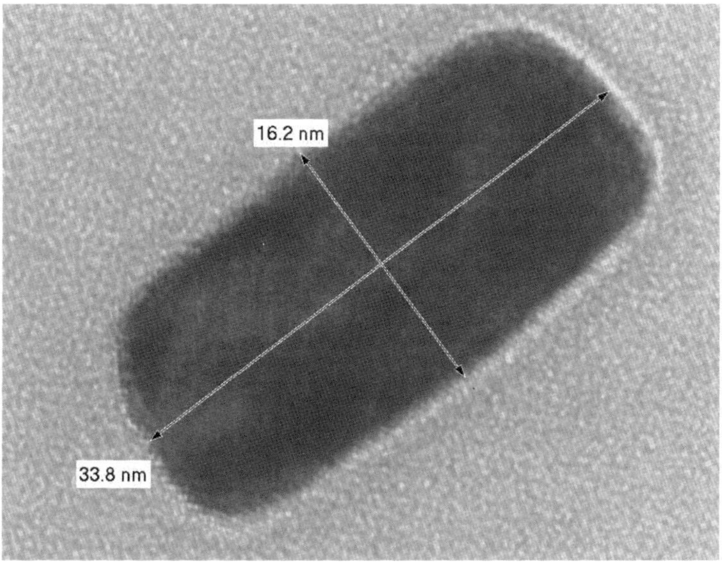

Fig. 2.14. TEM photograph of the non-smooth interfaces between the oxide region (surrounding) and silicon region (center), photo courtesy of A*STAR, IME, Singapore.

Surface scattering also reduces the mobilities of electrons and holes in silicon. In a power MOSFET, the carriers in the inversion layer is travelling in a direction nearly perpendicular to the silicon surface under strong inversion of the perpendicular gate electric field. As such carriers are traversed closer to the interface. Their mobility decreases due to the increased surface scattering at the rough interface between the gate dielectric and silicon. Figure 2.14 shows the high resolution transmission electron microscopic view of the interface between an oxide and silicon. Besides surface roughness scattering, carriers travelling close to the interface between gate oxide and silicon also suffer from additional scatterings due to oxide-fixed charges. Carrier mobility in the inversion layer depends not only on the orientation of the surface crystallographic plane, but also on the direction of the current flow. This is due to the anisotropy of the conductivity mass in silicon. Thus, the effect of different carrier scattering mechanisms is crucial to the operating characteristics of the power semiconductor devices.

In summary, the carrier mobility is affected by and as a function of the device operating temperature, the background doping concentration, the excess carrier injection level, the magnitude of field, surface scattering, the crystallographic plane, and the orientation of current flow.

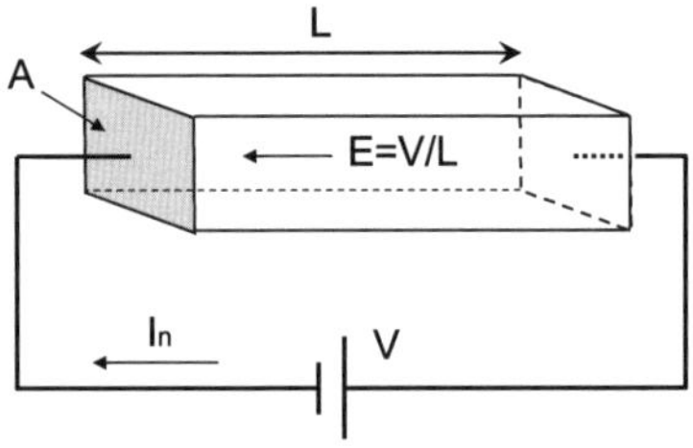

Fig. 2.15. Silicon under drift field.

The movement of electrons in the conduction band and holes in the valence band forms the current flow. This transport mechanism is known as carrier drift, and the resulting current is called the drift current. Consider a piece of silicon as shown in Fig. 2.15. The drift current due to n electrons per unit volume is given by

$$I_{\mathrm{n}} = -qAnv_{\mathrm{dn}} = qAn\mu_{\mathrm{n}}\hat{E}, \tag{2.32}$$

where A is the cross-sectional area. By analogy, the hole drift current can be written as

$$I_{\mathrm{p}} = qAp\,\mu_{\mathrm{p}}\hat{E}. \tag{2.33}$$

2.4.2. Carrier Diffusion

Carrier transport in semiconductor can also occur due to the differences in chemical potentials. The diffusion of electrons or holes results from their movement from higher concentration to lower concentration locations. Because electrons and holes are charged particles, their motion induces a current flow known as the diffusion current. From Fick's first law,

$$F = -D\frac{dN}{dx}, \tag{2.34}$$

where F is the carrier flux, D is the diffusion constant, and N is the carrier density function. The diffusion of electron induces electron diffusion current

$$I_{\mathrm{n}} = qAD_{\mathrm{n}}\frac{dn}{dx}, \tag{2.35}$$

where D_{n} is the diffusion constant for electrons. Similarly, the diffusion of holes induces hole diffusion current:

$$I_{\mathrm{p}} = -qAD_{\mathrm{p}}\frac{dp}{dx}, \tag{2.36}$$

where D_p is the diffusion constant for holes. The negative sign for the hole diffusion current indicates that hole diffusion current flows in the direction opposite to the gradient of hole distribution. The total electron current is the sum of both the drift and diffusion currents:

$$I_n = qA\left(\mu_n n\hat{E} + D_n\frac{dn}{dx}\right). \tag{2.37}$$

Similarly, the total hole current is

$$I_p = qA\left(\mu_p p\hat{E} - D_p\frac{dp}{dx}\right). \tag{2.38}$$

Thus, current flow in silicon comprises of electron drift and diffusion currents in the conduction band and the hole drift and diffusion currents in the valence band.

Under thermal equilibrium conditions, the net current in a nonuniformly doped semiconductor must be zero. Thus for n-type semiconductor, from Eq. (2.37),

$$q\mu_n n\hat{E} + qD_n\frac{dn}{dx} = 0. \tag{2.39}$$

The electric field within the nonuniformly doped semiconductor is directly proportional to the gradient of the intrinsic Fermi level, i.e.

$$\hat{E} = \frac{1}{q}\frac{dE_{fi}}{dx}. \tag{2.40}$$

From Eq. (2.17),

$$\frac{dn}{dx} = \frac{d}{dx}\left(n_i e^{\frac{(E_f - E_{fi})}{kT}}\right) = -\frac{n_i}{kT}e^{\frac{E_f - E_{fi}}{kT}}\frac{dE_{fi}}{dx} \tag{2.41}$$

since $dE_f/dx = 0$. Substituting Eqs. (2.40) and (2.41) into Eq. (2.39), we obtain

$$q\mu_n n\hat{E} + qD_n\left(-\frac{q}{kT}n\hat{E}\right) = 0. \tag{2.42}$$

Thus, for a nonuniformly doped semiconductor,

$$\frac{D_n}{\mu_n} = \frac{kT}{q}, \tag{2.43}$$

which is the Einstein relationship for electrons. Similarly,

$$\frac{D_p}{\mu_p} = \frac{kT}{q} \tag{2.44}$$

for holes. Thus, both the electron and hole diffusion constants are related to the carrier mobilities by the same Einstein relationship:

$$\frac{D_n}{\mu_n} = \frac{D_p}{\mu_p} = \frac{kT}{q}. \tag{2.45}$$

It should be noted that the Einstein relationship is also valid under non-equilibrium conditions.

2.4.3. Resistivity

From Eq. (2.32), the conductivity σ_n of a semiconductor due to electron drift can be expressed as

$$\sigma_n = q\mu_n n, \tag{2.46}$$

and from Eq. (2.33), the conductivity σ_p due to hole drift can also be defined as

$$\sigma_p = q\mu_p p. \tag{2.47}$$

The overall conductivity for both electron and hole carriers' drift within the semiconductor is

$$\sigma = q(\mu_n n + \mu_p p). \tag{2.48}$$

The resistivity of the material is defined as the reciprocal of its conductivity as

$$\rho = \frac{1}{q(\mu_n n + \mu_p p)}. \tag{2.49}$$

Thus, the resistivity of a semiconductor is controlled by the concentrations of the free electrons in the conduction band and the free holes in the valence band as well as their mobilities of these carriers. The curves on resistivity of n-type and p-type silicon are shown in Fig. 2.16 as a function of background doping level.

2.5. Bandgap Reduction

The magnitude of effective intrinsic carrier concentration in silicon varies under the condition of heavy doping concentration. This is due to the band structure of a heavily-doped silicon altered by (a) the formation of impurity band as the spacing between individual impurity atoms becomes smaller and the interaction between adjacent impurity atoms leads to a splitting of impurity levels, (b) the conduction and valence band edges no longer exhibit a parabolic shape which leads to the formation of band tails, and (c) the interaction between the free carriers and impurity atoms leads to a modification of the density of states at band edges, as shown in Fig. 2.17. The electrostatic interaction of the

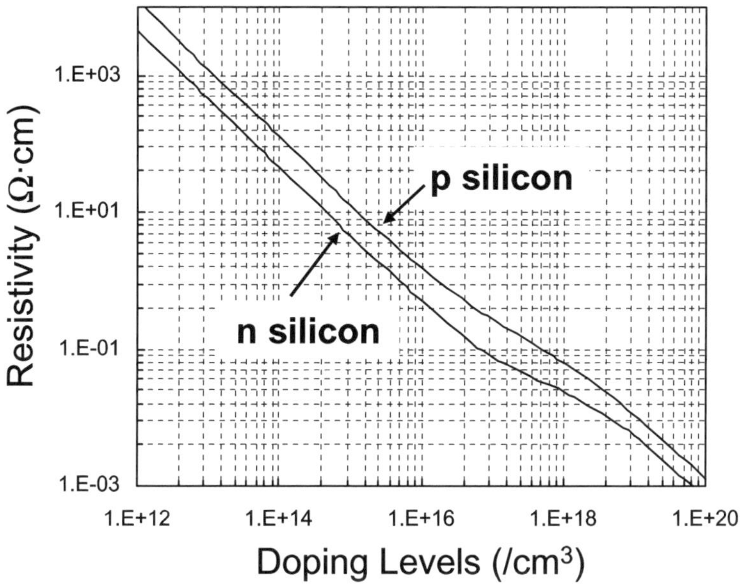

Fig. 2.16. Resistivities of n-type and p-type silicon at various doping levels at 300 K.

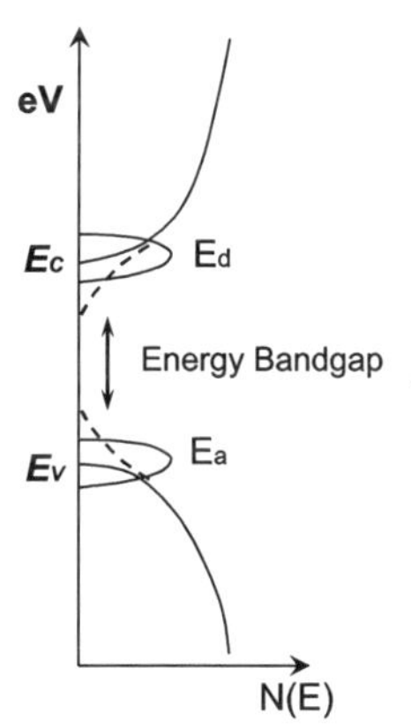

Fig. 2.17. Energy bandgap at high doping level.

minority carriers and the high concentration of the majority carriers leads to a reduction in the thermal energy required to create an electron–hole pair.

The electrostatic potential of the screened minority carrier as a function of distance r from the carrier, can be solved using Poisson's equation in spherical coordinates (Lanyon and Tuft, 1978)

$$\frac{1}{r^2}\frac{d}{dr}\left(r^2\frac{dV}{dr}\right) = +\frac{q}{\varepsilon_s}(n(r) - N_d), \tag{2.50}$$

where N_d is the donor concentration for n-silicon semiconductor. For doping concentration below degenerated level, $n(r)$ is related to the local potential, $V(r)$, by Boltzmann equation so that

$$\frac{1}{r^2}\frac{d}{dr}\left(r^2\frac{dV}{dr}\right) = \frac{q}{\varepsilon_s}N_d\left(e^{\frac{qV}{kT}} - 1\right). \tag{2.51}$$

For low excess electron concentration, Eq. (2.51) can be simplified using exponential series expansion to

$$\frac{1}{r^2}\frac{d}{dr}\left(r^2\frac{dV}{dr}\right) = \frac{q}{\varepsilon_s}N_d\frac{qV}{kT}. \tag{2.52}$$

The screened Coulombic potential is

$$V(r) = \frac{q}{4\pi\varepsilon_s}\frac{e^{-\frac{r}{r_s}}}{r}, \tag{2.53}$$

where the screening radius is given by

$$r_s = \sqrt{\frac{\varepsilon_s kT}{q^2 N_d}}. \tag{2.54}$$

Equation (2.52) can be solved by recognizing that a Bessel's differential equation for a real variable r with a solution of the form of Eq. (2.53) is

$$r^2\frac{d^2V}{dr^2} + 2r\frac{dV}{dr} + \frac{r^2}{r_s^2}V = 0. \tag{2.55}$$

The electric field distribution for the screened Coulombic field, $\hat{E}(r)$, is

$$\hat{E}(r) = -\frac{dV(r)}{dr} = \frac{q}{4\pi\varepsilon_s r}e^{-\frac{r}{r_s}}\left(\frac{1}{r} + \frac{1}{r_s}\right) \tag{2.56}$$

and the unscreened field distribution $\hat{E}_o(r)$ is

$$E_o(r) = \frac{q}{4\pi\varepsilon_s r^2}. \tag{2.57}$$

The bandgap reduction is the difference in electrostatic energy between the screened and unscreened cases

$$\Delta E_G = \frac{\varepsilon_s}{2}\int(\hat{E}_o^2(r) - \hat{E}^2(r))dr. \tag{2.58}$$

Using integration by parts, an expression for the bandgap narrowing in the units of eV (electron-Volt) can be found as

$$\Delta E_{\mathrm{G}} = \frac{3q}{16\pi\varepsilon_{\mathrm{s}}}\sqrt{\frac{q^2 N_{\mathrm{d}}}{\varepsilon_{\mathrm{s}} kT}}. \tag{2.59}$$

At room temperature, bandgap narrowing in silicon in the unit of eV is given by

$$\Delta E_{\mathrm{G}} = 2.2252 \times 10^{-2} \times \sqrt{N_{\mathrm{d}} \times 10^{-18}}, \tag{2.60}$$

where N_{d} is the donor concentration for n-type silicon. In silicon at room temperature, bandgap narrowing becomes significant at doping concentration above 10^{18} cm^{-3}. The mass-action law must be modified for bandgap narrowing as

$$n_{\mathrm{ie}}^2 = n_{\mathrm{i}}^2 e^{\frac{q\Delta E_{\mathrm{G}}}{kT}}. \tag{2.61}$$

At room temperature in silicon, the effective intrinsic carrier concentration will increase by a factor of over 70 times when the doping concentration increases from 10^{15} cm^{-3} to 10^{20} cm^{-3}. Figure 2.18 shows the effective intrinsic carrier density and the bandgap reduction as a function of background doping level at 300 K.

2.6. Carrier Recombination

Under thermal equilibrium, an electron in the valence band may acquire sufficient thermal energy to make the transition to the conduction band. A hole is left behind in the valence band after each successful electron transition. This thermal generation process is known as the electron–hole pair generation. In silicon, about 1.0×10^{10} cm^{-3} electron–hole pairs are generated at room temperature. At the same time, an electron may lose its energy and jump from the conduction band to the valence band. Such process is called the recombination. Under thermal equilibrium, the generation rate is equal to the recombination rate for the constant free carrier concentration.

In a semiconductor, carrier generation can also be accomplished by optical means. This requires an incident photon having an energy equal to or greater than the bandgap energy of the semiconductor. When the energy of the impinging photon is absorbed by an electron in the valence band, the electron is excited to the conduction band and a hole is created in the valence band. Excess carriers are generated by the impinging photons such that $n(T)p(T) > n_{\mathrm{i}}^2(T)$. Since for each electron generated by the impinging photon, a hole is left behind in the valence band as such, the space charge neutrality is preserved. This is equivalent to injecting both electrons and holes in equal amounts into the semiconductor.

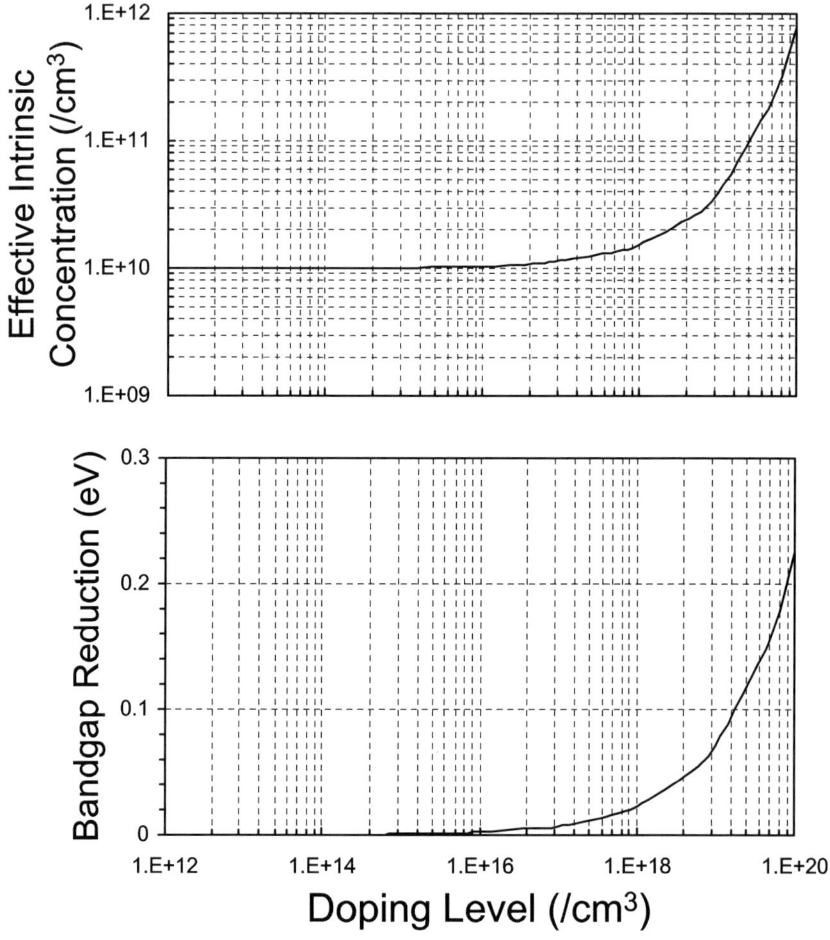

Fig. 2.18. The effective intrinsic carrier concentration (upper) and the amount of bandgap reduction (lower) for silicon under various doping levels at 300 K.

Or, carriers can also be injected by electrical means by forward biasing a semiconductor junction. Usually, this is done for single type of carriers, either electrons or holes, to be injected into one side of the junction according to the field direction. If the injected carrier density is less than the majority carrier (background) density, the injection is termed as low-level injection. Or, if the injected carrier density is greater or comparable to the majority carrier density, then the injection is termed as high-level injection. High-level injection reduces the carrier mobility as discussed in the preceding section.

The semiconductor returns to its equilibrium condition as the excess carriers are recombined or annihilated as a result of (a) an electron jumping directly from the conduction into the valence band (direct recombination process), or (b) an electron jumping from the conduction band and a hole jumping from the valence band into recombination centers (defect energy levels) located within the forbidden gap (indirect recombination process). The excess energy of the electrons must be dissipated by the emission of a photon without any change in its momentum (radiative recombination) or a phonon which involves both a change in its energy and momentum, or by the transmission of energy to a third particle that can be either an electron or a hole for Auger recombination, as shown in Fig. 2.19.

In direct recombination, for example, electrons, the rate of transition is given by

$$U = B(\Delta d)p_o, \tag{2.62}$$

where B is the statistical recombination probability coefficient, Δd is the excess electron density in the conduction band, and p_o is the number of holes in the valence band. From the above relationship, the minority carrier (e.g. the electron) lifetime can be defined as

$$\tau = \frac{\Delta d}{U} = \frac{1}{Bp_o}. \tag{2.63}$$

Carrier lifetime is the average time an excess carrier remains free before it recombines with the opposite carrier. The deviation of excess carrier (e.g. electron or hole) concentration in a semiconductor is then given by

$$\Delta d = \tau U. \tag{2.64}$$

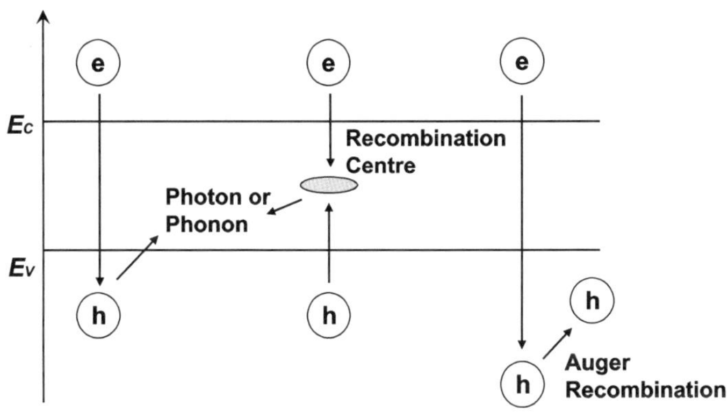

Fig. 2.19. Modes of carrier recombination processes in silicon.

In indirect recombination, carriers can interact with recombination trap centers located within the energy bandgap according to Shockley–Read–Hall (SRH) statistic (Hall, 1952; Schockley and Read, 1952) via (a) electron emission G_c, (b) electron capture, R_c, (c) hole emission, G_v, and (d) hole-capture, R_v. If G_c and R_c have a higher probability of occurring than G_v and R_v, the recombination centers act more as electron centers. Otherwise, they act more like hole centers. From another point of view, the centers act more as capture centers if R_c and R_v have a higher probability than G_c and G_v. Otherwise, they act more as generation centers. The rate of electron capture is expressed as

$$R_c = v_{th}\sigma_n n N_t(1 - f(E_t)), \tag{2.65}$$

where v_{th} is the carrier thermal velocity and is equal to $\sqrt{3kT/m}$, σ_n is the capture cross-section of the trap center for electron. The quantity of σ can be treated as a measure of how close the free carrier has to come to the physical vicinity of the center to be captured. $n = n_o + \Delta n$, N_t is the density of these centers and $f(E_t)$ is the occupancy factor for the trap center which is expressed by the Maxwell–Boltzmann distribution function:

$$f(E_t) = \frac{1}{e^{\frac{E_t - E_f}{kT}}}. \tag{2.66}$$

If the concentration of the recombination center is N_t, then the concentration of occupied centers is $N_t \times f(E_t)$ which denotes the probability of occupancy of a center by an electron. Similarly, $N_t \times (1 - f(E_t))$ represents the probability of unoccupancy, i.e. the center is ready to capture an electron. The rate of electron emission from the center into the conduction band G_c is proportional to the number of centers occupied:

$$G_c = e_n N_t f(E_t), \tag{2.67}$$

where e_n is the electron emission probability. The rate of valence band hole-capture is proportional to the number of centers occupied and the concentration of holes in the valence band:

$$R_v = v_{th}\sigma_p p N_t f(E_t), \tag{2.68}$$

where σ_p is the hole-capture cross-section of the trap center and $p = p_o + \Delta p$. The rate of hole emission from the center into the valence band is proportional to the number of centers unoccupied

$$G_v = e_p N_t(1 - f(E_t)), \tag{2.69}$$

where e_p is the hole emission probability. In the equilibrium state, the rate of electron capture is equal to the rate of electron emission. Thus, the electron emission probability can be calculated as

$$v_{th}\sigma_n n N_t(1 - f(E_t)) = e_n N_t f(E_t) \tag{2.70a}$$

or

$$e_n = v_{th}\sigma_n N_c e^{\frac{E_t - E_C}{kT}} = v_{th}\sigma_n n_i e \frac{E_t - E_i}{kT}. \tag{2.70b}$$

Similarly, if the rate of hole-capture is equal to the rate of hole emission in the steady-state, then

$$v_{th}\sigma_p p N_t f(E_t) = e_p N_t (1 - f(E_t)) \tag{2.71a}$$

or

$$e_p = v_{th}\sigma_p N_V e^{\frac{(E_V - E_t)}{kT}} = v_{th}\sigma_p n_i e \frac{E_i - E_t}{kT}. \tag{2.71b}$$

Under steady-state conditions, the principle of detailed balance requires that the rate by which electrons enter the conduction band is equal to the rate by which electrons leave the conduction band, and

$$R_c - G_c = R_v - G_v. \tag{2.72}$$

Solving for the trap occupancy factor energy level, we obtain

$$f(E_t) = \frac{\sigma_n N_c e^{\frac{E_t - E_c}{kT}} + \sigma_p N_v e^{\frac{E_v - E_t}{kT}}}{\sigma_n \left(n + N_c e^{\frac{E_t - E_c}{kT}}\right) + \sigma_p \left(p + N_v e^{\frac{E_v - E_t}{kT}}\right)}. \tag{2.73}$$

The steady-state net recombination rate, U, is equal to

$$U = R_c - G_c = R_v - G_v. \tag{2.74}$$

Thus,

$$U = \frac{\sigma_n \sigma_p v_{th} N_t (np - n_i^2)}{\sigma_n \left(n + N_c e^{\frac{E_t - E_c}{kT}}\right) + \sigma_p \left(p + N_v e^{\frac{E_v - E_t}{kT}}\right)}, \tag{2.75}$$

where $np = (n_o + \Delta d)(p_o + \Delta d)$, Δd represents the deviation of excess carrier density, either Δn or Δp, as defined in Eq. (2.64). It can be seen from Eq. (2.75) that the term $(np - n_i^2) > 0$ gives the nonequilibrium state and it also makes the recombination rate as a function of the carrier injection level. The minority hole lifetime in the n-silicon can be defined as

$$\tau_{po} = \frac{1}{\sigma_p v_{Tp} N_t}, \tag{2.76}$$

while the minority electron lifetime in p-silicon is

$$\tau_{no} = \frac{1}{\sigma_n v_{Tn} N_t}, \tag{2.77}$$

where σ_p and σ_n are the capture cross-sections for the electron and hole at the recombination center, v_{Tp} and v_{Tn} are the carrier thermal velocities for the hole and electron with the values of 4.3×10^6 cm/s and 5.2×10^6 cm/s, respectively at 300 K.

Then, the net steady-state recombination rate is

$$U = \frac{np - n_i^2}{\tau_{po}\left(n + N_c e^{\frac{E_t - E_c}{kT}}\right) + \tau_{no}\left(p + N_v e^{\frac{E_V - E_t}{kT}}\right)}. \tag{2.78}$$

Space-charge neutrality requires that the excess electron and hole concentrations are equal. Then, the carrier lifetime can be expressed as

$$\tau = \frac{\Delta d}{U} = \left[\frac{\tau_{po}\left(n_o + N_c e^{\frac{E_t - E_c}{kT}} + \Delta d\right) + \tau_{no}\left(p_o + N_v e^{\frac{E_V - E_t}{kT}} + \Delta d\right)}{n_o + p_o + \Delta d}\right]. \tag{2.79}$$

If we assume that both the capture probabilities and thermal velocities for electrons and holes are identical, then the expression can be simplified as (Grove, 1967)

$$U = \frac{cN_t(pn - n_i^2)}{n + p + 2n_i \cosh\left(\frac{E_t - E_i}{kT}\right)}, \tag{2.80}$$

where $c = v_{th} \cdot \sigma$. Considering an n-type silicon with the energy level of recombination center at the midband, i.e. $E_t = E_i$, it can be seen that the recombination rate will go down with higher n and p concentrations. The recombination rate is also lower if the trap energy level moves away from the mid-bandgap toward either the conduction band edge or the valence band edge. It is easy to understand that, for a recombination center to be in the middle of the bandgap, it renders the highest possibility for both the electron and the hole to reach the center. If the center moves toward one side, it will be easier for one type of the carrier to enter and leave the center, but much harder for another type of carrier, as the recombination process does need both types of carriers to arrive at the center. The rate of recombination will lower in the case of the centers away from the mid-bandgap.

Under the condition of low-level injection and the center at the mid-bandgap, the net recombination rate U for n-silicon is found to be

$$U = \frac{cN_t(n_n p_n - n_{no} p_{no})}{n_n + p_n + 2n_i} \tag{2.81}$$

or

$$U \approx cN_t(p_n - p_{no}) \tag{2.82}$$

for $n_n = n_{no}$ and $n_n \gg p_n + 2n_i$. So, the minority carrier (hole) lifetime, $\tau_p = \frac{\Delta p}{U}$, can be calculated as $\tau_p = \frac{1}{cN_t}$ for $p_n - p_{no} = \Delta p$. For the case, it can be seen that the minority carrier lifetime is inversely proportional to the trap density, the thermal velocity, and the capture cross-section area.

2.6.1. Carrier Lifetime

From the environment where carrier transport occurs, the lifetime can be categorized into two types, namely the injection (minority) carrier lifetime named SRH lifetime (Shockley and Read, 1952; Hall, 1952) and the space-charge generation lifetime (Shockley and Read, 1952; Sah *et al.*, 1957). The former is a measurement of carrier lifetime of minority carrier injected into or generated within a region where the opposite type of carrier claims to be the majority. The latter is related to the measurement of carrier generation within the reverse-biased space-charge region, and it determines the p–n junction leakage current. Both lifetimes are affected by the energy levels of recombination centers and their capture cross-section areas located at the silicon bandgap (Baliga, 1977).

The minority carrier lifetime is also a function of the injection level, ranging from below 1% of the doped background to over 100 times of that, besides other factors already mentioned earlier. From Eq. (2.79), under low-level injection, the injected carrier lifetime is relatively constant and is independent to Δn and can be given by the expression:

$$\tau_{LL} = \frac{\tau_{po}\left(n_o + N_C e^{\frac{E_t - E_C}{kT}}\right) + \tau_{no}\left(p_o + N_V e^{\frac{E_V - E_t}{kT}}\right)}{n_o + p_o}. \tag{2.83a}$$

For a heavily doped n-type silicon, i.e. $n_o \gg p_o$, the equation can be simplified as

$$\tau_{LL} = \tau_{po}\left(1 + e^{\frac{E_t - E_f}{kT}}\right) + \tau_{no} e^{\frac{2E_i - E_t - E_f}{kT}} \approx \tau_{po}\left(1 + e^{\frac{E_t - E_f}{kT}}\right). \tag{2.83b}$$

Similarly, for a heavily doped p-type silicon it can be expressed as

$$\tau_{LL} = \tau_{no}\left(1 + e^{\frac{E_f - E_t}{kT}}\right) + \tau_{po} e^{\frac{E_t + E_f - 2E_i}{kT}} \approx \tau_{no} + \tau_{po} e^{\frac{E_t + E_f - 2E_i}{kT}}. \tag{2.83c}$$

When the injection level increases, the minority carrier lifetime goes up quickly provided that the recombination center is located at the upper bandgap, i.e. closer or above the middle of the bandgap, where $E_t \geq 0.5\,\text{eV}$. For a higher level injection, the component of Δd in Eq. (2.79) cannot be ignored as that in the low-level injection condition. The lifetime varies and can be expressed by

$$\tau_{\Delta d\uparrow} = \frac{\tau_{\text{LL}} \times \dfrac{1+\Delta d(\tau_{\text{po}}+\tau_{\text{no}})}{\tau_{\text{po}}\left(n_{\text{o}}+N_{\text{C}}e^{\frac{E_{\text{t}}-E_{\text{C}}}{kT}}\right)+\tau_{\text{no}}\left(p_{\text{o}}+N_{\text{V}}e^{\frac{E_{\text{V}}-E_{\text{t}}}{kT}}\right)}}{1+\dfrac{\Delta d}{n_{\text{o}}+p_{\text{o}}}}. \tag{2.84a}$$

When the injection level goes to be extremely high, the high-level lifetime approaches the final value of

$$\tau_{\text{HL}} = \tau_{\text{po}} + \tau_{\text{no}}. \tag{2.84b}$$

The variation of lifetime with injection level is important to the operation of bipolar junction devices, where carriers need to have a longer high-level injection lifetime for on-state conduction and shorter low-level lifetime for fast switching of current tail during turn-off. This leads to the necessity of performing lifetime control by purposely creating defect levels in silicon at the device area where shorter carrier lifetime is needed for reducing the tail current at switching.

In a reverse-biased p–n junction, a leakage current will normally flow along the direction of the electric field. This is caused by the minority carriers diffusing into the depletion region from the edges, or the electron–hole pair generation within the region. For the minority carriers swept out of an n-type semiconductor, the rate of generation can be found from Eq. (2.78) by setting $p = 0$ as $p = p_{\text{o}} + n$ and $p_{\text{o}} = -\Delta n$,

$$-U_{\text{sp}} = \frac{n_{\text{i}}^2}{\tau_{\text{po}}\left(n + N_{\text{C}}e^{\frac{E_{\text{t}}-E_{\text{C}}}{kT}}\right) + \tau_{\text{no}}N_{\text{V}}e^{\frac{E_{\text{V}}-E_{\text{t}}}{kT}}}. \tag{2.85}$$

This corresponds to a "lifetime" of

$$\begin{aligned}\tau_{\text{sp}} &= \frac{\Delta n}{U_{\text{sp}}} = \frac{-p_{\text{o}}}{U_{\text{sp}}} = \frac{\tau_{\text{po}}\left(n_{\text{o}} + N_{\text{C}}e^{\frac{E_{\text{t}}-E_{\text{C}}}{kT}} - p_{\text{o}}\right) + \tau_{\text{no}}N_{\text{V}}e^{\frac{E_{\text{V}}-E_{\text{t}}}{kT}}}{n_{\text{o}}} \\ &\approx \tau_{\text{po}}e^{\frac{E_{\text{t}}-E_{\text{f}}}{kT}} + \tau_{\text{no}}e^{\frac{2E_{\text{i}}-E_{\text{f}}-E_{\text{t}}}{kT}}\end{aligned} \tag{2.86a}$$

for the space-charge region, where $n_{\text{o}} \approx p_{\text{o}}$. Similarly, for a p-type semiconductor, the space-charge generation lifetime can be derived as

$$\tau_{\text{sp}} \approx \tau_{\text{po}}e^{\frac{E_{\text{f}}+E_{\text{t}}-2E_{\text{i}}}{kT}} + \tau_{\text{no}}e^{\frac{E_{\text{f}}-E_{\text{t}}}{kT}}. \tag{2.86b}$$

Note that, as the depletion layer is penetrating in the said region, i.e. n for Eq. (2.86a) and p for Eq. (2.86b), the doping concentrations of these regions are normally low in power devices. By setting $E_f = E_{fi}$ for the depletion region, a set of curve on the space-charge generation lifetime against the recombination energy can be obtained (Sah *et al.*, 1957). With the assumption of uniform generation of carriers within the depletion region, it is preferred to have a large space-charge generation lifetime in order to maintain a low leakage current.

2.6.2. Carrier Lifetime Control

The carrier injection lifetime and the space-charge generation lifetime determine the property of power semiconductor devices for both on-state and off-state. The basic approach of providing lifetime control for bipolar devices, especially for high-voltage and high-current applications, is to manipulate the location of the recombination level within the energy bandgap, i.e. to create defects intentionally during wafer fabrication by various means. In principle, having the recombination level at the middle of the bandgap will result in a shorter low-level injection carrier lifetime for switching and for a larger high-level injection carrier lifetime at the on-state conduction, as shown in Figs. 2.20 and 2.21. This property is particularly needed for high-frequency operation of the power device. Figure 2.20 shows the variations of minority carrier injection lifetime ratios at different injection levels, different background doping levels, and under different lifetime control mechanisms in n silicon. In general, the lifetime rises when the injection level is higher although at different ratios. Note that, under lightly doped background, the platinum-controlled

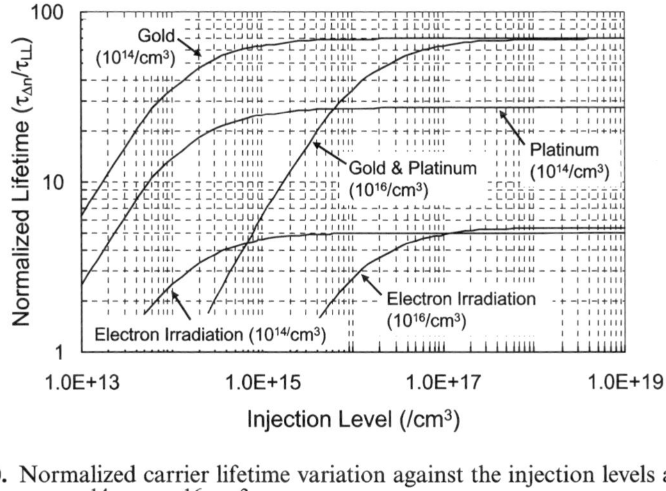

Fig. 2.20. Normalized carrier lifetime variation against the injection levels at doping background of 10^{14} and $10^{16}/cm^3$.

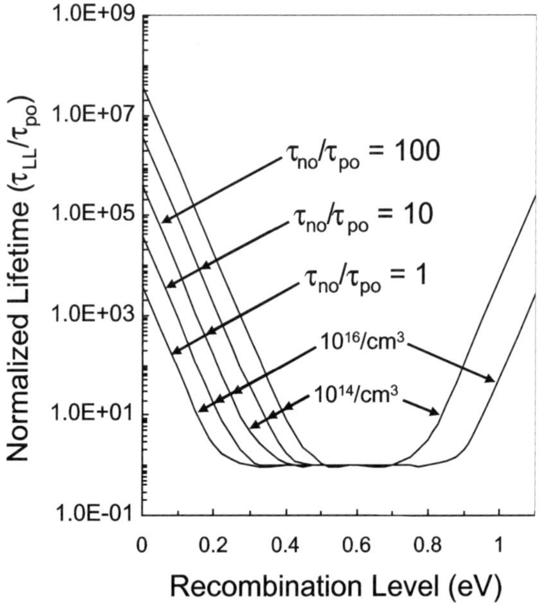

Fig. 2.21. Minority carrier low-level lifetime as a function of recombination levels at doping concentrations of 10^{14} and 10^{16}/cm^3.

low-level lifetime is higher; therefore the final ratio at high injection is different from that of the heavier-doped platinum-controlled n silicon. Figure 2.21 shows the variations of low-level lifetime as a function of recombination energy level. It is seen that, the low-level lifetime reaches its minimum where the recombination energy level locates around the mid-bandgap. The selectivity is more confined at lower doping background as shown. However, by having the mid-band recombination level, the space-charge generation lifetime will be lower, as shown in Fig. 2.22 and this will result in a higher leakage current. The situation becomes more specific in lightly doped silicon. This is logical as that, for the recombination center to be at the mid-bandgap, it is easier for electron–hole pair recombination and so as for generation. It is then preferable to have the level moving away from the mid-bandgap for high-voltage devices to minimize the leakage current. Hence, it is always a trade-off situation between on-state conduction and off-state blocking when choosing the right recombination level, e.g. between midband and band edge (Baliga and Krishna, 1977; Baliga, 1977).

In single-crystal silicon, the carrier lifetime can be as long as 1 ms, and it can be shortened during fabrication because of the defects caused by

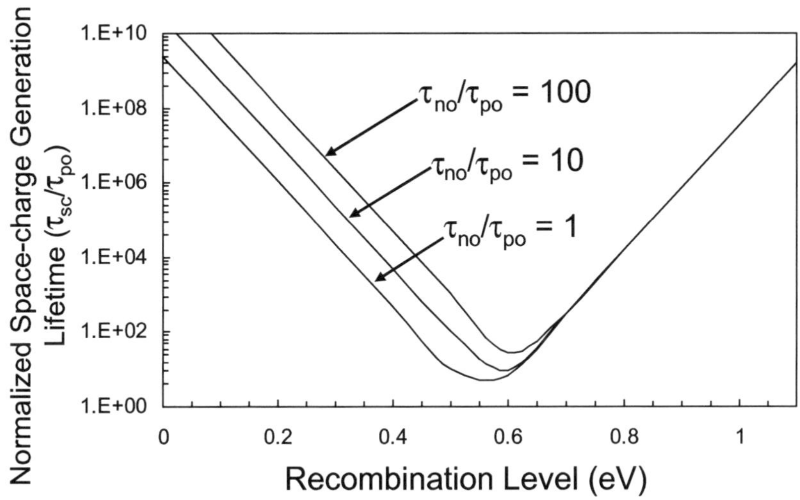

Fig. 2.22. Space-charge generation lifetime as a function of recombination level at 300 K in n-type silicon.

high-temperature processing, unwanted impurity diffusion from chemicals used in the etching and cleaning process, or from the residues in the equipment. By proper control of the temperature gradient, not exceeding 1°C/min, and proper passivation on silicon surface, the carrier lifetime of longer than 100μs can be preserved (Benda *et al.*, 1999). Preservation of carrier lifetime is needed for devices operating at low-frequency and high-current conditions. Otherwise, a properly controlled carrier lifetime reduction is usually performed, either globally or locally.

The optimization process starts with first finding the ratio of $\tau_{\mathrm{HL}}/\tau_{\mathrm{LL}}$ and $\tau_{\mathrm{SC}}/\tau_{\mathrm{LL}}$, and then to maximize both ratios by selecting a proper recombination level. By taking Eqs. (2.84b) and (2.83b) for n-type silicon, the expression of $\tau_{\mathrm{HL}}/\tau_{\mathrm{LL}}$ can be obtained

$$\frac{\tau_{\mathrm{HL}}}{\tau_{\mathrm{LL}}} = \frac{\tau_{\mathrm{po}} + \tau_{\mathrm{no}}}{\tau_{\mathrm{po}}\left(1 + e^{\frac{E_{\mathrm{t}}-E_{\mathrm{f}}}{kT}}\right) + \tau_{\mathrm{no}} e^{\frac{2E_{\mathrm{i}}-E_{\mathrm{t}}-E_{\mathrm{f}}}{kT}}}. \tag{2.87}$$

The ratio is also a function of doping concentration and temperature as indicated. Figure 2.23 shows the variations where the solid lines indicate temperature at 300 K and the dashed lines indicate the temperature at 400 K. It is seen that, the shoulder, i.e. the flat top region, becomes narrower when the background doping drops, and even much narrower when the temperature

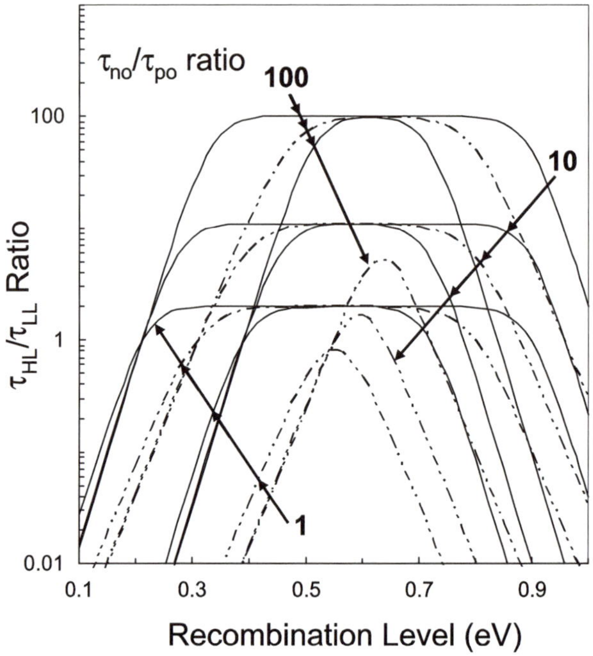

Fig. 2.23. Variations of high-level lifetime to low-level lifetime ratio against the recombination level in the cases of n-doping of $10^{16}/\text{cm}^3$ (wide shoulder) and $10^{13}/\text{cm}^3$ (narrow shoulder), and temperatures of 300 K (solid line) and 400 K (dashed line).

goes up. The selectivity of optimized energy level is more confined when the shoulder region is narrow. By finding $d(\tau_{HL}/\tau_{LL})/dt = 0$, the optimized recombination level can be found for Eq. (2.87),

$$E_t = E_i + \frac{kT}{2}\ln\left(\frac{\tau_{no}}{\tau_{po}}\right). \tag{2.88}$$

The optimized level is found to be a function of the electron–hole lifetime ratio and the temperature. It is independent from the background doping level, as E_f term disappears from the derivation. For the same approach, the ratio of space-charge generation lifetime to the low-level injection lifetime, e.g. for n-silicon, can be expressed as

$$\frac{\tau_{SP}}{\tau_{LL}} = \frac{\tau_{po}e^{\frac{E_t-E_i}{kT}} + \tau_{no}e^{\frac{E_i-E_t}{kT}}}{\tau_{po}\left(1+e^{\frac{E_t-E_f}{kT}}\right) + \tau_{no}e^{\frac{2E_i-E_t-E_f}{kT}}}. \tag{2.89}$$

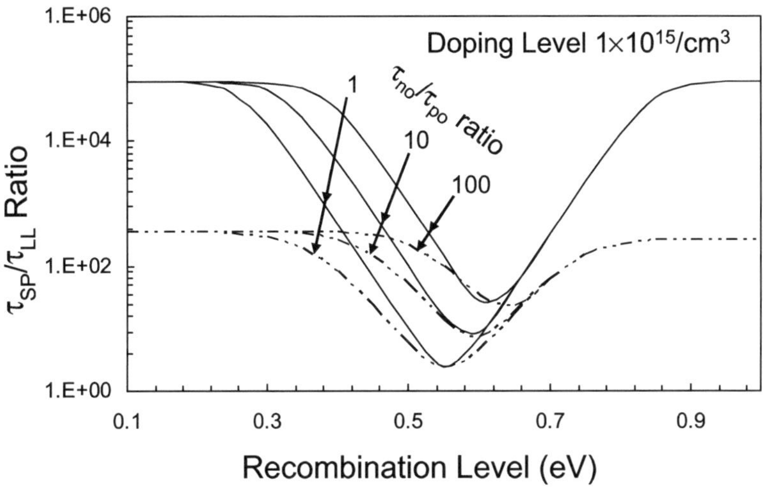

Fig. 2.24. Variation of space-charge generation lifetime to low-level injection lifetime ratio at 300 K (solid line) and 400 K (dashed line) at doping level of $10^{15}/\text{cm}^3$ in silicon.

The variation is plotted as shown in Fig. 2.24. As it can be seen, each curve has a minimum value close to the mid-bandgap. When the temperature goes up to 400 K, the ratio drops lower, and a lower ratio indicates a higher leakage current under space-charge depletion. To find the maximum value for the ratio, although it occurs at both sides of the valley banks, a recombination level equal to E_C can be substituted into Eq. (2.89), with a bit of approximation to obtain the maximum ratio. The maximum ratio can then be expressed as

$$\frac{\tau_{\text{SP}}}{\tau_{\text{LL}}} \approx e^{\frac{E_f - E_i}{kT}}. \tag{2.90}$$

In Eq. (2.90), the ratio is a function of both the doping level and the temperature. The equation indicates that, it is difficult to achieve both a low leakage current at lightly doped drift region with a small low-level injection lifetime, especially when the device temperature is high. Such a combined condition usually occurs in high-voltage semiconductor devices. Figure 2.24 indicates the same when the maximum ratio drops 340 times when temperature varies from 300 K to 400 K at a silicon doping level of $10^{15}/\text{cm}^3$.

In practice, to diffuse metal, such as copper in germanium, gold, platinum and iron in silicon creates the impurities defects in silicon and thus forms additional effective recombination levels at specific locations. Gold is the most frequently used element to create effective combination centers. It diffuses into silicon at a temperature range of 800–1000°C using a large thermal budget. An alternative is to use platinum due to the nature of its recombination centers lying away from mid-bandgap; thus the device property on leakage current can be improved. The limitations of using metal diffusion techniques are that, first, atoms of diffused metal may not be evenly distributed and this will depend on the process conditions. Second, the diffusion process needs to be done prior to device metallization (which forms the contacts). Therefore, it is not possible to test the device after lifetime control but prior to device metallization, and it is also not possible to alter the lifetime control by second metal diffusion if it is not correct after the device metallization. Normally, the diffused metal creates more than one trap energy level. For this situation, the behavior of the recombination process is more complicated than that of the single-level recombination process as analyzed earlier. However, due to difference in the capture cross-section, one can identify a majority recombination center out of others created by the metal diffusion. And, as a reasonable approximation, we then consider only one recombination center for simplicity in analysis. Another concern in the metal diffusion is that, when the concentration is comparable to the majority carrier concentration, the carrier removal effect occurs. It happens when each of the deep-lying energy levels associated with the metal atoms will remove one majority carrier from the conduction band (if for n-silicon) or from the valence band (if for p-silicon). A simple way of finding the majority carrier concentration is by subtracting the metal atom density from the dopant density, i.e. $p \approx N_A - N_t$. In this case, the resistivity of silicon will go up as a result of metal diffusion. Diffusion of metal atoms is normally used for large devices, such as diode, thyristor, and GTO (gate turn-off thyristor), lifetime control.

Another means of creating these levels is by high-energy particle bombardment, e.g. electron, proton implantation, alpha or gamma rays irradiation (Baliga and Su, 1977; Carlson *et al.*, 1977; Mogro-Campero *et al.*, 1986; Hallen and Bakowski, 1989). The induced lattice damage and vacancies form defect energy levels for recombination centers similar to other defect levels. Among them, the electron irradiation is the commonly used option due to the good control in defect density and homogeneous property throughout the silicon area. Research shows that the annealing process made by hydrogen will give a lower reverse leakage current but a higher on-state voltage drop compared to that by the nitrogen (Jo *et al.*, 2000). Gamma irradiation is able to penetrate deeper into silicon surface, so that post-metallization lifetime control can be performed. This overcomes the inconvenience occurred in the metal diffusion technique.

Proton implantation has a limited penetration depth but with the advantage of its precise depth to lifetime control. This is due to the fact that the density of the trap centers is proportional to the implantation dosage but inversely proportional to the level of proton energy. Therefore, much higher trap density can be found at a thin layer along the surface at the end of the implantation range (energy smaller than 1 MeV) for higher lattice damage at this localized layer. This enables a localization of high defect concentration, leaving most of the silicon bulk unaffected. Due to the accuracy in localization, the method is useful for power MOSFET and IGBT parallel cells' lifetime control. The shortcoming of this technique is that proton implantation needs to be performed under vacuum, and it is a costly process.

Another mature and more recent development on lifetime control is by helium implantation (Akiyama *et al.*, 1991; Konish *et al.*, 1996; Fang *et al.*, 2001). The high dose of helium implantation produce bubbles of density above 3.5×10^{20} He/cm^3 in silicon. Helium is not stable in the bubble form but evaporates through the silicon layer when under thermal treatment in nitrogen. The escape rate depends upon factors of the temperature, the He density, the distance from the surface, and the bubble size (Griffioen *et al.*, 1987). If the silicon is under a high temperature treatment, e.g. 1250°C, the He bubble tends to escape faster. And, this leaves behind the "void layer", or called voids, which is a localized lifetime control region. The region remains stable at a confined range even after further thermal treatment. The void density and diameter depend on the He density and the annealing temperature, whereas a higher He density will give a higher void density. The usage of the He implantation and high temperature voids layer formation enables the precise control of the localized region, where lifetime needs to be reduced. This is different from the proton implantation, where the carrier lifetime control is made in depth profile. Helium implantation gives localized lifetime control in lateral direction. As reported by Saggio *et al.*, 1998), lifetime control by the voids layer formation yields an on-state voltage drop of 2.1 V (at 100A/cm^2) unaffected with turn-off time of 70 ns reduced in comparison to that of 3.5 V with similar turn-off time by unlocalized lifetime control for a 600 V IGBT device at the n-drift region near the n^+-buffer.

A list of dominant levels for gold, platinum, and electron-irradiation induced recombination centers together with their localized coverage, allowed thermal budget, and capture cross-sections is shown in Table 2.5 (Milnes, 1973; Baliga and Su, 1977; Saggio *et al.*, 1998, Evwaraye and Baliga, 1977).

2.6.3. Auger Recombination

The Auger recombination process occurs when the energy of phonon released by SRH electron–hole recombination is received by the third particle for the subsequent carrier recombination. The process becomes obvious when there is

Table 2.5. Dominant deep energy levels for gold, platinum, electron/proton irradiation, and helium implantation induced recombination centers with capture cross-sections in silicon.

	Deep level position (eV)	Allowed thermal budget	Localised area (μm)	Capture cross section (cm^2)	
				Hole	Electron
Gold	$E_V + 0.56$ (for n-type) $E_V + 0.35$ (for p-type)	800°C for long hours	100	6.08×10^{-15}	7.21×10^{-17}
Platinum	$E_V + 0.8$ (for n-type) $E_V + 0.42$ (for p-type)			2.7×10^{-12}	3.2×10^{-14}
Argentum	$E_V + 0.83$ (for n-type) $E_V + 0.32$ (for p-type)			2.8×10^{-14}	1.2×10^{-15}
				1×10^{-15}	1×10^{-13}
Electron irradiation	$E_V + 0.71$ $E_V + 0.27$ (n-type)	350°C for 4 h	10–50	8.66×10^{-16}	1.62×10^{-16}
Proton implantation	—	—	10	—	—
Helium implantation	$E_V + 0.55$ (for n-type) $E_V + 0.45$ (for p-type)	—	3	3.0×10^{-15}	1.0×10^{-15}
		1250°C to form void layers, upto 5 h	0.2–0.5		

an extremely high-level carrier injection where high concentration of carriers of both electrons and holes exists, or when excessive carriers are injected into a heavily doped region. Measurement of Auger recombination was performed by Dziewior and Schmid (1977) for highly doped and highly excited silicon, and the Auger lifetime can be expressed as

$$\tau_{\mathrm{A}}^{n} = \frac{1}{2.8 \times 10^{-31} \times n^2} \tag{2.91a}$$

$$\tau_{\mathrm{A}}^{p} = \frac{1}{10^{-31} \times p^2}, \tag{2.91b}$$

where n and p are the concentrations of heavily doped n-type and p-type silicon, where Auger energy transfer is made to the electron and hole at the conduction band and valence band, respectively. And, for very high-level injection, the Auger lifetime is expressed as

$$\tau_{\mathrm{A}}^{\Delta C} = \frac{1}{3.4 \times 10^{-31} \times (\Delta C)^2}, \tag{2.91c}$$

where ΔC is the injected carrier concentration for cases when it is greater than $5 \times 10^{17}/\mathrm{cm}^3$. The graphs on Auger lifetime is shown in Fig. 2.25 for an easy reference.

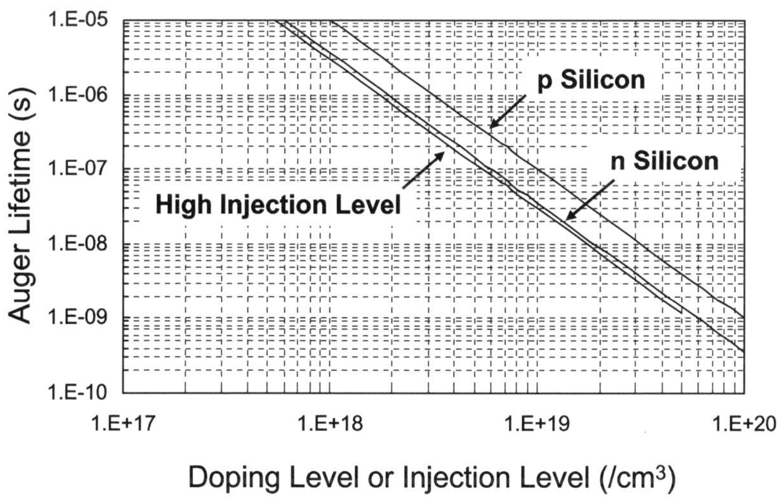

Fig. 2.25. Minority carrier Auger recombination lifetime as a function of the doping level or injection level in silicon.

The effective carrier lifetime is then calculated by combining both the SRH lifetime and the Auger lifetime by the following expression:

$$\tau_{\text{eff}} = \frac{1}{\frac{1}{\tau_{\text{SRH}}} + \frac{1}{\tau_A}}. \tag{2.92}$$

Therefore, for high injection level, the high-level injection lifetime cannot be a constant value as that of Eq. (2.84b) for SRH lifetime. From Eq. (2.92), the Auger lifetime which is smaller will dominate the minority carrier effective high-level lifetime.

2.7. Basic Equations in Semiconductor

Two important basic equations in semiconductor device physics are the carrier continuity equation and the Poisson's equation. The mechanisms of carrier transport, generation, and recombination in a semiconductor material give rise to a change in the carrier concentrations with time. The carrier continuity equation is derived from the principle of conservation of carriers which states that there must be a spatial and time continuity in the carrier concentrations. Thus, carriers (i.e. electrons or holes) cannot be arbitrarily created, or destroyed at a given time or location without mutual carrier interaction, but they must be transported to, created, or destroyed through interactions with carrier conservation law observed.

Consider a small incremental distance Δx in a semiconductor having unit cross-sectional area, as shown in Fig. 2.26. The currents, both incoming to a cross-section area and outgoing from the same area after an increment distance Δx are indicated. The continuity of the carrier flow requires that the incoming number of carriers entering Δx be equal to the net carrier change within the

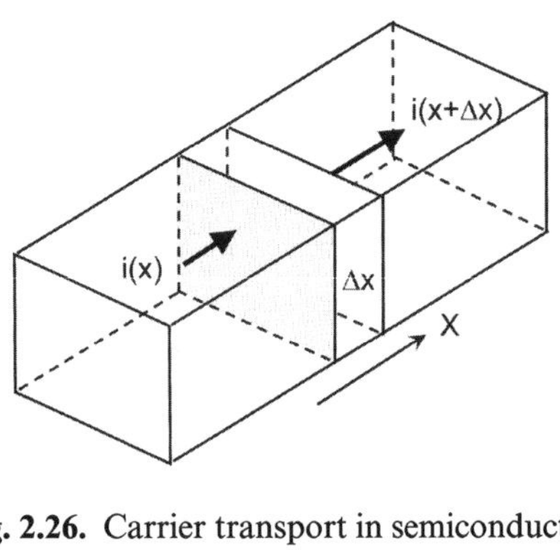

Fig. 2.26. Carrier transport in semiconductor.

increment plus carriers leaving the increment. The net quantity difference of incoming and outgoing carriers of the Δx region as a function of time is

$$\frac{\Delta C(x)}{\Delta t} = \frac{i(x)}{q} - \frac{i(x+\Delta x)}{q}, \tag{2.93}$$

where ΔC is the change of the carrier number within the increment, and i is the current. By adding $A\Delta x$ into Eq. (2.93), the variation of carrier density in terms of variation of current density can be obtained as

$$\frac{\Delta C(x)}{\Delta t A \Delta x} = \frac{\Delta d(x)}{\Delta t} = -\frac{\Delta i(x)}{qA\Delta x} = -\frac{\Delta j(x)}{q\Delta x}, \tag{2.94}$$

where A is the cross-sectional area, d is the carrier density, and j is the current density. This forms the basic equation of carrier continuity. It indicates that the carrier concentration in the confined increment can vary only if the incoming and outgoing current densities are different. However, the equation does not include variations from carrier generation and recombination processes occurring within the increment. To be more complete, carrier generation rate G and recombination rate U are added into Eq. (2.94) to yield

$$\frac{\Delta d(x)}{\Delta t} = G - R - \frac{\Delta j(x)}{q\Delta x}. \tag{2.95}$$

For hole carrier (in n-type silicon), the continuity equation can be written as

$$\frac{\partial p}{\partial t} = -\frac{1}{q}\frac{\partial J_{\mathrm{p}}}{\partial x} + G - \frac{p - p_{\mathrm{o}}}{\tau_{\mathrm{p}}}. \tag{2.96}$$

By analogy, the electron (in p-type silicon) continuity equation is

$$\frac{\partial n}{\partial t} = \frac{1}{q}\frac{\partial J_{\mathrm{n}}}{\partial x} + G - \frac{n - n_{\mathrm{o}}}{\tau_{\mathrm{n}}}. \tag{2.97}$$

These continuity equations can be useful in determining carrier transport actions and in deriving the on-state characteristics of semiconductor devices.

For Poisson's equation, it states that the gradient of the electric field in a space-charged region is directly proportional to the charge density in the region. Thus, it provides the relationship of the space-charged region in a semiconductor for the electric field, E, and is given by

$$\frac{d\hat{E}}{dx} = \frac{\rho}{K_{\mathrm{S}}\varepsilon_{\mathrm{o}}}, \tag{2.98}$$

where ρ is the net space-charge density, K_{s} is the dielectric constant of the semiconductor, and ε_{o} is the permittivity of free space. By integrating the

Poisson equation, the electric field distribution in the space-charged region can be found as:

$$\hat{E} = \int \frac{\rho}{K_s \varepsilon_o} dx. \tag{2.99}$$

The net space-charge in a semiconductor region is

$$\rho = q\left(p + N_d^+ - \left(n + N_a^-\right)\right), \tag{2.100}$$

where an ionized donor atom is positively charged and an ionized acceptor is negatively charged. Due to the electric field in the region, free carriers, such as p and n, are drifted away from the region. Therefore, the carrier densities can be approximated as $p \approx p_o$ and $n \approx n_o$ under the equilibrium state. Substituting the net space-charge into Eq. (2.98), we obtain

$$\frac{d\hat{E}}{dx} = \frac{q}{K_s \varepsilon_o}(p_o + N_d - (n_o + N_a)). \tag{2.101}$$

The above equation can then be integrated over the space-charge region to obtain the electric field distribution and subsequently for the voltage potential distribution. This equation is especially useful for device breakdown analysis to be described later in this chapter.

2.8. p–n Junction Electrostatics

When the regions of n-type and of p-type silicon are neighbored to each other, normally by junction diffusion process, electrons tend to diffuse from the n-region to the p-region and leave behind positive fixed charge of ionized donors N_d^+ at the vicinity of the junction. Similarly, holes diffuse from the p-region to the n-region, leaving behind negatively charged acceptor ions N_a^-. Consequently, a space-charge region is formed at the vicinity of the junction, as shown in Fig. 2.27. An electric field is then established within the space-charge region by the fixed charges of both sides. This field is in such a direction that further diffusion of electrons and holes is prevented. Therefore, the equilibrium condition at the junction is reached when the diffusion flux of carriers is neutralized by the drift flux resulting from the electric field. Under thermal equilibrium, there is no net carrier movement, except very small quantity of

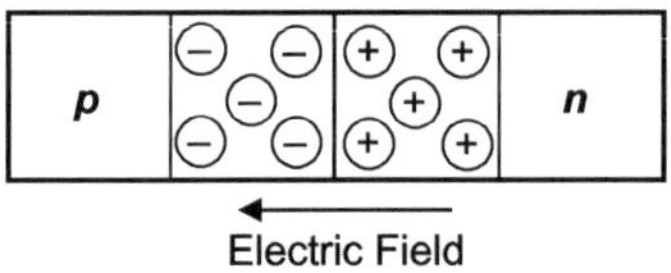

Fig. 2.27. p–n diode at the equilibrium zero-bias state.

electron–hole pairs generated within the region, and this can happen only when the probability of occupancy at a given energy level is the same everywhere. Consequently, the Fermi level must be the same everywhere along the p–n junction.

The relationship between the charge distribution and the electric field in the space-charge region is governed by Poisson's equation:

$$\frac{d\hat{E}}{dx} = -\frac{q}{\varepsilon_s}((n - p) - (N_d - N_a)). \tag{2.102}$$

At the n-side of the space-charge region, Poisson's equation is

$$\frac{d\hat{E}}{dx} = \frac{qN_d}{\varepsilon_s} \quad \text{for } 0 < x < x_n. \tag{2.103}$$

Similarly, at the p-side of the space-charge region, Poisson's equation is

$$\frac{d\hat{E}}{dx} = \frac{-qN_a}{\varepsilon_s} \quad \text{for } -x_p < x < 0. \tag{2.104}$$

Since the electric field is zero in the neutral regions at both edges of the depletion layer, we have

$$\hat{E}(x_n) = E(-x_p) = 0. \tag{2.105}$$

Integrating Eq. (2.104) for the p-side field,

$$\int_{E_p(-x_p)}^{E_p(x)} d\hat{E} = \int_{-x_p}^{x} \frac{-qN_a}{\varepsilon_s} dx \tag{2.106a}$$

or

$$\hat{E}_p(x) = \frac{-qN_a}{\varepsilon_s}(x_p + x). \tag{2.106b}$$

Similarly, the electric field on the n-side of the depletion layer is given by integrating Eq. (2.103):

$$\int_{\hat{E}_n(x)}^{\hat{E}_n(x_n)} d\hat{E} = \int_{x}^{x_n} \frac{qN_d}{\varepsilon_s} dx \tag{2.107a}$$

or

$$\hat{E}_n(x) = -\frac{qN_d}{\varepsilon_s}(x_n - x). \tag{2.107b}$$

As can be seen, the electric field is negative which denotes that the direction of field is at $-x$-direction on both sides of the depletion layer with the maximum value located at the metallurgical junction. Since the electric field is continuous at the metallurgical junction, $\hat{E}_p(x=0) = \hat{E}_n(x=0)$. Thus,

$$N_a x_p = N_d x_n, \tag{2.108}$$

which means that the negative charge on the p-side is exactly balanced by the positive charge on the n-side. It should be noted that the semiconductor as a whole is charge-neutral or quasi-neutral, despite the existence of the localized positive charges at the n-side of the space-charge region and localized negative charges at the p-side of the space-charge region.

The electric field is related to the electrostatic potential by

$$\hat{E} = -\frac{d\psi}{dx}. \tag{2.109}$$

From Eq. (2.106b)

$$\frac{d\psi}{dx} = \frac{qN_a}{\varepsilon_s}(x_p + x). \tag{2.110}$$

Integrating the above equation for the p-region,

$$\int_{\psi_p(-x_p)}^{\psi_p(x)} d\psi = \int_{-x_p}^{x} \frac{qN_a}{\varepsilon_s}(x_p + x)dx. \tag{2.111}$$

Thus,

$$\psi_n(x) = \frac{qN_a}{2\varepsilon_s}(x_p + x)^2 + \psi_{po}, \tag{2.112}$$

where the electrostatic potential at $x = -x_p$ is denoted as ψ_{po}. Similarly, for the n-side of the depletion layer,

$$\int_{\psi(x)}^{\psi_n(x_n)} d\psi = \int_{x}^{x_n} \frac{qN_d}{\varepsilon_s}(x_n - x)dx. \tag{2.113}$$

Thus,

$$\psi_p(x) = -\frac{qN_d}{2\varepsilon_s}(x_n - x)^2 + \psi_{no}, \tag{2.114}$$

where the electrostatic potential at $x = x_n$ is denoted as $\psi_n(x_n) = \psi_{no}$. Since the electrostatic potential must also be continuous at the p–n junction interface,

$$\psi_n(0) = \psi_p(0) \tag{2.115a}$$

or

$$\frac{qN_a}{2\varepsilon_s}x_p^2 + \frac{qN_d}{2\varepsilon_s}x_n^2 = \psi_{no} - \psi_{po} = \psi_o \tag{2.115b}$$

for the total reverse voltage ψ_o -applied across both regions. From Eqs. (2.115b) and (2.108), the depletion widths on the n-side and p-side of the junction can be found as

$$x_n = \sqrt{\frac{2\varepsilon_s \psi_o N_a}{qN_d(N_a + N_d)}} \tag{2.116}$$

and

$$x_p = \sqrt{\frac{2\varepsilon_s \psi_o N_d}{qN_a(N_a + N_d)}}. \tag{2.117}$$

The total depletion width is

$$x_d = x_n + x_p. \tag{2.118}$$

If the impurity concentration at one side of the junction is much larger than that at the other side, the junction is termed a one-sided or abrupt junction. The abrupt junction is a good approximation for a diffused junction having a shallow junction depth. For a one-sided p^+–n junction, the depletion width extends mainly into the lightly doped side of the junction:

$$x_d \approx x_n \approx \sqrt{\frac{2\varepsilon_s \psi_o}{qN_d}}. \tag{2.119}$$

Under thermal equilibrium, the energy-band diagram of a p^+–n junction is shown in Fig. 2.28(a). Both the intrinsic and Fermi levels are indicated. If a positive voltage V_f is applied to the p-side with respect to the n-side, the potential barrier at the p^+–n junction reduces to $\psi_o - V_f$, as shown in Fig. 2.28(b). The reduced potential barrier allows majority carriers to start to diffuse across the junction (then become the minority carriers at the other side) so that a visible current can flow through the forward-biased junction. If a negative voltage $-V_R$ is applied to the p-side with respect to the n-side, the potential barrier is increased to $\psi_o + V_R$, as shown in Fig. 2.28(c). The increased potential barrier continues to prevent carrier transport across the junction. Consequently, there is just a small leakage current and the junction is said to be heavily reverse-biased. Under the reverse-bias condition, the depletion layer width increases according to the total reverse voltage applied:

$$x_d = \sqrt{\frac{2\varepsilon_s(\psi_o + V_R)}{qN_d}}. \tag{2.120}$$

This unique property in the p–n junction provides control in the current flow based on the voltage applied across the junction. The term for such a control is called "current rectification". The junction remains functional as a rectifier till it breaks down by either high-field avalanche or Zener tunneling mechanisms.

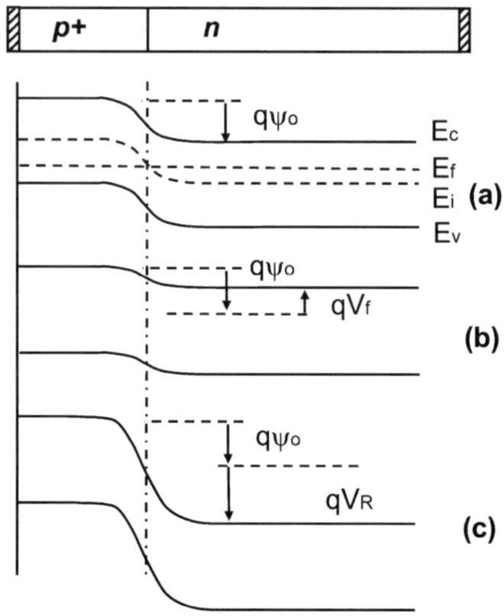

Fig. 2.28. Energy states of p–n junction under (a) zero, (b) positive, and (c) negative bias.

2.9. Junction Breakdown Phenomena

For power semiconductor devices, the breakdown voltage and current-handling capability are the two important parameters that determine the power ratings of the devices. For the p–n junction devices, the blocking voltage is supported entirely by its reverse-biased depletion region. As such, there exists the high electric field within the region. It is this electric field that sweeps out any carriers, either by diffusion from the adjoining quasi-neutral edges into the region or by the process of space-charge generation within the region. In silicon p–n junction, these mobile carriers are accelerated to a saturated drift velocity of 1×10^7 cm/s within the depletion region as the magnitude of the electric field reaches 1×10^5 V/cm. At higher electric fields, these carriers attain sufficient kinetic energy that their collisions with the atoms in the lattice can excite valence band electrons into the conduction band. This process of the generation of new electron–hole pairs is known as impact ionization, as shown schematically in Fig. 2.29. Since the electron–hole pairs created in the depletion layer by the impact ionization process can also undergo acceleration by the existing electric field, they can then participate in further impact ionization for more electron–hole generation. Thus, impact ionization is a multiplicative process that leads to the avalanche breakdown in a reverse-biased p–n junction. Avalanche breakdown is thus defined as the condition under which the impact ionization process attains an infinite rate. Thus, avalanche breakdown

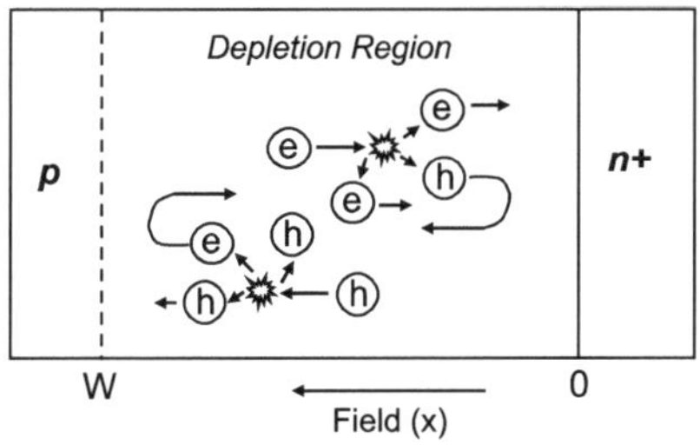

Fig. 2.29. Schematic diagram showing the electron and hole carriers' impact on ionization processes within the depletion region.

forms the fundamental limit for the maximum operating voltage of the power semiconductor devices.

The process can be characterized by the impact ionization coefficients of the electron and the hole. The impact ionization coefficient for electrons, α_{n}, is defined as the number of electron–hole pairs created by an electron traversing 1 cm through the depletion layer in the opposite direction of the electric field. It is related to the electric field by (Baliga, 1987)

$$\alpha_{\mathrm{n}} = 7 \times 10^5 \times e^{\frac{-1.23 \times 10^6}{\hat{E}}}. \tag{2.121}$$

On the other hand, the impact ionization coefficient for holes, α_{p}, is defined as the number of electron–hole pairs created by a hole traveling 1 cm along the direction of the electric field in the depletion region. It is related to the electric field by

$$\alpha_{\mathrm{p}} = 1.6 \times 10^6 e^{\frac{-2 \times 10^6}{\hat{E}}}. \tag{2.122}$$

Both coefficients increase very rapidly with increasing electric field. Approximations have been made to use only one ionization coefficient for simplicity in calculation. Various curves on the ionization coefficients are shown in Fig. 2.30 for easy reference (Fulop, 1967). Among these curves, two approximation curves are

$$\alpha = 9 \times 10^5 e^{\frac{-1.8 \times 10^6}{\hat{E}}} \tag{2.123a}$$

for approximation made by Maserjian, and

$$\alpha = 1.8 \times 10^{-35} \times \hat{E}^7 \tag{2.123b}$$

for approximation made by Fulop. From Fig. 2.30, it shows that the Fulop curve gives a better approximation for both electron and hole impact ionization coefficients. The condition under which the impact ionization achieves an

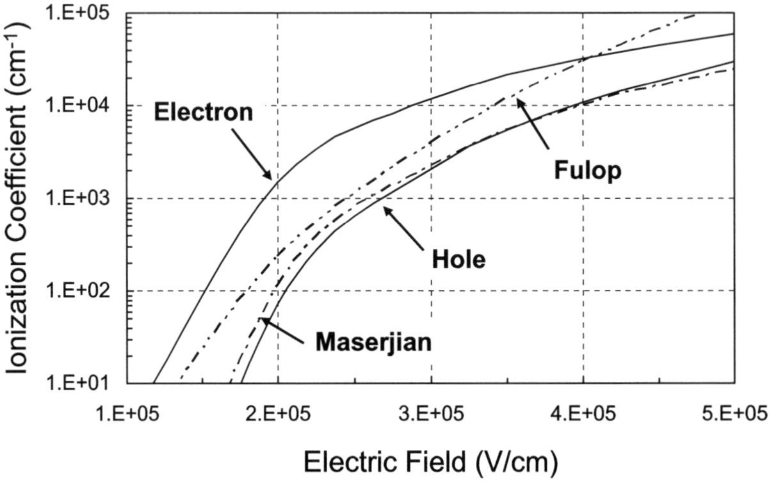

Fig. 2.30. Various ionization coefficients as a function of electric field.

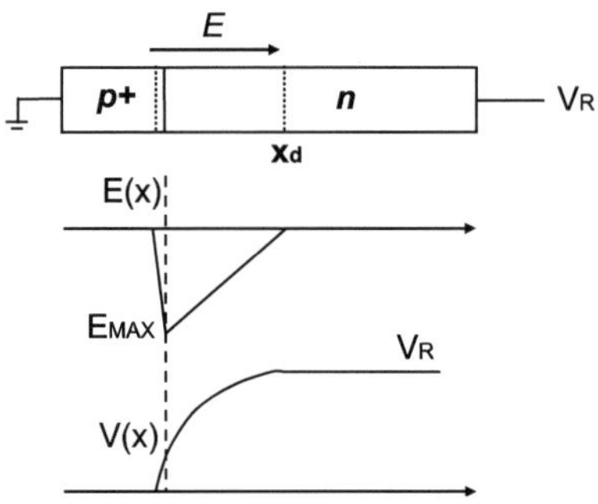

Fig. 2.31. Field and potential distributions of a reverse-biased p–n junction.

infinite rate determines the avalanche breakdown voltage of a reverse-biased p–n junction. Consider a p^+–n junction with a reverse-bias voltage applied across, as shown in Fig. 2.31. As discussed in the previous section, the depletion layer in the p^+–n junction extends mainly into the lightly doped n-side. Assume that an electron–hole pair exists within the depletion region due to space-charge generation. In the presence of the electric field established by the localized charges, the electron will be swept toward the n-region and the hole will be swept toward the p^+-region. In traversing a distance dx within the depletion region, the single hole will create $\alpha_p dx$ electron–hole pairs while the single electron will create $\alpha_n dx$ electron–hole pairs. The total number of electron–hole pairs generated in the depletion region due to the original electron–hole pair created by the space-charge generation at a distance x away from the p$^+$–n junction is given by

$$M(x) = 1 + \int_0^x \alpha_p M(x)dx + \int_x^{x_d} \alpha_n M(x)dx, \tag{2.124}$$

where x_d is the depletion layer width in the lightly doped n-side of the p$^+$–n junction. Performing the integration, the multiplication coefficient is

$$M(x) = \frac{e^{\int_0^x (\alpha_p - \alpha_n)dx}}{1 - \int_0^{x_d} \alpha_n e^{\int_0^x (\alpha_p - \alpha_n)dx} dx}. \tag{2.125}$$

The multiplication becomes infinity when the ionization integral becomes equal to 1:

$$\int_0^{x_d} \alpha_n e^{\int_0^x (\alpha_p - \alpha_n)dx} dx = 1. \tag{2.126}$$

For convenience, in further derivation, we use the single ionization coefficient (Fulop, 1967). Thus, for most applications, the ionization integral can be simplified to

$$\int_0^{x_d} 1.8 \times 10^{-35} \hat{E}^7 dx = 1. \tag{2.127}$$

Due to the strong dependence of the ionization coefficients on the electric field, the avalanche integral needs only to be performed for the reverse-biased junctions near the highest electric field region to characterize its breakdown performance.

Equation (2.127) treats the impact ionization coefficient independent from temperature variation. In fact, the coefficient is a function of temperature and becomes lower due to the reduction of carrier velocity and the shorter distance between impacts, e.g. lack of acceleration distance, at a higher temperature. It was shown that about 50% increase in breakdown voltage at the doping level of $10^{14}/\text{cm}^3$ can be obtained for temperature changes from 300 K to 500 K in silicon, and a higher breakdown ratio in germanium (Crowell and Sze, 1966).

Another mode of breakdown, normally occurring at the p$^+$–n$^+$ junction, is caused by Zener tunneling. Under extreme high-field situations, i.e. $\hat{E} > 10^6$ V/cm, the covalence bond between neighboring atoms can be distorted and even broken. The process creates free carriers when the covalence bond disappears. This phenomenon of quick generation of electron–hole pair under high field, without multiplication process, is called Zener breakdown. The word "tunneling" symbolizes the direct penetration of electrons through the energy bandgap without going through the recombination/generation centers.

2.9.1. Abrupt p^+–n Junction

Consider an abrupt p^+–n junction in which the doping concentration of the heavily doped side is much higher than the lightly doped side. In practice, these are the devices fabricated by performing a shallow acceptor diffusion into a lightly donor-doped substrate or epi-layer. As described in the previous section, the depletion layer extends primarily into the lightly doped side, as shown in Fig. 2.31. The analysis is pertained to the depletion region of the p^+–n junction (i.e. across the junction at the n-side), where it can be approximated as a parallel–plane abrupt junction. The electric field distribution within the depletion region in the lightly doped n-side is given by

$$\hat{E}(x) = -\frac{qN_\mathrm{d}}{\varepsilon_\mathrm{s}}(x_\mathrm{d} - x), \tag{2.128}$$

where the depletion width is approximated by x_d at the lightly doped n-region. The electrostatic potential distribution is

$$\psi(x) = \frac{qN_\mathrm{d}}{\varepsilon_\mathrm{s}}\left(x_\mathrm{d}x - \frac{x^2}{2}\right), \tag{2.129}$$

where the electrostatic potential at the junction is referenced to zero (in actual situation, it will be slightly larger than zero due to p^+-depletion). During the reverse-bias condition, the reverse voltage is supported entirely by the depletion region. Thus, the reverse-biased depletion width becomes

$$x_\mathrm{d} = \sqrt{\frac{2\varepsilon_\mathrm{s}(V_\mathrm{R} + \psi_\mathrm{o})}{qN_\mathrm{d}}} \approx \sqrt{\frac{2\varepsilon_\mathrm{s}V_\mathrm{R}}{qN_\mathrm{d}}}. \tag{2.130}$$

The maximum electric field occurs at the metallurgical junction and it is given by

$$\hat{E}_\mathrm{m} = -\sqrt{\frac{2qN_\mathrm{d}V_\mathrm{R}}{\varepsilon_\mathrm{s}}}. \tag{2.131}$$

Substituting the electric field distribution from Eq. (2.128) into the ionization integral of Eq. (2.127) for silicon, we obtain

$$\int_0^{x_\mathrm{d}} 1.8 \times 10^{-35}\left(\frac{qN_\mathrm{d}}{\varepsilon_\mathrm{s}}(x_\mathrm{d} - x)\right)^7 dx = 1. \tag{2.132}$$

Performing the integration, the depletion layer width at breakdown for the parallel–plane junction is

$$W_\mathrm{BR,PP} = 2.604 \times 10^{10} \times N_\mathrm{d}^{-0.875}. \tag{2.133}$$

To obtain the breakdown voltage, substituting the depletion layer width at breakdown into Eq. (2.129) at $x = x_d$, we obtain

$$V_{BR,PP} = 5.238 \times 10^{13} N_d^{-0.75}. \tag{2.134}$$

A multiplying coefficient of $(E_G/1.1)$ is added to Eq. (2.134) for breakdown voltage when dealing with semiconductor material with different bandgap energy from that of silicon (Sze and Gibbons, 1966). The maximum electric field at breakdown occurs at the junction ($x = 0$), and can be found by substituting Eq. (2.134) into Eq. (2.131) to yield

$$|\hat{E}_{MAX,PP}| = \sqrt{\frac{2qN_d(5.2381 \times 10^{13} \times N_d^{-0.75})}{\varepsilon_S}} = 4023.43 \times N_d^{0.125}. \tag{2.135}$$

The three preceding equations give a good estimation of planar junction breakdown width, field, and voltage.

2.9.2. Linearly Graded Junction

If the doping concentration of a p–n junction changes linearly from n-type to p-type throughout the entire length of the device, such p–n junction is called a linearly graded junction. A symmetrical distribution of localized charges is expected in this linearly graded p–n junction. Assuming that the rate of change of the doping concentration with distance is given by H, the charges in the depletion region are then given by

$$Q(x) = +qHx \quad \text{for } 0 < x < W \tag{2.136a}$$

at the n-side and

$$Q(x) = -qHx \quad \text{for } -W < x < 0 \tag{2.136b}$$

at the p-side. Poisson's equation for the linearly graded junction is

$$\frac{d\hat{E}}{dx} = \frac{qHx}{\varepsilon_S}. \tag{2.137}$$

Integrating the above equations yields

$$\hat{E}(x) = -\frac{qH}{2\varepsilon_S}(W^2 - x^2). \tag{2.138}$$

It can be seen that the electric field varies nonlinearly with distance for the linearly graded junction, as shown in Fig. 2.32. The maximum electric field

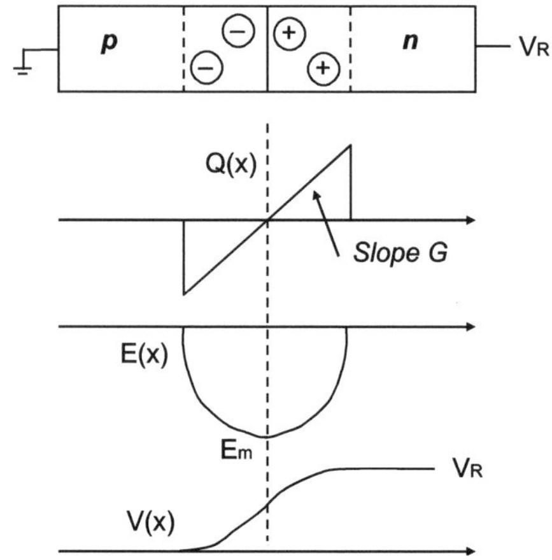

Fig. 2.32. Charge, field, and potential distribution of a linearly graded p–n junction.

occurs at the metallurgical junction, and is given by

$$\hat{E}_{\text{max}} = -\frac{qH}{2\varepsilon_s}W^2. \tag{2.139}$$

The electrostatic potential distribution in the depletion region can be found by integrating the electric field distribution:

$$\psi(x) = \frac{qH}{\varepsilon_s}\left(\frac{W^2x}{2} + \frac{W^3}{3} - \frac{x^3}{6}\right), \tag{2.140}$$

where the electrostatic potential at the edge of the depletion region at the p-side of the junction is referenced to zero, i.e. $\psi(-W) = 0$. The potential distribution is a cubic function of distance along the depletion region. During reverse-bias condition, the reverse voltage is supported entirely by the depletion region. Thus, the voltage at the end of the depletion region at the n-side shall be equal to the applied reverse-bias V_R. From the above equation, the reverse-biased depletion width can be found as

$$W = \left(\frac{3\varepsilon_s V_R}{2qH}\right)^{\frac{1}{3}}. \tag{2.141}$$

Substituting the electric field distribution of Eq. (2.138) into the ionization integral from Eq. (2.127) for silicon, the following is obtained:

$$\int_{-W}^{W} 1.85 \times 10^{-35}\left(\frac{qH}{2\varepsilon_s}(W^2 - x^2)\right)^7 dx = 1. \tag{2.142}$$

Performing the integration, the depletion width at breakdown is

$$W_{\text{LPP}} = 2W = 8.883 \times 10^5 \times H^{\frac{-7}{15}}. \qquad (2.143)$$

To obtain the breakdown voltage, substitute the depletion width at breakdown into Eq. (2.140)

$$V_{\text{BR,LPP}} = 9.022 \times 10^9 \times H^{\frac{-2}{5}}. \qquad (2.144)$$

The same multiplying coefficient of $(E_G/1.1)$ is added to Eq. (2.144) for breakdown voltage when dealing with a semiconductor material with different bandgap energy from that of silicon (Sze and Gibbons, 1966b). The critical electric field at breakdown occurs at the metallurgical junction and is found by substituting Eq. (2.143) into Eq. (2.139):

$$|\hat{E}_{\text{MAX,LPP}}| = 1.524 \times 10^4 \times H^{\frac{1}{15}}. \qquad (2.145)$$

Comparing the breakdown equation in the linearly graded p–n junction to that of the abrupt p^+–n junction, the linearly graded junction has a higher breakdown voltage than that of the abrupt junction for the same range of doping levels. This is because the linearly graded junction has a wider depletion width than that of the abrupt junction for voltage blocking.

2.10. Punchthrough Phenomenon

The junction punchthrough occurs when the depletion layer boundary moves from one junction and touches another junction as it expands. This can happen under two possible different device environments, namely within the open-base transistor structure (for example, the n^+–p–n^+) and within the lightly doped diode structure (for example, the n^+–p–p^+). The former, as described earlier, causes the breakdown, but the latter does not. Both structures are shown in Fig. 2.33 for a more clear illustration on the different kinds of punchthrough nature.

By looking at the open-base transistor, when the externally applied negative voltage is increased, the depletion width extends from the n^+–p junction toward the right side with the field distribution gradient maintained. Note that, the field gradient is a function of the background doping concentration and it shall maintain the same as long as the doping concentration of the p-region does not change. Once the field touches the p–n^+ junction, punchthrough occurs. The p–n^+ junction is not a reverse-biased junction as that of the n^+–p junction, but rather it is a forward junction. Electrons from the n^+-emitter will enter the forward-biased junction and drift (due to the presence of field) through the center depletion width to arrive at the n^+-collector to the left. In this case, the depletion region is no longer short of free carriers, and the continuous current

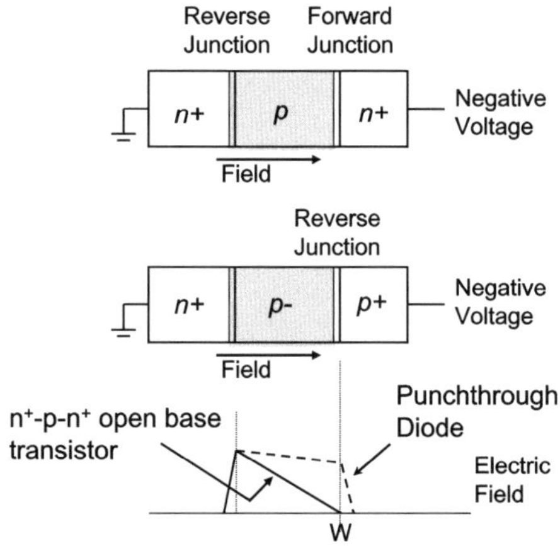

Fig. 2.33. Depletion layer punches through at open-base transistor and punchthrough diode.

flow indicates that the punchthrough has caused the junction breakdown. For a junction to break down, both the avalanche and punchthrough limits are to be considered and whichever is lower will set the breakdown voltage for the device rating. Figure 2.34 indicates both limits as a function of base doping concentration and base length (or called punchthrough length).

For the punchthrough diode structure, both junctions are reverse-biased. As such, no hole or electron carriers will be able to diffuse through and drift across the depletion region even when punchthrough occurs. However, the depletion punchthrough does alter the shape (not the gradient) of the field distribution profile. The rule of having triangular shape of the field distribution profile ceases to be valid after the punchthrough occurs. After its occurrence, the field profile cannot sizeably extend further to the right side within the p^+-region (because of the sudden drop of the filed profile in the p^+-region); therefore, when the external voltage increases, the field distribution profile will go upward with a fixed (or confined) base width equal to the p^--drift width. The slope of the field profile remains the same. As the rule of maximum avalanche field remains valid at the main junction, it is then favorable to have a lower doping concentration (marked as p^-) for the drift region. This enables a much flattened field distribution profile and early punchthrough before the avalanche sets in. The flattened field distribution profile will eventually form a rectangle-like shape which will support about twice the voltage for the same width of the drift region for a non-punchthrough structure.

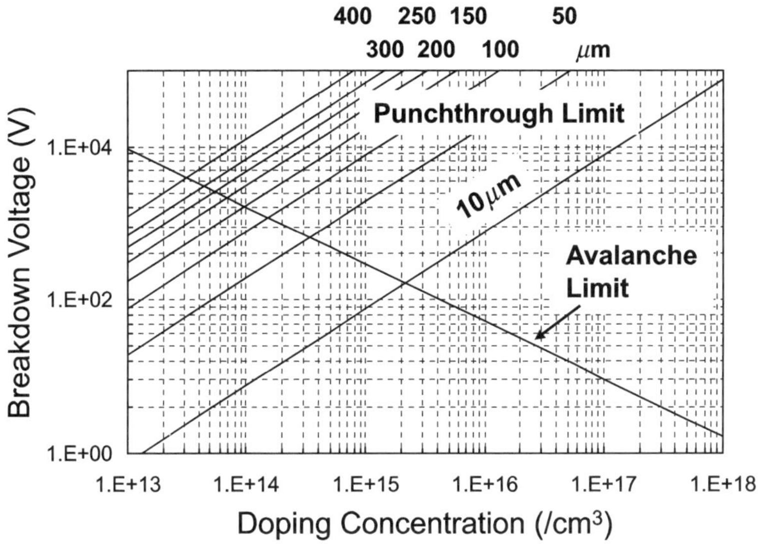

Fig. 2.34. Breakdown limits on avalanche and punchthrough in silicon.

For the punchthrough diode with a finite p^--layer width as shown in Fig. 2.33, the depletion region lies mainly in this layer and the reverse-blocking voltage is determined by both the critical breakdown field and the width of the p^--layer. The Poisson equation is

$$\frac{d\hat{E}}{dx} = -\frac{qN_A}{\varepsilon_S}. \tag{2.146}$$

Integrating the above equation within the p^- region,

$$\int_0^x d\hat{E} = \int_0^x -\frac{qN_A}{\varepsilon_S} dx = \hat{E}(x). \tag{2.147}$$

Thus,

$$\hat{E}(x) = -\frac{qN_A}{\varepsilon_S} x + \hat{E}_C, \tag{2.148}$$

where $E(x = 0) = \hat{E}_C$ which is the maximum critical field at the junction. This distribution of the electric field is shown in Fig. 2.33 as dashed line. Note that the magnitude of the electric field decreases linearly from the main (n^+–p^-) junction to the p^-–p^+ junction. The potential distribution is

$$\int_0^W d\psi(x) = -\int_0^W \hat{E}(x)dx. \tag{2.149}$$

F35

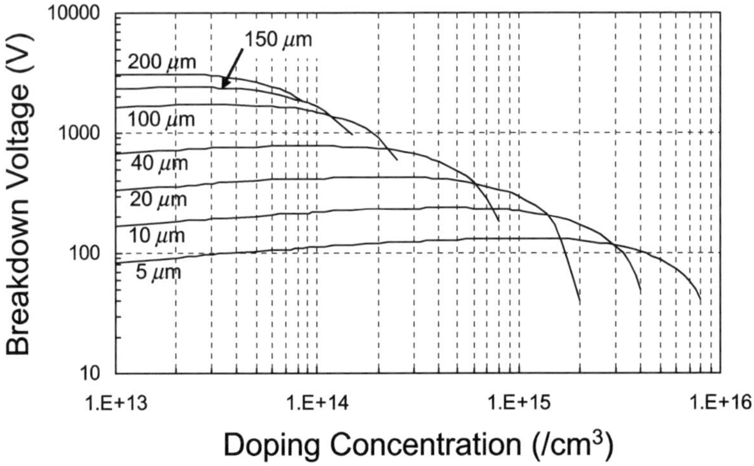

Fig. 2.35. Breakdown voltage of a punchthrough diode at different drift-region doping levels and lengths.

Performing the integration with $E(x)$ from Eq. (2.148) yields

$$V(W) = -\left(\hat{E}_{\mathrm{c}}W - \frac{qN_{\mathrm{A}}}{\varepsilon_{\mathrm{S}}}\frac{W^2}{2}\right), \tag{2.150}$$

where the negative sign indicates a negative voltage applied. By comparing with the n^+–p diode, the punchthrough diode would have a higher breakdown voltage (close to double) for the same base width. This is because the field profile is no longer a triangular shape as that in the abrupt junction. A set of curves as shown in Fig. 2.35 can be derived by using Eq. (2.150). It is important to understand the depletion punchthrough concept as it forms the foundation in studying new high-voltage structures, such as the superjunction structure to be discussed in a forthcoming chapter.

2.11. Junction Termination

In practical high-voltage power semiconductor devices, their breakdown voltages are limited by the edge terminations and they can be as low as 10% of the achievable breakdown voltages at the bulk parallel–plane junctions (Kao and Wolley, 1967). Since dopants diffuse both vertically and laterally from the diffusion window during the diffusion process, cylindrical junctions are created at the sidewalls while spherical junctions are created at the sharp corners as shown in Fig. 2.36. Normally, the lateral diffusion is less than the vertical diffusion, e.g. 85% of the value. However, for the simplicity of analysis,

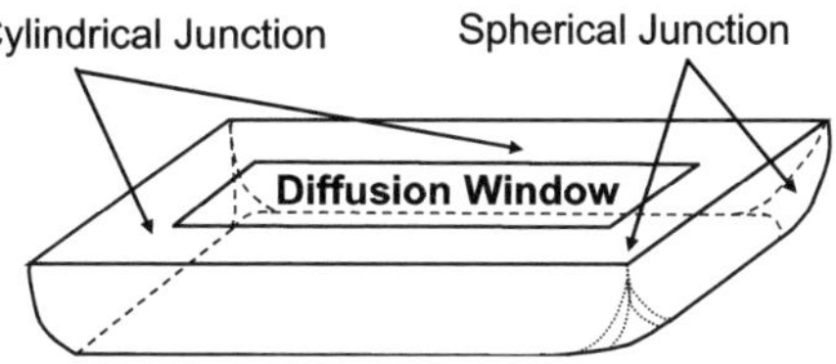

Fig. 2.36. Cylindrical junction and spherical junction formed under the diffusion window.

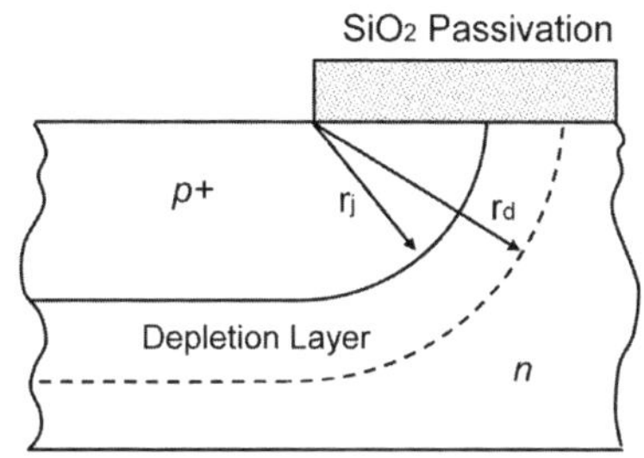

Fig. 2.37. Lateral junction diffusion and cylindrical depletion layer.

it is assumed that lateral diffusion has the same length as the vertical junction depth. Electric field crowding occurs at both the cylindrical and spherical sharp corners because the requirement for charge balance between the two sides of the junction need to be fulfilled, and also because the charge density at one side of the junction is more dense than the other side of the junction. As such, the practical device breakdown voltage of the power semiconductor device is actually limited by the sharp corner breakdown rather than the planar breakdown as described earlier (Sze and Gibbons, 1966a). Methods to alter depletion layer curvature such as using floating field rings, field-plates, bevelled-edge terminations, etch contour terminations, and the junction extension by ion implantation are often employed in high-voltage power semiconductor devices to reduce the field crowding. The presence of surface charge also has a strong influence on the shape of depletion layer near the surface of the reverse-biased junction.

2.11.1. Cylindrical Junction

The cylindrical junction is formed at the sidewalls of the planar diffused junction due to dopant lateral diffusion, as shown in Fig. 2.37. For a p^+–n junction, the Poisson equation in cylindrical coordinates is given by

$$\frac{1}{r}\frac{d(r\hat{E})}{dr} = \frac{qN_\mathrm{d}}{\varepsilon_\mathrm{s}}, \tag{2.151}$$

where $E(r)$ is the electric field distribution along a radius r extending into the depletion layer in the lightly doped n-region, as shown in Fig. 2.37. Using the boundary condition that the electric field at the depletion layer edge in the lightly doped n-region is zero, i.e. $E(r_d) = 0$, the electric field distribution within the cylindrical depletion region is given by

$$\hat{E}(r) = \frac{qN_d}{2\varepsilon_s}\left(\frac{r_d^2 - r_j^2}{r}\right). \tag{2.152}$$

The maximum electric field occurs at the metallurgical junction:

$$\hat{E}_{MAX,CY}(r_j) = \frac{qN_d}{2\varepsilon_s}\left(\frac{r_d^2 - r_j^2}{r_j}\right). \tag{2.153}$$

It should be noted that the maximum electric field in the cylindrical junction is higher than that in the parallel–plane junction approximately by a factor of $\frac{r_d}{2r_j}$. The potential distribution is found by integrating Eq. (2.152) to be

$$V(r) = \frac{qN_d}{2\varepsilon_s}\left[\left(\frac{r_j^2 - r^2}{2}\right) + r_d^2 \ln\left(\frac{r}{r_j}\right)\right]. \tag{2.154}$$

It has been shown that as the junction depth, r_j, increases, the breakdown voltage of the cylindrical junction increases and begins to approach the higher breakdown voltage of the parallel–plane junction. This can be intuitively understood by looking at Fig. 2.38 for two different junction depths. As it can be seen, there is a higher degree of electric field crowding occurring in the shallow p^+–n junction compared to that in the deeper p^+–n junction. This is because the negative charges in the p^+-diffused region, at which the electric field lines terminate, are located closer to the surface instead of being spread out along a wider junction area as in the case of the deep p^+–n junction. Therefore, the enhanced crowding of the electric field in the shallow junction is responsible for its lower breakdown voltage compared to that in the deep junction.

Figure 2.39 shows the breakdown voltage at the cylindrical junction for the lightly doped silicon if the same maximum avalanche field at the junction is set. These curves are calculated by first finding the maximum allowable avalanche field at the junction. With the maximum field obtained, the depletion widths of cylindrical or spherical curvatures at various junction depths can be calculated. By knowing the doping level, junction depth, and depletion width, the voltage at breakdown can then be obtained. It is seen that for doping levels between 10^{13} and $10^{15}/\text{cm}^3$, the breakdown voltage does not change much for junction depth below 2 μm.

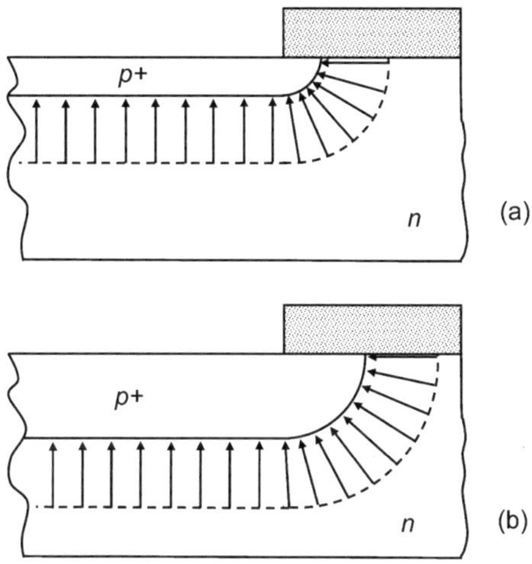

Fig. 2.38. Shallow junction in (a) has more crowded field lines at the cylindrical junction than that in (b).

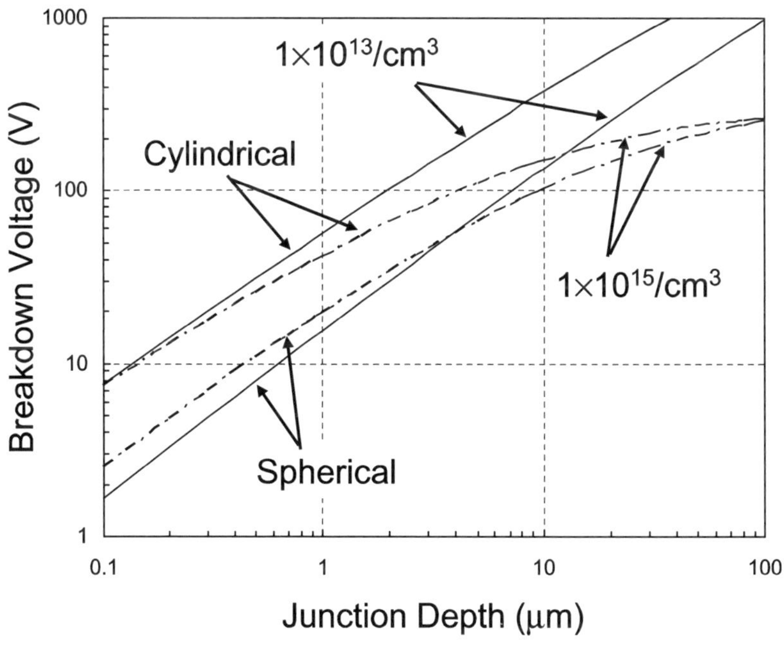

Fig. 2.39. Breakdown voltage at various junction depths for cylindrical and spherical junction at doping levels of 10^{13} and $10^{15}/cm^3$ in silicon.

2.11.2. Spherical Junction

The spherical junctions are formed at the sharp corners as the junctions take the form of a one-quarter of a spheroid with the radius equal to the distance of the lateral diffusion. To simplify the analysis, the distance of the lateral diffusion is again taken to be approximately equal to the junction depth. For a p^+–n junction, Poisson's equation in spherical coordinates is

$$\frac{1}{r^2}\frac{d(r^2\hat{E})}{dr} = \frac{qN_\mathrm{d}}{\varepsilon_\mathrm{s}}. \tag{2.155}$$

The electric field distribution can be found by integrating the Poisson's equation and using the boundary condition that the electric field is zero at the edge of the depletion layer, i.e. $E(r_\mathrm{d}) = 0$:

$$\hat{E}(r) = \frac{qN_\mathrm{d}}{3\varepsilon_\mathrm{s}}\left(\frac{r_\mathrm{d}^3 - r^3}{r^2}\right). \tag{2.156}$$

The maximum electric field for the spherical junction occurs at the metallurgical junction and is given by

$$\hat{E}_{\mathrm{MAX,SP}}(r_\mathrm{j}) = \frac{qN_\mathrm{d}}{3\varepsilon_\mathrm{s}}\left(\frac{r_\mathrm{d}^3 - r_\mathrm{j}^3}{r_\mathrm{j}^2}\right). \tag{2.157}$$

The maximum electric field in the spherical junction is higher than that in the parallel–plane junction by a factor of $\frac{1}{3}\left(\frac{r_\mathrm{d}}{r_\mathrm{j}}\right)^2$. It is higher than that in the case of cylindrical junction by a factor of $\frac{2}{3}\left(\frac{r_\mathrm{d}}{r_\mathrm{j}}\right)$. The potential distribution can be found by integrating Eq. (2.156):

$$V(r) = \frac{qN_\mathrm{d}}{3\varepsilon_\mathrm{s}}\left[\left(\frac{r_\mathrm{j}^2 - r^2}{2}\right) + r_\mathrm{d}^3\left(\frac{1}{r_\mathrm{j}} - \frac{1}{r}\right)\right]. \tag{2.158}$$

The breakdown voltage for a spherical junction is much lower than that of the cylindrical junction. As such, device designers strive to avoid sharp corners by rounding them off in high-voltage power devices and integrated circuits. Usually, the radius of curvature of the rounded corner should be several times larger than the depletion width at breakdown to be effective. The curves on the spherical junction breakdown can also be found in Fig. 2.39 calculated by the above equations.

2.11.3. Floating Field Ring

To ease the field crowding situation, floating field rings are often employed to divert the depletion charge away from the planar diffused main junction in

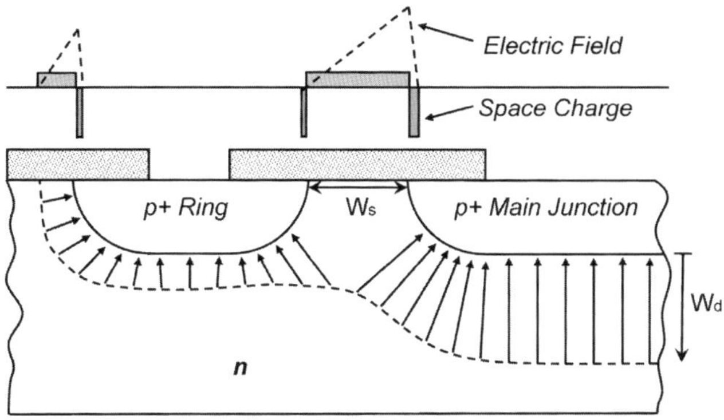

Fig. 2.40. Floating field ring and the depletion charge diversion for less field crowding.

order to reduce the electric field crowding at the main junction, as shown in Fig. 2.40. The ring has the similar effect in reducing the surface electric field (Kao and Wolley, 1967), and is indicated in the same figure. The figure shows the space-charge distribution, and the field is diverted from the original one large triangle to a later two smaller triangles for a lower field magnitude. It seems to be that the length of the ring, which is the distance separating two field triangles, need not be very large but be large enough to have a visible field diversion. Floating field rings are often fabricated simultaneously with the main junction to have a similar junction depth. The distance of the ring away from the main junction is critical in determining the effectiveness of field diversion (Adler and Temple, 1977). At low reverse-bias, the depletion layer of the main junction does not extend to the region which hosts the floating field rings. The floating field ring then acquires the same potential of the lightly doped p-region, normally zero. As the reverse-bias on the main junction increases, its depletion layer widens and it touches the floating field ring. Once punchthrough occurs, the floating field ring assumes the potential of the punchthrough point in the depletion region, i.e. a field-bypassed region, to extend the depletion profile away from the main junction. A simple analysis made by one-dimensional field here can be used for the floating field ring voltage calculation. For a floating field ring with a small width, its potential is related to the main junction by

$$V_{\mathrm{ffr}} = \frac{qN_{\mathrm{a}}}{\varepsilon_{\mathrm{s}}}\left(W_{\mathrm{d}}W_{\mathrm{s}} - \frac{W_{\mathrm{s}}^2}{2}\right) = \sqrt{\frac{2qN_{\mathrm{a}}W_{\mathrm{s}}^2V_r}{\varepsilon_{\mathrm{s}}}} - \frac{qN_{\mathrm{a}}}{2\varepsilon_{\mathrm{s}}}W_{\mathrm{s}}^2, \qquad (2.159)$$

where W_{d} is the depletion width of the main junction and W_s is the spacing between the main junction and the floating field ring. As can be seen, the potential of the floating field ring varies in value as a function of the square root of the applied potential of the main junction. However, this analysis may not be accurate for the actual two-dimensional field situation at the corner of the diffusion window. The optimal spacing between the floating field ring and the main junction for the single ring case is about 0.24 times of the depletion layer width at breakdown on the lightly doped side (10^{13}–10^{14}/cm^3) of the parallel–plane junction (Adler and Temple, 1977). At this optimal spacing, the breakdown voltage has been shown to increase by nearly a factor of 2 over the cylindrical planar diffused junction without floating field ring. It is hoped that at the optimized distance, the impact ionization process will occur at both junctions, i.e. the main junction and the field ring junction, at the same time by avalanche breakdown. Normally, for a practical reduction of field crowding, multiple field rings are used, e.g. three to five, to have a better result for higher breakdown voltage. By properly choosing the spacings between the rings, the field rings can be used as a field divider to raise the breakdown voltage closer to the bulk junction.

2.11.4. Etched Contour Termination

Another way of diverting the field contour is by etched contour, especially for large current devices on wafer form. The moat is typically chemically etched using either a wet isotropic etchant or a dry plasma etch process. The principle behind this is to etch away part of the silicon at the lightly doped side near the junction, so the depletion contour can be manipulated to be similar to that of having a floating field ring. In Fig. 2.41, both the original depletion contour before etching and the actual depletion contour after etching are shown. As part of the silicon is removed, to maintain the charge balance between both sides of the junction, the depletion boundary of the lightly doped side has to extend to the far side, resulting in a lower field strength along the region near the contour edge, and a less field crowding at the corner of the cylindrical junction (Temple and Adler, 1976). For a one-dimensional view, it is a fair approximation to assume that both areas A and B of the space-charge region

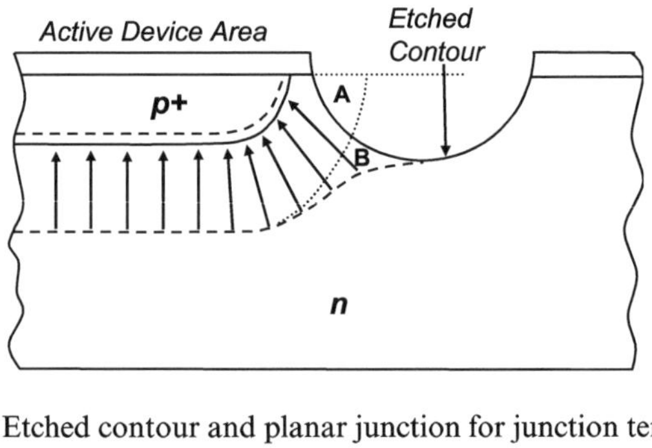

Fig. 2.41. Etched contour and planar junction for junction termination.

are equal. Area A is located much closer to the junction curvature compared to area B. Therefore, it is clear that by diverting the charge away from the junction curvature, both the field strength and the degree of crowding can be reduced.

Note that, the etched contour mentioned here is done at one side of the junction to reduce field crowding at the junction curvature. The etched contour can also be done by cutting through the junction to form a bevelled-edge-like surface for field reduction.

2.11.5. Bevelled Edge Termination

Besides reducing the field crowding at the junction curvature to avoid junction breakdown, it is also of concern that, for large power semiconductor devices such as thyristor, the breakdown may also occur at the silicon surfaces, either at the top surface or at the side surface. It is understood that the surface breakdown occurs at a field level which is substantially lower than the corresponding value for bulk breakdown. The obvious reason can be due to the defects at the surface caused by a process which creates localized high-field spots leading to breakdown. As there is no more silicon beyond the surface, the bevelled edge termination technique is used to shape the surface, so that the depletion contours on each side of the junction can be diverted in a way to reduce the field strength near the surface (Adler and Temple, 1978; Cornu, 1973; Cornu *et al.*, 1974). However, the approach does consume certain amount of silicon and it is not suitable for junction termination of small devices.

Both positive-bevelled and negative-bevelled edges are shown in Fig. 2.42. These bevelled edges, when carefully chosen, can be used to effectively improve breakdown voltage of a p^+–n junction. The positive-bevelled junction [Fig. 2.42(a)], is the one in which the depletion area decreases when proceeding from the highly doped side to the lightly doped side. The selective removal of silicon at the lightly doped side of the depletion near the surface forces

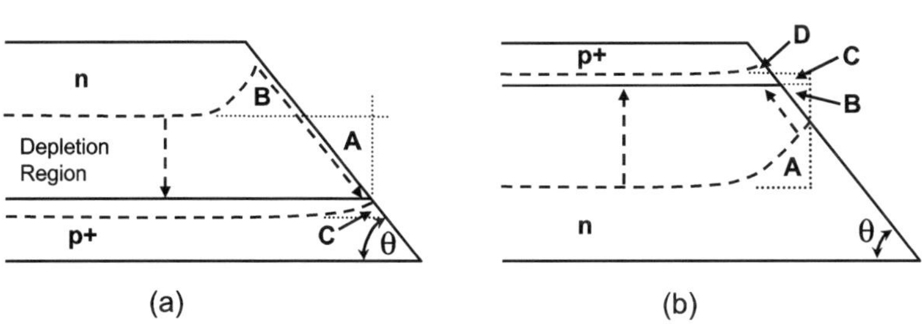

(a) (b)

Fig. 2.42. The junction termination by (a) positive-bevelled edge and (b) negative-bevelled edge.

the depletion layer on the lightly doped side to expand, resulting in a significant reduction in the maximum surface electric field. For the simplicity of one-dimensional analysis, it is assumed that the charge in area A is equal to the charge in area B plus the charge in area C, for the charges to be balanced at each side of the junction. Therefore, when the bevelled angle θ is small, area A increases and this forces area B to increase as well. By looking at the field lines at the surface and in the bulk, i.e. the dashed lines, it can be seen that the depletion width is stretched because of the bevelled edge, and this in turn will reduce the field strength along the surface, because the same voltage is seen across the junction but for a larger junction width.

The negatively bevelled junction, as shown in Fig. 2.42(b), has a charge relationship of area A+area B = area C−area D. However, the charge density of area A is much lesser than area D; this makes area A usually much larger than area D. In this case, the filed line, i.e. the dashed lines, along the surface is usually shorter than that in the bulk. Only in the case where the bevelled angle is reduced to be very small (below 10°), this creates a long stretch of area B along the surface for a lower surface field. Approximately, a positive-bevelled angle of 45° will reduce the maximum electric field to below 40% of that without bevelled angle. For the same amount of field reduction, the negatively bevelled angle would be around 4° (Baliga, 1987) which definitely consumes a larger area than that of the positively bevelled edge. Figure 2.43 gives the curve of

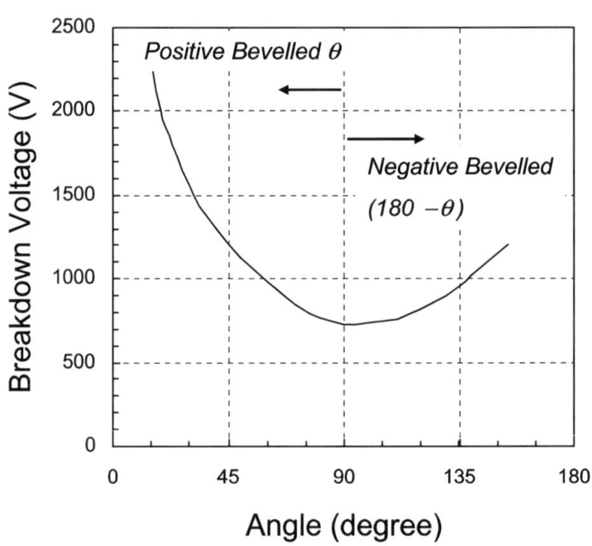

Fig. 2.43. Breakdown voltage variation as a function of bevelled angle (0–90° positive bevel, 90–180° negative bevel) for base doping level between 10^{13}–10^{14}/cm^3.

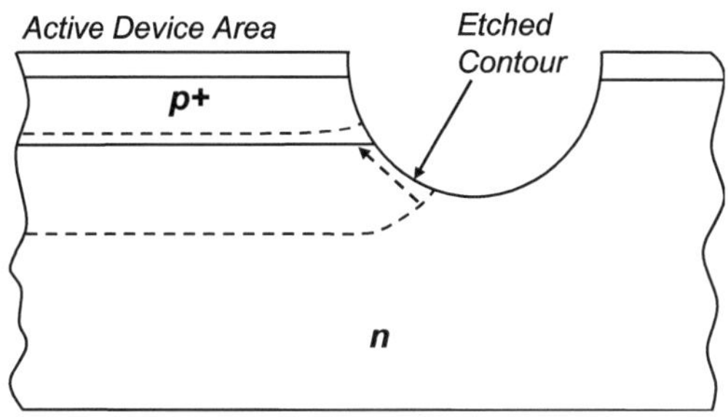

Fig. 2.44. The etched contour with negatively bevelled edge.

breakdown voltage at different bevelled angles (CSIHPDPIC, 1996). The bevelled angle θ is read as it is in the diagram when it is positively bevelled, and as $180 - \theta$ when it is negatively bevelled. For example, for the negatively bevelled angle of 45° [refer to Fig. 2.42(b)], it reads 135° in Fig. 2.43.

Similar bevelled contour can be formed by chemical etching, as shown in Fig. 2.44. Comparing this structure with Fig. 2.41, the difference between these two is derived the etched contour sidewall. In Fig. 2.41, the etched contour does not touch the diffused junction, and the main focus is to reduce the field crowding at the junction curvature. There is a field appearing along the contour-etched sidewall, but not high enough to be critical. A higher field appears at the top surface where the junction is located compared to that at the sidewall surface. In Fig. 2.44, when the etched contour touches the diffused junction, the sidewall surface field is then of concern as the diffused junction disappears from the top junction. The structure shown is a negatively bevelled junction and it will result in a higher surface field.

The bevelled surface needs to be treated properly to remove the contamination and damage after the process of sandblasting and grinding, and to cover the surface with passivation material.

2.11.6. Field Plate

The field plate as shown in Fig. 2.45(a) is a thin oxide layer surrounding the junction termination with metal contact deposited on top. The field plate is applied at the same potential as the p^+-layer to create a field of similar nature as that of the p^+-layer but without a physical space-charge junction in silicon. This oxide field diverts the depletion contour stretched away from the junction to create a similar field reduction effect as that of using the field rings or etched contour terminations. To be effective, the oxide layer thickness shall be thin enough for sufficient field to occur in silicon. For high voltage devices, thin

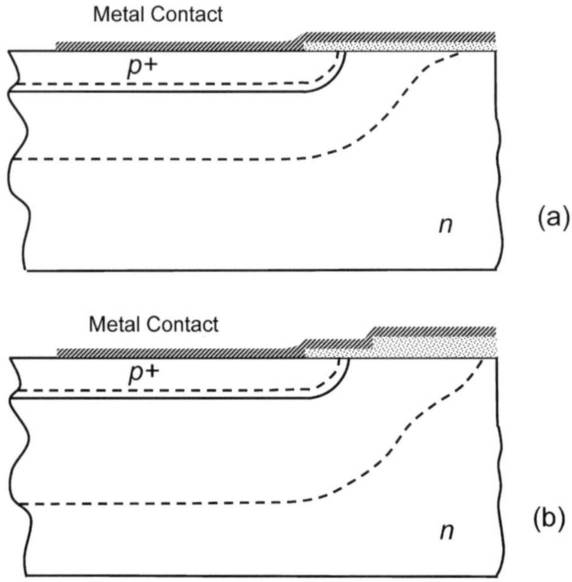

Fig. 2.45. Field plates with (a) homogeneous oxide thickness and (b) stacked oxide layers.

oxide may face the problem of breakdown. In this case, stepped oxide layers, as shown in Fig. 2.45(b), can be used to avoid the problem (Benda *et al.*, 1999). Since the field plate and the floating field ring have similar nature and effect in field reduction, they can be used together to bring down the field crowding to a minimum level.

2.11.7. Junction Termination Extension

Junction termination extension is made by ion implantation to the end of the junction for the formation of a thin and lightly doped layer around the perimeter of the diffusion junction, as shown in Fig. 2.46. The technique can be applied for planar junction or at etched contour junction as shown. The extended thin layer provides extension on depletion layer similar to that of a field plate to reduce the field crowding at the main junction (Temple, 1983; Temple and Tantraporn, 1986; Stengl and Gosele, 1985). Multiple junction termination extension layers with different doping concentrations, in a technique called Variation of Lateral Doping (VLD), can be used to obtain better results in field reduction, both at the main junction curvature and at the surface junction. The technique is considered to be more effective and space-saving than that of using the floating field rings.

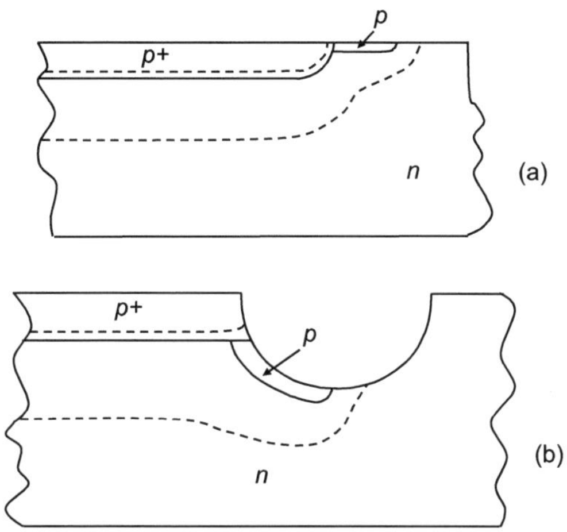

Fig. 2.46. Junction termination extension at (a) planar junction and (b) etched contour junction.

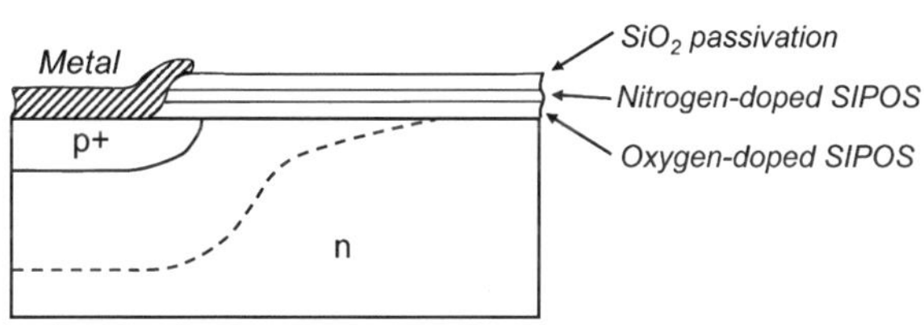

Fig. 2.47. Planar junction with SIPOS termination/passivation.

2.11.8. SIPOS (Semi-insulating Polycrystalline Silicon) Termination

SIPOS films are chemically vapor-deposited polycrystalline silicon doped with oxygen or nitrogen atoms. The passivation films normally are of three layers (Matsushita *et al.*, 1976; Stockmerier and Lilja, 1991), as shown in Fig. 2.47. The bottom layer is oxygen-doped SIPOS film with proper control of oxygen content for the film resistivity. This semi-insulating film serves two purposes: (a) to screen any undesired charge on silicon surface to avoid localized breakdown, and (b) to allow current flowing in SIPOS in order to form a linear potential gradient at the surface to relieve field crowding near the planar junction, so that the depletion boundary is diverted away from the junction. The film resistivity shall be low enough to allow (a) and (b) to be effective, but not too low to avoid high leakage current. Note that, the film resistance is also a function of temperature. When at high temperature, the leakage current goes higher. The second layer of the film is nitrogen-doped SIPOS to prevent sodium ions contamination; normally this is a very thin layer. The final layer is the SiO_2 or Si_3N_4 passivation that helps in the protection from dielectric breakdown.

2.12. Summary

In this chapter, the fundamental concept on power semiconductor physics and junction electrostatics including termination techniques were reviewed. This included the carrier transport physics, junction electrostatics, and the breakdown phenomena. It is hoped that by understanding these device physics, it will bring a more in-depth understanding on the analyses and utilization of power semiconductor devices. The same knowledge is also useful in working on power microelectronic device design and analyses.

References

M. S. Adler and V. A. K. Temple, Theory and breakdown voltage of planar junctions with a single field limiting ring, *IEEE Trans. Electron Devices* **24**, 107–113 (1977).

H. Akiyama, M. Harada, H. Konoh and Y. Akasaka, Partial lifetime control in IGBT by helium irradiation through mask patterns, *Int. Symp. Power Semiconductor Devices ICs* 187–191 (1991).

M. S. Adler and V. A. K. Temple, Maximum surface and bulk electric fields at breakdown for planar and beveled devices, *IEEE Trans. Electron Devices* **25**, 1266–1270 (1978).

B. J. Baliga and S. Krishna, Optimization of recombination levels and their capture cross-section in power rectifiers and thyristors, *Solid State Electron.* **20**, 225–232 (1977).

B. J. Baliga, Technological constraints upon properties of deep levels used for lifetime control in the fabrication of power rectifiers and thyristors, *Solid State Electron.* **20**, 1029–1032 (1977).

B. J. Baliga and E. Su, Comparison of gold, platinum, and electron irradiation for controlling lifetime in power rectifiers, *IEEE Trans. Electron. Devices* **24**, 685–688 (1977).

B. J. Baliga, High voltage device termination techniques — A comparative review, *IEE Proc. I: Commun. Speech Vision* **129**, 173–179 (1982).

B. J. Baliga, *Modern Power Devices*, Chap. 3 (John Wiley & Sons, New York, 1987).

A. Bar-Ler, *Semiconductors and Electronic Devices*, 2nd edn. (Prentice Hall International, 1984), p. 15.

V. Benda, J. Gowar and D. A. Grant, *Power Semiconductor Devices: Theory and Applications* (John Wiley & Sons, England, 1999), pp. 114–126.

R. O. Carlson, Y. S. Sun and H. B. Assalit, Lifetime control in silicon power devices by electron or gamma irradiation, *IEEE Trans. Electron Devices* **24**, 1103–1108 (1977).

J. B. Casady, A. K. Agarwal, L. B. Rowland, S. Seshadri and R. R. Siergiej, 4H-SiC power devices: Comparative overview of UMOS, DMOS and GTO device structures, *Power Semiconductor Material Devices*, Material Research Society Symposium Proceedings **483**, 27–38 (1998).

W. J. Choyke and G. Pensl, Physical properties of SiC, *Mater. Res. Soc. Bull.*, 25 (March, 1997).

Committee of specialist investigators on high performance power devices and power IC (CSIHPDPIC), *Power Semiconductor Device and Power IC Handbook* (Corona Publishing Co., Ltd, Japan, 1996), p. 54.

J. A. Cooper Jr, M. R. Melloch, R. Singh, A. Agarwal and J. W. Palmour, Status and prospects for SiC power MOSFETs, *IEEE Trans. Electron Devices* **40**, 658–664 (2002).

C. R. Crowell and S. M. Sze, Temperature dependence of avalanche multiplication in semiconductors, *Appl. Phys. Lett.* **9**, 242–244 (1966).

J. Cornu, Field distribution near the surface of beveled p–n junctions at high voltage devices, *IEEE Trans. Electron Devices* **20**, 347–352 (1973).

J. Cornu, S. Schweitzer and O. Kuhn, Double positive bevel: A better edge contour for high voltage devices, *IEEE Trans. Electron Devices* **21**, 181–184 (1974).

J. Dziewior and W. Schmid, Auger coefficients for highly doped and highly excited silicon, *Appl. Phys. Lett.* **31**, 346–348 (1977).

L. F. Eastman and U. K. Mishra, The toughest transistor yet, *IEEE Spectrum* 28–33 (May, 2002).

A. O. Evwaraye and B. J. Baliga, The dominant recombination centers in electron-irradiated semiconductor devices, *J. Electrochem. Soc.* **124**, 913–916 (1977).

J. Fang, Z. Li, H. Li and J. Yang, High speed LIGBT with localized lifetime control by using high dose and low energy helium implantation, *6th Solid-State Integrated Circuit Technol. Conf.*, 166–169 (2001).

W. Fulop, Calculation of avalanche breakdown of silicon p–n junctions, *Solid-State Electron.* **10**, 39–43 (1967).

C. C. Griffioen, J. H. Evans, P. C. De Jong and A. Van Veen, Helium desorption/permeation from bubbles in silicon: A novel method of void production, *Nucl. Instr. Meth. Phys. Sec. B* **27**, 417–420 (1987).

A. S. Grove, *Physics and Technology of Semiconductor Devices* (John Wiley & Sons, New York, 1967), pp. 133–134.

R. N. Hall, electron–hole generation in germanium, *Phys. Rev.* **87**, 387 (1952).

A. Hallen and M. Bakowski, Combined proton and electron irradiation for improved GTO thyristors, *Solid State Electron.* **32**, 1033–1037 (1989).

J. Jo, J. Park, H. J. Kim, S. H. Lee, Z. Y. Shen and Y. Nishihara, Lifetime control by low energy electron irradiation and hydrogen annealing, *Device Res. Conf.* 61–62 (2000).

Y. C. Kao and E. D. Wolley, High voltage planar p–n junctions, *Proc. IEEE* **55**, 1409–1414 (1967).

Y. Konish, Y. Onishi, S. Momota and K. Kakurai, Optimized local lifetime control for the superior IGBTs, *Int. Symp. Power Semiconductor Devices ICs* 335–338 (1996).

H. P. D. Lanyon and R. A. Tuft, Bandgap narrowing in heavily doped silicon, *IEEE Int. Electron Devices Meet. Digest* 316–319 (1978).

T. Matsushita, T. Aoki, T. Ohtsu, H. Yamoto, H. Hayashi, M. Okayama and Y. Kawana, Highly reliable high-voltage transistors by use of the SIPOS process, *IEEE Trans. Electron Devices* **23**, 826–830 (1976).

A. G. Milnes, *Deep Impurities in Semiconductors*, Chap. 2 (Wiley, New York, 1973).

A. Mogro-Campero, R. P. Love, M. F. Chang and R. Dyer, Localized lifetime control in insulated-gate transistors by proton implantation, *IEEE Trans. Electron Devices* **33**, 1667–1671 (1986).

N. Ramungul, T. P. Chow, M. Ghezzo, J. Kretchmer and W. A Hennessy, A fully planarized 6H-SiC UMOS insulated gate bipolar transistor, *Device Res. Conf.* 56–57 (1996).

C. T. Sah, R. N. Noyce and W. Shockley, Carrier generation and recombination in p–n junction and p–n junction characteristics, *Proc. IRE* **45**, 1228–1243 (1957).

M. Saggio, V. Raineri, F. Frisina and E. Rimini, Localized lifetime control in silicon bipolar power devices by voids induced by He ion implantation, *Power Semiconductor Material and Devices*, Material Research Society Symposium Proceedings **483**, 45–50 (1998).

W. Shockley and W. T. Read, Statistics of the recombination of holes and electrons, *Phys. Rev.* **87**, 835–842 (1952).

R. Stengl and U. Gosele, Variation of lateral doping — A new concept to avoid high voltage breakdown of planar junctions, *IEEE Int. Electron Devices Meet. Digest*, 154–157 (1985).

T. Stockmerier and K. Lilja, SIPOS-passivation for high voltage power devices with planar junction termination, *Int. Symp. Power Semiconductor Devices and ICs* 145–148 (1991).

S. M. Sze and G. Gibbons, Effect of junction curvature on breakdown voltage in semiconductors, *Solid State Electronics* **9**, 831–845 (1966a).

S. M. Sze and G. Gibbons, Avalanche breakdown voltages of abrupt and linearly graded p–n junction in Ge, Si, GaAs and GaP, *Appl. Phys. Lett.* **8**, 111–113 (1966b).

V. A. K. Temple and M. S. Adler, The theory and application of a simple etch contour for near ideal breakdown in plane and planar p–n junction, *IEEE Trans. Electron Devices* **23**, 950–955 (1976).

V. A. K. Temple, Increased avalanche breakdown voltage and controlled surface electric fields using junction termination extension (JTE) technique, *IEEE Trans. Electron Devices* **30**, 954–957 (1983).

V. A. K. Temple and W. Tantraporn, Junction termination extension for near-ideal breakdown voltage in p–n junctions, *IEEE Trans. Electron Devices* **33**, 1601–1608 (1986).

3

Bipolar Junction Diode

3.1. Introduction

In Chapter 2, it has given brief description on the p–n junction as a two-terminal semiconductor bipolar device functioning for rectification. The junction acts as a switch which allows current to flow under forward bias condition and blocks the current by a much higher potential barrier under reverse bias condition. So, a simple p–n junction diode is serving as an uncontrollable switch. In a sense, the behavior of the semiconductor diode is preset by the internal p–n junction, rather than by or with any combined influence of external control signal. The diode structure can be complemented by the field-controlled mechanism, called field-controlled diode, or similar variations to create the ability of external control (Baliga, 1981; Thapar and Baliga, 1997). Bear in mind that, the semiconductor diode can also be used for applications other than current switching, e.g. used as a small signal attenuator with no current switching. Or, for some applications, the diode is fabricated in the integrated circuits serving as a capacitor under reverse bias condition. In this case, the diode will not carry current, but serve as a voltage-controlled junction capacitor. Nevertheless, for the main applications of power semiconductor diodes described here, the main properties remain on the switching behaviors and power related issues, such as the current conduction, blocking and reverse recovery.

There are various ways to form a diode structure on silicon and the variations mainly due to the process differences and the need for integration with other devices on the same silicon die. Sometimes, the diode structure is formed as a parasitic component integrated with another device, such as MOSFET device has a built-in body diode. Figure 3.1 shows various diode structures formed by different fabrication processes. Among them, Fig. 3.1(a) shows a simple vertical p–n junction diode by diffusion process. Figure 3.1(b) is the

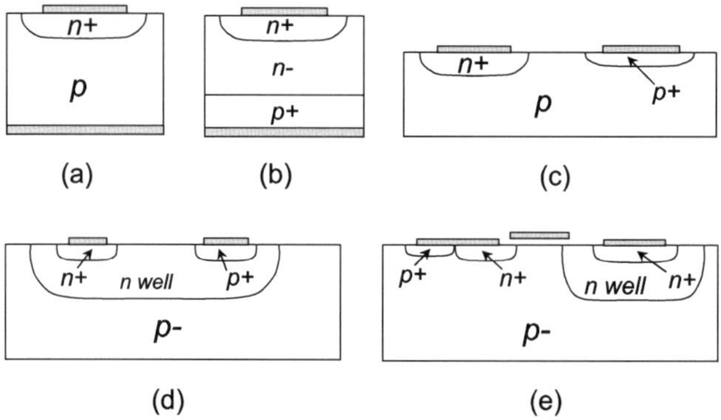

Fig. 3.1. Various diode structures: (a) vertical p–n diode, (b) vertical p–i–n diode, (c) lateral p–n diode, (d) lateral p–i–n diode, and (e) MOSFET body diode.

vertical p–i–n diode using the similar fabrication process as in (a), but made on the p^+–n^- epi-wafer. Figure 3.1(c) gives the lateral p–n diode where the p^+ layer is very shallow made for ohmic contact purpose. The lateral p–n diode is suitable for integration. Figure 3.1(d) gives the lateral p–i–n diode using the n-well in CMOS process as the drift region. Figure 3.1(e) is a lateral LDD n-MOSFET structure with p-body shorted to the source (left n^+ contact). The body diode is then formed between the p-body and the drain contact (right n^+ contact) with the n-well drift region in between. The body diode may conduct during the fly-wheeling process used in power converters.

The behaviors of the junction diode device are determined by two basic elements, namely the junction and the semiconductor doping within the device. The junction behaves like a check-valve, which allows a certain type of carriers to go through but not the other type of the carriers in the same direction of flow. To describe the concept more clearly, let us take a p^+–n junction as an example. If the junction is reverse biased, then the electric field is running from the n-side to the p^+-side. Under such a condition, the junction will allow electron carriers to move from p^+-side to n-side, or equally capable for hole carriers to move from n-side to p^+-side. And at the same time, the electron carriers are prevented to move from n-side to p^+-side, nor can the hole carriers flow from the p^+-side to n-side. The concept seems to be trivial for one single junction under steady-state, but it becomes rather important when dealing with a multiple-junction device or under transient condition when electron and hole carriers are mixed at different regions and a reverse-biased junction does carry current flow. The second element affecting the behaviors is the doping level. If an area is used for carrier emission, it then needs to have a higher concentration of doping for a better emission efficiency to boost up the current density. Otherwise, for a low emission efficiency it means not enough number of carriers and, in term, the device has a high on-state voltage drop. If the area serves to sustain high field, then a lower doping level drift-region shall be used to raise the breakdown voltage. The high resistivity of a lightly doped region at off-state may not be the same resistivity during on-state current conduction depending on the amount of excess carriers flowing into this region. If a large amount of excess carriers are injected into the lightly doped region, then the injection creates an effect called the ***conductivity modulation*** for the situation that more free carriers are now available within the region and the resistivity of the region becomes much lower for the conduction period.

In modern power electronic circuits, the diode device plays a major role in current rectification and energy fly-wheeling. A fast switching diode leads to a lower switching loss and in turn a higher system efficiency. Besides the switching speed, other desired characteristics such as low forward voltage drop, low leakage current, and high temperature capability are also desired. Power Schottky barrier diodes made of metal-silicon junction are generally used for high-speed switching applications due to its less amount of excess carrier storage compared to the conventional diodes. However, the Schottky barrier diode has a limited role above 100 V due to its higher reverse leakage current. The p^+–n^-–n^+ (or called p–i–n) diodes can be used for high-voltage switching and rectification purpose. However, the switching speed is relatively slower for the recovery time needed to clear up the excess carriers in the long drift region. High voltage GaAs power diodes are now commercially available for high-speed and high temperature switching applications for its larger bandgap compared with that of silicon. Diodes made of SiC material are still in the laboratory development stage and hopefully to be available soon in the market.

In this chapter, the device physics of power p^+–n^-–n^+ and Schottky barrier diodes are reviewed and discussed. An experimental investigation on the high temperature characteristics of a silicon p^+–n^-–n^+ diode is accompanied to completely cover the steady-state silicon diode properties. Diodes of GaAs and SiC are briefly described. This is followed by switching behaviors of the junction diodes. New devices, such as the field-controlled diode, MPS (Merged p–(i)–n and Schottky) diode and synchronous rectifier are also described at the end of the chapter.

3.2. Basic Junction Diode Theory

3.2.1. Forward Conduction

The current flow in a p–n junction diode is determined by the diffusion and subsequent recombination of the minority carriers injected into regions on either side of a forward-biased junction, as shown in Fig. 3.2. The current

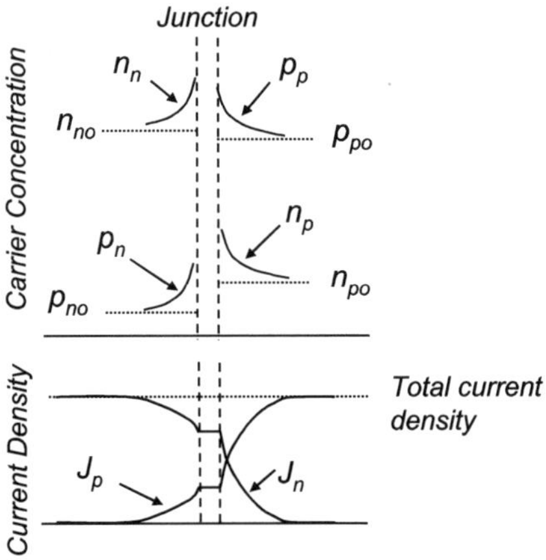

Fig. 3.2. Distribution of carrier densities and current densities across a forward-biased junction.

flow is a consequence of carrier movement following the Schockley's diffusion approximation (Schockley, 1949). At low-level injection, the minority-carrier densities at the edges of the space-charge layer with the application of a forward bias, V_f, are given by

$$n_p = n_{p0} e^{\frac{qV_f}{kT}}, \tag{3.1}$$

$$p_n = p_{n0} e^{\frac{qV_f}{kT}}, \tag{3.2}$$

where n_{p0} and p_{n0} are the equilibrium minority carrier concentrations in the p and n regions respectively, k is the Boltzmann's constant, and T is the temperature. Assume that both p and n sides are uniformly doped and sufficiently long in length that all the injected minority carriers recombine before they reach the contact. For generation in the space-charge region is too small to be considered, the time-independent hole diffusion equation for the n-side of the junction is

$$D_p \frac{d^2 p_n(x)}{dx^2} - \frac{\Delta p_n(x)}{\tau_p} = 0, \tag{3.3}$$

where τ_p is the hole minority-carrier lifetime, D_p is the hole carrier diffusion coefficient and the excess minority carrier at the n-side, $\Delta p_n(x)$, is defined as

$$\Delta p_n(x) = p_n(x) - p_{n0} \tag{3.4}$$

The general solution to Eq. (3.3) is in the form of

$$\Delta p_n(x) = A e^{-\frac{x}{\sqrt{D_p \tau_p}}} + B e^{\frac{x}{\sqrt{D_p \tau_p}}}, \tag{3.5}$$

where A and B are the constants to be determined using the appropriate boundary conditions. Since all injected holes are recombined before reaching the external contact, the boundary condition at $x = \infty$ is

$$p_{\rm n}(x=\infty) = p_{\rm n0}. \tag{3.6}$$

At the edge of the depletion layer at the n-side of the junction where $x = 0$, the injected minority-carrier (hole) density is given by Eq. (3.2)

$$p_{\rm n}(x=0) = p_{\rm n0}e^{\frac{qV_{\rm f}}{kT}}. \tag{3.7}$$

The constant B is zero in order to satisfy the boundary condition in Eq. (3.6). Thus, the distribution of the injected hole is found to be

$$\Delta p_{\rm n}(x) = p_{\rm n0}\left(e^{\frac{qV_{\rm f}}{kT}} - 1\right)e^{-\frac{x}{L_{\rm p}}}, \tag{3.8}$$

where

$$L_{\rm p} = \sqrt{D_{\rm p}\tau_{\rm p}} \tag{3.9}$$

is the hole diffusion length in the n region. Note that the minority carriers decay exponentially from the initial injected density given by Eq. (3.2) to the equilibrium concentration of $p_{\rm n0}$ near the external contact. The hole diffusion current density is

$$J_{\rm p}(x) = -qD_{\rm p}\frac{dp_{\rm n}(x)}{dx} = \frac{qD_{\rm p}p_{\rm n0}}{L_{\rm p}}\left(e^{\frac{qV_{\rm f}}{kT}} - 1\right)e^{-\frac{x}{L_{\rm p}}}. \tag{3.10}$$

At the edge of the junction depletion where the minority hole carriers entering the n region, the hole diffusion current density is

$$J_{\rm p}(0) = \frac{qD_{\rm p}p_{\rm n0}}{L_{\rm p}}\left(e^{\frac{qV_{\rm f}}{kT}} - 1\right). \tag{3.11}$$

Similarly, the electron distribution at the p-side of the junction is

$$\Delta n_{\rm p}(x) = n_{\rm p0}\left(e^{\frac{qV_{\rm f}}{kT}} - 1\right)e^{-\frac{x'}{L_{\rm n}}} \tag{3.12}$$

and the electron diffusion current density is

$$J_{\rm n}(x) = \frac{qD_{\rm n}n_{\rm p0}}{L_{\rm n}}\left(e^{\frac{qV_{\rm f}}{kT}} - 1\right)e^{-\frac{x'}{L_{\rm n}}}, \tag{3.13}$$

where

$$L_{\rm n} = \sqrt{D_{\rm n}\tau_{\rm n}} \tag{3.14}$$

is the minority electron diffusion length in the p-region, and x' is the spatial dimension at the p-region with $x' = 0$ located at the edge of the depletion layer. The total current density is obtained by adding the minority carrier current density components at the junction space-charge layer edge, i.e. $x = x' = 0$. Therefore,

$$J_\mathrm{f} = q\left(\frac{D_\mathrm{p} p_\mathrm{n0}}{L_\mathrm{p}} + \frac{D_\mathrm{n} n_\mathrm{p0}}{L_\mathrm{n}}\right)\left(e^{\frac{qV_\mathrm{f}}{kT}} - 1\right) = J_0\left(e^{\frac{qV_\mathrm{f}}{kT}} - 1\right), \tag{3.15}$$

where J_0 is the saturation current density and is given by

$$J_0 = q\left(\frac{D_\mathrm{p} p_\mathrm{n0}}{L_\mathrm{p}} + \frac{D_\mathrm{n} n_\mathrm{p0}}{L_\mathrm{n}}\right). \tag{3.16}$$

Figure 3.2 also shows the minority-carrier diffusion currents for a forward-biased p–n junction.

3.2.2. Short-Base Diode

In a narrow-base or called short-base diode, the length of the lightly doped side of a p^+–n diode is close to or less than the minority-carrier diffusion length. Hence, all the remaining minority carriers will be recombined at the contact. The boundary condition of Eq. (3.6) must be modified to

$$p_\mathrm{n}(x = W_\mathrm{n}) = p_\mathrm{n0} \tag{3.17a}$$

or

$$\Delta p_\mathrm{n}(x = W_\mathrm{n}) = 0, \tag{3.17b}$$

where W_n is the n-base width. Applying the new set of boundary conditions gives

$$\Delta p_\mathrm{n}(x = 0) = \Delta p_\mathrm{n}(0) = A + B, \tag{3.18}$$

$$\Delta p_\mathrm{n}(x = W_\mathrm{n}) = Ae^{-\frac{W_\mathrm{n}}{L_\mathrm{p}}} + Be^{\frac{W_\mathrm{n}}{L_\mathrm{p}}} = 0. \tag{3.19}$$

Solving for constants A and B and substituting back into Eq. (3.5) yields,

$$\Delta p_\mathrm{n}(x) = \Delta p_\mathrm{n}(0)\left(\frac{e^{\frac{W_\mathrm{n}-x}{L_\mathrm{p}}} - e^{\frac{-(W_\mathrm{n}-x)}{L_\mathrm{p}}}}{e^{\frac{W_\mathrm{n}}{L_\mathrm{p}}} - e^{-\frac{W_\mathrm{n}}{L_\mathrm{p}}}}\right). \tag{3.20}$$

Thus, the distribution of the injected hole carrier is

$$\Delta p_\mathrm{n}(x) = \Delta p_\mathrm{n}(0)\frac{\sinh\left(\frac{W_\mathrm{n}-x}{L_\mathrm{p}}\right)}{\sinh\left(\frac{W_\mathrm{n}}{L_\mathrm{p}}\right)}. \tag{3.21}$$

The hole current density can be found as

$$J_p(x) = \frac{qD_p p_{n0}}{L_p}\left(e^{\frac{qV_f}{kT}} - 1\right)\frac{\cosh\left(\frac{W_n - x}{L_p}\right)}{\sinh\left(\frac{W_n}{L_p}\right)}, \tag{3.22a}$$

$$J_p(0) = \frac{qD_p p_{n0}}{L_p}\left(e^{\frac{qV_f}{kT}} - 1\right)\frac{1}{\tanh\left(\frac{W_n}{L_p}\right)} \tag{3.22b}$$

which is independent of x. If W_n/L_p is much less than one, the sinh( ) terms in Eq. (3.21) can be replaced in approximation by their arguments, and the excess carrier concentration becomes

$$\Delta p_n(x) = \Delta p_n(0)\left(1 - \frac{x}{W_n}\right). \tag{3.23}$$

Note that the carrier concentration becomes a linear function of position as a consequence of negligible thermal generation–recombination in a region much shorter than a diffusion length. In this case, the hole current density is

$$J_p(x) = \frac{qD_p p_{n0}}{W_n}\left(e^{\frac{qV_f}{kT}} - 1\right). \tag{3.24}$$

Note that the hole current density is now controlled by the width of the lightly doped n-region rather than the diffusion length as in the long diode. Figure 3.3 gives the selected I–V curves on diode forward conduction with and without the end-region recombination and higher current is observed on the short-base diode.

3.2.3. Junction Capacitance

When the diode junction is reverse-biased, the space-charge region behaves like a variable capacitor with the capacitance controlled by the applied voltage. The capacitance per square area is defined as

$$C \equiv \frac{dQ}{dV} = \frac{d(qN_d x_n)}{dV_R}, \tag{3.25a}$$

where, for abrupt junction

$$x_n = \sqrt{\frac{2\varepsilon_s N_a(V_R + \psi_0)}{qN_d(N_a + N_d)}}, \tag{3.25b}$$

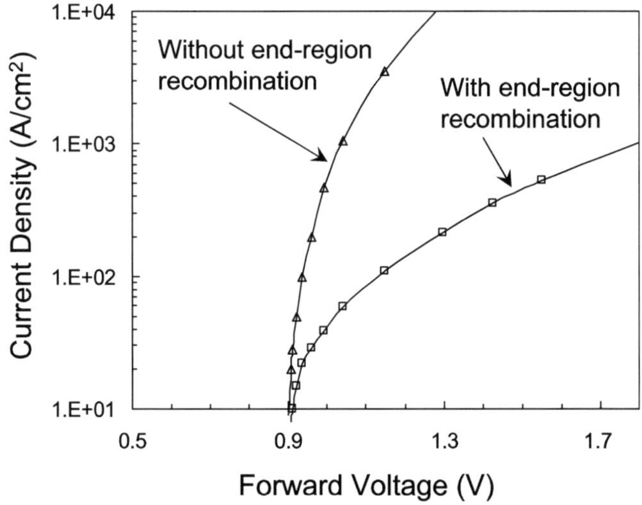

Fig. 3.3. Calculated diode forward current conduction with and without the end-region recombination.

where ψ_0 is the built-in junction potential barrier. Combine Eqs. (3.25a) and (3.25b), the junction capacitance can be calculated as

$$C = \sqrt{\frac{q\varepsilon_s N_a N_d}{2(N_a + N_d)(V_R + \psi_0)}}. \tag{3.26}$$

From the equation, it can be seen that the capacitance changes nonlinearly with applied voltage and becomes smaller when the voltage increases. For an abrupt p^+–n junction, the doping level of N_a is much greater than N_d, so Eq. (3.26) can be simplified as

$$C \approx \sqrt{\frac{q\varepsilon_s N_d}{2(V_R + \psi_0)}}. \tag{3.27}$$

So, a lower doping of N_d will increase the depletion width quickly and subsequently a lower capacitance appears. If the junction is a linearly graded junction, then Eq. (3.25b) needs to be changed to Eq. (3.28) in order to calculate the correct amount of space-charge

$$x_n = \left(\frac{3\varepsilon_s(V_R + \psi_0)}{qH}\right)^{\frac{1}{3}}, \tag{3.28}$$

where H is the doping gradient as described previously in Chapter 2. The junction capacitance can now be calculated as

$$C = \frac{dQ}{dV} = \frac{d\left(\frac{1}{2}qHx_{\mathrm{n}}^2\right)}{dV} = 0.436 \times \left(\frac{Hq\varepsilon_{\mathrm{s}}^2}{V_{\mathrm{R}} + \psi_0}\right)^{\frac{1}{3}}. \tag{3.29}$$

Similar trend of lower capacitance with higher voltage is observed in Eq. (3.29) but with different proportion with voltage compared to that of the abrupt junction. In Eq. (3.29), the junction capacitance decreases more moderately.

SPICE, an acronym for Simulation Program with Integrated Circuit Emphasis, is a general-purpose circuit analysis program widely used in the simulation of electronic circuits. It was developed by the Electronic Research Laboratory at the University of California, Berkeley in the late 1960s and was released to the public in 1972. Over the years, SPICE has gone through many upgrades. SPICE became very popular when MicroSim in 1984 introduced a personal-computer (PC) version of SPICE known as PSpice (MicroSim, 1984). In SPICE, the total capacitance of an abrupt $\mathrm{n^+}$–p junction diode consists of both a junction capacitance and a storage capacitance modeled by

$$C_{\mathrm{t}}(V_{\mathrm{d}}) = \frac{q}{NkT}T_{\mathrm{t}}I_{\mathrm{s}}e^{\frac{qV_{\mathrm{d}}}{nkT}} + C_{\mathrm{j0}}\left(1 - \frac{V_{\mathrm{d}}}{V_{\mathrm{j}}}\right)^{-0.5}. \tag{3.30}$$

In the above equation, n is the emission coefficient and it falls within the range of 1–2. T_{t} is the transit time and it can be approximated by the minority electron lifetime in the lightly doped p-region of the $\mathrm{n^+}$–p junction diode. V_{j} is the junction potential and is given as

$$V_{\mathrm{j}} = \frac{kT}{q}\ln\left(\frac{N_{\mathrm{a}}N_{\mathrm{d}}}{n_{\mathrm{i}}^2}\right), \tag{3.31}$$

where N_{d} and N_{a} are the donor concentration in the $\mathrm{n^+}$ region and acceptor concentration in the p-region, respectively. I_{s} is the saturation current and is given by

$$I_{\mathrm{s}} = qA\left(\frac{D_{\mathrm{p}}p_{\mathrm{n0}}}{L_{\mathrm{p}}} + \frac{D_{\mathrm{n}}n_{\mathrm{p0}}}{L_{\mathrm{n}}}\right), \tag{3.32}$$

where p_{n0} and n_{p0} are the minority hole concentration in the $\mathrm{n^+}$ region and minority electron concentration in the p-region. D_{p} and D_{n} are the diffusion coefficients for holes in the $\mathrm{n^+}$ region and electrons in the p-region, respectively. L_{p} and L_{n} are the diffusion lengths for holes in the $\mathrm{n^+}$ region and electrons in the p-region, respectively. C_{j0} is the zero-bias junction capacitance and is given as

$$C_{j0} = \frac{\varepsilon_s A}{\sqrt{\frac{2\varepsilon_s}{q} V_j \left(\frac{1}{N_a} + \frac{1}{N_d}\right)}}, \tag{3.33}$$

where ε_s is the permittivity of silicon.

3.3. High-Voltage p$^+$–n$^-$–n$^+$ Diode

The p$^+$–n$^-$–n$^+$ diode as shown Fig. 3.4 consists of three semiconductor regions with two high–low junctions, the p$^+$–n$^-$ and n$^+$–n$^-$ junctions. The diode is sometimes called the p–i–n diode or punchthrough diode for its lightly doped base region and the easy punchthrough nature. The doping and thickness of the n$^-$ region are controlled to achieve the desired reverse blocking voltage as described in Chapter 2. Due to the relatively low doping concentration of the n$^-$ region, the injected carrier concentration often exceeds the base background concentration. As such, the high-level carrier injection occurs while the injected hole density is much greater than the background concentration of the n$^-$ layer. Similar situation occurs for electron carriers as charge quasi-neutral constraint in the n$^-$ region requires the condition that the concentrations of the injected holes must be equal to that of the electrons in the base region, such that

$$n(x) = p(x) \quad \text{where } -d \leq x \leq d. \tag{3.34}$$

It should be noted that the concentrations of the excess holes and electrons injected from both p$^+$ and n$^+$ regions can become several orders much higher than the background concentration of n$^-$. As such, the conductivity in this layer changes a lot and becomes much higher than it was. This phenomenon

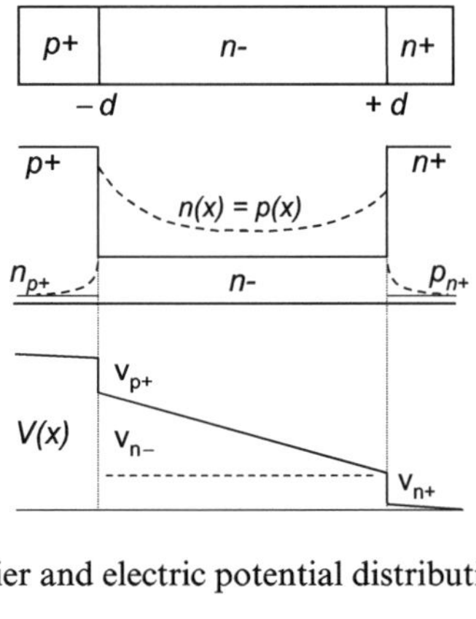

Fig. 3.4. Excess carrier and electric potential distributions in a p–i–n diode.

is known as the ***conductivity modulation***. The conductivity modulation is the mechanism for the reduction of the on-state voltage drop in the p^+–n^-–n^+ diode. Such a property makes the lightly doped n^- region used for high-voltage blocking layer now a less burden for on-state current conduction. However, the degree of conductivity modulation is proportional to the amount of the excess carriers in the base region. When a p–i–n diode conducts under heavy conductivity modulation at high-current conduction, the amount of excess storage charge is very large. This makes it difficult to quickly turn-off large current as more time is needed to clear up the amount of excess carrier storage. Therefore, the outcome of conductivity modulation makes the turn-off time as a function of its on-state current.

3.3.1. Forward Conduction

For the silicon p–i–n diode, the forward conduction can be briefly divided into four different regions, namely the tiny current recombination region, the low-level diffusion current region, the medium to high-level injection region, and the very high current ohmic region, as depicted in Fig. 3.5. For the first region, the current is proportional to $e^{\frac{qV_f}{2kT}}$. However, it is too small, e.g. a few orders of the reverse leakage current, to be important in diode forward conduction. The second region represents a low-level carrier injection and

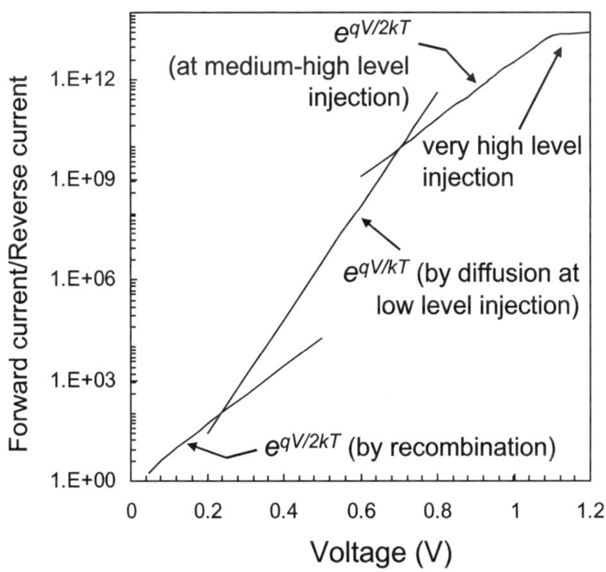

Fig. 3.5. Conduction modes of p–i–n diode under different forward bias levels.

diffusion to the lightly doped base region. This is described by Shockley's theory of the p–n junction (Shockley, 1949) as stated in the previous section. The forward current is proportional to $e^{\frac{qV_f}{kT}}$. The third region begins at medium to high-level injection. The diode current is approximately determined by the recombination process of electrons and holes in the base region, whereas there is very little carrier injection across the heavily doped borders to p^+ and n^+ regions. The forward current is proportional to $e^{\frac{qV_f}{2kT}}$ as described by the Hall's theory (Hall, 1952). The fourth region represents the very high current region where the carrier concentration in the base region is extremely high filled with both electrons and holes. Diffusion process occurs at both the lightly doped region and the heavily doped n^+ and p^+ regions. The diffusion current at the heavily doped regions can no longer be ignored and shall be included as part of the total current (Rose, 1964; Herlet, 1968). Conductivity modulation saturates at this point and the diode behaves more like an ohmic device now.

The diode conduction behavior at low-level forward current was described early. For the medium level of forward current, the mechanism of forward current conduction in the p^+–n^-–n^+ diode consists predominantly of the recombination of holes and electrons in the n^- layer, as shown in Fig. 3.4. The recombination of holes in the n^+ region and the recombination of electrons in the p^+ region are significant only at high-level of forward current conduction, i.e. the third region. The current density in the n^- region under steady-state condition is given by

$$J_{n^-} = q\int_{-d}^{d} \frac{c(x)}{\tau_a} dx = \frac{2qd\bar{c}}{\tau_a}, \tag{3.35}$$

where $c(x)$ is the distribution of excess carrier density, $\bar{c}$ is the average excess carrier density, τ_a is the amipolar carrier lifetime equal to $\tau_{p0} + \tau_{n0}$, and d is half of the n^- layer width, i.e. $-d$ to $+d$ covering the entire n^- region. The high-level lifetime is used here to indicate a large amount of excess carriers, both hole and electron, diffusing into the base region for recombination process to occur. Since the free carrier density is directly proportional to the current density, the voltage drop in the n^- layer will then be treated independent of the current density. The free carrier density distribution in the drift region can be found by solving the carrier continuity equation:

$$\frac{d^2c(x)}{dx^2} - \frac{c(x)}{D_a\tau_a} = 0, \tag{3.36}$$

where D_a is the ambipolar diffusion coefficient given by

$$D_a = \frac{2D_nD_p}{D_n + D_p} = \frac{L_a^2}{\tau_a} \tag{3.37}$$

and L_a is the ambipolar diffusion length. Equation (3.36) can rewritten as

$$\frac{d^2c(x)}{dx^2} = \frac{c(x)}{L_a^2}. \tag{3.38}$$

The general solution to Eq. (3.38) is

$$c(x) = K_1 \cosh\left(\frac{x}{L_a}\right) + K_2 \sinh\left(\frac{x}{L_a}\right). \tag{3.39}$$

The boundary condition at the n^+–n^- junction can be determined by recognizing that the current transport is made entirely by electrons. Thus, the hole current is almost zero at $x = +d$:

$$J_p(+d) = q\mu_p p(x = +d)\hat{E}(x = +d) - qD_p \left.\frac{dp}{dx}\right|_{x=+d} \approx 0, \tag{3.40}$$

where $\hat{E}(x)$ is the electric field in the base region. From the above equation and the Einstein relationship, the electric field at the n^+–n^- (i.e. $x = +d$) junction is

$$\hat{E}(+d) = \frac{kT}{qp(x = +d)} \left.\frac{dp}{dx}\right|_{x=+d}. \tag{3.41}$$

The electron current density at $x = +d$ is given by

$$J_n(+d) = q\mu_n n(x = +d)\hat{E}(x = +d) + qD_n \left.\frac{dn}{dx}\right|_{x=+d}. \tag{3.42}$$

Substituting Eqs. (3.34) and (3.41) into Eq. (3.42), the current density at $x = +d$ can be obtained as

$$J_{n^-} = J_n(+d) = 2qD_n \left.\frac{dn}{dx}\right|_{x=+d}. \tag{3.43}$$

Similarly, the boundary condition at the p^+–n^- junction with $J_n(-d) = 0$ yields

$$J_{n^-} = J_p(-d) = -2qD_p \left.\frac{dn}{dx}\right|_{x=-d}. \tag{3.44}$$

The solution of Eq. (3.39) with the above boundary conditions is given by (Herlet, 1968)

$$n(x) = p(x) = \frac{J_{n^-} \times \tau_a}{2qL_a}\left(\frac{\cosh\left(\frac{x}{L_a}\right)}{\sinh\left(\frac{d}{L_a}\right)} - B\frac{\sinh\left(\frac{x}{L_a}\right)}{\cosh\left(\frac{d}{L_a}\right)}\right), \tag{3.45}$$

where

$$B = \frac{\frac{\mu_n}{\mu_p} - 1}{\frac{\mu_n}{\mu_p} + 1}. \tag{3.46}$$

Thus, the concentration of electrons and holes in the n^- region is proportional to the current density J. For the condition if $\mu_n = \mu_p$, then $B = 0$. In this case, $n(x)$ and $p(x)$ both will have a symmetrical distribution, and the boundary values at $x = +d$ and $x = -d$ are equal. If μ_n is not exactly equal to μ_p, then an asymmetrical carrier density distribution, due to the second term in Eq. (3.45), is obtained. For example, if $\mu_n > \mu_p$, the carrier concentration is higher at the p^+–n^- junction, i.e. $x = -d$ and lower at the n^+–n^- junction, i.e. $x = +d$, as depicted in Fig. 3.4.

For a fixed current density, the average value $\overline{c}$ of the carrier concentration distribution is independent of the value B. This is because by integrating Eq. (3.45), only the first symmetrical term gives a contribution to the average value. Thus,

$$\overline{c} = \frac{\tau_a J_{n-}}{2dq}. \tag{3.47}$$

The concentrations of electrons and holes, for $c(x) = n(x) = p(x)$, in the neutral n^- region at two junctions are (Herlet, 1968)

$$c(-d) = \overline{c}\frac{d}{L_a}\coth\left(\frac{d}{L_a}\right)\left(1 + B\tanh^2\left(\frac{d}{L_a}\right)\right), \tag{3.48}$$

$$c(+d) = \overline{c}\frac{d}{L_a}\coth\left(\frac{d}{L_a}\right)\left(1 - B\tanh^2\left(\frac{d}{L_a}\right)\right). \tag{3.49}$$

Equations (3.48) and (3.49) can then be used to obtain the current density at the edges of lightly doped n base by substituting them into Eqs. (3.43) and (3.44). The diode voltage drop comprises of contributions of the voltage drops across the space-charge regions at both the p^+–n^- and n^+–n^- junctions, and the resistive voltage drop across the middle n^- region. The resistive drop plays an important role in voltage variation by conductivity modulation. The voltage drop across the middle n^- layer, V_{n^-}, can be found by integrating the electric field $\hat{E}_{n^-}(x)$ over the n^- region,

$$V_{n^-} = \int_{-d}^{+d} \hat{E}_{n^-}(x)dx \tag{3.50}$$

and, the current density in the n^- region is

$$J_{n^-} = q(\mu_n n(x) + \mu_p p(x))\hat{E}_{n^-}(x) + \frac{kT}{q}\left(\mu_n\frac{dn}{dx} - \mu_p\frac{dp}{dx}\right). \tag{3.51}$$

So, the electric field $\hat{E}_{n^-}(x)$ can be obtained from the total current density of Eq. (3.51). Taking into account that $n(x) = p(x)$ and $dn/dx = dp/dx$, the electric field in the n^- layer is

$$\hat{E}_{n^-}(x) = \frac{J_{n^-}}{q(\mu_n + \mu_p)n(x)} - B\frac{kT}{qn(x)}\frac{dn}{dx}. \tag{3.52}$$

It can be seen that the electric field comprises of an *ohmic component* proportional to the current density and another component which depends upon the carrier concentration distribution. Carrying out the integration in Eq. (3.50), the voltage across the n^- layer is (Herlet, 1968)

$$\begin{aligned} V_{n^-} = \frac{kT}{q}&\left(\frac{J}{J_{n^-}}\frac{8b}{(b+1)^2}\frac{\sinh\left(\frac{d}{L_a}\right)}{\sqrt{1 - B^2\tanh^2\left(\frac{d}{L_a}\right)}}\right. \\ &\times \tanh^{-1}\left(\sqrt{1 - B^2\tanh^2\left(\frac{d}{L_a}\right)}\sinh\left(\frac{d}{L_a}\right)\right) \\ &\left. + B\cdot\ln\left(\frac{1 + B\tanh^2\left(\frac{d}{L_a}\right)}{1 - B\tanh^2\left(\frac{d}{L_a}\right)}\right)\right), \end{aligned} \tag{3.53}$$

where b is the ratio of μ_n/μ_p. As it indicates, the voltage drop across the n^- region is independent of its current density, e.g. $J = J_{n^-}$ for the second region in Fig. 3.4, as treated and it is a function of d/L_a and B only. The voltage drops across the p^+–n^- and n^+–n^- junctions also contribute to the diode forward voltage drop. According to Boltzmann's law, the injected minority carrier densities at these junctions are:

$$p(-d) = p_{n0}e^{\frac{qV_{p^+}}{kT}} \tag{3.54}$$

$$n(+d) = n_{n0}e^{\frac{qV_{n^+}}{kT}}. \tag{3.55}$$

The voltage drops are calculated as

$$V_{p^+} = \frac{kT}{q}\ln\left(\frac{p(x=-d)}{p_{n0}}\right) = \frac{kT}{q}\ln\left(\frac{p(x=-d)N_{d,n^-}}{n_i^2}\right), \tag{3.56}$$

$$V_{n^+} = \frac{kT}{q}\ln\left(\frac{n(x=+d)}{N_{d,n^-}}\right). \tag{3.57}$$

The total voltage drops across the two high–low junctions are

$$V_{\mathrm{p}^+} + V_{\mathrm{n}^+} = \frac{kT}{q} \ln\left(\frac{n(x=+d)p(x=-d)}{n_\mathrm{i}^2}\right). \tag{3.58}$$

Combining Eqs. (3.53) and (3.58), the total voltage drop across the diode can be obtained. Equation (3.58) can be written in a form by replacing n and p from Eqs. (3.48) and (3.49) as

$$V_{\mathrm{p}^+} + V_{\mathrm{n}^+} = \frac{2kT}{q} \ln\left(\frac{\tau_\mathrm{a} J_{\mathrm{n}^-} \sqrt{1 - B^2 \tanh^4(d/L_\mathrm{a})}}{2qn_\mathrm{i} L_\mathrm{a} \tanh(d/L_\mathrm{a})}\right). \tag{3.59}$$

Hence, the forward current density at medium to high-level injection by the recombination of holes and electrons in the n^- layer without end-region recombination at the n^+ and p^+ regions is

$$J_{\mathrm{n}^-} = \frac{2qn_\mathrm{i} D_\mathrm{a}}{d} F\left(\frac{d}{L_\mathrm{a}}\right) e^{\frac{q(V_{\mathrm{p}^+} + V_{\mathrm{n}^+} + V_{\mathrm{n}^-})}{2kT}}, \tag{3.60a}$$

where the function of $F()$ is

$$F\left(\frac{d}{L_\mathrm{a}}\right) = \frac{\frac{d}{L_\mathrm{a}} \tanh\left(\frac{d}{L_\mathrm{a}}\right)}{\sqrt{1 - B^2 \tanh^4\left(\frac{d}{L_\mathrm{a}}\right)}} e^{\frac{-qV_{\mathrm{n}^-}}{2kT}}. \tag{3.60b}$$

Thus, at medium current densities when high-level injection prevails in the n^- layer, the forward current density varies as a function of $e^{\frac{qV_\mathrm{f}}{2kT}}$, comparing to that of the low-level injection where the forward current density varies according to $e^{\frac{qV_\mathrm{f}}{kT}}$ as that of a regular p–n junction diode.

At a very high current density, the current–voltage characteristics is found to deviate from the derived exponential behavior. This is due to the significant recombination in the n^+ and p^+ regions as well as a reduction in the diffusion length by carrier–carrier scattering. For such a case, the forward current density must include these recombination components in the p^+ and n^+ regions:

$$J_\mathrm{f} = J_{\mathrm{n}^-} + J_{\mathrm{p}^+} + J_{\mathrm{n}^+}. \tag{3.61}$$

Assuming that the width of the n^+ and p^+ regions are less than the carrier diffusion lengths, the current densities in these regions can be derived from low-level injection theory since the minority-carrier density is much lower than the high doped background levels. Thus,

$$J_{p^+} = \frac{qD_{n,p^+}n_{0,p^+}}{L_{n,p^+}\tanh\left(\frac{W_{p^+}}{L_{n,p^+}}\right)}\left(e^{\frac{qV_{p^+}}{kT}} - 1\right), \tag{3.62}$$

$$J_{n^+} = \frac{qD_{p,n^+}p_{0,n^+}}{L_{p,n^+}\tanh\left(\frac{W_{n^+}}{L_{p,n^+}}\right)}\left(e^{\frac{qV_{n^+}}{kT}} - 1\right), \tag{3.63}$$

where W_{p^+} and W_{n^+} are the thickness for the p^+ and n^+ regions. For a high–low junction, the injected carriers on either side of the junction are related by (Sze, 1981):

$$\frac{p_{p^+}(x=-d)}{p_{n^-}(x=-d)} = \frac{n_{n^-}(x=-d)}{n_{p^+}(x=-d)}, \tag{3.64}$$

where p_{p^+} and n_{p^+} are the hole and electron concentrations on the p^+ side of the junction, p_{n^-} and n_{n^-} are the hole and electron concentrations on the n^- side of the junction. Due to low-level injection level on the heavily doped sides:

$$p_{p^+}(-d) = p_{0,p^+}. \tag{3.65}$$

The injected electron concentration in the p^+ region is related to the voltage across this junction:

$$n_{p^+}(-d) = n_{0,p^+}e^{\frac{qV_{p^+}}{kT}}. \tag{3.66}$$

Thus, Eq. (3.64) can be re-written as

$$n(-d)p(-d) = p_{0,p^+}n_{0,p^+}e^{\frac{qV_{p^+}}{kT}} = n_{ie,p^+}^2 e^{\frac{qV_{p^+}}{kT}}, \tag{3.67}$$

where n_{ie,p^+} is the effective intrinsic carrier concentration in the p^+ region. Using the charge neutrality condition, i.e. $n(x) = p(x)$ in the base region:

$$e^{\frac{qV_{p^+}}{kT}} = \left(\frac{n(x=-d)}{n_{ie,p^+}}\right)^2. \tag{3.68}$$

Substituting Eq. (3.68) into Eq. (3.62) yields

$$J_{p^+} \approx \frac{qD_{n,p^+}n_{0,p^+}}{L_{n,p^+}\tanh\left(\frac{W_{p^+}}{L_{n,p^+}}\right)}\left(\frac{n(x=-d)}{n_{ie,p^+}}\right)^2. \tag{3.69}$$

Similarly,

$$J_{n^+} \approx \frac{qD_{p,n^+}p_{0,n^+}}{L_{p,n^+}\tanh\left(\frac{W_{n^+}}{L_{p,n^+}}\right)}\left(\frac{n(x=+d)}{n_{ie,n^+}}\right)^2. \tag{3.70}$$

From Eq. (3.58), it indicates that the total voltage drop of both junctions are proportional to the average carrier concentration as $\ln(\bar{c})$. For high-level injection, the voltage drop can now be related to $\ln(\sqrt{J})$, which has a very weakly variation, by looking at Eqs. (3.69) and (3.70). For approximation, the total voltage drop of both junctions can be assumed constant (Herlet, 1968). As for the voltage drop across the base region, Eq. (3.53) is no longer independent of forward current as the total current J is not the same as the base diffusion/recombination current J_{n-}. Instead, it increases strongly with the forward current in a ratio of approximately $V \propto \sqrt{J}$. This phenomenon of deviation from the exponential function can be seen as the saturation of conductivity modulation, whereas the increase of carrier density in the base region does not have dominant effect on the reduction of resistivity. We can then treat this I–V region as ohmic. It should be noted that, the combined effect of carrier–carrier scattering and Auger recombination also contributes to a higher base-region voltage.

3.3.2. Reverse Blocking

As discussed in Chapter 2, the reverse blocking voltage of a p^+–n^-–n^+ diode is supported by the lightly doped n^- layer. As such, the width and doping level of the n^- layer determine the maximum reverse blocking voltage of the diode. The limited ability of silicon junction diode to block current flow in reverse direction under elevated temperature sets its maximum operating temperature. There are two possible mechanisms responsible for current flow under reverse bias conditions. The first mechanism is by space-charge generation of electron–hole pairs within the space-charge region. Under reverse bias conditions, electron–hole pair generated within the space-charge region will be swept out of this region by the existence of an intense electric field within this region. If the generation process is assumed to be uniform over the space-charge region, the space-charge generation current is given by (Sah *et al.*, 1957)

$$J_{sc} = \frac{qW_d n_i}{\tau_{SP}}, \tag{3.71}$$

where W_d is the depletion layer thickness in the n^- layer. Since the depletion width increases with reverse bias, the space-charge generation leakage current increases with a higher reverse bias. The space-charge generation leakage current is also a strong function of temperature due to the thermal energy available for the generation of electron–hole pairs within the space-charge region. This is because, in Eq. (3.71), n_i increases with temperature.

The second carrier mechanism responsible for the generation of reverse leakage current comes from electron–hole pairs generated near the depletion edges of the reverse-biased junction. These carriers diffuse into the edges of the depletion region and are swept across the depletion region by the electric field. These diffusion leakage currents are

$$J_{\mathrm{i,n^-}} = \frac{qD_{\mathrm{p}}n_{\mathrm{i}}^2}{L_{\mathrm{p}}N_{\mathrm{d,n^-}}} \tag{3.72}$$

from the n$^-$ region, and

$$J_{\mathrm{i,p^+}} = \frac{qD_{\mathrm{n}}n_{\mathrm{i}}^2}{L_{\mathrm{n}}N_{\mathrm{a,p^+}}} \tag{3.73}$$

from the p$^+$ region. Note that the main depletion region for the p–i–n diode is located at p$^+$–n$^-$ junction. The diffusion component of the reverse leakage current increases rapidly with temperature due to the term of n_{i}^2 dependence. As such, the diffusion component of the reverse leakage current dominates at temperatures above 100°C. The total leakage current is the sum of the components in Eqs. (3.71)–(3.73).

3.3.3. Temperature Effect

The temperature effect on diode current conduction is described by looking at fabricated diode devices. The epitaxial p$^+$–n$^-$–n$^+$ silicon diode fabricated with an n$^+$ diffusion on a lightly doped n-epitaxial layer of concentration of 1×10^{16} cm^{-3} on top of an p$^+$ (100) substrate of concentration of 1×10^{18} cm^{-3}, and the p$^+$–n junction diode with similar active area fabricated by diffusing phosphorus into the boron-doped (100) silicon substrates are measured to investigate the temperature effect (Ang, 1995). Both diode structures are shown in Fig. 3.6. For the p–i–n structure, the high–low junction is formed

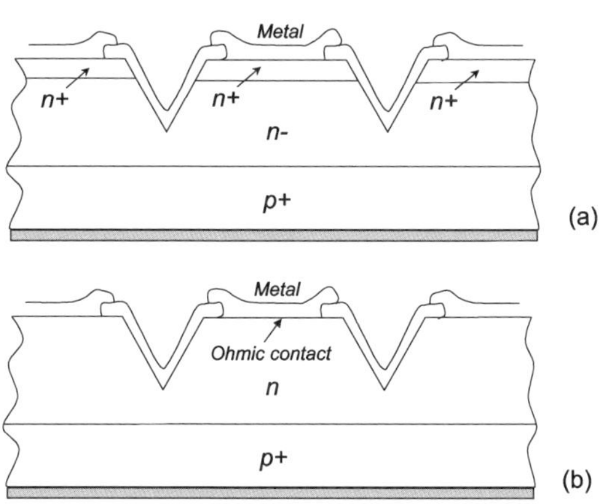

Fig. 3.6. Cross-section views of (a) the p–i–n diode and (b) p–n diode fabricated on silicon.

at the diffused n^+/n-epitaxial layer with an n-epitaxial/p^+ substrate junction. The V-groove is made to form isolated islands for phosphorus implantation for the n^+ emitter on the top. The V-groove trenches are not needed for the p^+–n diode, however, they are made to have the similar device active region. The p^+–n junction was formed epitaxially (flat junction), while the high–low junction was formed by diffusing an n^+ layer into the lightly doped n-epitaxial layer. For the p^+–n diode, the phosphorus-diffused junction depth is 2 μm and its doping profile is selected to yield a similar breakdown voltage of 27 V as that of the p^+–n^-–n^+ diode.

Figure 3.7 shows the diode quality factor for forward conduction under low bias condition as a function of temperature for the epitaxial p^+–n^-–n^+ silicon diode. The diode quality factor at room temperature is defined as

$$n = \frac{q}{kT}\frac{\Delta V_{\mathrm{f}}}{\Delta(\ln I_{\mathrm{f}})} \tag{3.74}$$

and it has a value of about 1.13. Carrier diffusion is the predominant current conduction mechanism for low-level injection at room temperature. The deviation of the diode quality factor from unity reveals that a small recombination current exists in addition to the predominant diffusion current. The diode quality factor at low forward bias condition reduces to unity at higher temperature as shown. This is because the diffusion component of the forward current increases at a faster rate than the recombination component at

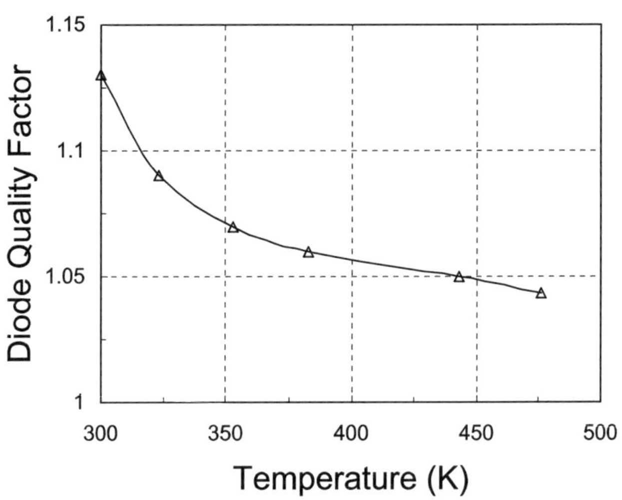

Fig. 3.7. Measured p–i–n diode quality factor under low forward-based condition as a function of temperature.

high temperature. The intrinsic carrier concentration increases with temperature and thus, the diffusion current increases more rapidly with temperature as it is proportional to the square of the intrinsic carrier concentration, whereas the recombination current increases linearly with higher intrinsic carrier concentration. Overall, the forward current at low bias voltage increases approximately as $e^{(-\frac{E_G - qV_f}{kT})}$ with temperature variation. A semi-logarithmic plot of the forward current versus temperature as a function of applied bias as shown in Fig. 3.8 can be used to explain the current conduction mechanism. The figure shows that the forward current under higher biasing voltage departs from the line and bends down at high temperature region. This phenomenon can be explained by the concept of activation energy. The activation energies for the forward current at different bias voltage are tabulated in Table 3.1.

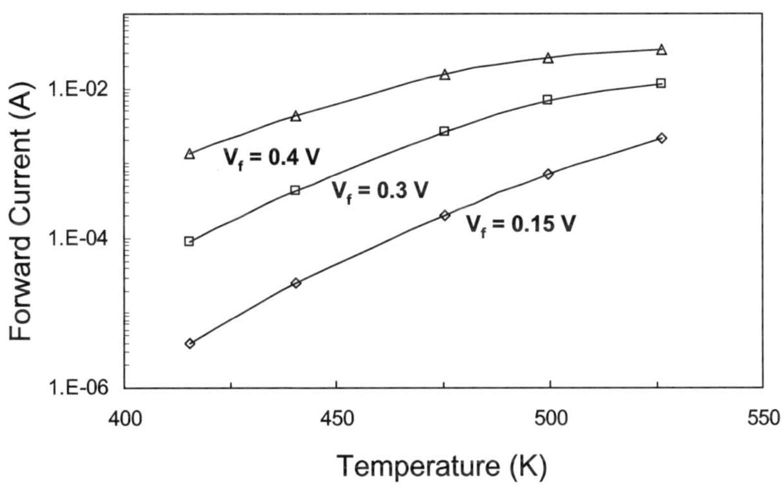

Fig. 3.8. Measured forward current versus temperature at various bias voltage for a p–i–n diode.

Table 3.1. Activation energies for forward current at different bias voltage.

Forward bias voltage (V)	**Activation energy (eV)**
0.1	1.02
0.2	0.95
0.3	0.86
0.4	0.73

The sum of the activation energy and the applied bias ($\times q$) are approximately in the range of 1.12–1.16 eV, which is around the bandgap energy of silicon at 300 K. Thus, the forward conduction is controlled by the temperature activated process. The forward current at a given temperature of T_2, can be expressed as

$$I_{f,T_2} \approx I_{f,T_1} e^{\frac{1}{k}\left(\frac{qV_{f,T_2}-E_{G,T_2}}{T_2}-\frac{qV_{f,T_1}-E_{G,T_1}}{T_1}\right)} \tag{3.75}$$

when the current at T_1, i.e. I_{f,T_1}, is known. The temperature dependence of the bandgap energy in eV is given by (Sze, 1981)

$$E_G = 1.17 - \frac{4.73 \times 10^{-4} T^2}{T + 636}. \tag{3.76}$$

The high-level injection dominates the current–voltage characteristics beyond 0.7 V at room temperature, and also at the lower voltage for higher temperature. According to Singh and Jain (1982), the onset of high-level injection occurs when the forward voltage at the p–n junction is higher than a threshold value given by

$$V_{j0} = \frac{2kT}{q} \ln\left(\frac{n_{n^-}}{n_i}\right) \tag{3.77}$$

where n_{n^-} is the doping concentration in the lightly doped n region. For the p^+–n^-–n^+ diode, V_{j0} at 300 K is 0.698 V. The linear curve of onset voltage versus temperature is shown in Fig. 3.9. For a higher temperature, a higher n_i makes V_{j0} to reduce for the high-level injection to occur at a lower forward voltage.

The forward characteristics of the p^+–n^-–n^+ diode is also influenced by the presence of the high–low n^+–n^- junction. The high–low junction forms potential barrier which reduces the number of minority carriers reaching the cathode. At 300 K, the n^+–n^- high–low junction for the measured diode induces a potential barrier of (Hauser and Dunbar, 1975)

$$\Phi_{hl} = kT \ln\left(\frac{n_{n^+}}{n_{n^-}}\right) \approx 0.0259 \times \ln\left(\frac{10^{19}}{10^{16}}\right) = 0.179\,\text{eV} \tag{3.78}$$

to the holes injected from the p^+ region. This barrier effectively reduces the hole current component. Thus, the difference in minority carrier concentration going from the lightly doped n-epitaxial region to the n^+ diffused region results in hole accumulation in the lightly doped n layer. Under high-level injection, i.e. when the forward voltage is much greater than the onset voltage, the p^+–n^-–n^+ diode operates under deep bipolar mode where the n^+ cathode injects electrons and the p^+ anode injects holes into the lightly doped

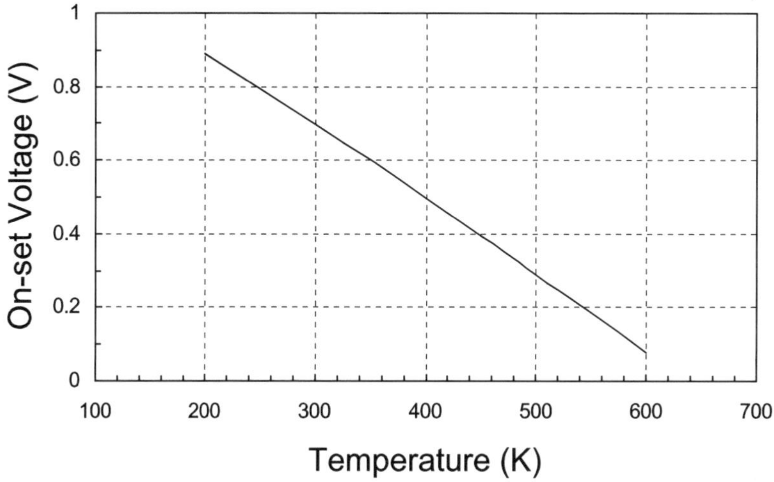

Fig. 3.9. The onset voltage of p–i–n diode as a function of temperature.

n region. This leads to conductivity modulation in the lightly doped n region and compensates the effect of high–low junction by high forward voltage.

The reverse characteristics of the p^+–n^-–n^+ and p^+–n diodes are distinctively different. The p^+–n^-–n^+ diode, it maintains a relatively stable leakage current at constant temperature when reverse-bias voltage increases, while the p^+–n diode exhibits a more significant change in leakage current with reverse-bias voltage. This is partially due to that the electrical field in p^+–n^-–n^+ diode raises rather slowly after punchthrough compared to that in the p^+–n diode when reverse-bias voltage is changed. The low field in p–i–n diode gives a much lower leakage current compared to that of a p–n diode. Both diodes have the leakage current increase when temperature goes higher. By analysis, the leakage current in an epitaxial layer with a minority-carrier diffusion length much greater than its thickness, i.e. $L_p \ll W_{n^-}$, consists of the electron and hole leakage currents (diffusion component) are given by

$$I_{\mathrm{e,leak,diff}} = qA\frac{D_n n_i^2}{L_n N_{a,p^+}} \tag{3.79}$$

and

$$I_{\mathrm{h,leak,diff},} = qA\frac{D_p n_i^2}{W_{n^+} N_{d,n^+}}, \tag{3.80}$$

where D_p and D_n are the diffusion coefficients of holes and electrons, W_{n^+} is the width of the n^+ region, and $N_{d,n+}$ is the dopant concentration in the

n^+ emitter region. In the above equation, n_i, D_n, and D_p are all temperature dependent. It should be noted that Eqs. (3.79) and (3.80) are valid for the p^+–n diode as well.

The difference in the reverse characteristics of the p^+–n^-–n^+ and p^+–n diodes, besides the dependency of electric field, is also attributed to the factor that the contribution of the generation current to the leakage current is negligible in the case of the former and is significantly higher in the case of the latter. The space-charge generation current is given as (Kao and Wolley, 1967)

$$I_G \propto \frac{n_i(V_\Phi + V_R)^{0.5}}{\tau_{SP}}, \tag{3.81}$$

where V_Φ is the junction built-in potential, V_R is the voltage by the external reverse bias, and τ_{SC} is the space-charge generation lifetime. The equation explains that the leakage current changes with reverse voltage and the temperature. When the temperature goes up, the intrinsic carrier density increases and the space-charge generation lifetime reduces, both help to bring the leakage current higher. The temperature dependence of n_i can be expressed as (Sze, 1981)

$$n_i^2 \approx 6.06 \times 10^{32} T^3 e^{-\frac{E_G}{kT}}. \tag{3.82}$$

The space-charge generation lifetime in the lightly doped drift region of the p^+–n^-–n^+ diode is expected to be higher than that of the p^+–n diode. Therefore, the reverse leakage current is expected to be higher in the p^+–n diode than that of the p^+–n^-–n^+ diode.

From the temperature dependence, the leakage current (diffusion component) can be expressed as (Ang, 1995)

$$I_{\text{leak,diff}} \propto T^{3+\frac{\alpha}{2}} e^{-\frac{1.17-\frac{4.73\times 10^{-4}T^2}{T+636}}{kT}}, \tag{3.83}$$

where α is the temperature coefficient, T^α, for D_p and D_n. Generally, the effect of the reduction in bandgap energy on the diffusion current is more pronounced and the temperature dependence of D_p or D_n can be neglected in the approximation. Thus,

$$I_{\text{leak,diff}} \propto e^{-\frac{E_G}{kT}}. \tag{3.84}$$

Figure 3.10 shows a typical leakage current versus reciprocal temperature plot for the p^+–n^-–n^+ diode. A straight line can be drawn through the data points above 323 K. The slope of this plot is

$$E_G = -k\frac{\Delta \ln I_{\text{leak}}}{\Delta(\frac{1}{T})} = 1.13\,\text{eV} \tag{3.85}$$

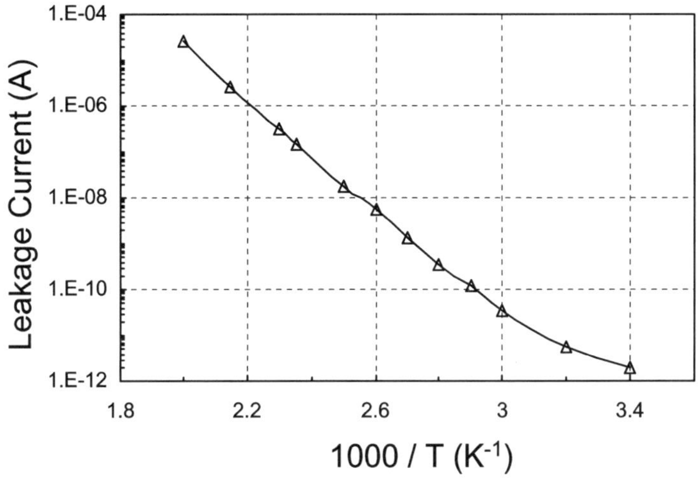

Fig. 3.10. Measured leakage current as a function of the reciprocal temperature for the p–i–n diode.

which is approximately equal to the bandgap energy of the silicon at 300 K. Thus, the reverse current increases exponentially with temperature by several orders of magnitude as the operating temperature increases from 300 K to 533 K. The measured reverse leakage current at 300 K is an order of magnitude higher than that predicted by the ideal Shockley equation, which assumes that the reverse current consists of only diffusion current. Cappelletti *et al.* (1985) reported a similar phenomenon for the p–n junction diodes fabricated on silicon followed by an anneal at 650°C for 4 hours. From Singh and Singh (1990), the reverse leakage current for large surface recombination velocities at the high–low junction can be higher as much as by a factor of $\coth(W_{n^-}/L_p)$ than that of an ideal Shockley diode. In a practical junction diode, the reverse saturation current consists of a diffusion current in the quasi-neutral region and also a generation current in the depletion region. The intrinsic carrier density at room temperature is low, as such, the generation current contributes a significant portion of the reverse leakage current (Green, 1990). As the operating temperature goes higher, the intrinsic carrier concentration increases and the diffusion current gradually dominates at higher temperatures.

3.4. Schottky Barrier Diode

Historically, the metal–semiconductor diode was the first practical semiconductor junction device. Braun discovered the metal–semiconductor junction in 1874. Early stage rectifiers were fabricated simply by pressing a metallic whisker onto a semiconductor surface, forming point-contact diodes. However, the characteristics of such point-contact rectifiers were not consistent. Later on, Schottky barrier diodes are formed by evaporating or sputtering metallic contact onto a semiconductor. When a metal and an n-type semiconductor are intimately joined, electrons from the n-type semiconductor having higher energy diffuse across the metal–semiconductor barrier to the metal side until the Fermi level of the metal–semiconductor system is aligned in thermal equilibrium as shown in Fig. 3.11(a). The migration of the electrons from the semiconductor to the metal leaves behind ionized donors as fixed positive charges creating a depletion layer similar to that of the p^+–n junction. The depletion of electrons in the vicinity of the interface produces an upward band bending near the semiconductor surface so that a built-in potential ψ_0 is established. This built-in potential accounts for the difference between the Fermi

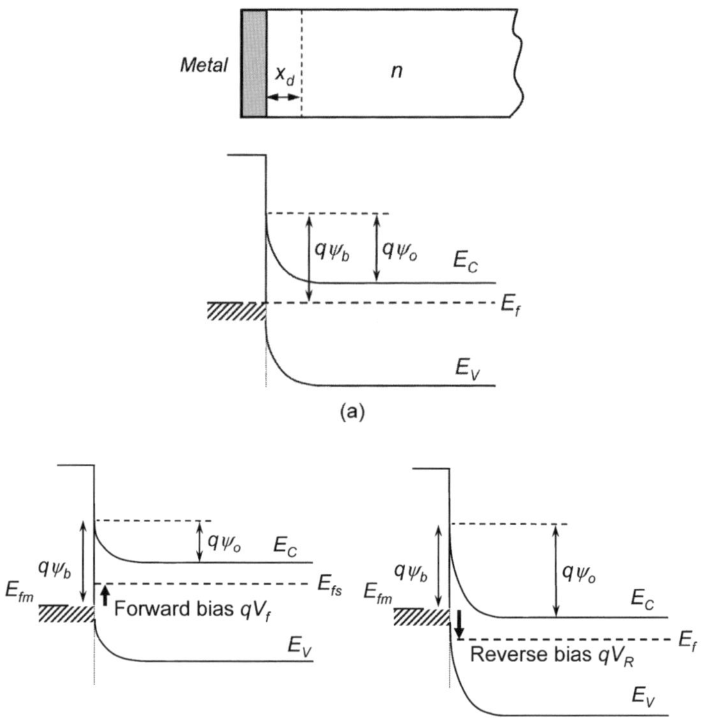

Fig. 3.11. (a) Schottky diode and the associated energy band diagram, (b) energy band diagram under forward bias, and (c) under reverse bias.

levels of the metal and semiconductor. Thus,

$$q\psi_0 = \phi_m - \phi_s \tag{3.86}$$

where ϕ_m and ϕ_s are the work functions (in eV) for the metal and semiconductor, respectively. The work function is in general defined as the work required to bring an electron from the Fermi level of the material to the vacuum level. The semiconductor work function ϕ_s and the electron affinity are related by

$$\phi_s = \chi_s + (E_c - E_{fs}), \tag{3.87}$$

where χ_s is the electron affinity (≈ 4.05 eV) which specifies the energy required to release an electron from the bottom of the conduction band to the vacuum level. E_{fs} is the fermi energy level in silicon. The built-in potential ψ_0 as described in Eq. (3.86) in semiconductor is supported by a space-charge layer with a width of x_d. The depletion width x_d is given by

$$x_d = \sqrt{\frac{2\varepsilon_s \psi_0}{qN_d}}, \tag{3.88}$$

where N_d is the dopant concentration in the n-type semiconductor. A potential barrier due to the discontinuity of allowed energy states between the metal and the n-semiconductor is formed at the interface with its barrier height given by

$$q\psi_b = \phi_m - \chi_s = (\phi_m - \phi_s) + (E_c - E_{fs}). \tag{3.89}$$

The barrier heights for n-type silicon depend on both the silicon doping level and the type of metal used to form the Schottky contact. A list of barrier heights for different metal contact is listed in Table 3.2 for lightly doped planar substrate (Kawakami *et al.*, 1985). The value of barrier height is calculated from the diode saturation current as that described in Eq. (3.91).

For a brief estimation, a Schottky diode with barrier height of 0.7 V, the forward voltage drop is about 0.35 V for 1 mA of forward current conduction.

3.4.1. Forward Conduction

If a positive bias V_f is applied to the metal with respect to the n-type semiconductor, the built-in potential is reduced to $\psi_0 - V_f$ as shown in Fig. 3.11(b). The reduction of the built-in potential height makes it easier for electrons in the semiconductor to move to the metal side. This forms the forward conduction condition. During forward bias, the predominant conduction mechanism is governed by the thermionic emission of electrons (Crowell and Sze, 1966) and the forward current density is given by

$$J_f = A^* T^2 e^{\frac{-q(\psi_0 - V_f)}{kT}} = J_s e^{\frac{qV_f}{kT}}, \tag{3.90}$$

Table 3.2. Barrier heights for different metal to n-type silicon Schottky contact.

Metal	ψ_b(V)	Metal	ψ_b(V)
Sc	0.30	Ni	0.63
Er	0.33	Fe	0.64
Y	0.38	W	0.65
Ti	0.48	Ta	0.66
Hf	0.49	Mo	0.67
Mn	0.49	Pd	0.70
Zr	0.51	Rh	0.73
C	0.55	Be	0.75
V	0.55	Re	0.75
Nb	0.59	Ir	0.84
Cr	0.60	Pt	0.85
Co	0.62	B	0.95
Al	0.69	Au	0.79

where A^* is the effective Richardson constant and the saturation current density is

$$J_s = A^* T^2 e^{\frac{-q\psi_0}{kT}}. \tag{3.91}$$

For electric field between 10^4 and 10^5 V/cm, the effective Richardson constants for electrons in silicon and gallium arsenide are 112 A cm^{-2} K^{-2} and 140 A cm^{-2} K^{-2}, respectively. For hole in silicon, the effective Richardson constant is 32 A cm^{-2} K^{-2} (Crowell, 1965; Andrew and Lepselter, 1970). It should be noted that only the majority carriers are involved in the conduction mechanism for metal–silicon Schottky barrier diode, as such, it is a majority carrier device. Because of this nature, the bipolar conductivity modulation behavior does not occur during the forward conduction in the conventional Schottky barrier diode.

For a high-voltage Schottky barrier diode, the n-silicon layer usually consists of a lightly doped n-type epitaxial layer and a heavily doped n$^+$ emitter as shown in Fig. 3.12 The thickness and doping of the lightly doped n-layer are adjusted to yield the desired reverse blocking voltage. Due to the absence of conductivity modulation in this lightly doped n-layer, the n-layer contributes to a much higher resistive voltage during forward conduction. Using Eq. (3.90), the total forward voltage drop of the Schottky barrier diode including the resistive drop in the drift region is

$$V_f = \frac{kT}{q} \ln\left(\frac{J_f}{J_s}\right) + J_f R_t \tag{3.92}$$

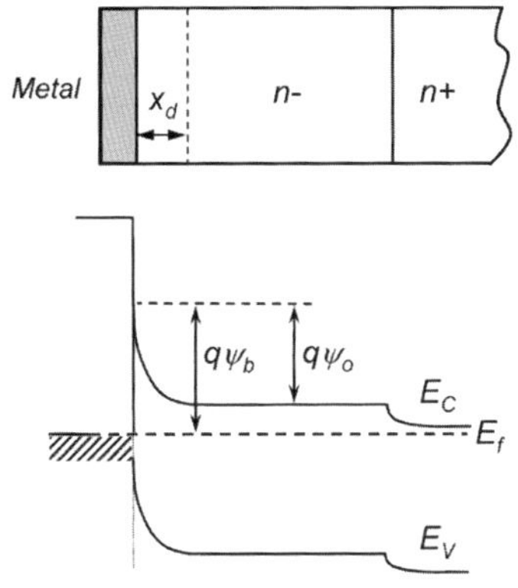

Fig. 3.12. Schottky barrier diode with drift region and n^+ emitter.

where R_t is the total specific resistance (in $\Omega\,cm^2$) contributed by the lightly doped n-region, the n^+ region and the metal contacts. The lightly doped n-region gives the major contribution and the specific resistance can be calculated by the following:

$$R_{n^-} = \frac{L_{n^-}}{q\mu_n N_d}. \tag{3.93}$$

In an actual device, the forward voltage drop of a Schottky barrier diode rated at a 50 V reverse blocking voltage is between 0.5 V and 0.6 V at a forward current density of 120 A/cm^2. For a comparable p–i–n junction diode, the forward voltage drop would be in the range of 0.8–0.9 V for the same current density. It should be noted that ordinary Schottky barrier diodes do not inject minority carriers and the resistive voltage drop at higher forward current conduction will be large.

Schottky barrier diodes can also be made with higher minority carrier barrier height, e.g. greater than 0.85 eV by a shallow implantation of opposite conductivity type of dopant, e.g. metal–n–p^-–p^+, to the epitaxial layer (Shannon, 1976; Wu, 1981), with the energy band structure is shown in Fig. 3.13. For such kind of high barrier Schottky diode, it does inject a moderate amount of minority carriers from n-layer into the lightly doped drift region under high-level carrier injection condition (Elfsten and Tove, 1985). Therefore, the conductivity modulation now occurs and the forward current–voltage characteristics are significantly altered from the classical exponential relationship. The typical forward I–V characteristics for the high-barrier Schottky diode has three distinctive regions of operations, namely a low-level injection, a high-level injection, and a transition region in between. The curve moves away from that of the conventional Schottky diode and entering into the transition region at

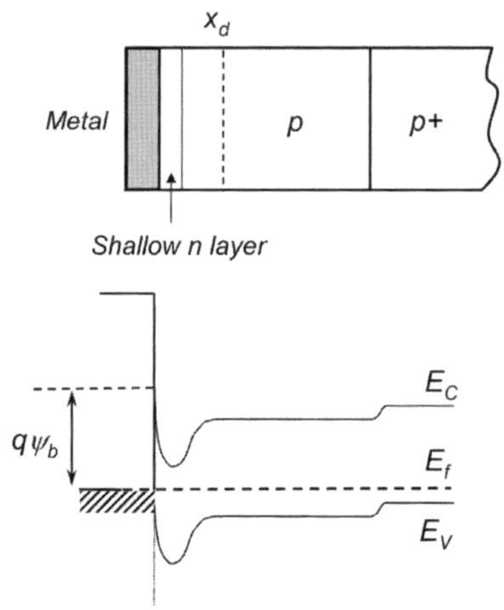

Fig. 3.13. High barrier Schottky diode by shallow n layer implantation and the energy state diagram.

higher injection level. Figure 3.14 shows the forward *J–V* curve to explain such a phenomenon. Note that, due to the transition region, the curve is not a smooth monotone curve but with a corner indicating the mode of conduction switching from the low-level junction to the high-level injection. Detailed modeling of the conduction characteristics can be found in Ng *et al.* (1990).

Contrarily, if the same type of dopant as the substrate but higher concentration, e.g. metal–n–n^-–n^+, is used for the shallow implantation, then the barrier height will be reduced. Approximately, a reduction in barrier height of 0.1–0.2 eV can occur if an implantation dose of 10^{13} cm^{-2} is used.

3.4.2. Reverse Blocking

When a negative voltage, V_R, is applied to the metal with respect to the n-semiconductor, the built-in potential height is increased and the reverse voltage is supported across the depletion layer in the semiconductor, as shown in Fig. 3.11(c). The depletion width at reverse bias is increased according to

$$x_d = \sqrt{\frac{2\varepsilon_s(\psi_0 + V_R)}{qN_d}}. \qquad (3.94)$$

The blocking voltage in a Schottky diode is about one-third of that in an abrupt p–n junction diode. The leakage current of the Schottky diode consists of the space-charge generation current in the depletion region, the diffusion current

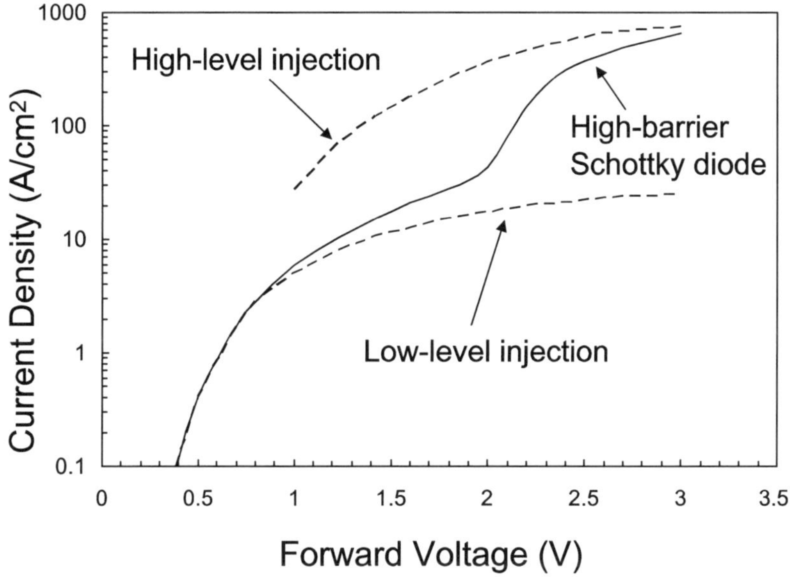

Fig. 3.14. Forward conduction curve of the high-barrier Schottky diode.

from the semiconductor neutral region, and the thermionic current across the barrier. From Eq. (3.90), the reverse current density is

$$J_{\mathrm{R}} = J_{\mathrm{s}}\left(e^{\frac{-qV_{\mathrm{R}}}{kT}} - 1\right) = -J_{\mathrm{s}}. \tag{3.95}$$

which is equal to the saturation current density. Due to T^2 dependence and the lower barrier height, the reverse current of the Schottky diode is generally larger than that of the p^+–n^-–n^+ diode. Even worse, under reverse-biased, the barrier height is found to be reduced (Sze, 1981) due to image force lowering. The amount of reduction of barrier height in eV is given by (Muller and Kamins, 1986)

$$\Delta q\psi_{\mathrm{b}} = \sqrt{\frac{q^3 \hat{E}_{\mathrm{m}}}{4\pi\varepsilon_{\mathrm{s}}}}, \tag{3.96a}$$

where the maximum field at the silicon-metal junction is

$$\hat{E}_{\mathrm{m}} = \sqrt{\frac{2qN_{\mathrm{d}}(V_{\mathrm{R}} + \psi_0)}{\varepsilon_{\mathrm{s}}}}. \tag{3.96b}$$

The amount of barrier height reduction (in eV) as a function of reverse bias voltage is shown in Fig. 3.15 for silicon background doping level of 10^{16} cm^{-3}.

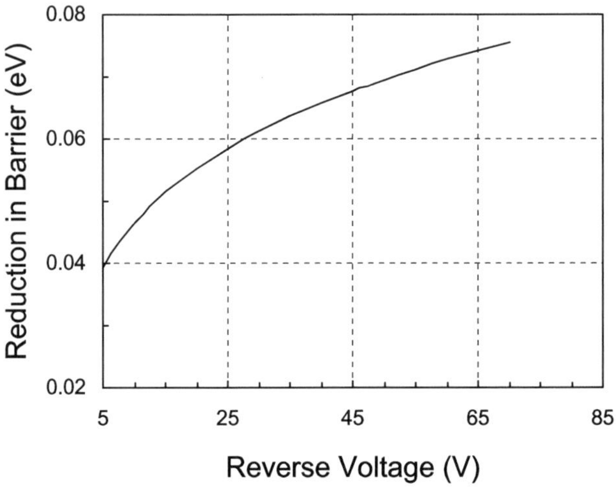

Fig. 3.15. Reduction in Schottky barrier height (ϕ_b) as a function of reverse bias voltage at silicon doping of 10^{16} cm^{-3}.

Hence, the reverse current density is modified by including the barrier height reduction to yield

$$J_r = A^* T^2 e^{\frac{-q(\psi_0 - \Delta\psi_b)}{kT}}, \tag{3.97}$$

where $\Delta\psi_b$ is the amount of barrier height lowering ($q\Delta\psi_b = \phi_b$). As the barrier is lower, the reverse leakage current gets higher. This makes the Schottky barrier less effective in blocking high voltage. Figure 3.16 gives the calculated curves for leakage current at various barrier heights and temperatures using the equations mentioned and the Richardson constant for silicon.

For a pause, to review what has been described for the p–i–n junction diode and the Schottky barrier diode, some short comparisons are made here. The forward current conduction in a metal–silicon Schottky barrier diode is made by majority carriers, whereas in a p–i–n junction diode it is by both majority and minority carriers. When the p–i–n diode is switched from the forward conduction to the reverse blocking, the switching speed is limited by the time needed to remove the excess carriers in the n-drift region. In a Schottky barrier diode, there is very little excess carrier storage, its switching speed is much higher and hence limited by the external circuit elements. For this reason, Schottky barrier diodes are ideal for high-frequency and fast-switching applications.

The reverse saturation current in a Schottky barrier diode is much larger than that of a p–i–n diode as additional amount of thermionic current density

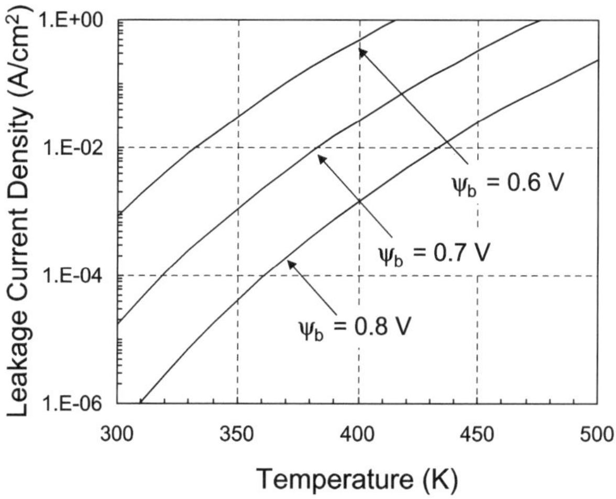

Fig. 3.16. Leakage current density as a function of temperature and barrier heights.

is added to the diffusion current and space-charge generation current. But, the forward-voltage drop will be less in a Schottky barrier diode for a smaller barrier height than in a p–i–n diode for the same current density. The typical cut-in voltage of a Pd/n^- silicon Schottky barrier diode is about 0.3 V compared to the silicon p–i–n diode of about 0.7 V at low forward current density. Also, the temperature dependence of Schottky barrier diodes and p–i–n diodes under forward bias is different.

3.5. Ohmic Contact

Ohmic contact is needed for semiconductor devices to be in good connection with external circuits without suffering a high interface voltage drop. As stated earlier, metal–semiconductor contact forms the Schottky potential barrier and a junction voltage drop exists during current conduction. Thermionic emission forms the carrier transport mechanism at the metal–semiconductor interface. One way to make the contact more ohmic is to lower the potential barrier, by using various kinds of metals for contact, so that the interface voltage drop can be reduced to a negligible amount. Near ohmic contacts can be formed by metal contacts on p-type silicon for a low barrier height. For the n-type metal contacts, such as normally by aluminum, gold, and platinum, the potential barrier heights remain between 0.7 eV and 0.9 eV and are independent from silicon doping concentration. To raise current conduction beyond thermionic emission, another mode of carrier transport mechanism is needed besides the thermionic emission. Yu (1970) described that by metal contact with heavily

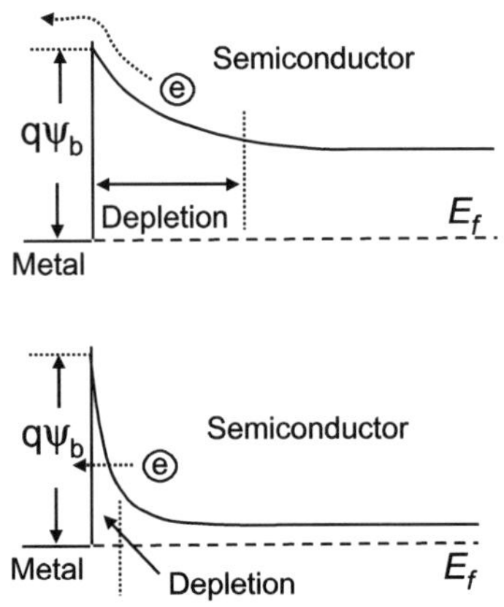

Fig. 3.17. Band diagrams for different Schottky contacts with lightly doped n-type silicon (upper) and heavily doped n-type silicon (lower).

doped n-type silicon, the carrier tunneling transport occurs from the semiconductor to the metal at the interface. Figure 3.17 depicts the band diagrams of metal–semiconductor interface at low and high n-type doping concentrations. For the lightly doped case, the current transport is made by the thermionic emission across the barrier height $q\psi_b$ as normally occurring for a Schottky diode. For heavily doped case, the depletion layer is much thinner and the field tunneling processes can occur by electrons travelling from the Fermi level in the semiconductor into the metal. This current transport process is named the field emission. A probability factor is defined as

$$E_{pf} = \frac{qh}{4\pi}\sqrt{\frac{2N_d}{m_e \varepsilon_s}} \tag{3.98}$$

to indicate the probability of tunneling process, where $m_e (= 9.109 \times 10^{-31}$ kg) is the electron rest mass, N_d is the n-type silicon doping and $h (= 6.626 \times 10^{-34}$ J·s) is the Planck's constant. If E_{pf} is much less than kT/q, the transport mechanism is mainly dominant by the thermionic emission. Otherwise, when much greater than kT/q, it is then dominant by the field emission. For E_{pf} values fall in between, the current transport is made by both mechanisms. As the n-silicon doping changing from the lower end to the higher end, E_{pf} value increases as shown in Fig. 3.18. The specific contact resistance can be found by the following definition (Yu, 1970):

$$R_c = \left.\frac{dV}{dJ}\right|_{V \to 0} \tag{3.99a}$$

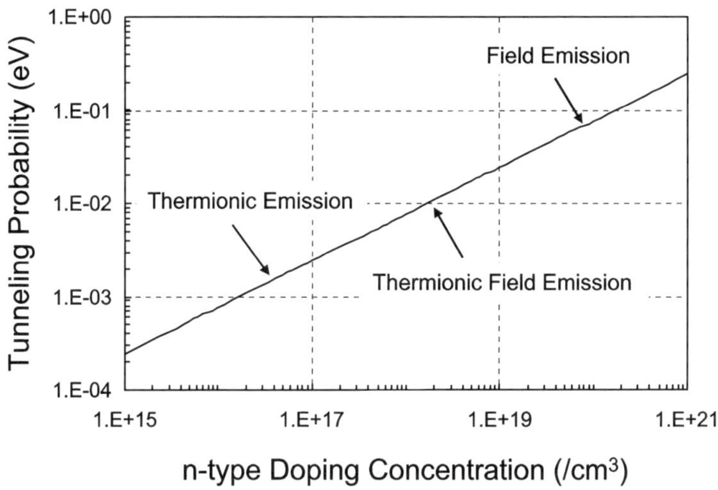

Fig. 3.18. The tunneling probability as a function of n-type doping concentration.

and is found to be

$$\frac{k}{qA^*T}e^{\frac{q\psi_b}{kT}} \quad \text{for thermionic emission region, or}$$

$$R_c = \frac{k^2}{qA^*\sqrt{\pi(q\psi_b + E_c - E_f)E_{pf}}} \times \cosh\left(\frac{E_{pf}}{kT}\right)\sqrt{\coth\left(\frac{E_{pf}}{kT}\right)}\,e^{\frac{q\psi_b + E_c - E_f}{E_{pf}\coth\left(\frac{E_{pf}}{kT}\right)} - \frac{E_c - E_f}{kT}}$$

for thermionic field emission region, or

$$\left(\frac{qA^*\pi T}{k\sin\left(\frac{\pi kT\ln\left(\frac{4q\psi_b}{E_c - E_f}\right)}{2E_{pf}}\right)}e^{\frac{-q\psi_b}{E_{pf}}} - \frac{4A^*qE_{pf}^2}{\left(k\ln\left(\frac{4q\psi_b}{E_c - E_f}\right)\right)^2}e^{\frac{-q\psi_b}{E_{pf}} - \frac{E_c - E_f}{2E_{pf}\ln\left(\frac{4q\psi_b}{E_c - E_f}\right)}}\right)^{-1}$$

for field emission region. (3.99b)

The specific contact resistance can be calculated by Eq. (3.99b) for three of the emission regions. The resistance curves are shown in Fig. 3.19 for potential

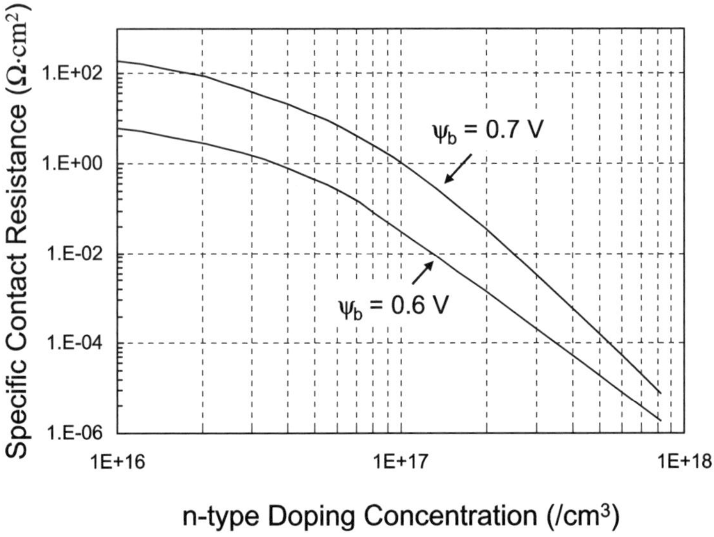

Fig. 3.19. Specific contact resistance as a function of n-type doping concentration for metals of different barrier heights.

barriers of 0.6 eV and 0.7 eV. It can be seen from the curves that the contact resistance drops significantly when the doping level increases from $10^{16}\,cm^{-3}$ to $10^{18}\,cm^{-3}$. The reduction in contact resistance indicates that the interface behaves like an ohmic contact, although the potential barrier remains exist. Hence, it is practical to always have a thin layer of n^+ implant at the metal contact, as shown in Fig. 3.20, to reduce the contact resistance to that of close to the ideal ohmic contact.

3.6. GaAs and SiC Power Diodes

The gallium arsenide power diode has significant advantages over the silicon counterpart due to the relatively higher bandgap energy of the gallium arsenide material. It has an energy bandgap of 1.43 eV which is approximately 0.32 eV larger than that of silicon. The significant advantages over silicon diodes are in terms of losses, breakdown voltage, switching frequency, and operating temperature range. Silicon Schottky barrier diodes have large leakage current and switching losses that make them impractical above 200 V. On the other hand, GaAs Schottky barrier diodes provide a higher breakdown voltage. However, a larger bandgap in GaAs also causes a higher forward voltage drop, but the impact of this disadvantage is mitigated because GaAs has an electron mobility, that is, five times higher than that of silicon.

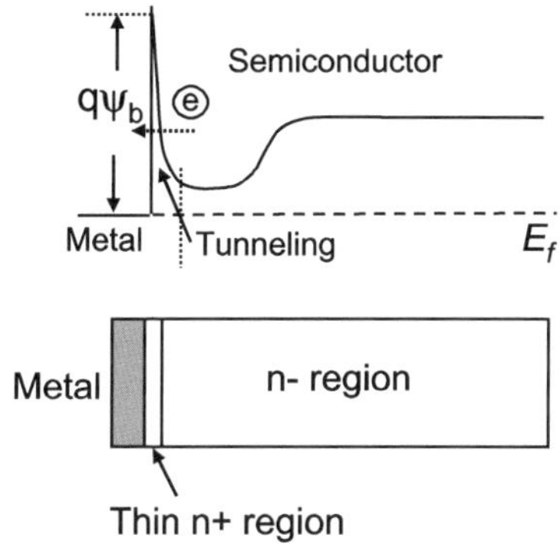

Fig. 3.20. A thin layer of n^+ is added to form the ohmic contact for n-type silicon–metal interface.

For 200 V and above, GaAs Schottky diode is favorable compared to the silicon one. GaAs Schottky barrier power diodes can be fabricated by forming dual-layer metallization of titanium-based contact on n–n^+ substrate for semiconductor–metal barrier and an aluminum top metal as the electrical contact. GaAs Schottky barrier diode with a breakdown voltage of 250 V and a current handling capability of 10 A are commercially available (Delaney *et al.*, 1995) for high-frequency power rectification applications. These diodes exhibit lower reverse-recovery time and less stored charge when compared to silicon Schottky barrier diodes of similar ratings. In addition, the softer reverse recovery and lower peak reverse current of the GaAs Schottky diodes are essential for reducing transition stresses on other switching devices, such as the power MOSFET and IGBT, during high-speed switching. The soft recovery of the GaAs Schottky barrier diodes is particular useful during inductive spike snubbering.

The GaAs p^+–n–n^+ (Ashkinazi *et al.*, 1995) diodes can also be made. The 0.31 eV difference in bandgap energy yields three orders of magnitude smaller in leakage current for the GaAs p^+–n–n^+ diodes than that of the silicon p–i–n diodes. GaAs p^+–n–n^+ diodes can be fabricated using a liquid phase epitaxial technique to grow the lightly doped n layer on heavily Zn doped (up to $2 \times 10^{19}\ cm^{-3}$)$p^+$ GaAs substrate. The ohmic contact (i.e. the n^+ layer) on the lightly doped n layer was formed by AuGeNi doping. The thickness of the n^+ layer or ohmic contact is about 1 μm. The resultant doping profile is a p^+–i–n–n^+ structure due to the presence of the near-fully compensated I region between the p^+ substrate and the lightly doped n-epitaxial layer. The presence of the near-fully compensated I layer yields a forward voltage and reverse recovery time comparable to those of silicon power diodes. However, the reverse recovery time, peak reverse-recovery current, and reverse-recovery charge of silicon diodes triple or quadruple at 200°C from their room temperature values. In comparison, the device parameters of GaAs junction diodes remain more stable within the high temperature range.

When high operating temperature, high-power, high-frequency and radiation hardened applications are needed, SiC diode becomes a good candidate (Singh *et al.*, 2002). The choice of SiC among other high bandgap materials, such as aluminum nitride (AlN), gallium nitride (GaN), boron nitride (BN), and diamond, is because of two main reasons, namely the availability of SiC bulk wafer and the relatively mature ability in fabrication processes. In comparison with GaAs, the SiC material has a higher thermal conductivity of up to 10 times, of about double in bandgap and with similar saturation velocity as that in GaAs. Table 3.3 list some of the basic 3C-, 4H-, and 6H-SiC properties (Casady and Johnson, 1996).

The diode can be made as either a majority carrier device of a Schottky diode or a bipolar carrier device of a p–n diode. For majority carrier device, the drift (lightly doped) region is always a concern during on-state as insufficient amount of conductivity modulation normally occurs. One way to reduce the voltage drop in the drift region is to raise the background doping so as to reduce the resistivity in this region, however the trade-off is that the peak electric field at the junction will also increase when the background doping increases. For SiC device, the critical electric field is about 10 times higher than that of silicon. This gives the field allowance for background doping to be raised. The

Table 3.3. Some of the SiC properties.

	3C-SiC	4H-SiC	6H-SiC
Bandgap (eV) @ low K	2.40	3.26	3.02
Critical E field (MV/cm)	2.12	2.2	2.5
Thermal conductivity @ 300 K (W cm^{-1} K^{-1})	3.2	3.7	4.9
Intrinsic concentration @ 300 K (cm^{-3})	1.5×10^{-1}	5×10^{-9}	1.6×10^{-6}
Saturation speed parallel to c-axis (cm/s)	—	2×10^{7}	2×10^{7}
Electron mobility (cm^2 V^{-1} s^{-1})	800	1000	400
Hole mobility (cm^2 V^{-1} s^{-1})	40	115	101
Dielectric constant	9.72	—	9.66

empirically derived relationship on the critical electric field as a function of background doping for 4H-SiC is (Konstantinov *et al.*, 1997)

$$E_{\text{BR}} = \frac{2.49 \times 10^6}{1 - 0.25 \log\left(\frac{N_d}{10^{16}}\right)}. \tag{3.100}$$

For example, a 25 kV SiC punchthrough diode would require around 200 μm of drift region doped at 5×10^{13} cm^{-3} or less with a high-level lifetime of larger than 20 μs. If the silicon material is used for the same 25 kV diode, then it will need a 2 mm drift region with a doping level of below 10^{12} cm^{-3} and a lifetime of at least 400 μs (Johnson *et al.*, 2001). This comparison shows the advantage of using SiC as the diode material for high-voltage application. However, the material properties are different from silicon, and these lead to certain difficulties in fabrication process. Detailed descriptions on the SiC process can be found in Casady and Johnson (1996) Singh *et al.* (2002), Tucker *et al.* (2001) and they are summarized here. Dopants for SiC material include nitrogen (N) for n-type doping, and aluminum (Al), boron (B), beryllium (Be), and gallium (Ga) for p-type doping. Undoped SiC is normally n-type due to the residual nitrogen from the material growth process. Unlike silicon, the donor and acceptor levels are not very close to the conduction and valance bands. For n-type, the activation energies are as large as 45 meV for 3C-SiC, 100 meV for 4H-SiC, and 150 meV for 6H-SiC. Note that, for hexagon structures, they have two donor energy levels, namely one for the cubic site and the other one for hexagonal site. The hexagonal site activation energy is normally smaller than that of the cubic site. For p-type, aluminum is the commonly used dopant and the energy level is about 200 meV for all types of polytypes. Due to the large activation levels, SiC devices usually have a partial thermal elevation and yield a lower concentration of free carriers at room temperature. It is then important to know the temperature ranges of the operation in order to determine the dopant levels to be used for SiC devices.

Two possible approaches can be used in fabrication process to add dopants into the SiC material. One is by the epitaxial growth process where nitrogen and carbon are competing for the carbon sites and aluminum and silicon are competing for silicon sites. Doping levels range from 10^{14} to above 10^{19} can be achieved by the epitaxial growth. Another technique is by high temperature multiple ion implantations. The commonly used dopants in implantation are aluminum (Al) and boron (B) for p-type, and nitrogen (N) and phosphorus (P) for n-type. The reason for high temperature is to reduce the lattice damage during the implantation and also to reduce the requirement for subsequent annealing. Typical implantation temperatures are 500°C for N, 400°C for Al, and 800°C for P (Johnson *et al.*, 2001). Table 3.4 gives a sample of implantation

Table 3.4. Multiple ion-implantations for diode p-body and n^+ cathode.

Al implantation schedule for p-body Energy/Dose	N implantation for n^+ cathode Energy/Dose
$1\,MeV/2.0 \times 10^{12}\,cm^{-2}$	$145\,keV/2.2 \times 10^{14}\,cm^{-2}$
$600\,keV/2.5 \times 10^{12}\,cm^{-2}$	$90\,keV/1.4 \times 10^{14}\,cm^{-2}$
$360\,keV/2.0 \times 10^{12}\,cm^{-2}$	
$200\,keV/1.4 \times 10^{12}\,cm^{-2}$	
$90\,keV/8.0 \times 10^{11}\,cm^{-2}$	

schedule for producing the p-body and n^+ cathode for a 4H-SiC planar junction diode (Tucker *et al.*, 2001).

The annealing conditions are important for reordering the crystal and for diffusing the dopants into SiC, especially needed when the same implant area is used for channel formation as well. The annealing temperature is normally high to obtain good properties in electrical activation for conductivity and to eliminate defects.

Etching of SiC can be made by high temperature molten salts, such as NaOH–KOH for wet-etching, or by fluorinated gases of the reactive ion dry etching process where $Si + 4F \rightarrow SiF_4$. The etch rates are usually slow, e.g. as low as 300 Å/minutes, compared to the silicon process of normally 2000 Å/minute or more.

The SiO_2 growth on SiC varies in quality depending on both surface's orientation and whether the surface is silicon-facing or carbon-facing. The oxidation temperature is usually around 1100°C. The hexagonal crystal structure of 4H- and 6H-SiC causes the properties of electron mobility, saturation velocity, impact ionization coefficients, and the thermal oxidation rate anisotropic. The quality of SiO_2 growth is very important in SiC devices mainly for two reasons, namely to sustain the off-state field stress and to serve as the gate oxide for channel formation. As mentioned earlier, the breakdown field in SiC ($2\text{–}3 \times 10^6\,V/cm$) is much higher than that of silicon and this approaches the similar breakdown stress as that of SiO_2. The worse situation is that, SiC has a larger dielectric constant of 9.5–10, compared to SiO_2 of 3.9. Therefore, under normal situation, the field within SiO_2 will be about 2.5 times higher than that in SiC. The oxide region may have an early breakdown under high field stress, especially at tight corners near the main junction and high temperature. The second concern on oxide growth is on the interface property, especially grown on the etched sidewall. Zhu *et al.* (2001) reported that the carrier effective mobility drops from $150\,cm^2/Vs$ to $100\,cm^2/Vs$, i.e. a 33% drop for the silicon horizontal channel on top of the surface versus the vertical channel by sidewall oxide, although the mobility reduction mechanisms are not entirely the same for SiC vertical channel. The situation gets worse in SiC when realized that the post-oxidation hydrogen anneal does not reduce the interface state densities and the effective mobility can drop below 25 cm^2/Vs at temperature above 475 K under the current fabrication capability. Current researches aim to improve the situation by growing high-quality oxide with low interface state density and trap density. The oxide interface density state can be obtained through the *C*–*V* measurement at high temperature, e.g. >600 K.

Ohmic contact is an another important aspect in SiC device fabrication. Alloys of Al/Ti are used for p-type ohmic contact by rapid thermal annealing with a best resistivity of below 0.01 mΩ cm^2 obtained for 4H-SiC. For high temperature p-type contact, instead of aluminum, boron carbide (B_4C) can be used. As for the n-type contact, it can be formed by nickel silicide (Ni_2Si) or nickel chromium (NiCr) to have a lowest resistivity of below 0.005mΩ cm^2. The contacts formed are generally stable at the temperature ranges below 400°C.

SiC diodes operating up to 8.6 kV and up to 970 K in temperature have been fabricated by Singh *et al.*, 2002; Alexandrov *et al.*, 2001; Tucker *et al.*, 2001) and others (Casady and Johnson, 1996; Johnson *et al.*, 2001). The measured forward voltage drop at 100 A/cm^2 for the 8.6 kV diode is 7.1 V with a reverse leakage current below 1 μA at 300 K. The diode also gives a faster recovery speed and a lower reverse-recover-recovery current compared to that of the silicon counterpart. The reverse-recovery charge is less sensitive in temperature increment, about 1–2 in ratio, compared to that of the silicon diode at temperature range of 300–450 K.

3.7. Switching Characteristics

Switching characteristics of a power silicon diode are determined by its intrinsic properties as well as the external circuit to which it is connected. For low voltage applications, Schottky barrier diodes are preferred over p–i–n diodes for the fast switching speed. Turn-off times of Schottky barrier diodes are faster than those of junction diodes since they are majority-carrier devices with less excess carrier storage compared to the p–i–n diodes. To overcome the shortcoming of Schottky diode large leakage current at high temperature and high voltage, a new kind of diode, called the merged p–(i)–n/Schottky or MPS diode (Baliga, 1987; Ang, 1989), was made suitable for high frequency and medium voltage applications. Its switching property is comparable to that of the Schottky barrier diode while its reverse-bias characteristic is similar to that of the p–i–n junction diode. Such a combined device will be described in more detail later in this chapter.

Heat generation from conduction or switching losses is another major concern for semiconductor devices. For example, a low forward on-state voltage drop is desired for device used for high-power, low frequency switching to minimize the conduction loss. Germanium power diodes, with a forward on-state voltage drop of about 0.3 V, are commercially available for high-power, low frequency use. The major drawback of germanium power diodes is again the high reverse leakage current due to its low bandgap energy of 0.67 eV at room temperature. For high frequency or low duty cycle applications, e.g. above 100 kHz, the forward and reverse recovery characteristics of a switching diode become important and they are directly linked to the switching loss as a dominant component in comparison with conduction loss.

3.7.1. Turn-on Transient

An overshoot voltage transient in forward conduction is observed across a p–(i)–n diode during the turn-on process as shown in Fig. 3.21. This overshoot arises from the existence of the parasitic inductance and the high resistive n-drift region as it is normally lightly doped. Conductivity modulation is expected to occur in this region to bring down the forward conduction voltage. However, such a modulation process takes finite time to occur and to complete as large amount of carriers of both types need to enter the region in order to build up the conductivity. The turn-on transient then exists between the duration of time period when the n-drift region undergoes from nil to full conductivity modulation, and it is called the *forward recovery* transient (Ko, 1961; Kano and Reich, 1964). In brief, the forward recovery phenomenon of

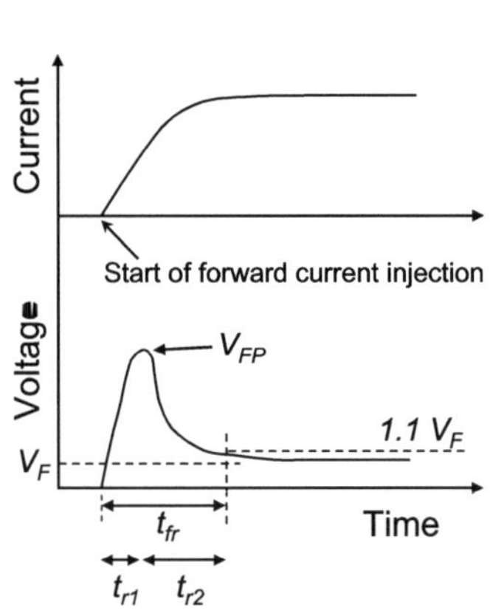

Fig. 3.21. Current and voltage waveforms showing the diode forward recovery phenomenon.

a diode occurs in a finite duration of time taken for a diode to go from the non-conducting state to a fully conducting state.

At quiescence state before turn-on, the junction capacitance serves to clamp the voltage across the diode to zero voltage. When the diode junction becomes forward biased, the injection of excess carriers commences. For a p^+–n^-–n^+ diode, both holes from p^+ emitter and electrons from n^+ emitter are injected into the n-drift region by lowering of barrier potentials. The diode current is the sum of both hole and electron currents as described earlier. Initially, the diode resistance is dominated by the ohmic resistance of the lightly doped n-drift region. However, as the concentration of injected excess carriers in the lightly doped drift region increases to the point that their concentration is comparable to the background, conductivity modulation begins and the associated ohmic voltage drop decreases. When a step-like current is injected into the diode, its forward voltage rises to a maximum value of $V_{\mathrm{fp,max}}$ which can be many orders of magnitude larger than the steady-state forward voltage drop V_{f}. The magnitude of $V_{\mathrm{fp,max}}$ is dependent upon the resistivity of the n^- region and the associated parasitic inductance at the wire-bond. As the diode current reaches its steady-state value at the end of t_{r1}, the parasitic inductive effect disappears. The concentration of injected excess carriers continues to increase until it reaches its steady-state value of diode current I_{f}. After t_{r1}, the voltage across the diode decays towards its steady-state forward voltage drop determined by ohmic components of the diode. It reaches steady-state at time t_{r2}. The diode forward recovery time t_{fr} is defined as the time it takes from the initial turn-on point to the point for the diode to attain the normal forward voltage drop. Thus,

$$t_{\mathrm{fr}} = t_{\mathrm{r1}} + t_{\mathrm{r2}}. \tag{3.101}$$

It should be noted that the diode forward recovery time t_{fr} is governed, not only by intrinsic properties of the diode, but also by the external circuit to which the diode is connected, as the rate of change in bulk resistance is a function of the magnitude of the applied current. The higher the magnitude of the applied current is, the shorter time the bulk resistance modulation process will need to come to its steady-state.

Besides the wire bind inductance, in quantity, the total voltage drop across the diode is comprised of the voltage drop across the junction and the bulk resistance. The increase in the junction voltage is needed for the high-level injection of excess carriers into the lightly doped n region. As the injected carrier density increases, the junction voltage increases with time toward its steady-state value. The excess carrier distribution and junction voltage of p–(i)–n diode at steady-state have been described earlier. At the recovery period, the conductivity of the lightly doped n-region increases for higher injected carrier density. Therefore, the junction voltage component increases with time and the bulk resistive voltage reduces with time during the forward recovery period. Overall, the diode behaves inductively by giving a voltage overshoot for large injected current density, capacitively for small current density giving an overdamped voltage response, and oscillatory response at intermediate current density (Ko, 1961). Using p^+–n diode as an example, the change of the diode bulk resistance, by hole carrier injection, during the forward recovery is given by:

$$R\left(\frac{t}{\tau_p}, J_d\right) \approx R_0\left[1 - \frac{1}{x_{n^-}}\ln\left(\frac{1 + J_d\mathrm{erf}\left(\sqrt{\frac{t}{\tau_p}}\right)}{1 + J_d\mathrm{erf}\left(\sqrt{\frac{t}{\tau_p}}\right)e^{-x_{n^-}}}\right)\right], \tag{3.102}$$

where

$$J_d = \frac{\left(1 + \frac{\mu_n}{\mu_p}\right)p_n(x = 0, t = \infty)}{p_{n0} + \frac{\mu_n}{\mu_p}n_{n0}}. \tag{3.103}$$

x_{n^-} is the normalized n^- base width (base width divided by L_p), R_0 is the base resistance without conductivity modulation, J_d is the diode current density and erf( ) is the error function. $x = 0$ refers to the position of junction. From the equation, it is seen that the modulation is made of a function of injected current density, the hole carrier lifetime, and the normalized base width. The expression is used for p^+–n diode where the hole injection as a minority carrier modulates the resistivity of the n^- base (drift) region. Similar expression can be made for n^+–p diode where the electron carrier injection is dominant in the drift region. For a p–i–n diode, the total resistance modulation is the combined effects of both carrier injections. The forward recovery period is usually below 0.5 μs as indicated by simulation and experimental works (Liang and Gosbell, 1990).

Due to the absence of the conductivity modulation in the lightly doped n^- region of a power Schottky diode of low barrier height, there is practically very brief forward recovery overshoot. Voltage overshoot is due primarily to the parasitic inductance in the bond wires and the change of junction voltage in Schottky barrier diode. For high barrier height Schottky diode, conductivity modulation does occur at high-level carrier injection and the forward recovery phenomenon is then noticeable.

3.7.2. Turn-off Transient

Turn-off characteristics of the p–(i)–n switching diode are of major concern in the power electronic circuits due to its longer recovery period (a few microseconds) and larger voltage and current overshoots. The bipolar device contains a large amount of excess carriers during forward conduction, especially under high-level injection. When the diode is directly applied a reverse bias immediately from its forward bias, these carriers need to be cleared in order for the diode to resume its reverse blocking state. Hence, the diode still remains at conduction state (but, current in the reverse direction) at the initial time period when the reverse bias voltage is applied. The process of changing diode from the full conduction state to the full blocking state is called the *reverse-recovery* process. During the recovery transient, the excess carriers removal process occurs, either via one side of the lightly doped region for n^+–p or p^+–n diodes, or via both boundaries of the base region for p–i–n diode (Moll *et al.*, 1962; Benda and Spenke, 1967; Berz, 1979).

When the reverse bias is directly applied, the orientation of the field within the device is then changed. As described earlier, hole carriers can travel through a reverse-biased junction from n-side to p-side, and electron carriers travel from p-side to n-side. Reverse current is now flowing through the device till the time that all the excess carriers are cleared from the lightly doped region. However, it is seldom to have the reverse voltage applied directly across the diode for recovery. Part of the reason is that the recovery will be too spiky and large current and voltage stress is induced. Also, stray inductance normally exists in the circuit connection. The recovery process is usually done via an external resistor or an inductor. Two or three stages are shown in Fig. 3.22 during the reverse recovery. At stage one, the excess carriers are swept out by the field

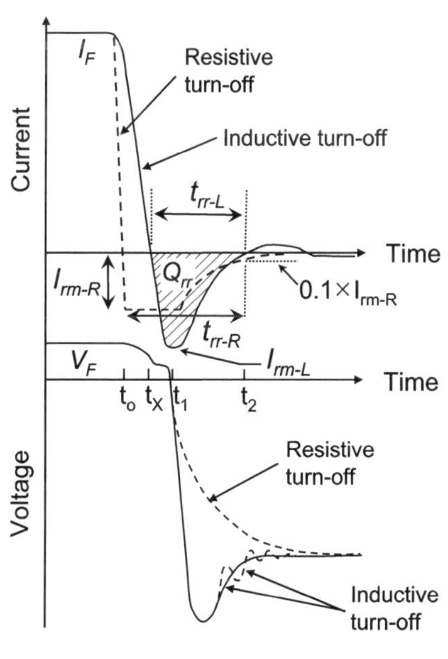

Fig. 3.22. Waveforms of diode under reverse recovery by inductive and resistive turn-off, note that the definitions of turn-off time t_{rr} are different for these two.

and the current waveform is limited by the external component connected to the diode. If it is connected to a resistor, then a step recovery current can be expected as shown in dashed line. Or, if it is connected to an inductor, which is the usual case, then a current ramp is expected, as shown in solid line. Note that, the excess carriers are not swept out from the lightly doped base region immediately as the current is ramping down and remains positive, i.e. positive carrier injection remains but weaker during the initial phase of the inductive reverse recovery. The concentration of excess carriers is dropping slowly due to less current injection and by the recombination process in the lightly doped base region. For the current ramp, the rate of change is determined by the magnitude of the applied reverse bias V_R and the total inductance L in the diode and external circuit, as

$$\frac{di_d(t)}{dt} = -\frac{V_R}{L}. \tag{3.104}$$

Or, if it is for resistive step-recovery, the recovery current is $-V_R/R$ during the period of between t_0 and t_1. If we define that $Q(t)$ is the excess charge density (cm^{-2}) in the lightly doped region at any time, using p^+–n diode as an example then for one-dimensional approximate,

$$Q(t) = \int_0^{W_{n^-}} (p_n - p_{n0})dx, \tag{3.105}$$

where W_{n^-} is the n-drift width. From the charge control analysis, the variation of excess charge is directly linked to continuity equation with the current flow by the following:

$$\frac{dQ(t)}{dt} + \frac{Q(t)}{\tau_F} = \frac{i_p(t)}{A} = j_p(t), \tag{3.106}$$

where A is the cross-section area. j_p is the forward hole current density. For p^+–n diode, the hole current is approximately equal to the diode current. τ_F is the effective forward lifetime for carriers and can be measured and expressed as (Tien and Hu, 1988)

$$\tau_F \approx \sqrt{(t_1 - t_X)(t_2 - t_1)}, \tag{3.107}$$

where t_X is the cross-point where the diode current drops to zero. Equation (3.106) can be revised to include the current induced in junction capacitance during the later stage, i.e. between the period of t_1 and t_2, when the reverse voltage builds up, as

$$\frac{dQ(t)}{dt} + \frac{Q(t)}{\tau_F} = j_p(t) + C_j\frac{dv_d(t)}{dt}, \tag{3.108}$$

where j_p is approximately equal to j_d, C_j is the junction capacitance per cross-section area. For inductive recovery, the current density changes linearly with time, the excess charge density between the time period t_0 and t_1 can be calculated as

$$Q(t) \approx \frac{V_R \tau_F}{L}\left(1 - e^{-\frac{t}{\tau_F}}\right) + \left(I_F - \frac{V_R}{L}t\right)\tau_F \qquad (3.109)$$

where I_F is the steady-state current before turn-off. Assume that the current linearly drops within the period and the capacitive current is very small and ignored. Figure 3.23 depicts the change of charge distribution in the lightly doped drift region. At time t_1, the hole concentration at the depletion region edge at p^+/n^- junction reaches zero. However, only a major portion of the stored charge, said Q_1 is removed from the n^- region and the leftover charge Q_2 is still in the base region. The reverse-recovery current now decreases quickly since insufficient holes carriers at the edge of the depleted p^+/n^- junction. The depletion region widens rapidly as time goes by and hence, the diode begins to regain its ability to support a reverse voltage. Between the period of t_1 and t_2, the reverse current varies quickly and, for inductive turn-off, this induces a large reverse voltage across the diode. During the period, the high field and expanded depletion region make the removal of excess charge much more effective. The rapid transient sometimes brings the circuit into ringing state where the circuit inductance and the junction capacitance have the exchange of energy flow. Equation (3.108) can be used to calculate the voltage across the diode by setting $Q(t) = j_d(t) \times \tau_R$ (Liang and Gosbell, 1990) as $Q(t)$ is no longer a charge-control variable.

$$v(t) = \frac{1}{C_j}\int\left(\tau_R \frac{dj_d(t)}{dt} + \left(\frac{\tau_R}{\tau_F} + 1\right) j_d(t)\right) dt, \qquad (3.110)$$

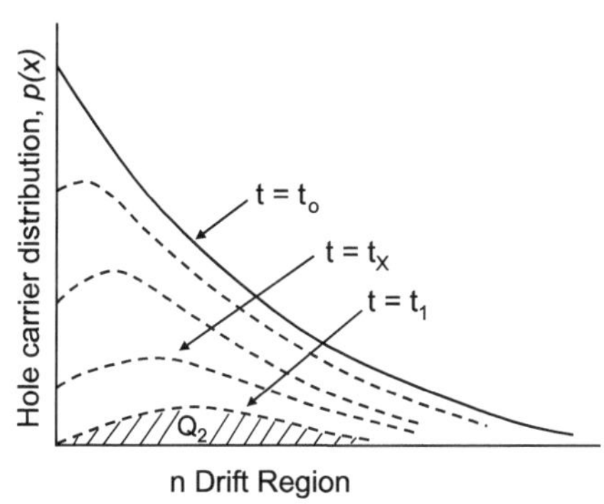

Fig. 3.23. Variation of hole carrier concentration in the n drift region during diode turn-off.

where τ_R is the effective reverse lifetime. The ratio of τ_F/τ_R is about 3–4 for most of the power diodes. The junction capacitance C_j is a function of the reverse voltage during turn-off, however for simplicity in calculation, a constant average value can be used. The overshoot voltage V_{RM} is normally higher than the external reverse voltage, V_R. At time t_2, the reverse-recovery current reduces to zero and the diode now blocks the reverse voltage V_R. The reverse-recovery time t_{rr} is defined as the time period where negative current is flowing, as shown in Fig. 3.22. Another simple way of calculating the reverse recovery time is by relating the charge removal during the period if the reverse-recovery charge Q_{rr} is known and assume current linearity (Fig 3.22).

$$Q_{rr} = \frac{1}{2} J_{rr} t_{rr}, \tag{3.111}$$

where J_{rr} is the reverse peak current density. The reverse-recovery time can be reduced by reducing the carrier lifetime in the n^- region for faster recombination. The softness factor of the reverse-recovery waveform is defined as $S = t_b/t_a$ where t_a and t_b are the time period shown in Fig. 3.24. The larger the value of S, the softer is the reverse recovery of the diode. Normally, there is a trade-off between the turn-off softness for less EMI/EMC interference, and the turn-off losses. Figures 3.25 and 3.26 give several reference curves to indicate the turn-off time and reverse-recovery charge are the functions of rate of current change (V/L) at different steady-state current ratings for a typical 300 V, 30 A power diode.

3.8. MPS (Merged p–i–n/Schottky) Diode

As described earlier, the Schotty diode has a lower barrier height for low voltage on-state conduction and fast switching for majority carrier conduction. The disadvantages of the Schottky diode are its high leakage current under high voltage and high temperature applications, and also its limited current carrying capability due to lack of conductivity modulation. In the p–i–n diode,

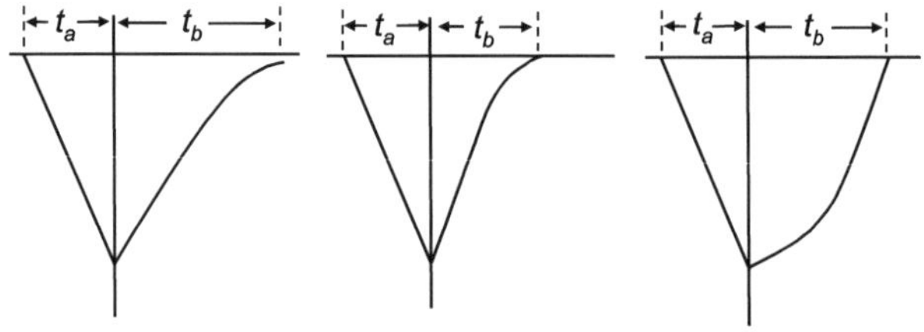

Fig. 3.24. Diode recovery currents with different levels of softness.

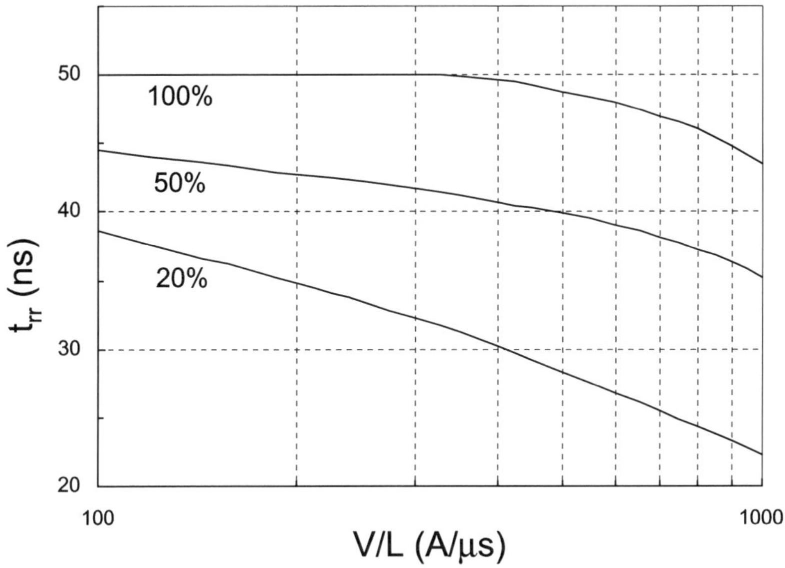

Fig. 3.25. Variation of turn-off time as a function of rate of current change at diode turn-off.

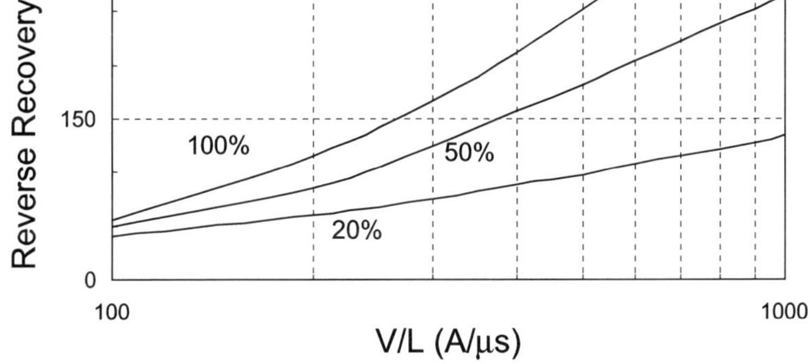

Fig. 3.26. Variation of reverse-recovery charge as a function of rate of current change at diode turn-off.

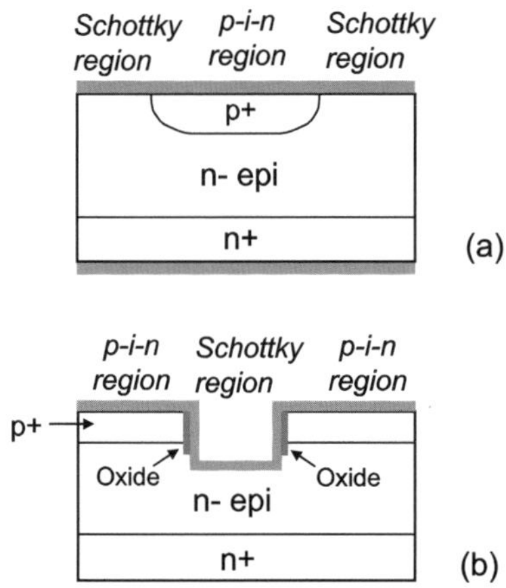

Fig. 3.27. MPS diodes with (a) top contact and (b) trench contact structures.

the advantages are mainly the low leakage current and high current carrying capability. But, the disadvantage is the slow switching speed and higher onset voltage for conduction. It is then a natural thinking to merge these two diodes together on the same silicon die to form a new kind of device, called *Merged p–i–n/SchottkyDiode* or MPS diode (Baliga, 1987; Ang, 1989). Figure 3.27 depicts the general structures of the MPS diode and Fig. 3.28 gives the typical on-state *J–V* curves for the MPS diode and its p–i–n and Schottky current components. The curve for Schottky component is drawn with the assumption that p–i–n diode is inactive. However, in the actual MPS conduction, the p–i–n diode is active above 0.7 V. Therefore, the MPS current is not the sum of these two currents, but higher due to the conductivity modulation for the entire n-drift region. As it can be seen, the MPS diode has an onset voltage similar to that of the Schottky diode and with a high-level current density similar to that of the p–i–n diode. The percentage of Schottky versus p–i–n on the MPS surface depends on the trade-off among the on-state voltage drop, switching time, and reverse leakage current.

3.9. Smart-Power Integrated Synchronous Rectifier

Modern power converter integrated circuits used in portable devices and fabricated by the (Bi)CMOS-compatible process need to have the nature of compactness in size and high conversion efficiency in order to prolong the usage time between consecutive charging needs. Both the supply and output voltages can be as low as one or two stages of cell-voltage, e.g. 1.2–2.4 V. A key element used in this conversion task for low output voltages is the *synchronous rectifier*

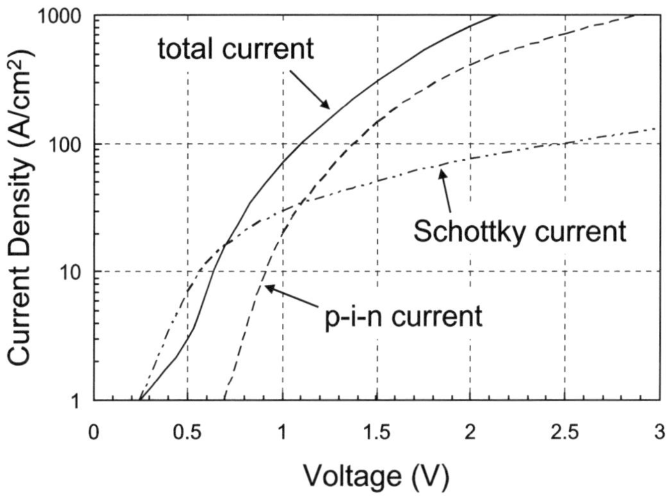

Fig. 3.28. The total current by both Schottky and p–i–n current components.

(SR), which is a very low on-state voltage rectifier, typically below 0.1 V drop at rated current, to reduce the output rectification loss to be lower than that of the Schottky or junction diode. An SR is formed by a MOSFET with an inherent body diode in parallel and so driven as to perform the function of a diode. Figure 3.29 shows the efficiency of a typical dc/dc converter system with and without using the synchronous rectifier as a function of the system output voltage. The difference becomes significant when the output voltage is below 2 V.

According to the gate drive configurations, the SRs can be categorized into three different types, namely the external-driven, the self-driven, and the smart-power synchronous rectifiers. In the external-driven configuration, the gate drive signal for the SR needs to be generated in synchronous with the main power MOSFET to avoid short circuit situation. Normally, a comfortable deadtime of more than 50 ns is placed between gate signals to the upper and lower MOSFETs to avoid short-circuit during switching transients. Figure 3.30 shows the circuit configuration and gate drive signals. A major advantage of such an SR is that it can be used for different dc/dc converter topologies. Also, the MOSFET is effectively driven to operate at the low-voltage linear region during the entire rectification period. This will keep the rectification efficiency not being affected by the system operating conditions, such as the variations of input/output voltage ratio and switching duty cycle. However, the tedious part is that a properly controlled gate drive circuitry with briefly regulated power

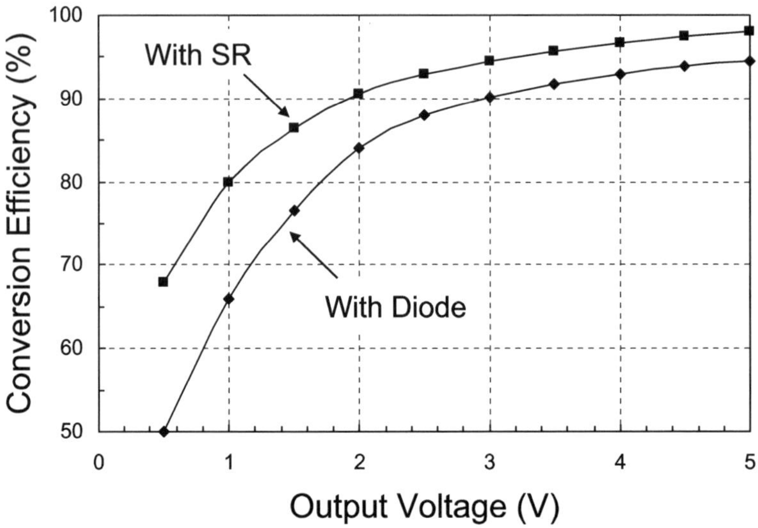

Fig. 3.29. Conversion efficiencies with synchronous rectifier and with diode of a dc/dc converter.

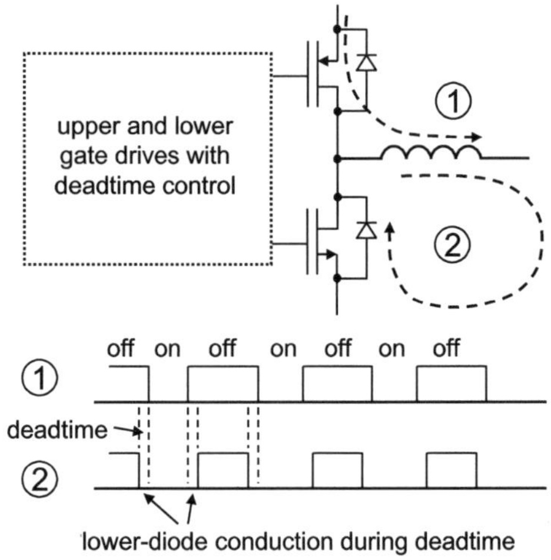

Fig. 3.30. Converter output stage (upper and lower MOSFETs) and the synchronous gate drive signals.

supply needs to be built externally. The self-driven SR uses one of the transformer tap as the gate drive. The advantage of having such a configuration is its simplicity in gate drive. However, the advantage is that the gate drive voltage is not constant but rounded at the corner of the switching edges. Also, the magnitude of the gate drive tends to vary with input/output voltage ratio and switching duty cycle. The use of synchronous rectifier has long known to be a possible solution but sometimes avoided due to its associated complexities and potential pitfalls. To overcome this problem, a smart-power integrated synchronous rectifier can be a good candidate for low-voltage switching solution (Liang *et al.*, 1995; Pan *et al.*, 1999).

There are three important parameters related to the operation of a synchronous rectifier, namely the product of on-state resistance and the input capacitance, i.e. the loss factor, the reverse-recovery time of the body diode, and the breakdown voltage. The power loss in a synchronous rectifier can be calculated as

$$P_{\mathrm{SR,loss}} = I_{\mathrm{rms}}^2 \times \frac{R_{\mathrm{on,sp}}}{A} + C_{\mathrm{in,sp}} \times V_{\mathrm{G}}^2 \times A \times f, \tag{3.112}$$

where A is the device area, $R_{\mathrm{on,sp}}$ is the specific on-state resistance, $C_{\mathrm{in,sp}}$ is the specific gate input capacitance, I_{rms} is the gate voltage to turn on the synchronous rectifier, and f is the switching frequency. The factor of duty cycle, if necessary, can also be added into the first term of Eq. (3.112) to identify the on-state conduction loss more appropriately. By increasing the device area, the conduction loss is reduced whereas the gate switching loss is increased. The minimum value of power loss can be obtained mathematically as follows (Baliga, 1989).

$$P_{\mathrm{SR,loss,min}} = 2 \times I_{\mathrm{rms}} \times V_{\mathrm{G}} \times \sqrt{f \times R_{\mathrm{on,sp}} \times C_{\mathrm{in,sp}}}. \tag{3.113}$$

The product of $R_{\mathrm{on,sp}}$ and $C_{\mathrm{in,sp}}$ can be considered as the loss factor for the synchronous rectifier. It can also be seen that, by keeping the gate voltage lower, if still sufficiently high to keep the on-state resistance low, the total loss can also be reduced. A lower threshold MOSFET can also provide a lower on-state conduction resistance, but with the penalty of a high input gate capacitance. Another way to lower the on-state resistance is to reduce the conduction channel length to a minimum, say below 0.5 μm. The shortcoming of having a shorter channel length is its lower drain-to-source breakdown voltage.

The body diode is a part of the power MOSFET structure, for example, as shown in Fig. 3.1(e) for lateral structure and in Fig. 3.31 for vertical structure, where the reverse recovery of the diode can be a concern. The body

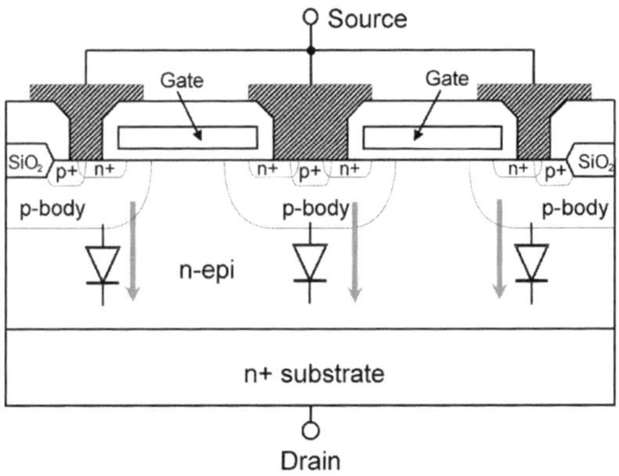

Fig. 3.31. MOSFET body diode and the current conduction from source to drain during deadtime.

diode enters its conduction state during the deadtime period when the lower-MOSFET is holding back from its on-state conduction. Such a deadtime period is needed to avoid short-circuit current spikes during switching period if the upper p-MOSFET and the lower n-MOSFET are both switched at the same time. The amount of deadtime to be inserted between them is dependent on the amount of switching current, the MOSFET capacitances and the gate drive capability. Understanding that, a minimum deadtime period will bring the maximum benefit of using synchronous rectifier. For minimum deadtime during high-frequency operation, this will keep the diode conduction stay at low injection level, i.e. with less charge storage, and the reverse recovery phenomenon will be less significant.

Figure 3.32 gives the schematic circuit diagram of a bipolar-MOSFET smart-power integrated synchronous rectifier. The circuit can be viewed in three stages, namely the voltage supply stage at the left, the diode-BJT flip-flop stage in the middle, and the power MOSFET with body diode at the right. The voltage supply is usually fed from the converter output which has a stable dc voltage. If this is not possible, an alternative is to have a diode (D1) to be in series with a small capacitor to form a simple power supply, as shown in the dashed line. A small current is flowing through resistor R2 to reach the transistor base and diode anode. The diode and the transistor are used to form a bi-stable flip-flop whose conduction state is determined by the external voltage bias across anode and cathode. For example, if the anode voltage potential is higher than that of cathode, this reduces the base current in the transistor compared to the diode (D2) current as the base–emitter junction is less forward biased. The transistor operating point then moves away from

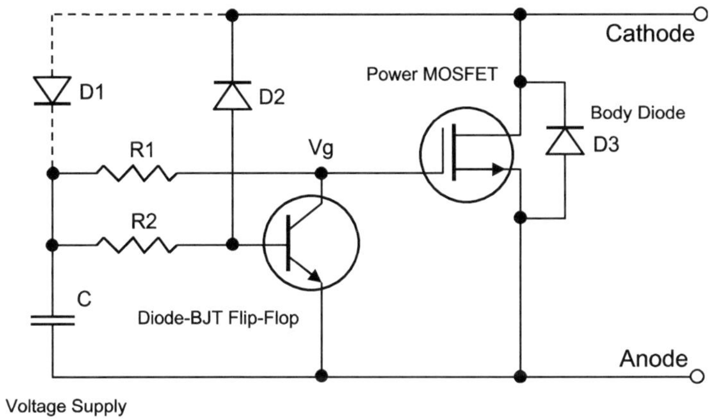

Fig. 3.32. Schematic circuit diagram of a bipolar-MOSFET synchronous rectifier.

the saturation region to have a higher collector voltage to the gate of power MOSFET. When the collector voltage is sufficiently high, the power MOSFET turns on to divert the main current from its body diode (D3) to the MOSFET channel. Contrarily, if the cathode voltage potential is higher than that of the anode, the transistor base current remains high. This makes the transistor to stay in the saturation region and the MOSFET gate voltage is below the threshold voltage. The device is at its blocking state, for both the MOSFET and the body diode. The bi-stable flip-flop ensures that the power MOSFET can only be turned on when the terminal voltage is at the correct state. This protection eliminates the requirement of deadtime adoption as normally needed in the conventional synchronous rectifier.

Figure 3.33 gives the device profile using the vertical UMOS as an example. The p+ sinker and buried layer provide the junction isolation between the low-power control circuit and the high-power UMOS structure. Partial silicon-on-insulator platform can also be used, and indeed a better option, for the needed isolation (Cai *et al.*, 2001). The resistors R1 and R2 can be made by surface layer polysilicon deposition or formed by well-resistor as their resistance values are not necessary to be precise.

Figure 3.34 shows the data curves from process and device simulations on the single device cell for the waveforms of gate voltage, synchronous rectifier current, and the body diode current as a function of terminal bias voltage. It can be seen that the rectification transition occurs at about the zero voltage. The MOSFET gate voltage rises quickly and, at 0.05 V terminal voltage, the gate voltage reaches its steady-state value. Below 0.5 V, the body diode current is almost zero. The single device cell was turned off by varying the terminal

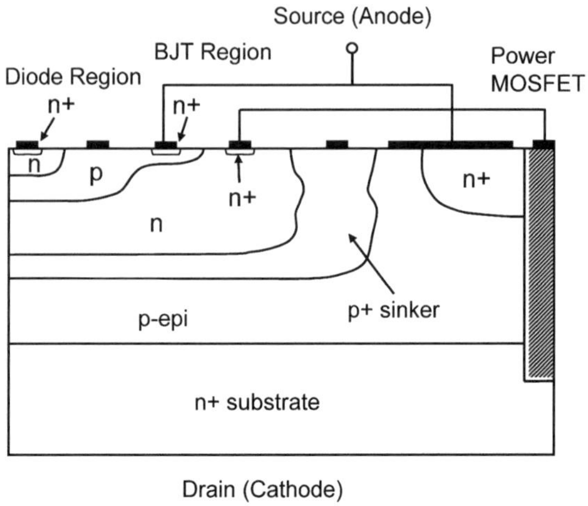

Fig. 3.33. Device profile of the smart-power integrated synchronous rectifier.

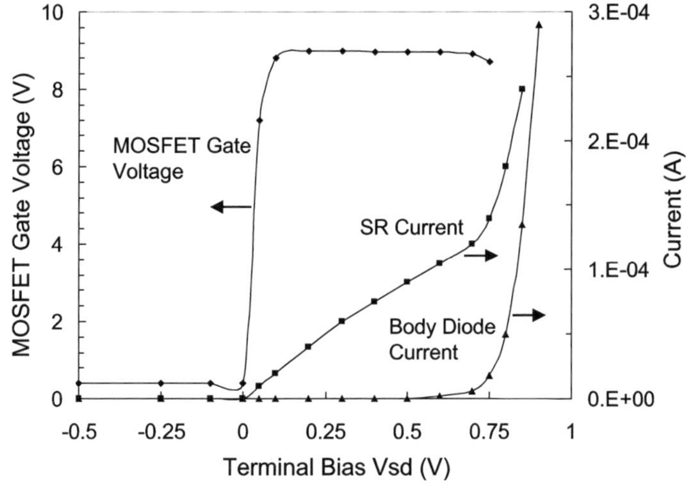

Fig. 3.34. Data curves on MOSFET gate voltage, synchronous rectifier current and body diode current for the smart power integrated synchronous rectifier by process and device simulations.

voltage from on-state conduction to reverse blocking both for the body diode (without MOSFET connected) and for the synchronous rectifier. Figure 3.35 shows the terminal voltage and reverse-recovery current waveforms. It can be seen that, the body diode alone has a normal reverse-recovery overshoot and time, while the synchronous rectifier has a much less recovery phenomenon.

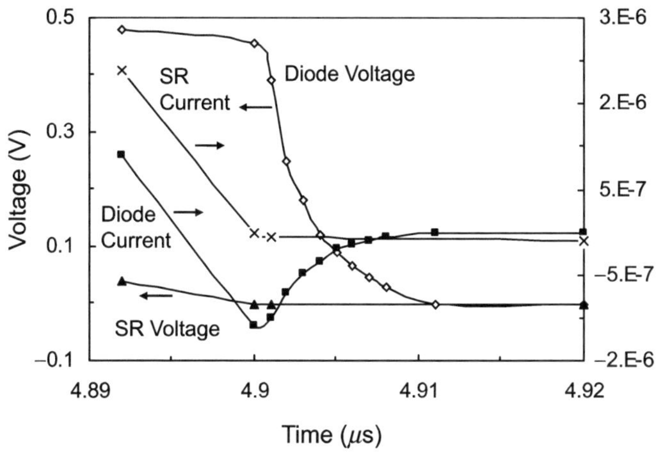

Fig. 3.35. Voltage and current waveforms of body diode (without MOSFET connected) and synchronous rectifier during turn-off by simulation.

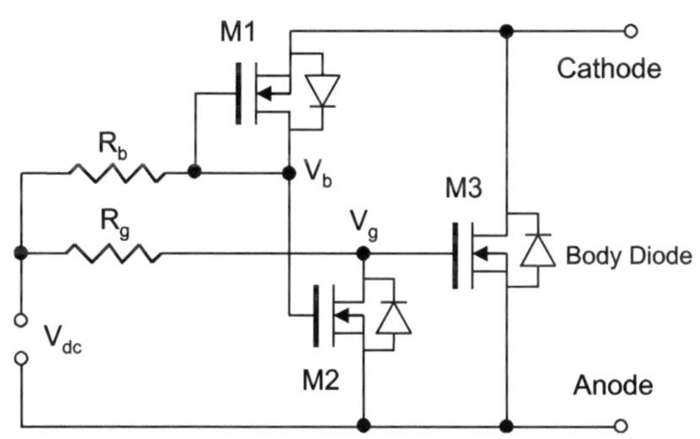

Fig. 3.36. Schematic diagram of MOSFET-based synchronous rectifier.

The smart power integration can also be modified for CMOS-compatible process where the bipolar transistor is removed as shown in Fig. 3.36 (Liang *et al.*, 2001). The description of how the synchronous rectifier works can be made similarly in forward and reverse bias conditions. For this configuration, the supply voltage V_{dc} needs to match the threshold voltage of MOSFET M1. Therefore, the threshold voltage tuning for MOSFET M1 shall be part of the process steps. When the anode voltage is higher than the cathode voltage, MOSFET M1 will be turned on. The gate voltage of M2 becomes lower than its threshold voltage, and it is turned off. The voltage V_g at M3 gate is then pulled up to turn on MOSFET M3. Once MOSFET M3 is turned on, it diverts the forward conduction current from the body diode to the MOSFET channel to reduce the on-state voltage drop. For the reverse bias condition, i.e. the cathode voltage is higher than the anode voltage, MOSFET M1 is turned off, and MOSFET M2 is turned on to pull down the voltage V_g at M3 gate. In this case, the device is at its blocking state.

3.10. Summary

In this chapter, the properties and operations of power junction diodes were discussed. These included the forward conduction, reverse blocking and switching behaviors; junction capacitance; temperature dependency; ohmic contact; Schottky barrier diode; MPS diode, and smart power synchronous rectifiers.

References

P. Alexandrov, W. Wright, M. Pan, M. Weiner, L. Jiao and J. H. Zhao, Demonstration of a 140 V, 800 V, fast recover 4H-SiC P-i-N/Schottky barrier (MPS) diode, *Int. Semiconductor Devices Res. Symp.* 13–16 (2001).

J. M. Andrew and M. P. Lepselter, Reverse current–voltage characteristics of metal-silicide Schottky diodes, *Solid State Electron.* **13**, 1011–1023 (1970).

S. S. Ang, High-temperature characteristics of epitaxial high–low junction n^+–n–p^+ silicon diodes, *Microelectron. J.* **26**, 375–382 (1995).

S. S. Ang, A V-groove Schottky/pn diode, *Microelectron. J.* **20**, 7–12 (1989).

G. Ashkinazi, J. Lorch and M. Nathan, Ultra-fast GaAs power diodes provides dynamic characteristics with better temperature stability than silicon diodes, *PCIM Mag.*, 10–16 (April, 1995).

B. J. Baliga, Barrier-controlled current conduction in field controlled thyristor, *Solid State Electron.* **24**, 617–620 (1981).

B. J. Analysis of a high-voltage merged p–i–n/Schottky (MPS) rectifier, *IEEE Electron Device Lett.* **8**(8), 407–409 (1987).

B. J. Baliga, Power semiconductor device figure of merit for high-frequency applications, *IEEE Electron Device Lett.* **10**, 455–457 (1989).

H. Benda and E. Spenke, Reverse recovery process in silicon power rectifiers, *Proc. IEEE* **55**, 1331–1354 (1967).

F. Berz, Step recovery of p–i–n diodes, *Solid State Electron.* **22**, 927–932 (1979).

J. Cai, C. Ren, Y. C. Liang, N. Balasubramanian and J. K. O. Sin, A partial SOI technology for single-chip RF power amplifiers, *IEEE Int. Electron Devices Meeting*, 891–894 (2001).

P. Cappelletti, G. F. Cerofolini and M. L. Polignano, Current–voltage characteristics of ideal silicon diodes in the range of 300–400 K, *J. Appl. Phys.* **57**(2), 646–647 (1985).
J. B. Casady and R. W. Johnson, Status of silicon carbine (SiC) as a wide-bandgap semiconductor for high temperature applications: A review, *Solid State Electron.* **39**, 1409–1422 (1996).
C. R. Crowell, The Richardson constant for thermionic emission in Schottky barrier diodes, *Solid State Electron.* **8**, 395–399 (1965).
C. R. Crowell and S. M. Sze, Current transport in metal–semiconductor barriers, *Solid State Electron.* **9**, 1035–1048 (1966).
S. Delaney, A. Salih and C. Lee, GaAs diodes improve efficiency of 500 kHz DC–DC converter, *PCIM Ventura Mag.* **21**(8), 10–11 (1995).
B. Elfsten and P. A. Tove, Calculation of charge distributions and minority-carrier injection ratio for high-barrier Schottky diodes, *Solid State Electron.* **28**, 721–727 (1985).
M. A. Green, Intrinsic concentration, effective densities of states, and effective mass in silicon, *J. Appl. Phys.* **67**(6), 2944–2954 (1990).
R. N. Hall, Power rectifiers and transistors, *Proc. IRE* **40**, 1512–1518 (1952).
A. Herlet, The forward characteristics of silicon power rectifiers at high current densities, *Solid State Electron.* **11**, 717–742 (1968).
J. R. Hauser and P. M. Dunbar, Minority carrier reflecting properties of semiconductor high–low junctions, *Solid-State Electron.* **18**, 715–716 (1975).
C. M. Johnson, N. G. Wright, M. J. Uren, K. P. Hilton, M. Rahimo, D. A. Hinchley, A. P. Knights, D. J. Morrison, A. B. Horsfall, S. Ortolland and A. G. O'Neill, Recent progress and current issues in SiC semiconductor Device for power applications, *IEE Proc. Circuits, Devices Syst.* **148**(2), 101–108 (2001).
K. Kano and H. J. Reich, Forward transient behavior of p–n junction diodes at high injection levels, *IEEE Trans. Electron Devices* **11**, 515–523 (1964).
Y. C. Kao and E. D. Wolley, High-voltage planar p–n junctions, *Proc. IEEE* **55**, 1409–1414 (1967).
T. Kawakami, Y. Amemiya and Y. Mizushima, Schottky diode design for low-voltage rectification, *Solid State Electron.* **28**, 885–891 (1985).
W. H. Ko, The forward transient behavior of semiconductor junction diodes, *Solid State Electron.* **3**, 59–69 (1961).
A. O. Konstantinov, Q. Wahab, N. Nordell and U. Lindefelt, Ionization rates and critical fields in 4H silicon carbide, *Appl. Phys. lett.* **71**, 90–92 (1997).

Y. C. Liang and V. J. Gosbell, Diode forward and reverse recovery model for power electronic SPICE simulations, *IEEE Trans. Power Electron.* **5**(3), 346–356 (1990).

Y. C. Liang, R. Oruganti and T. B. Oh, Design consideration of power MOSFET for high frequency synchronous rectification, *IEEE Trans. Power Electron.* **10**, 388–395 (1995).

Y. C. Liang, C. Y. Lim and G. S. Samudra, MOS-controlled smart power synchronous rectifier, *IEEE Region 10 Conf.*, 480–482 (2001).

J. L. Moll, S. Krakauer and R. Shen, P–N junction charge-storage diodes, *Proc. IRE* **50**, 43–53 (1962).

R. S. Muller and T. I. Kamins, *Device Electronics for Integrated Circuits* (John Wiley & Sons, 1986), p. 140.

W. T. Ng, S. Liang and C. A. T. Salama, Schottky barrier diode characteristics under high-level injection, *Solid State Electron.* **33**, 39–46 (1990).

H. L. Pan, Y. C. Liang and R. Oruganti, Design of smart power synchronous rectifier, *IEEE Trans. Power Electron.* **14**, 308–315 (1999).

A. Rose, Comparative anatomy of models for double injection of electrons and holes into solids, *J. Appl. Phys.* **35**(9), 2664–2678 (1964).

C. T. Sah, R. N. Noyce and W. Shockley, Carrier generation and recombination in p–n junctions and p–n junction characteristics, *Proc. IRE* **45**, 1228–1243 (1957).

W. Shockley, The theory of p–n junction in semiconductors and p–n junction transistors, *Bell Syst. Technol. J.* **28**, 435–489 (1949).

J. M. Shannon, Control of Schottky barrier height using highly doped surface layers, *Solid State Electron.* **24**, 537–543 (1976).

R. Singh, J. A. Cooper Jr., M. R. Melloch, T. P. Chow and J. W. Palmour, SiC power Schottky and PiN diodes, *IEEE Trans. Electron Devices* **49**, 665–672 (2002).

S. N. Singh and P. K. Singh, A new expression for minority carrier surface recombination velocity at low-high junction of an n^+–p–p^+ silicon diode, *Solid State Electron.* **33**, 968–970 (1990).

S. N. Singh and G. C. Jain, Effect of the back-surface field on the open-circuit voltages of p^+–n–n^+ and n^+-p–p^+ silicon solar cells, *Solar Cells* **5**, 143–172 (1982).

S. M. Sze, *Physics of Semiconductor Device* (Wiley, New York 1981), 15–19 and p. 90.

N. Thapar and B. J. Baliga, The accumulation channel driven bipolar transistor (ACBT), *IEEE Electron Device Lett.* **18**(5), 178–180 (1997).

B. Tien and C. Hu, Determination of carrier lifetime from rectifier ramp recovery waveform, *IEEE Electron Device Lett.* **9**, 553–555 (1988).

J. B. Tucker, M. V. Rao, N. A. Papanicolaou, J. Mittereder, A. Elasser, A. W. Clock, M. Ghezzo, O. W. Holland and K. A. Jones, Characteristics of planar n–p junction diode made by double-implantations into 4H-SiC, *IEEE Trans. Electron Devices* **48**, 2665–2670 (2001).

C.-Y. Wu, Barrier height enhancement of the Schottky barrier diode using a thin uniformly-doped surface layer, *Solid State Electron.* **24**, 857–862 (1981).

A. Y. C. Yu, Electron tunneling and contact resistance of metal-silicon contact barriers, *Solid State Electron.* **13**, 239–247 (1970).

Y. Zhu, Y. C. Liang, S. Xu, P-D. Foo and J. K. O. Sin, Folded gate LDMOS transistor with low on-resistance and high transconductance, *IEEE Trans. Electron Devices* **48**, 2917–2928 (2001).

4

Power Metal–Oxide–Semiconductor Field-Effect Transistor

4.1. Introduction

The operational principle of the metal–oxide–semiconductor field-effect transistor (MOSFET, or sometimes called MOS in short) devices originates from the field-effect concept by Lilienfeld in the 1930s. When the industry eventually learned how to grow the good-quality gate dielectric with acceptable charges, the n-MOS technology and the later CMOS technology took off in a fast path. Today, MOS device technology, namely CMOS or BiCMOS process, becomes prevalent and dominant in integrated circuit (IC) development. At the same time, power MOSFET devices began to emerge and were introduced in 1970s. So far, many structures for power MOSFET devices have been proposed, such as the double-diffusion MOSFET (or DMOS), V-groove MOSFET (or VMOS), and trenched-gate UMOS. Due to many processing difficulties related to the gate dielectrics and the enhanced electric field at the tip of the V-groove (Holmes and Salama, 1973), the VMOS structure is seldom used in production. Instead, the DMOS structure is a preferred choice. For the UMOS structure, it is seen as the preferred choice to provide a low on-state resistance. The introduction of power MOSFET brought about the prediction of the extinction of the bipolar power transistor. However, this prediction did not happen immediately as cost and reliability plagued the power MOSFET. The bipolar power transistor continued to be used in many high-voltage and high-current applications. Only in the mid-1980s, the bipolar power transistor finally met the most critical rival with the introduction of the insulated-gate bipolar transistor (IGBT), which will be described in Chapter 5.

The high gate impedance is a primary feature of the power MOSFET that greatly simplifies its gate drive circuitry. The negative temperature coefficient of the drain current, i.e. the drain current reduces when temperature goes up, in power MOSFET also provides an additional advantage over their bipolar counterparts. In a sense, the bipolar transistor is very susceptible to thermal run-away but not the case for power MOSFET. However, the power MOSFET suffers a higher on-state conduction loss as compared to the bipolar power transistor of similar die sizes because of the absence of conductivity modulation in power MOSFET. In spite of this, the trend of higher switching frequency to increase power density in power electronic system continues, therefore, the choice of using power MOSFET over conventional power bipolar switching transistor is evident.

In this chapter, the operations of the power MOSFET are described. This begins with the basic MOS physics necessary for the understanding of the device. This is then followed by the derivation of current–voltage characteristics. Various power MOSFET structures and their switching characteristics are mentioned along with the presentations of experimental switching waveforms under different operating and load conditions. Gate driver circuits for power MOSFET in various configurations are also discussed. Finally, more advanced gate structures, such as the dummy-gate and folded-gate MOSFET devices and the radio frequency (RF) MOSFET device on the partial silicon on insulator (SOI) platform are introduced.

4.2. Basic MOS Physics

The basic MOSFET structure consists of a field-controlled channel region upon which majority carriers flow between the drain and source. It is thus important to understand the basic properties of the metal–oxide–semiconductor structure that controls the conductance of the channel by the application of a gate voltage. Figure 4.1 shows the MOS capacitor on p-type silicon along with its energy band diagram in the thermal equilibrium state.

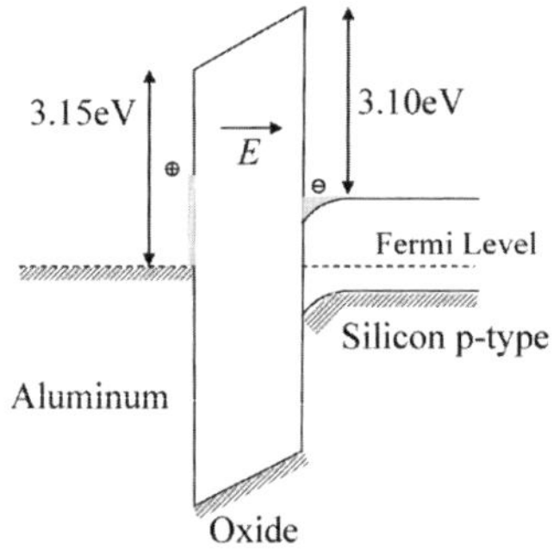

Fig. 4.1. The energy band diagram of the metal–oxide–silicon at the thermal equilibrium state.

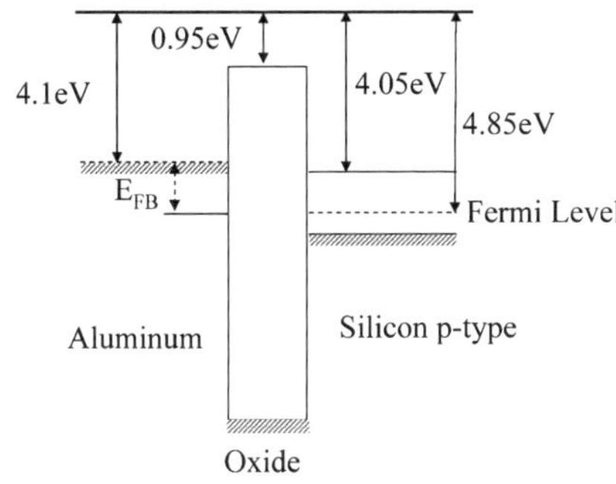

Fig. 4.2. Energy band diagram of the metal–oxide–silicon under the flat-bed state.

The Fermi level of metal is at the top of the filled band. For aluminum, it is around 4.1 eV away from the vacuum level. For silicon, the Fermi level is a function of the doping type and concentration. The edge of conduction band is around 4.05 eV away from the vacuum level, so the Fermi level is calculated as (5.17 eV $-E_f$) away from the vacuum level (refer to Fig. 4.2). The Fermi levels of metal and silicon are aligned to the same level by charge transfer between them. For p-type silicon, the Fermi level of metal is higher than that of the silicon, therefore negative charge is transferred from the metal to the silicon. The potential difference which corresponds to the initial difference in Fermi levels is now across the oxide and the depletion region at the silicon interface. The depletion of mobile carriers, i.e. holes in p-type silicon, at the silicon–oxide interface forms the space charge as that of the junction barrier. The density of the induced charge Q_S, in this region is given by the Gauss law:

$$Q_S = -qN_A x_{d0} = -\varepsilon_S \hat{E}_S, \tag{4.1}$$

where N_A is the dopant concentration, x_{d0} is the depletion width in silicon at thermal equilibrium, $\hat{E}_S$ is the electric field at the semiconductor surface, and ε_S is the dielectric constant of silicon. The potential difference between metal and silicon is distributed to the oxide voltage drop, V_{ox}, and the semiconductor surface potential, Ψ_S. The symbol Ψ (in V) is used to represent the potential and $q \times \Psi$ (in eV) is used to relate the potential in the energy band. The MOS system has charge storage and field associated as those of a capacitor.

The application of a gate voltage V_G across the MOS capacitor establishes an additional electric field between the gate electrode and the underlying semiconductor. This causes the system to depart from its thermal equilibrium state (Fig. 4.1). Depending on the polarity and magnitude of the gate voltage, the underlying silicon can accumulate different type and different amount of charge, and the MOS system functions in several possible states, namely the flat-band, the accumulation, the depletion, and the inversion states. They are described below using the p-type silicon as an example.

4.2.1. Flat-Band State

When the applied gate voltage serves to exactly compensate the difference in metal–silicon work functions, the net electric field in the oxide and the silicon interface is neutralized. Under such a condition, the energy-band is no longer bending downwards at the silicon–oxide interface but becomes flat as shown in Fig. 4.2. The gate voltage V_G is

$$V_G - V_B = V_{FB} = \Psi_M - \Psi_S, \tag{4.2}$$

where V_B is the substrate reference voltage, Ψ_M and Ψ_s are the potential difference between the Fermi level and the vacuum level. As Ψ_M is less than Ψ_s for the case of p-type silicon, the voltage V_{FB} to be applied at the gate is negative.

4.2.2. Accumulation State

When a greater negative gate voltage is applied, the energy-band starts to move upwards at the interface and a positive charge region by hole accumulation is formed near the surface as shown in Fig. 4.3. The resulting negative surface potential induces an upward band bending at the silicon surface. When looking at the energy-band, a larger $E_i - E_f$ near the silicon surface indicates that there is a higher hole density at the surface compared to that in the bulk. Consequently, the surface conductivity is increased.

4.2.3. Depletion State

When a small positive voltage is applied to the gate electrode with respect to the p-silicon substrate, the resulting positive surface potential induces the

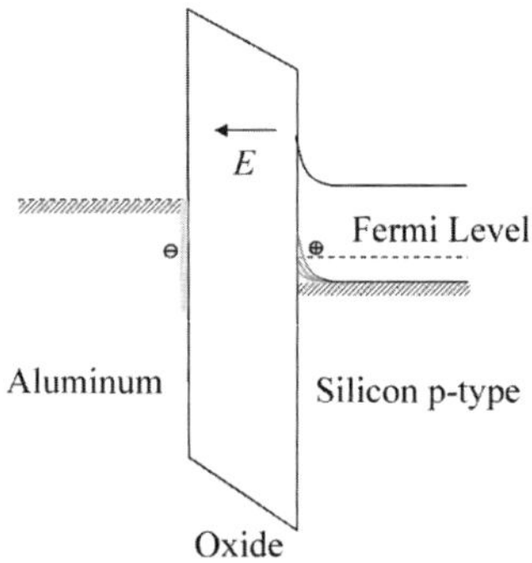

Fig. 4.3. Energy band diagram of the metal–oxide–silicon under the accumulation state.

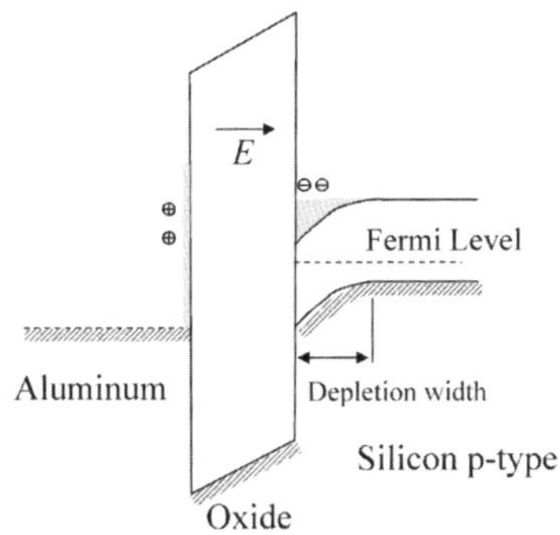

Fig. 4.4. Energy band diagram of the metal–oxide–silicon under the depletion state.

energy-band bending further downwards and a negative space-charge region is formed at the surface as shown in Fig. 4.4. A smaller $E_i - E_f$ at the surface indicates that holes are depleted from the vicinity of the oxide–silicon interface. The space charge in the depletion region which is the only bulk charge in silicon is given by

$$Q_B = Q_d = -qN_A x_d, \tag{4.3}$$

where x_d is the width of the space-charge layer and is greater than the thermal equilibrium value of x_{d0}. Using the depletion approximation, and by setting $V_B = 0$, the semiconductor surface potential can be found to be

$$\Psi_S = \frac{qN_A x_d^2}{2\varepsilon_s}. \tag{4.4}$$

And, the potential distribution in the silicon along the depletion region is given by

$$\Psi(x) = \Psi_S - \int_0^x \hat{E}_s(x)dx = \Psi_S\left(1 - \frac{x}{x_d}\right)^2 \tag{4.5}$$

which is similar to that of an n^+–p abrupt junction diode with a reference voltage at p-substrate set to zero.

4.2.4. Inversion State

If a large positive voltage is applied to the gate electrode with respect to the p-silicon substrate, the resulting large positive surface potential causes the intrinsic level to bend downwards and crossover the constant silicon Fermi level near the silicon surface. This indicates that an inversion is formed at the surface in which the electron density is larger than the hole density. The negative charge region at the silicon surface now consists of the inversion-layer charge, Q_i, and the depletion-layer charge, Q_d, as shown in Fig. 4.5. If the inversion-layer

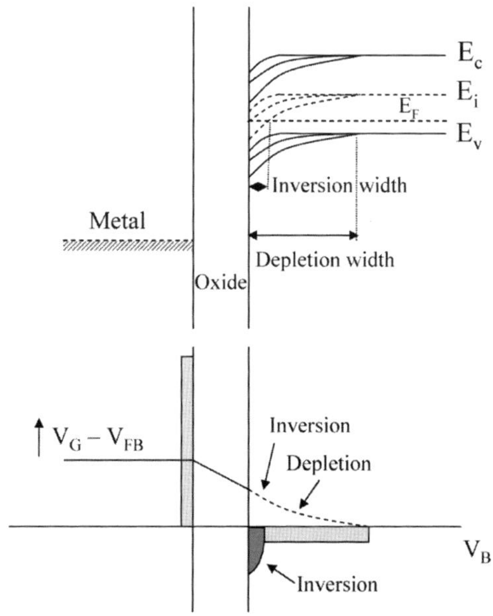

Fig. 4.5. Energy band diagram of the metal–oxide–silicon moving from the depletion state to the inversion state by increasing the gate bias.

charge is small compared to the depletion-layer charge, the surface is said to be in weak inversion. From the energy-band diagram, E_i bends slightly below the Fermi level at weak inversion. As the surface band bending further, the inversion-layer charge becomes greater than the depletion-layer charge. This is the onset condition of the strong inversion. At strong inversion, by setting $V_B = 0$, the space-charge-layer width will be at its maximum and is given by (Muller and Kamins, 1986):

$$x_{dm} = \sqrt{\frac{2\varepsilon_s(2|\Psi_f - \Psi_i|)}{qN_A}}, \tag{4.6}$$

where $2 \times |\Psi_f - \Psi_i|$ indicates the amount of potential (also indicated in the energy-band by the quantity of $2 \times q \times |\Psi_f - \Psi_i|$ in eV) in bending the energy-band in strong inversion. Ψ_f is the Fermi level potential and Ψ_i is the intrinsic level potential, both are in volts. It should be noted that all additional induced charges will be accumulated in the inversion layer after the occurrence of strong inversion and the charge within the depletion layer will remain approximately constant at the equilibrium state. Thus, the space-charge-layer width remains at x_{dm}, even with further (and slow) increase of the gate voltage.

The depletion charge Q_d (in negative) remains constant at

$$Q_d = -qN_A x_{dm} = -\sqrt{4qN_A\varepsilon_s|\Psi_f - \Psi_i|}. \tag{4.7}$$

4.2.5. MOS Capacitance

The MOS capacitor consists of an oxide capacitance and a semiconductor capacitance in series manner. The oxide capacitance per unit area is at a constant quantity and is related to the gate dielectric thickness by

$$C_{ox} = \frac{\varepsilon_{ox}}{d_{ox}}, \tag{4.8}$$

where ε_{ox} is the oxide dielectric constant and d_{ox} is the oxide thickness. At the same time, the semiconductor capacitance per unit area is a function of the depletion-layer width which is a function of the gate voltage

$$C_s = \frac{\varepsilon_s}{x_d}. \tag{4.9}$$

A small positive gate voltage around V_{FB} is applied at the gate, x_d can be treated approximately to be the extrinsic Debye length L_D. The semiconductor capacitance can be calculated as (Muller and Kamins, 1986):

$$C_s = \frac{\varepsilon_s}{L_D} = \frac{\varepsilon_s}{\sqrt{\frac{\varepsilon_s kT}{qN_A}}}. \tag{4.10}$$

When a higher gate voltage is applied, the depletion width is widened and the semiconductor capacitance becomes smaller. Hence, during depletion state, the MOS capacitance is a function of the applied gate voltage. When the depletion width varies in the range less than the maximum width x_{dm} as described in Eq. (4.6) before the onset of strong inversion, the total gate capacitance can be calculated by putting the oxide capacitance and the semiconductor depletion capacitance in series manner as

$$C(V_G) = \frac{C_{ox}}{1 + \sqrt{\frac{2C_{ox}^2(V_G - V_B - V_{FB} - V_{ox})}{\varepsilon_s qN_A}}}, \tag{4.11}$$

where V_{ox} is the voltage drop across oxide region, V_B is the substrate reference voltage and is normally zero when connected to ground. Note that, V_{FB} is negative for p substrate. Equation (4.11) gives the capacitance variation with gate voltage of the range between V_{FB} and that of onset of inversion state. When the surface potential ($V_G - V_{FB} - V_{ox} - V_B$) reaches $2(\Psi_f - \Psi_i)$, the depletion width reaches its maximum value of x_{dm}. For a higher gate voltage, the accumulation charge, which is mobile, increases and the depletion charge,

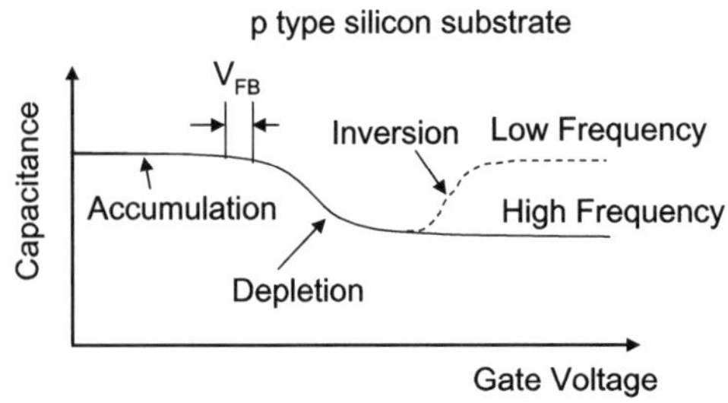

Fig. 4.6. Variation of capacitance of a MOS capacitor with different applied gate bias.

which is fixed, is kept constant. The inversion layer becomes dominant and the semiconductor capacitance becomes larger. The total MOS capacitance increases to eventually approach the value of C_{ox}. However, if the gate voltage varies in a high-frequency manner, the accumulation charge, with carrier lifetime in microseconds, may not be able to vary quick enough to follow the gate voltage. In this case, the depletion width may go beyond x_{dm}, called deep depletion, and the semiconductor capacitance will be kept at a low value. Figure 4.6 gives the variation of MOS capacitance with gate voltage.

The above analyses were done with the assumption that there is no charge in the dielectric layer. Within the gate dielectric, such as the oxide, there are always fixed charges residing within the oxide and at the interface of the oxide near semiconductor. Oxide charge exists in the forms of fixed charge, trapped charge, and mobile charge. The charge, depending on its polarity, creates extra accumulation or depletion effects on the semiconductor surface. Besides the polarity, the distribution of the oxide charge, e.g. the distance from the semiconductor interface, also determines the degree of influence on the semiconductor surface state. The closer the oxide charge to the semiconductor surface is, the stronger the influence will be. For the modern fabrication process to form the thermally grown gate oxide, the charge per unit area can be well controlled to be below 10^{10} cm^{-2}. The low oxide charge provides less variation on the interface state away from the above-mentioned condition. The flat band voltage, V_{FB}, is now modified by the existence of oxide charge to be

$$V_{\text{FB}} = V_{\text{FBO}} - \frac{(d_{\text{ox}} - d)Q_{\text{ox}}}{C_{\text{ox}} d_{\text{ox}}} \tag{4.12a}$$

when the oxide charge Q_{ox} can be quantized at a particular fixed location distanced away, measured as d, from the semiconductor interface. V_{FBO} is the flat band voltage without any oxide charge as in Eq. (4.2). If the charge is distributed all over in the oxide layer, then the flat band voltage can be adjusted as

$$V_{\text{FB}} = V_{\text{FBO}} - \frac{1}{C_{\text{ox}}} \int_{d_{\text{ox}}}^{0} \frac{(d_{\text{ox}} - x)}{d_{\text{ox}}} \rho(x) dx, \tag{4.12b}$$

where $\rho(x)$ is the oxide charge density distribution with distance x away from the semiconductor interface.

4.2.6. Threshold Voltage

The electron carriers in the inversion layer (and also the hole carriers in the accumulation layer but not discussed here) are mobile in the p-semiconductor under gate voltage bias. If there are nearby contacts, one at each end of the mobile-carrier layer called drain and source contacts, it is possible to run current through the thin layer and the semiconductor surface forms a conduction channel. In order to create a good conduction channel between the source and drain contacts, the semiconductor surface must be strongly inverted. The threshold voltage V_T is defined as the gate voltage required in creating a visible inversion in the channel region

$$V_T = 2(|\Psi_f - \Psi_i|) + V_{FB} + V_B + \frac{\sqrt{4qN_A\varepsilon_s|\Psi_f - \Psi_i|}}{C_{ox}} - \frac{Q_{ox}}{C_{ox}}. \tag{4.13}$$

Equation (4.13) is valid when both drain and source contacts are connected to the substrate potential, i.e. $V_D - V_B = 0$. The oxide charge influence becomes obvious where neutron irradiation process is applied in the MOSFET fabrication, resulted in positive charge residue. Following the radiation, the n-channel power MOSFET can have a negative threshold voltage (Baliga, 1987). Figure 4.7 shows the depletion region when a drain voltage is

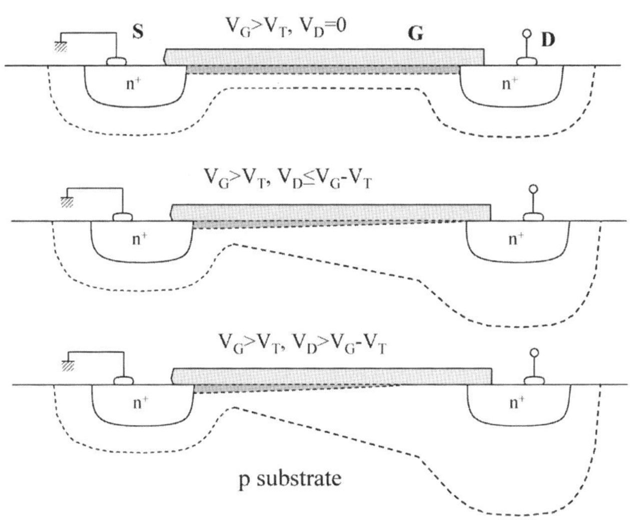

Fig. 4.7. Depletion and inversion layers of a MOSFET at different applied drain voltage.

applied. It is noticed that the depletion width near the drain area is wider than other places under the gate. As the negative charge induced by the gate bias is shared between the depletion and inversion layers, a wider depletion layer, means less accumulation charge, can be formed in area near the drain. With the reduction of accumulation charge, the onset of strong inversion occurs at the surface potential of $2 \times |\Psi_f - \Psi_i| + V_C - V_B$ instead of $2 \times |\Psi_f - \Psi_i|$ as stated earlier, where $V_C - V_B$ is the voltage between the semiconductor surface and the substrate across the channel region and it varies from drain to source. At the drain location, $V_C = V_D$ and at the source location, $V_C = V_B$. The threshold voltage is then adjusted to be the following to include the effect on the influence of wider depletion layer.

$$V_T = 2|\Psi_f - \Psi_i| + V_{FB} + V_C + \frac{\sqrt{2qN_A\varepsilon_s(2|\Psi_f - \Psi_i| + V_C - V_B)}}{C_{ox}} - \frac{Q_{ox}}{C_{ox}}. \tag{4.14}$$

The maximum depletion width is adjusted to be

$$x_{dm} = \sqrt{\frac{2\varepsilon_s(2|\Psi_f - \Psi_i| + V_C - V_B)}{qN_A}} \tag{4.15}$$

with the influence of applied drain voltage. Consequently, the amount of the depletion charge is also adjusted to be

$$Q_d = -\sqrt{2qN_A\varepsilon_s(2|\Psi_f - \Psi_i| + V_C - V_B)} \tag{4.16}$$

which is larger than that of without the drain voltage applied. These equations indicate that the substrate voltage also plays a role in the depletion charge formation and the threshold voltage value. For a negative substrate voltage applied at the p-substrate, it extends the depletion region and reduces the accumulation charge density. In a sense, the threshold voltage is seen to be increased to a more positive value. Such effect is called the body effect or the back-gate effect in published literature.

From the point of view of device profiles, the threshold voltage of the MOSFET is affected by the body doping concentration, the oxide charge, the oxide thickness, and the body bias.

4.3. Static Characteristics

Figure 4.8(a) shows the basic n-channel MOSFET structure. Typically, the n-channel MOSFET is fabricated by diffusing or implanting phosphorus into a p-type silicon substrate to form the drain and source. For power MOSFET, as shown in Fig. 4.8(b), a lightly doped n-drift region is present between the n^+

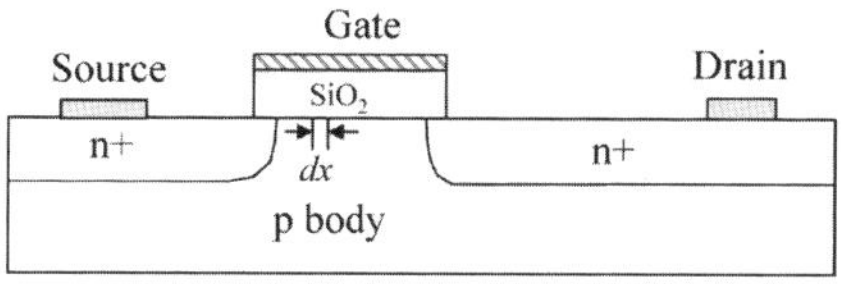

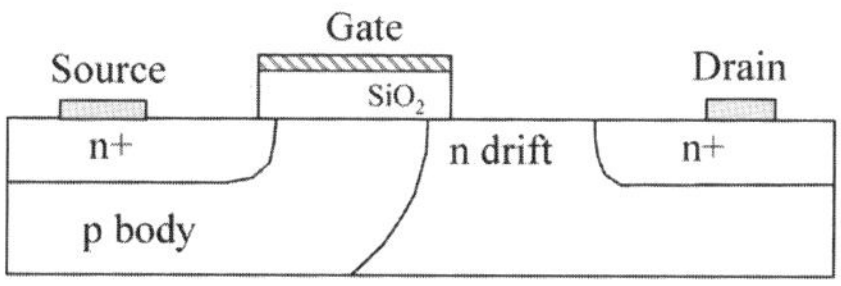

Fig. 4.8. (a) Plain n-MOSFET structure and (b) power n-MOSFET structure with the n-drift region added for higher voltage rating.

drain and the channel region to sustain a higher blocking voltage. The channel is formed below a silicon dioxide gate dielectric with a metal, polysilicon, or polycide gate electrode. With a small negative voltage applied to the gate electrode with respect to the p-substrate, the channel is accumulated with holes. As such, no current flows in this MOSFET. This is the cut-off region of operation. The device can be viewed as a two back-to-back p–n junction diodes or an open-base transistor.

With a small positive voltage applied to the gate electrode with respect to p-substrate, a low conductivity n-channel is formed between the source and drain. The source is defined as the region where electrons come from, and the drain as the region where electrons go to. The device is now under subthreshold region of operation. For power MOSFET, this is not an important operating characteristic as the main purpose of having the power MOSFET is to serve as a low-resistance conduction switch. When a large positive voltage of greater than the threshold voltage V_T is applied to the gate electrode with respect to p-substrate, the semiconductor surface is strongly inverted. As such, a high conductivity layer with mobile electron carriers named "n-channel" is formed between the source and the drain. With a positive voltage applied to the drain with respect to the source, electrons can drift from the source to the drain through the n-channel. The magnitude of the current flow depends on the magnitude of the applied drain-to-source voltage, V_{DS}. At low V_{DS}, the conductive n-channel is not distorted and the drain current is proportional to the V_{DS} in an almost linear manner. This is called the linear region of operation. At high V_{DS}, the channel near the drain pinches off (refer to Fig. 4.7) and the drain current is kept almost constant with further increasing V_{DS}. This is appropriately called the saturation region of operation.

Besides the lateral structure, the three common vertical structures are the V-groove MOS (VMOS), doubled-diffused MOS (DMOS), and UMOS shown in Fig. 4.9(a)–(c). The vertical structure with common substrate drain is meant for handling large current and blocking high voltage. This feature not only allows for more source area but also ease the electric field crowding at the gate. The forward blocking capability is provided by the p–n junction at the p-body and the light-doped n^- drift region. The p-body region is shorted to the source by the source metal in order to establish a fixed p-body potential during device operation. The body-source short also forms the antiparallel diode (p-body to n^--drift region) for fly-wheeling current conduction during inductive switching. For MOSFET devices of a high forward blocking voltage would require the drift region to be not only with a lower doping concentration but also with a larger thickness. In a sense, the on-state conduction resistance becomes higher as contributed by the drift region and the channel voltage drop becomes less dominant in the overall on-state voltage.

The V-groove in the VMOS structure is formed by an orientation-dependent etch. The channel region for this structure is formed along the wall of the V-groove which is made by an angle of 54.7° against the device surface on (100) silicon substrate. The VMOS structure was the first commercial power structure. However, due to the many processing difficulties and crowded electric field at the tip of the V-groove, this structure was abandoned later on. In the DMOS structure, the n^+ source and p-body are formed by a double-diffusion process through a common mask opening defined by the edge of the polysilicon gate. The difference in the lateral diffusion between the p-body and n^+ source regions defined the channel region. The UMOS structure has a U-shaped groove formed for the gate region by silicon reactive ion etching similar to the trench etching in the dynamic random access memory (DRAM) process. The UMOS will have the highest channel density of the three structures which allows for significant reduction in the on-resistance of the power MOSFET.

4.3.1. Linear Region Operation

Consider the n-channel MOSFET shown in Fig. 4.8(a) with a gate voltage of greater than the threshold level and a small voltage of V_{DS} applied to the drain with respect to the source. With the source shorted to the p-body ($V_S = V_B$), the induced channel (inversion) charge is given as (Muller and Kamins, 1986)

$$\begin{aligned} Q_I &= -C_{ox}(V_G - 2|\Psi_f - \Psi_i| - V_{FB} - V_C) \\ &\quad + \sqrt{2qN_A\varepsilon_s(2|\Psi_f - \Psi_i| + V_C - V_S)} \\ &\approx -C_{ox}(V_G - V_T - (V_C - V_S)). \end{aligned} \tag{4.17}$$

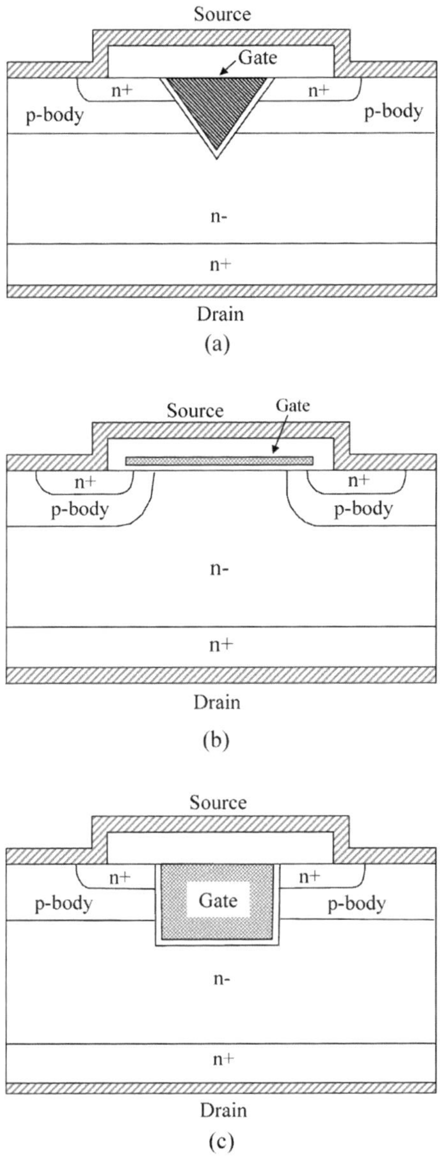

Fig. 4.9. Structures of (a) VMOS, (b) DMOS, and (c) UMOS.

The approximation is made by the assumption that the depletion charge is kept at constant at different V_C values, i.e. the x_{dm} is constant. The potential variation (ohmic voltage drop) along the (n) channel can be written as

$$I_D\, dR = I_D \frac{dx}{-W\mu_n Q_I} = dV_C \approx I_D \frac{dx}{-W\mu_n C_{ox}(V_G - V_T - (V_C - V_S))}, \tag{4.18}$$

where I_D is the drain current, W is the channel width (the dimension out of the paper), μ_n is the electron mobility along the channel, and dR is the incremental resistance along the channel. Equation (4.18) can be rewritten as

$$\begin{aligned} I_D \int_0^L dx &= -W\mu_n \int_{V_S}^{V_D} Q_I dV_C \\ &\approx -W\mu_n C_{ox} \int_{V_S}^{V_D} (V_G - V_T - (V_C - V_S))dV_C. \end{aligned} \tag{4.19}$$

By integrating both sides, the drain current can be found as

$$\begin{aligned} I_D &= \frac{\mu_n W}{L}\left(C_{ox}\left(V_G - V_{FB} - 2|\Psi_f - \Psi_i| - \frac{1}{2}V_{DS}\right)V_{DS} \right. \\ &\quad \left. - \frac{2}{3}((2|\Psi_f - \Psi_i| + V_{DS})^{1.5} - (2|\Psi_f - \Psi_i|)^{1.5})\sqrt{2\varepsilon_s q N_A}\right) \\ &\approx \frac{\mu_n W C_{ox}}{L}\left((V_G - V_T)V_{DS} - \frac{V_{DS}^2}{2}\right). \end{aligned} \tag{4.20}$$

The approximation is made by ignoring the dependency of vertical field at different drain voltage, i.e. by setting the threshold voltage a constant value.

For a small drain voltage applied, the above assumption can be valid and the drain current can be further simplified to be as

$$I_D \approx C_{ox}\mu_n \frac{W}{L}((V_G - V_T)V_{DS}) \tag{4.21}$$

which is a linear relationship between V_{DS} and I_D. The I–V curves plotted by Eqs. (4.20) and (4.21) for the linear region of operation are shown in Fig. 4.10. A channel resistance at low V_{DS} can be defined from Eq. (4.21) as

$$R_{ch} = \frac{L}{W\mu_n C_{ox}(V_G - V_T)}. \tag{4.22}$$

For a power MOSFET, the conduction resistance is more than just the channel resistance as described in the above equations. The total on-state resistance

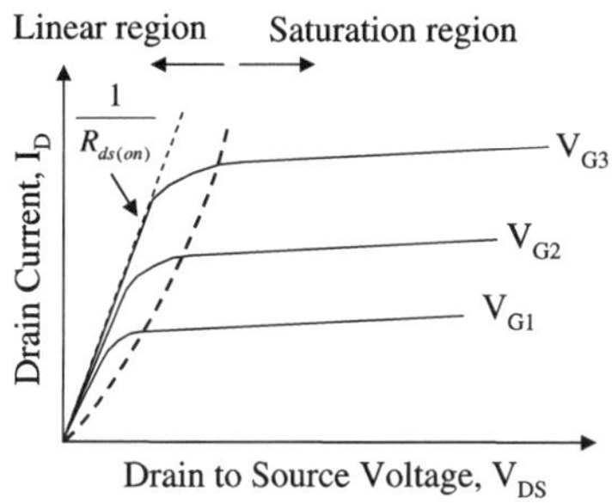

Fig. 4.10. MOSFET *I–V* curves at different gate voltages.

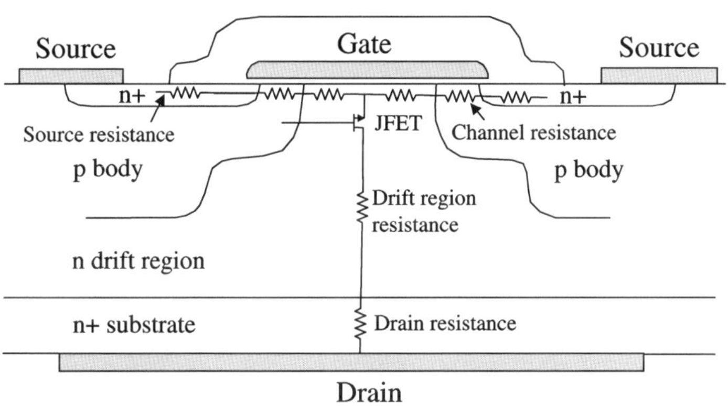

Fig. 4.11. Distribution and location of conduction resistances along the current path.

$R_{ds,on}$ includes the channel resistance (R_{ch}), substrate resistance (R_{sub}), JFET region resistance (R_{JFET}), drift-region resistance (R_d), the source diffusion resistance (R_{n+}), and the drift-region accumulation resistance (R_{dA}) as shown in Fig. 4.11. The n-channel JFET region is formed by the depletion layers in the drift region between the p-base diffusions. Of these resistive components, the JFET, channel, and especially the drift-region resistance are the significant contributions to the on-resistance in the DMOS structure. The UMOS structure does not have the JFET region and as such has a lower on-state resistance. Since both the drift-region resistance and channel resistance are inversely proportional to the carrier mobility, the on-resistance of the p-channel power MOSFET is larger by about 2.5 times than that of the n-channel counterpart with a similar device area, due to the difference between the electron and hole mobilities. The on-state resistance is an important parameter as it determines the maximum current handling capability and power dissipation of the power

MOSFET. To minimize $R_{ds,on}$, the gate voltage should be large enough for a given drain current to maintain the operation in the linear region. However, by increasing the gate voltage beyond the optimum value, due to surface scattering effect, it has the diminishing effect on lowering the $R_{ds,on}$, especially for high voltage devices (Ang, 1995).

In the linear region of operation, a gain parameter or transconductance at low V_{DS} can be defined as the ratio of the change in drain current corresponding to a change in gate voltage

$$g_m = \frac{dI_D}{dV_G} = \mu_n C_{ox} \frac{W}{L} V_{DS}. \tag{4.23}$$

The transconductance indicates the current-carrying capability of the device. Equation (4.23) can be applied in laboratory measurement to find the effective channel mobility at different gate voltage by knowing the oxide capacitance and device dimensions.

4.3.2. Saturation Region Operation

As the drain-to-source voltage V_{DS} increases, the influence of depletion charge variation along the channel region becomes obvious. Along the channel, the quantity of accumulation charge becomes weaker and the depletion charge becomes stronger when looking from the source region toward the drain region. In the complete saturation point, at sufficiently large drain voltage, it can be considered that the surface potential induced by gate bias is "neutralized" near the point of full-saturation. The inversion layer will then disappear at the point of the channel and is now seen pinched off. Further increase of the drain-to-source voltage would not increase the drain current but just to extend the pinched-off region longer toward source region (refer to Fig. 4.7). It is then called that the drain current (and the MOSFET channel) is saturated. The onset of current saturation can be calculated by setting the inversion charge of Eq. (4.17) at drain to be equal to zero:

$$\begin{aligned} Q_I = {} & -C_{ox}(V_G - 2|\Psi_f - \Psi_i| - V_{FB} - V_{DS,Sat}) \\ & + \sqrt{2qN_A\varepsilon_s(2|\Psi_f - \Psi_i| + V_{DS,Sat})} = 0. \end{aligned} \tag{4.24}$$

Solving the equation, the onset saturated drain voltage $V_{DS,Sat}$ can be found as

$$\begin{aligned} V_{DS,Sat} &= V_G - V_{FB} - 2|\Psi_f - \Psi_i| + \frac{\varepsilon_s q N_A}{C_{ox}^2}\left(1 - \sqrt{1 + \frac{2C_{ox}^2}{\varepsilon_s q N_A}(V_G - V_{FB})}\right) \\ &\approx V_G - V_T. \end{aligned} \tag{4.25}$$

Substituting the above quantity into Eq. (4.20) yields

$$\begin{aligned} I_{D,Sat} = {} & \frac{\mu_{ns}W}{L}\left(\frac{1}{2}C_{ox}\left((V_G - V_{FB} - 2|\Psi_f - \Psi_i|)^2 \right.\right. \\ & \left. - \left(\frac{\varepsilon_s q N_A}{C_{ox}^2}\right)^2\left(1 - \frac{2C_{ox}^2(V_G - V_{FB})}{\varepsilon_s q N_A}\right)\right) \\ & - \frac{2}{3}\sqrt{2\varepsilon_s q N_A}\left(\left(V_G - V_{FB} + \frac{\varepsilon_s q N_A}{C_{ox}^2}\sqrt{1 - \frac{2C_{ox}^2(V_G - V_{FB})}{\varepsilon_s q N_A}}\right)^{1.5}\right. \\ & \left.\left. - (2|\Psi_f - \Psi_i|)^{1.5}\right)\right). \end{aligned} \tag{4.26}$$

Equation (4.26) can be derived in approximation by neglecting the depletion-charge variation to receive a simple expression as

$$I_{D,Sat} = \frac{1}{2}\frac{\mu_{ns}C_{ox}W}{L}(V_G - V_T)^2 \approx \frac{1}{2}\frac{\mu_{ns}C_{ox}W}{L}V_{DS,Sat}^2, \tag{4.27}$$

where μ_{ns} is the effective surface mobility during saturation which is lower than that of the linear operation mode. The approximation in Eq. (4.27) is made at the onset saturated point occurring near the knee region, i.e. $V_{DS} \approx (V_G - V_T)$ where the I–V curves start to bend to become flat. As it can be seen, the drain current varies approximately as the squared of the onset saturated drain-to-source voltage $V_{DS,Sat}$ at the knee points. For a fixed gate voltage, there is very little variation in drain current beyond the onset saturation point with further increase of the drain-to-source voltage as Eq. (4.26) is independent of the drain voltage (if without considering the channel reduction by drain-body depletion at high voltage). The transconductance (or called the small-signal current gain) of the MOSFET at saturation operation can be found as

$$g_{ms} = \frac{dI_D}{dV_G} = \frac{\mu_{ns}C_{ox}W}{L}(V_G - V_T). \tag{4.28}$$

4.3.3. Mobility Degradation

Carrier mobilities in the channel are influenced by several degradation factors, namely by the surface scattering, the high-field saturation, and the substrate bias effects. The surface scattering occurs when carriers are driven by the vertical field (when gate voltage applied) toward the microscopic rough oxide–semiconductor interface. When they hit the surface, they rebound. This forms the zigzag profile of traveling path and in scattering pattern, which makes the effective carrier mobilities lower. At high gate voltage, the influence of vertical field becomes obvious, and as much as 30% of reduction in drain current may occur (Veendrick, 1992). The second factor is due to the carrier velocity saturation under high lateral field between source and drain as mentioned in Chapter 2. The third factor is by the substrate bias effect which influences the accumulation charge formation. For example, a negative substrate bias on p-substrate diminishes the crowding of accumulation charge and which, in turn, helps to raise the carrier mobility in the channel.

4.3.4. Forward Blocking

The power MOSFET structure consists of a parasitic n^+(source)–p(base)–n(drift) transistor where the p-base region is shorted to the n^+ source to keep the parasitic BJT inactive. In the forward blocking mode, the drain voltage is supported across the p-base/n-drift region junction. The p-base doping is normally much higher than that of the n-drift region. This enables the depletion region extends mainly in the n-drift region. The breakdown voltage can be calculated similar to that of the p–n^-–n^+ punchthrough structure described in Chapter 2. There will be still the depletion region although small in the p-base region. For a power MOSFET, the channel length is relatively long to prevent the p-base punchthrough at high voltage. One reason of allowing so is that the on-state conduction loss is mainly dominated by the drift region resistance. The channel resistance plays a relatively smaller role for device voltage rated higher than a few hundred volts.

4.4. Switching Characteristics

The switching speed of a power MOSFET is faster than that of a power bipolar switching transistor due to the absence of minority (excess) carrier storage in the device drift region. The transit-time-limited switching speed of power MOSFET is generally less than 0.1 ns. However, in practice, switching in power MOSFET is limited by parasitic capacitances in device structure and inductances in the chip wire-bond and external circuit. The switching speed is also determined by the gate driver capability in supplying current to charge the device input capacitance. The input capacitance is the sum of gate-to-source capacitance C_{GS} and gate-to-drain capacitance C_{GD}. The capacitance C_{GD} is sometimes called the Miller feedback capacitance C_{RSS} due to the Miller effect when drain voltage varies. Besides the input capacitance, the MOSFET also contains the output capacitance C_{OSS} which comprises of the drain-to-gate capacitance C_{DG} and the drain-to-source capacitance C_{DS}.

4.4.1. Turn-on Transient

To switch on a power MOSFET, a step voltage V_P from a voltage-driver or a constant current I_G from a current-driver is applied through the gate resistance R_G to charge the gate-to-source capacitance C_{GS} and the gate-to-drain capacitance C_{GD} (Fig. 4.12). Figure 4.13 shows the waveforms by the current drive. As long as the gate voltage is below the threshold voltage V_T, no visible drain current will flow in the channel. A delay time T_o, is defined as the period taken for the gate voltage to reach the threshold level. During the period, the input capacitance is the sum of C_{GS} and C_{GD} capacitance without Miller effect. If it is for a voltage-driver, the gate voltage rises exponentially according to (Clemente and Pelly, 1981; Baliga, 1987):

$$V_{GS} = V_P(1 - e^{-\frac{t}{R_G(C_{GS}+C_{GD})}}). \tag{4.29a}$$

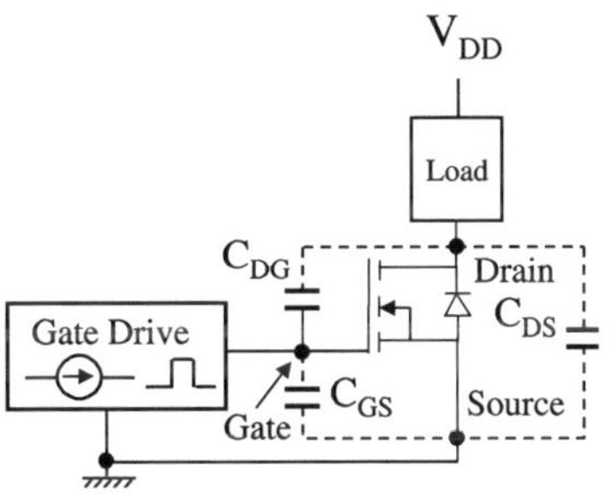

Fig. 4.12. MOSFET parasitic capacitances.

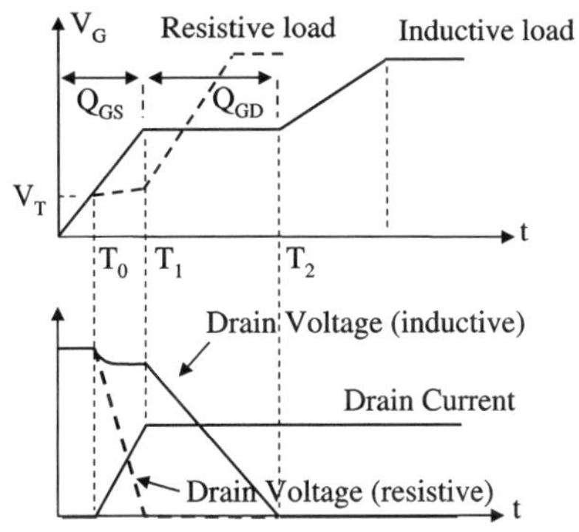

Fig. 4.13. Turn-on waveforms of a MOSFET in the circuit of Fig. 4.12 by gate drive current.

And if for the current-driver, the gate voltage changes according to

$$V_{GS} = \frac{t}{C_{GS} + C_{GD}} I_G. \tag{4.29b}$$

For the voltage-driver, the delay time T_o can be calculated as

$$T_o = R_G(C_{GS} + C_{GD}) \ln\left(\frac{V_P}{V_P - V_T}\right) \tag{4.30a}$$

and

$$T_o = \frac{V_T(C_{GS} + C_{GD})}{I_G} \tag{4.30b}$$

for the current-driver. When the gate-to-source voltage V_{GS} reaches the threshold level, the device begins to conduct current and operates in the saturation region. It is so because the drain voltage is high and the gate voltage is low at the initial turn-on point. The drain current and the gate voltage continue to increase in proportional with the device transconductance in the following expression

$$I_D = g_{ms}(V_{GS} - V_T), \tag{4.31}$$

where g_{ms} is the transconductance as described in Eq. (4.28). The drain voltage may or may not fall depending on the external circuit configuration. If it is for a resistive load, the drain voltage will start to fall when the drain current rises, as shown in the dashed line. Otherwise, for an inductive load, the drain voltage is more or less the same as the supply voltage as shown in the solid line. In a sense, the device will function in the saturation mode for a longer period for the case of inductive (constant) load-current turn-on. As such, the turn-on loss is higher in the inductive load case compared to the resistive counterpart. For the inductive load, when the drain current reaches the inductive load current I_L, the diode ceases to conduct and the drain voltage is now free to move. The gate-to-source voltage can be expressed in approximate as

$$V_{GS} = V_T + \frac{I_L}{g_{ms}}. \tag{4.32}$$

For the resistive load, the drain voltage falls earlier once the gate-to-source voltage exceeds the threshold level. And for the inductive load, the drain voltage starts to fall when the device carries the whole inductor current. Regardless the timing of occurrence, when the drain voltage falls, Miller effect occurs. This will draw additional current to charge the Miller-effect-influenced gate-to-drain capacitance C_{GD}.

For the inductive load and the voltage-driver, the gate-to-source voltage V_{GS} remains constant after period T_1 because of the constant drain current.

The gate current now flows into C_{GD} and can be expressed by

$$I_G = \frac{V_P - V_T - \frac{I_L}{g_{ms}}}{R_G}. \tag{4.33}$$

From Kirchhoff's voltage law, the drain-to-source voltage is

$$V_{DS} = V_{DG} + V_{GS}. \tag{4.34}$$

For a constant V_{GS}, the rate of change of the drain-to-source voltage during the period of T_2 is

$$\frac{dV_{DS}}{dt} = \frac{dV_{DG}}{dt} = -\frac{I_G}{C_{GD}} = -\frac{V_P - V_T - \frac{I_L}{g_{ms}}}{R_G C_{GD}}. \tag{4.35}$$

Equation (4.35) is also valid for the case of current-driver by setting the current I_G a constant value equal to the gate drive current. The drain-to-source voltage can be found as

$$V_{DS} = V_{DD} - \frac{I_G}{C_{GD}}t = V_{DD} - \frac{V_P - V_T - \frac{I_L}{g_{ms}}}{R_G C_{GD}}t, \tag{4.36}$$

where V_{DD} is the supply voltage. Thus, the drain voltage decreases from the supply voltage linearly with time during the T_2 interval. Toward the end of period, the device enters the linear operation mode, i.e. ($V_{GS} - V_T > V_{DS}$), the gate-to-drain capacitance C_{GD} becomes higher and this slows down the V_{DS} drop.

For the resistive load, the charging of the C_{GS} and the Miller capacitor C_{GD} are occurring at the same time. The variation of gate voltage is calculated as

$$\frac{dV_{GS}}{dt} = \frac{I_G}{C_{GD}(1 + g_{ms}R_L) + C_{GS}}. \tag{4.37}$$

And, the variation of drain voltage can be calculated as

$$\frac{dV_{DS}}{dt} = -\frac{g_{ms}I_G R_L}{C_{GD}(1 + g_{ms}R_L) + C_{GS}}. \tag{4.38}$$

4.4.2. Turn-off Transient

The MOSFET can be switched off by removing the applied gate voltage (connected to ground potential) or by applying a small negative voltage through the gate resistance. The turn-off process, as shown in Fig. 4.14 is

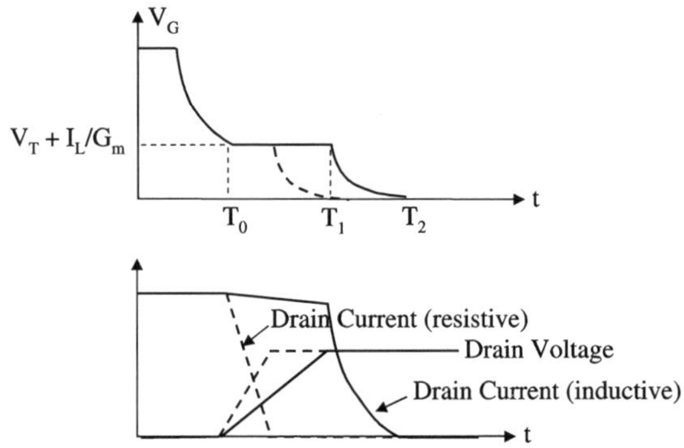

Fig. 4.14. Turn-off waveforms of a MOSFET.

essentially an opposite process as that of turn-on. The gate-to-source voltage decreases exponentially with time as a result of the discharging of its gate-to-source capacitance C_{GS} through R_G. Thus,

$$V_{GS} = V_P e^{-\frac{t}{R_G C_{GS}}}. \tag{4.39}$$

When it reaches the value of $V_T + I_L/g_m$, the drain voltage starts to rise. The period T_o can be calculated as (Ramshow, 1993):

$$T_O = R_G C_{GS} \ln\left(\frac{V_G}{V_T + \frac{I_L}{g_m}}\right). \tag{4.40}$$

For the resistive load, after T_o period the drain current starts to fall following the rise of drain voltage. However, for the inductive load, the drain current remains at pre-turn-off value I_L. The freewheeling diode D remains off until the drain-to-source voltage reaches V_{DD}. As the drain current is at the same level, the gate-to-source voltage V_{GS} will remain at constant. The gate current I_G will flow through the Miller capacitor C_{GD}. During this period,

$$I_G = -\frac{V_{GS}}{R_G} = -\frac{V_T + \frac{I_L}{g_m}}{R_G}. \tag{4.41}$$

The negative sign indicates that current is flowing out of gate. For a high-current power MOSFET, the discharge current is quite sizable and the charge can be recovered for partial gate drive supply. The rate of change of the drain-to-source voltage can then be calculated.

$$\frac{dV_{DS}}{dt} = \frac{dV_{DG}}{dt} = \frac{I_G}{C_{GD}}. \tag{4.42}$$

And, the drain-to-source voltage is

$$V_{DS} = V_{DS(on)} + \frac{I_G}{C_{GD}}t, \tag{4.43}$$

where $V_{DS(on)}$ is the on-state voltage. The drain-to-source voltage V_{DS} continues to increase linearly to the steady-state voltage V_{DD} during the T_1 interval. Now, the device enters the saturation mode of operation. The interval of T_1 is found as

$$T_1 = \frac{(V_{DD} - V_{DS(on)})R_G C_{GD}}{\frac{I_L}{g_m} + V_T}. \tag{4.44}$$

The freewheeling diode switches on once the drain voltage reaches the supply voltage. There may be overshoot on drain voltage depending on the stray inductance in the circuit. At the same time, the gate voltage is no longer clamped. Consequently, drain current starts to fall to zero during the interval of T_2. The gate-to-source voltage reduces exponentially according to

$$V_{GS} = \left(\frac{I_L}{g_m} + V_T\right) e^{-\frac{t}{R_G(C_{GS}+C_{GD})}} \tag{4.45}$$

with an exponential decrease in the drain current given by

$$I_D = g_m \left(\left(\frac{I_L}{g_m} + V_T\right) e^{-\frac{t}{R_G(C_{GS}+C_{GD})}} - V_T\right). \tag{4.46}$$

This interval ends when the gate-to-source voltage reaches the threshold voltage V_T and the drain current is reduced to zero. The duration for this interval is

$$T_2 = R_G(C_{GS} + C_{GD}) \ln\left(\frac{I_L}{g_m V_T} + 1\right). \tag{4.47}$$

The gate-to-source voltage continues to decrease to zero. For a resistive load, the change of drain current occurs at the same time as that of the drain voltage. The change of gate voltage is

$$\frac{dV_{GS}}{dt} = -\frac{I_G}{C_{GS} + (1 + g_m R_L)C_{GD}} \tag{4.48}$$

and that of the drain voltage is

$$\frac{dV_{DS}}{dt} = \frac{g_m R_L I_G}{C_{GS} + (1 + g_m R_L)C_{GD}}. \tag{4.49}$$

When V_{GS} reaches below the threshold voltage, the drain current drops to zero and the drain voltage reaches the supply voltage.

$$I_D = \frac{V_{DD}}{R_L} - \frac{g_m I_G}{C_{GS} + (1 + g_m R_L)C_{GD}}t. \tag{4.50}$$

During inductive switching, voltage switching spike occurs at the drain terminal of the MOSFET due to either the turn-on delay of the fly-wheeling diode or the stray inductance in the circuitry. Hence, the MOSFET device must be selected such that it can withstand switching spikes. Or else, a resistor–capacitor snubber bypass can be placed across the drain and the source terminals. However, the use of a snubber circuit reduces the speed of circuit operation and, at the same time, it increases the power dissipation. Second, the in-rush current which flows during the initial turn on of certain converters is another consideration. This is likely to occur when the gate drive is insufficient to maintain a fully inverted channel to carry the in-rush current. It is necessary to ensure that the initial current in-rush does not cause excess power dissipation.

4.4.3. Gate Charge

The total gate charge required in each switching is normally specified for gate drive in order to establish a desired drain current under given circuit applications (Barkhordarian, 2002). It is normally independent of temperature variation. When a constant gate current I_{G} is applied to the gate circuit, both the C_{GS} and C_{GD} are charged at a constant rate and the gate voltage is rising accordingly

$$\frac{dV_{\mathrm{G}}}{dt} = \frac{I_{\mathrm{G}}}{C_{\mathrm{GS}} + C_{\mathrm{GD}}}. \tag{4.51}$$

At the initial period, the gate voltage keeps increasing linearly as long as both capacitance values are constant. When the gate voltage reaches the threshold level V_{T}, the device enters its conduction region and its drain current starts to increase with V_{G} accordingly. The increase V_{G} in continues linearly till the drain current reaches the predetermined value set by the external inductive load element. When this happens, drain current reaches the steady-state (limited by the constant load current). The gate voltage needed to support the drain current is

$$V_{\mathrm{G}} = V_{\mathrm{T}} + \sqrt{\frac{2I_{\mathrm{o}}}{\mu_{\mathrm{ns}} C_{\mathrm{ox}} \frac{W}{L}}}, \tag{4.52}$$

where I_{o} is the inductive load current which can be considered as a constant current source. The device is operating at the saturation mode as the drain voltage remains the same as the supply voltage at this moment. The drain voltage starts to fall after this point as shown earlier in Fig. 4.13. When the drain voltage changes, the Miller effect starts to maneuver the charging current through the gate–drain capacitance. The gate current is now charging the capacitance

C_{GD} as the gate voltage is kept constant, and the drain voltage reduces at a rate according to gate charging current,

$$\frac{dV_{DS}}{dt} = -\frac{I_G}{C_{GD}}. \tag{4.53}$$

The capacitance C_{GD} does not change much along the way when drain voltage is falling as long as the transistor remains in the saturation mode. Once the drain voltage has fallen to its on-state value, the transistor enters the linear conduction mode, and the capacitance C_{GD} increases to a higher value. Because of the higher C_{GD} value, the trend of drain voltage falling becomes slower. After the drain voltage is stable, the gate current reduces to zero and gate voltage is stable at voltage V_P.

The gate–source charge during the period T_1 can be calculated as

$$Q_{GS} = I_G \times T_1 \times \frac{C_{GS}}{C_{GS} + C_{GD}}. \tag{4.54}$$

Similarly, the gate–drain charge during the periods T_1 and T_2 is

$$Q_{GD} = I_G \times T_1 \times \frac{C_{GD}}{C_{GS} + C_{GD}} + I_G \times T_2. \tag{4.55}$$

The total gate charge divided by the sum of both the gate–source and gate–drain charges is the minimum charge required to bring the device to the linear on-state conduction. In practice, a slightly larger gate charge is used to ensure a proper turn-on. The gate charge information in association with the input capacitance ($C_{GS} + C_{GD}$) can be found in the device manufacturer's databook. The charge information is used to find the maximum current requirement for the needed turn-on delay time in gate drive circuit design.

4.4.4. High-Frequency Operation

For power amplifier application at high-frequency, the device operation is usually limited by the charging and discharging of the input capacitance which consists of the gate-to-source capacitance C_{GS} and the gate-to-drain capacitance C_{GD},

$$C_{in} = C_{GS} + (1 + g_m R_L)C_{GD}. \tag{4.56}$$

The maximum small-signal operating frequency of the power MOSFET can be defined as the frequency at which the input gate current becomes equal to the output drain current. The input gate current is given by

$$i_G = 2\pi f C_{in} v_G, \tag{4.57}$$

where the input gate signal voltage is v_G. The small-signal drain current can be found as the product of its transconductance and the gate signal voltage,

$$i_D = g_m v_G. \tag{4.58}$$

Equating Eqs. (4.57) and (4.58) to find the maximum frequency of operation:

$$2\pi f_{max} C_{in} v_G = g_m v_G \tag{4.59a}$$

or,

$$f_{max} = \frac{g_m}{2\pi C_{in}}. \tag{4.59b}$$

Thus, a large gate current is generally required to achieve this maximum operating frequency of the power MOSFET.

4.4.5. Parasitic Body Diode

An integral p–n^-–n^+ diode is inherently formed by the p-body, the epitaxial n^- drift region, and the n^+ drain in the power MOSFET structure as shown in Fig. 4.11. In normal forward operations, this integral body-drain diode is reverse biased. The body-drain diode becomes conductive when the power MOSFET is reverse biased or it is under the inductive fly-wheeling operation where a current flowing from source to drain. The current handling capability of this integral body-drain diode is comparable to that of the power MOSFET itself. If the device functions as a synchronous rectifier, the gate can be biased during the diode conduction to provide the MOSFET operation in the third-quadrant where negative current conduction is utilized.

The body-drain diode must exhibit a fast reverse recovery and a low reverse recovery charge. Since the minority carrier lifetime in the epitaxial n^- region is generally high, the reverse recovery characteristic of the body-drain diode is generally not suitable for fast switching applications. As such, electron irradiation is used to introduce recombination centers in the n^- drift region to reduce the minority carrier lifetime. Gold or platinum doping for deep recombination centers can also be used for the same purpose. Unfortunately, this technique also increases the MOSFET on-state voltage.

4.5. *dv/dt* Limit

The rate of drain-to-source voltage change, *dv/dt*, during forward blocking state or switching states has significant influence on the device blocking and recovery capabilities. There are many ways, normally or abnormally, to bring the MOSFET devices into conduction state. The normal way is to apply a gate bias to form the conduction channel. The abnormal ways are namely to raise the drain voltage to the avalanche level to cause breakdown conduction, or to fluctuate the drain voltage in a fast transient manner to induce capacitive current to bias the emitter junction of the open-base transistor. A high *dv/dt* also creates failure in turn-off due to the large discharge current from the Miller capacitor.

The *dv/dt* effects during blocking and switching are described here. For an open-base transistor, the depletion layer at one of the junction sustains the blocking voltage. In MOSFET structure, the depletion region sits at the junction between the p-base and the n-drift regions. The higher the drain voltage is, the wider the depletion region becomes. This is the junction capacitance which plays a major role in the voltage-transient conduction. The capacitive current induced by the fluctuation of the drain voltage flows through the p-base to reach the source contact, as shown in Fig. 4.15. If the current is large enough, the emitter junction may be forward-biased. When this occurs, the open-base transistor becomes conductive. The equation to relate the drain voltage change to the conduction threshold can be expressed as

$$C_J \frac{dv_D}{dt} + v_D \frac{dC_J}{dt} = \frac{0.7}{R_P}, \tag{4.60}$$

where R_P is the p-base region resistance and 0.7 V is the nominal voltage for a forward p–n junction. The second voltage-transient induced turn-on occurs during the switching from on-state to the off-state. Recall that, the Miller capacitance discharges during the turn-off process when the drain voltage starts to rise from the on-state level to the supply voltage level. During the transient, the drain to gate Miller capacitance C_{DG} plays a major role to bias the gate. When drain voltage rises quickly, a large current is flowing through the capacitance. The same current will flow through the gate resistor R_G and the gate to source capacitance C_{GS}. When the gate voltage goes above the threshold level, the

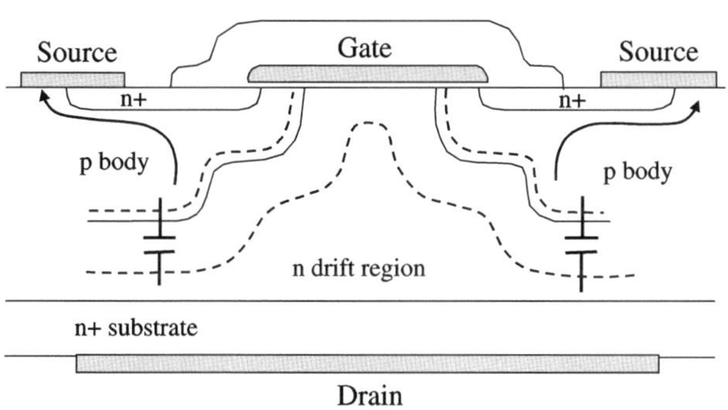

Fig. 4.15. *dv/dt* current flowing through p-body to forward bias p-body/n$^+$ junction.

conduction channel will then be formed. And, the device enters the conduction state. So, the turn-off is failed.

The relationship on the change of drain voltage to the gate voltage variation can be expressed as

$$C_{GD}\frac{dV_{DG}}{dt} = \frac{V_G}{R_G} + C_{GS}\frac{dV_G}{dt}. \tag{4.61}$$

To reduce the risk of turn-off failure, one can reduce the gate resistance to raise the drain voltage *dv/dt* limit as indicated by the equation.

4.6. Dummy-Gated Structure

The vertical MOSFET structure is a suitable candidate for high-voltage and high-power applications. For high-frequency operation, such as used for the RF base station, it has an inferior frequency-response capability due to the high parasitic feedback capacitance C_{rss} between the drain and the gate (Trivedi *et al.*, 1999). The drain-to-gate feedback capacitance C_{rss} consists of two parts, namely the gate-substrate overlap capacitance C_{ox} and the silicon bulk capacitance C_{Si}. The C_{rss} is defined as the series combination of C_{ox} and C_{Si}:

$$C_{rss} = \frac{C_{ox} \times C_{Si}}{C_{ox} + C_{Si}}. \tag{4.62}$$

The overlap capacitance C_{ox} is a function of the gate/substrate overlap length L_{OV} and the oxide thickness; while the bulk depletion capacitance C_{Si} is related to the bulk doping concentration, the junction area, and the applied drain bias. As the drain voltage varies, so do the depletion charge and the depletion capacitance.

C_{ox} is almost constant for a particular structure. With the increasing drain voltage, C_{Si} decreases due to the expanded depletion width at the reverse-biased p-body junction. To get a lower C_{rss}, both C_{OX} and C_{Si} must be minimized. For that, the gate structure in the conventional Radio Frequency Vertical Double-diffusion MOS (RF VDMOS) devices is normally split into two portions to just cover the channel regions, as shown in Fig. 4.16. Such an approach will reduce the feedback capacitance C_{rss} by a reduction of the gate-to-drain overlap area. However, the approach also brings several disadvantages, namely a higher silicon bulk capacitance by shallower substrate depletion; a higher on-state resistance by removing the top accumulation layer; and a lower breakdown voltage by electric field crowding on the silicon surface at the gate-end (Darwish and Board, 1984).

An approach made by (Xu *et al.*, 2001) introduced a dummy-gate located between two active gates to reduce the parasitic feedback capacitance but

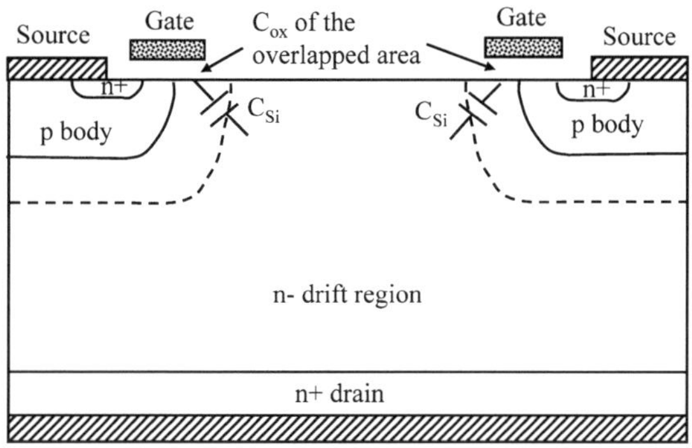

Fig. 4.16. The split-gate VDMOS structure with identified C_{ox} and C_{Si}.

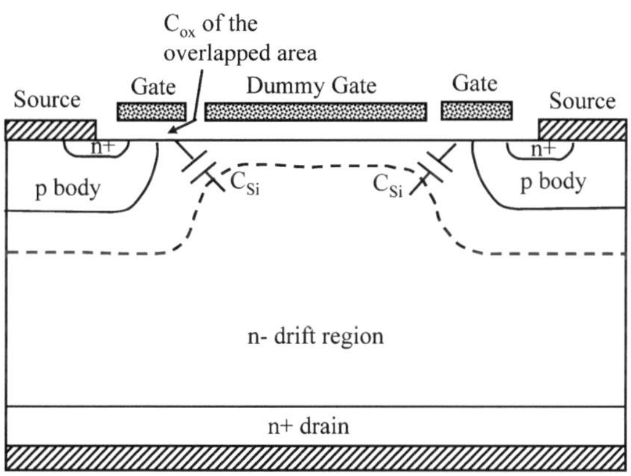

Fig. 4.17. The dummy-gate VDMOS structure with low-C_{Si}.

without those disadvantages caused by the split-gate approach, as shown in Fig. 4.17. The dummy-gated approach of splitting the gate into two parts just covering the active channel regions reduces the gate overlap capacitance. In addition, the dummy gate is shorted to the source. This enables the formation of depletion layer under the dummy gate to pinch off the depletion layers between two p-body regions when the device is at its low-voltage forward blocking state.

For the same blocking voltage rating, the dummy-gate structure allows the MOSFET to have a shorter channel and a larger gate to drain overlap area to minimize the on-state resistance. Hence, the transconductance gain can be improved, leading to a higher RF performance for the power device. Experimental results showed that with the dummy-gate a 51% lower on the feedback capacitance, a 21% lower on the on-state resistance, a 100% increase in output resistance, and a higher and linear transconductance are achieved (Xu *et al.*, 2001). Furthermore, the safe operating area (SOA) of the device, which is limited by the turning-on of the parasitic transistor, is improved. This allows a higher power density to be handled by the dummy-gate power MOSFET device.

4.7. Folded Gate Structure

For low-voltage (<40 V) LDMOS for power integrated circuits, the desirable performance is to have a low on-state resistance and a high transconductance. To reduce the on-state resistance, copper was utilized for lower metallization resistance (Kobori *et al.*, 1999), and both the RESURF mechanism and NTOP technique (Efland *et al.*, 1996; Merchant *et al.*, 1997, 1999) were applied to reduce the drift region resistance. With these improvements, the channel resistance becomes an increasingly important factor.

The channel resistance at low-V_{ds} region is determined by the channel length L, effective channel carrier mobility μ_{eff}, the gate oxide capacitance C_{ox} per unit area, the channel width W, and the gate voltage overdrive, $(V_{gs}-V_{th})$. The channel resistance R_{ch} can be optimized by reducing the channel length L and by increasing the effective channel electron mobility μ_{eff}. However, for a given device technology, the minimum channel length is dictated by the amount of channel charge needed and to avoid the premature channel punchthrough. In process, the channel length is determined by the difference between the lateral diffusions of the p-body and n^+ source. To get a short channel, the process control window needs to be very narrow. As to the channel carrier mobility, it is limited by surface scattering to about $500\,V/cm^2$ for n-channel silicon MOSFET with a gate threshold voltage, V_{th}, of about 2 V (Shenai, 1990). To get a low on-resistance, a wider channel width is needed. However, this will consume more chip area due to its planar nature. This leads to a higher cost and other side parasitic effects.

The folded gate structure, as shown in Fig. 4.18, can reduce this limitation. The structure, being different from the conventional LDMOS, provides double channel density with only one extra mask added into the LDMOS process flow. The structure is made on the corrugated substrate which leads to an enlarged surface on substrate (Blanchard, 1983). This corrugated substrate can be made by trenching the silicon surface before the gate formation. This makes the gate region to be folded. By folding the gate, the channel density is increased. Experimentally, it achieved a reduction in specific on-resistance by 45% and at the same time an increase in transconductance by 64% (Zhu *et al.*, 2001).

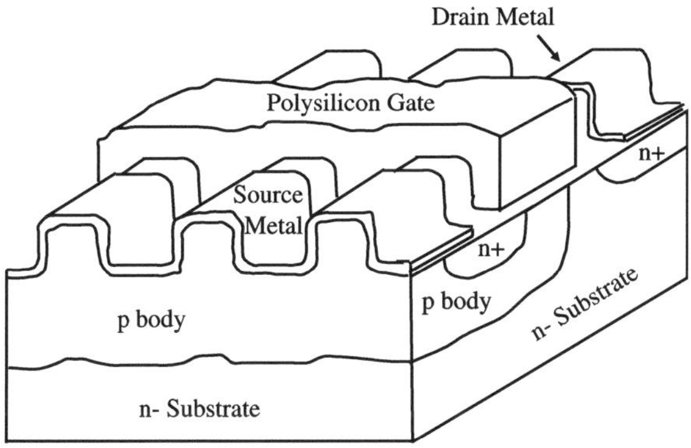

Fig. 4.18. The schematic structure of the folded-gate LDMOS.

4.8. Lateral Radio Frequency (RF) Power MOSFET

Power MOSFET devices are capable of operating at high frequency for narrow-band and high-gain signal amplification at the base station and for mobile portable platforms. For base station application, the concern is mainly on the higher RF current gain. The aspect on integration with other circuit components for space-saving is not a great concern. Thus, the vertical structure with heavily doped substrate as the common drain is a preferred choice to reduce the on-state resistance and also to lower the stray inductance without the drain metal route. As to the application on mobile platform, such as cellular phones and personal communication systems, the integration aspect becomes important in order to save the space and packaging cost. Hence, lateral structure is chosen to form devices for such an application.

The structure can be formed by the double diffused process which utilizes the same mask to self-align the channel position. The channel length is determined by the difference in lateral diffusion in p-body and source emitter, normally between 0.25 and 0.8μm for various voltage of operation. Besides the requirement in lowering the on-state resistance under the system voltage rating, the power MOSFET needs to have low parasitic capacitance as well. Equation (4.59) shows that the maximum operation frequency is a function of the input capacitance. For that, the gate–drain capacitance, C_{GD}, as part of the input capacitance, can be reduced by three approaches, namely to have a thicker gate oxide, to reduce the gate and drift-region overlap length, and to low the doping concentrations of the drift region. The thicker gate oxide, serving as the first approach to lower the gate–drain capacitance, will also increase the threshold voltage and lower the transconductance gain. The second approach on reduction in gate/drift overlap length requires a good control in process precision, and the third approach on reduction in drift region concentration will bring the on-state resistance higher.

To maintain a higher transconductance gain and lower threshold voltage, it is necessary for the gate oxide to be thinner. However, the disadvantage, besides the high gate–drain capacitance, is also that a thinner gate oxide will encourage the oxide breakdown near the gate/drift overlap region due to the field crowding. Research was carried out to have an uneven gate oxide thickness, such as the stepped gate oxide (Lin *et al.*, 1995) and the graded gate oxide (Xu *et al.*, 2000), in order to obtain a thin oxide at most of the channel region and a thicker oxide at the gate/drift overlap region to reduce the gate–drain capacitance and to avoid oxide breakdown.

4.8.1. Graded Gate

The graded gate structure enables a thin gate oxide in the channel region and has the gate oxide increasing gradually at the gate/drift overlap region, as shown in Fig. 4.19. The graded gate is formed by the oxidation process after gate polysilicon patterning. In conventional LDMOS process, the p-body is formed through the high-temperature drive-in process. During the drive-in process nitrogen is normally used as the medium gas. For the graded gate process, oxygen is used during the drive-in. This causes an oxidation on the exposed polysilicon surface at both sides leading to a shrunken polysilicon gate, followed by the oxidation processes at the drain and source regions. Bird beaks are formed at both sides of the gate region and this lifts the polysilicon gate at the edge. Hence, the thin gate oxide remains in the channel region and a thicker oxide is formed at the gate/drift overlap region. This provides a high

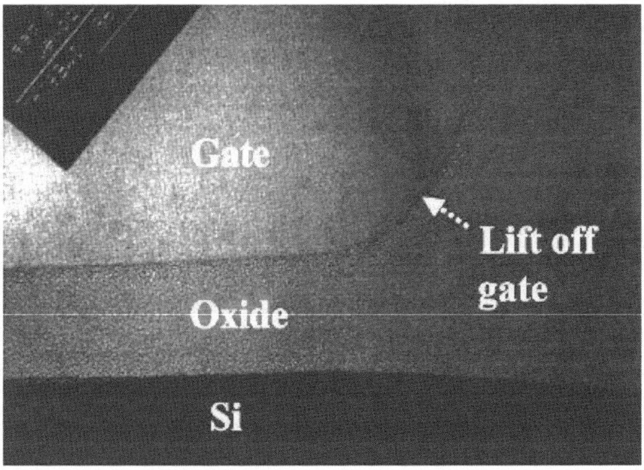

Fig. 4.19. SEM photo of the graded-gate structure (Xu *et al.*, 2000).

transconductance and the gate–drain capacitance is reduced. The graded gate structure provides a lower electric field at the gate–drift region. Using this technology, the gate oxide can be made to be below 150 Å in the channel region and up to 1000 Å at the gate–drift overlap region. A sufficient high breakdown voltage can be realized by the technology to obtain a very thin gate oxide.

4.8.2. Stepped Lateral Double Diffusion

The gate-to-drain capacitance can be further reduced when the approach of having a lower drift region doping is taken. By lowering the drift region doping, the silicon junction near the gate–drift overlap region can be depleted more quickly to have a lower junction capacitance. A lower doping at the gate–drift junction region brings another advantage of relief in hot-carrier degradation when the device is used as an amplifier with gate and drain both biased at high voltage. The degradation is due to the local high electric field and hot-electron injection into the gate oxide to cause the current-carrying capability to degrade over time.

On the other hand, the drift region has a great influence on the on-state conductivity and it is not a preferred choice to lower the doping level for the entire drift region. In CMOS process, the Halo angle counter-implantation is normally carried out at the gate-edge region under gate oxide spacer of 0.05 to 0.1μm in width to locally lower the drain region concentration. This helps to reduce the field crowding at the corner region. For power MOSFET devices, the similar technique can be used in the drift region but longer in width, so a stepped LDD structure which has lower doping near the gate/drift junction region and higher doping for the rest of the drift region can be created by an additional mask. This stepped LDD approach provides the choice for lower silicon junction capacitance C_{Si} without largely increasing the on-state resistance.

4.8.3. Partial Silicon-on-Insulator Platform

The drain-to-source output capacitance C_{OSS}, which comprises of the drain-to-substrate capacitance C_{DS} and the drain-to-gate capacitance C_{GD}, has a great influence on the performance of the RF power amplifier in terms of the power added efficiency and the harmonic distortion. The power delivered to the load of R_{L} can be expressed as (Trivedi, 1999):

$$P_{\mathrm{out}} = \frac{V_{\mathrm{in}}^2 g_{\mathrm{m}}^2 R_{\mathrm{L}}}{2(\omega^2 C_{\mathrm{OSS}}^2 R_{\mathrm{L}}^2 + 1)}. \tag{4.63}$$

The equation indicates that the output power of the LDMOS is a function of the output capacitance, especially at a higher frequency. Figure 4.20 shows

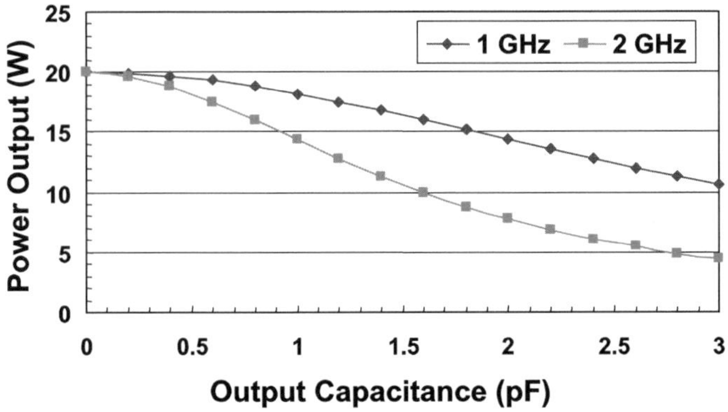

Fig. 4.20. RF output power as a function of the LDMOS amplifier output capacitance at 1 GHz (diamond points) and 2 GHz (square points).

the relationship between the power output of an RF power amplifier and output capacitance at different frequencies. It can be seen, by keeping the output capacitance low, that the amplifier will have a higher power-added efficiency with less sensitive to frequency variation. For RF power amplifier, the output matching circuitry is normally added for maximum power transfer and to reduce the harmonic contents. The concern is on distortion factor due to the handicap situation of having a fixed output circuitry to try to match a voltage-controlled variable MOSFET output capacitance. In large-signal operation, the output capacitance of a bulk MOSFET is a function of the drain voltage bias fluctuation. For a drain–substrate abrupt junction, the junction capacitance, similar to what was mentioned in Chapter 3 on a p–n junction, can be expressed as

$$C_{\mathrm{J,DS}} = \sqrt{\frac{q\varepsilon_{\mathrm{S}} N_{\mathrm{A}} N_{\mathrm{D}}}{2(N_{\mathrm{A}} + N_{\mathrm{D}})(V_{\mathrm{R}} + \psi_{\mathrm{o}})}}, \tag{4.64}$$

where N_A is the dopant acceptor concentration, N_D is the dopant donor concentration, V_R is the reverse-bias voltage across the junction, and ψ_o is the junction barrier potential. This equation indicates that the capacitance is a function of the drain bias voltage. Although drain/substrate junctions are usually not abrupt, the capacitance is still strongly influenced by the voltage bias. When the device biasing point varies or it operates under a large voltage swing at output, the capacitance value varies. The output matching becomes complicated and difficult when the output capacitance varies in a large range.

For a bulk power LDMOS, the drain region is relatively bigger for the need of low contact resistance and high current handling capability. Thus, the C_{DS}

capacitance contributes in the large extend to the overall output capacitance. One possible solution is that the SOI wafer can be utilized to fabricate the device to reduce the output capacitance and also to keep it at a relatively constant value. The approach does solve the mentioned problem electrically but raise another concern in thermal stress due to low thermal conductivity of the oxide buried layer, especially when it is thick, under the epilayer. Due to the nature of high-power application, the device is expected to generate extensive amount of heat. Simulation results show that the hot-spot temperature near the channel region may rise from 70°C to 200°C in comparison with device fabricated on bulk wafer. Due to the reduction in the mobility of carriers at high temperature, the on-resistance is expected to increase and the transconductance is expected to decrease. Another concern on SOI devices is that the source contact cannot be grounded to the substrate via the p^+ sinker connectivity normally made for the bulk RF LDMOS. Without the backside contact, additional metal route for source contact is necessary. This makes the common lead inductance higher. One approach can be made to remedy the situation by removing (etching) part of the epilayer silicon and buried oxide to provide a better thermal conductivity. However, for the substrate buried underneath the oxide, it is unsuitable for the formation of gate channel, partially due to the high doping level and partially due to the rough surface after removing the epilayer and buried oxide. The channel region should remain on the SOI platform.

To keep the benefit of having low-C_{DS} of SOI structure and to avert the overheating in the device, a more versatile partial SOI platform of buried oxide layer just below the drain region is needed, as shown in Fig. 4.21. The structure shall provide three advantages, namely to greatly lower the drain–substrate capacitance; to raise the drain–substrate breakdown, and to reduce the thermal stress of allowing heat-flow at the channel region to go through the normal bulk substrate. These properties provide great advantage in high-frequency power applications.

In Fig. 4.22, it is shown that the C_{DS} value in bulk structure at zero bias is 28 pf and in partial SOI structure is 10 pf (for the same gate width), a reduction of 64.3%. In partial SOI structure, the C_{DS} curve is rather flat with very little change with bias voltage. This characteristic is good for RF circuit application for a constant C_{DS} output impedance matching to provide a stable output characteristic.

4.8.4. Partial SOI Platform Formation

Technological solutions for the formation of partial SOI substrate include wafer bonding techniques, selective epitaxy, SIMOX with localized oxygen implantation, and uniform zone melting re-crystallization technique (Nguyen *et al.*, 1992; Bussmann *et al.*, 1991; Dilhac *et al.*, 1996). Ren *et al.* (2000) and

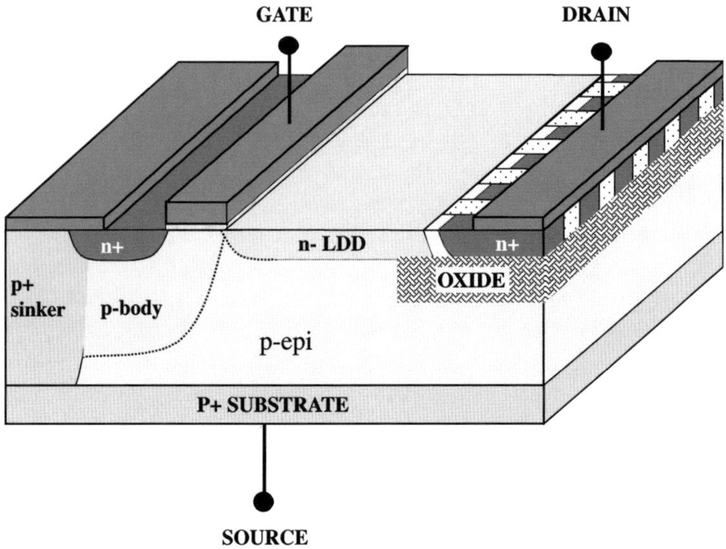

Fig. 4.21. RF LDMOS on the partial SOI platform.

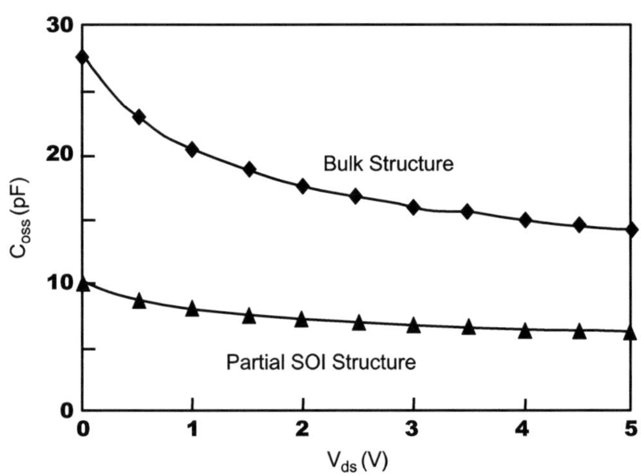

Fig. 4.22. Measurement of the output capacitance of the LDMOS on the bulk silicon wafer and the partial SOI platform.

Liang *et al.* (2000) proposed the partial-SOI formation process based on direct thermal oxidation and pinch-off principle. The process is fully compatible to the standard CMOS process. Besides, the oxide thickness is not limited by the normal thermal oxidation constraint (below 2 μm) and it can be formed from a few microns to tens of microns in thickness. The oxide platform can be selectively located in any area within a silicon bulk wafer. Thus the technology has the flexibility to place an SOI platform anywhere it may be needed, without affecting other area of bulk silicon.

The process sequence and SEM cross-sections of the partial-SOI platform formation are illustrated in Figs. 4.23(a)–(d) and 4.24(a) and (b), respectively, and described here (Ren *et al.*, 2002). The process utilizes a series of narrow and deep trenches. Part of the trench sidewall toward the top is covered by the nitride and the remaining lower part of the sidewall is exposed. Then, these trenches undergo the thermal oxidation process. Due to the nitride coverage at the top part of the sidewall, only the lower part of the trench is oxidized to form the thick layer of buried oxide. By the oxide pinch-off formation, the oxide thickness can be very much thicker than that of growth by the surface oxidation. The process steps can be combined with standard device processes

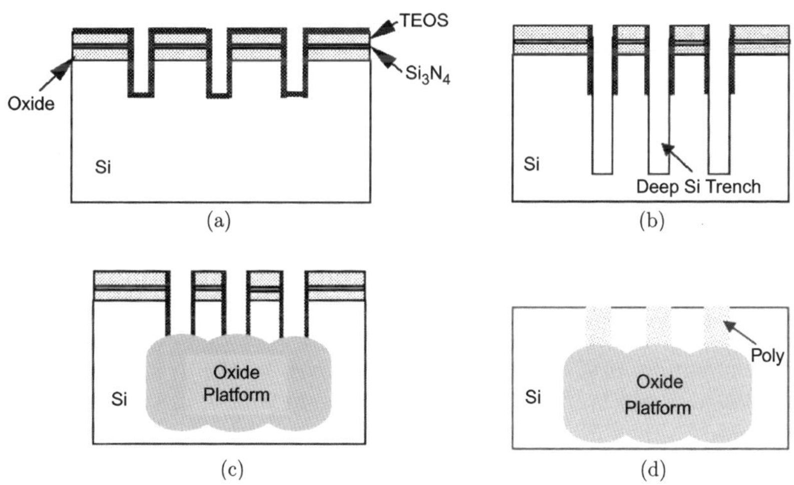

Fig. 4.23. Key fabrication steps in the partial SOI technology: (a) Si oxidation, Si_3N_4 deposition and TEOS (SiO_2) deposition. RIE shallow silicon trench, thin SiO_2 thermal growth, and the Si_3N_4 layer deposition. (b) Blanket RIE to remove Si_3N_4 and thin SiO_2 at the top surface and at the bottom of the trench. Deep Si trench formed by RIE. (c) Partial-oxide substrate formation using wet oxidation process. Oxide pinch-off in deep trenches, and the oxide layer grown is in the lateral direction. (d) Sidewall Si_3N_4 and SiO_2 removed. Poly Si refill and planarization.

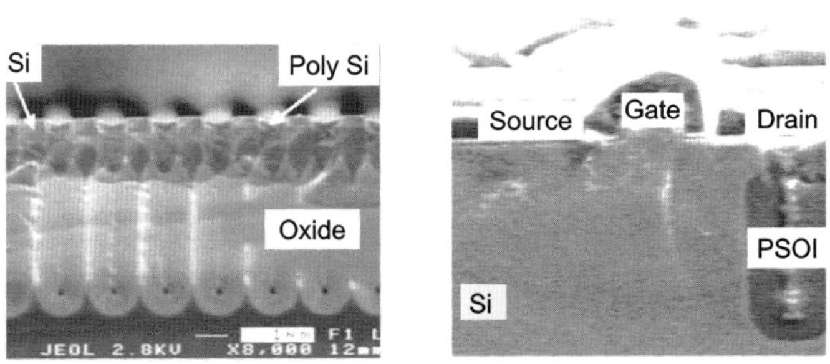

(a) (b)

Fig. 4.24. SEM photo of (a) the partial SOI platform and (b) the LDMOS on the partial SOI platform.

with one additional process of LPCVD TEOS (SiO_2) deposition added as the hard mask for Si_3N_4 and Si etching.

By trench window patterning and etching, the initial silicon trench of 0.5–1.0 μm depth is formed by reactive ion etching (RIE). The depth of the silicon trench is constraint by the Si epilayer thickness on the topside of wafer. A thin layer of SiO_2 is thermally grown on all surfaces, and then a Si_3N_4 layer is deposited. Blanket RIE is employed to remove Si_3N_4 and the thin oxide layers from all horizontal surfaces both on the top and on the bottom of the trench. There will be a SiO_2/Si_3N_4 spacer left along the sidewalls of the trench. The Si_3N_4 layer on sidewall of the trench will be used as the protection mask during thermal oxidation process. A deeper Si trench is then followed by RIE blank etching. During etching the sidewall nitride spacer may be slightly etched away. The depth of the new Si trench varies by the requirement of buried oxide thickness. As stated, the key feature on trench sidewall is that the top portion of the sidewall remains covered by the protective nitride spacers. The thermal oxidation is utilized to form oxide at the exposed sidewalls. The oxide grows laterally both inwards and outwards away from the trench sidewalls. Oxidation process continues until the newly formed oxide layers pinch off, both inside the trenches and within the surrounding silicon area. After that, the top-sidewall Si_3N_4 and SiO_2 are removed by wet etching and the original silicon trenches are restored. The trenches are filled with polysilicon followed by the planarizing etch to obtain a flat surface.

A layer of 5.0 μm, 10 Ω cm, p-type doped (100) silicon epilayer on a 0.02 Ω cm, boron-doped silicon substrate is used to go through the process. The RF power LDMOS is formed by first having the p^+ sinker implant/drive-in and partial-SOI substrate formation as described. Then, the thick field oxide

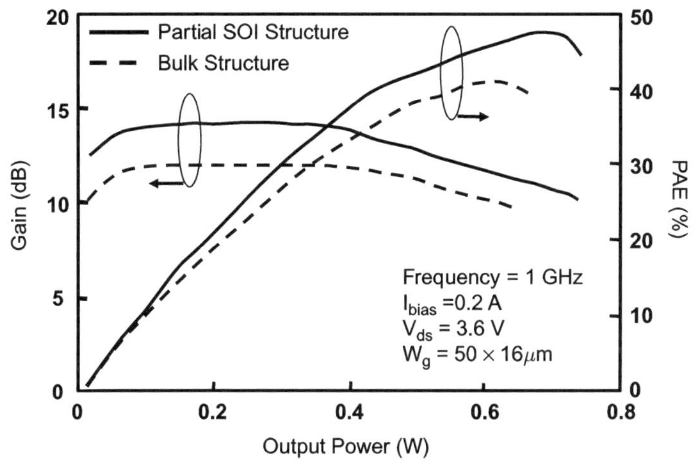

Fig. 4.25. Measured power gain and power added efficiency versus output power at 1.0 GHz.

was patterned to open the active area. A layer of 175 Å gate oxide is grown and approximately 6000 Å of LPCVD polysilicon was deposited and doped to obtain a sheet resistance of 10 Ω/cm^2. The layer was then patterned to define the gate. The channel was formed by double diffusions of the p-body and the n^+ source. Finally, contact to gate and drain are made of 1.7μm layer of Al-Si-Cu metal followed by a forming gas anneal at 420°C for 30 minutes.

Fabricated samples of conventional and P-SOI MOSFETs of the same device dimensions were measured to show the difference in characteristics, such as the leakage current, output capacitance, power-aided efficiency (Fig. 4.25), and second harmonic suppression (Fig. 4.26).

4.9. Parallel and Series Operations

Paralleling the power MOSFETs can be employed to increase the current handling capability for a lower $R_{ds(on)}$, to ease the packaging constraints by having a smaller die dimension, and to improve the thermal performance. When the operating temperature increases, the $R_{ds(on)}$ of the power MOSFET will increase due to the reduction of its carrier mobility. Thus, an equalization of $R_{ds(on)}$ and a good thermal coupling among MOSFETs are essential for the steady-state parallel operation. It is understood that, during linear-mode ($V_{GS} - V_{th} > V_{DS}$) operation, the $R_{ds(on)}$ increases with higher drain current, the $R_{ds(on)}$ of the particular power MOSFET that carriers a higher current will become higher, thereby reducing the disparity of $R_{ds(on)}$ among the

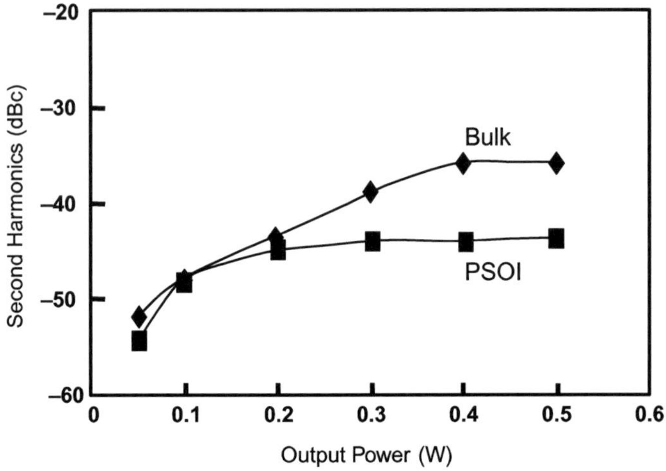

Fig. 4.26. Measured second harmonic power versus output power of power amplifier module.

paralleled devices. However, the amount of mismatch in $R_{ds(on)}$ of the paralleled devices is correctable only to a limited amount by thermal enforcement and is more suitable for device having large temperature coefficient. Another approach to equalize the current sharing is to insert a small source resistance to each of the device to dynamically adjust the gate-to-source voltage when the device current varies. The disadvantage is of course, the higher on-state resistance.

During dynamic switching, the current sharing in the device saturation region of operation is influenced by factors, such as the threshold voltage and transconductance that have little influence on static current sharing in the fully conducting linear region. To improve dynamic current sharing, the source inductance should also be as equal as possible to balance the division of the load current at high switching speed. In a sense, the layout of the paralleled devices, including the packaging wire inductance, should be as symmetrical and equal as possible. Another concern is on the high-frequency parasitic ringing during transient operation when the devices pass through the saturation region during switching (Siliconix, 1984). It is generally attributed to the common-mode or differential-mode oscillations. The former occurs at high-frequency high-gain devices coupled with the internal and external circuit parasitics. The latter occurs when paralleled power MOSFETs operate in push–pull mode as an oscillator where the tuned circuit comprises of the parasitic inductance of the interconnecting wire and the parasitic capacitance of the MOSFETs (Kassakian and Lau, 1984). The frequency of oscillation can range from 1 MHz to 300 MHz. It should be noted that the peak gate voltage generated by such an oscillation can be sufficiently large to damage the gate oxide. The oscillation can be prevented by minimizing the parasitic inductance and capacitance in all interconnects. The inclusion of the series gate resistor and ferrite-bead inductor helps to mitigate this differential oscillation through damping. However, the gate resistors will also slow down the switching speed of the power MOSFETs.

Power MOSFETs can also be connected in series to increase their voltage ratings. They are connected in the source-follower configuration. Static voltage sharing is aided by the drain-to-gate resistors while the dynamic voltage sharing is aided by the drain-to-source capacitors. The disadvantages are that a higher gate voltage is required to ensure the proper series operation, and additional switching loss is anticipated due to the external drain-to-source capacitors. Generally, there is no obvious advantage in cost and circuit performance of series operation of power MOSFETs when a device of the required voltage rating is available, e.g. the recently available superjunction power MOSFET. The power MOSFET with voltage rating of 1000 V is commercially available.

4.10. Gate Drive Circuits

To switch a power MOSFET between the blocking state and the conducting state, its gate-to-source voltage must be varied to enable the transfer of charge in and out of the gate. The characteristics of the gate drive circuit determine the switching performance of the device. During on-state bias, the magnitude of the gate voltage must be sufficiently high to yield the desired drain current. In most the power MOSFETs, the maximum gate-to-source voltage is specified to be 20 V to avoid stressing and damaging the gate dielectric. The minimum gate-to-source voltage required to ensure that the MOSFET remains fully-on while carrying the desired drain current can be found from its I_D versus V_{DS} transfer characteristics. Normally, the gate voltage is increased above the minimum required value in order to decrease $R_{ds,on}$ for a low on-state $V_{DS,on}$ for minimum power dissipation. However, the shortcoming is that, the turn-off delay time increases as the on-state gate voltage increases. The off-state gate voltage must be sufficiently lower than the threshold voltage to provide adequate noise immunity. However, even a gate voltage of 0 V may not provide sufficient noise immunity for the switching of inductive loads especially when the drain current falls rapidly. The fast transient in drain voltage may therefore induce a voltage higher than the threshold value at gate through the Miller capacitance C_{GD} coupling. In this case, a low-impedance gate drive circuit is usually used to hold down the gate voltage and to ensure it to be below the threshold value during turn-off transient. Besides the gate drive resistance, the low conductivity of the polysilicon gate also plays a part in boosting the gate voltage. For this, it may be necessary to apply a negative bias to the gate to prevent spurious turn-on of the power MOSFET. However, similarly, the negative gate voltage prolongs the turn-on delay time of the power MOSFET during turn-on process.

In some applications, an asymmetric gate drive circuit is required to achieve a fast turn-on and slow turn-off or vice versa. The asymmetric gate drive can be implemented using a diode in parallel with a resistor connected to the gate electrode. The slow turn-on configuration is used to reduce the current surge in the power MOSFET induced by the reverse recovery charge of the diode clamped across an inductive load. The fast turn-off is desired to minimize the turn-off switching losses.

A different gate drive configuration is required for the high-side configuration which is needed in many switching converter topologies such as the push–pull or full-bridge switching converters. The high-side power MOSFET can be either an n-channel or a p-channel device. Since the electron mobility is about three times higher than that of holes, the n-channel power MOSFET will have higher current handling capability than their p-channel counterparts for the same die size. Thus, n-channel power MOSFET is usually cheaper than the p-channel device for the same voltage and current ratings. However, the gate drive voltage needs to be higher than the drain voltage if n-MOSFET is used. This is an additional requirement as the drain voltage is normally the highest voltage in the system. On the other hand, the gate drive for the high-side p-channel power MOSFET is relatively simple. To switch on a high-side p-channel power MOSFET, the gate voltage needs to be pulled down to a few volts below the threshold voltage via a transistor. Since the cost of the gate driver circuit is fixed, the choice would be more economical to have the p-channel power MOSFET in low-current applications with minimum gate drive. The usage of n-channel power MOSFET with a high-side gate drive circuit will be more suitable when the current requirement increases above a certain level.

Practically, bootstrapping, charge pumping, and level-shifting techniques are used to obtain high-side gate drive signal (Murari *et al.*, 1996). Figure 4.27 shows a simple bootstrap circuit. The bootstrap capacitor C_B is charged up by the auxiliary supply V_A during the period when the lower MOSFET is turned on. In the lower MOSFET on-state period, the voltage at the output terminal, V_O is close to the ground potential, and the capacitor is charged close to the level of auxiliary supply V_{DC}. When the lower MOSFET turns off, the output voltage will go up to a higher level. The bootstrap diode D is now reverse-biased when the voltage, V_{CB} (with respect to the ground), is higher than the auxiliary

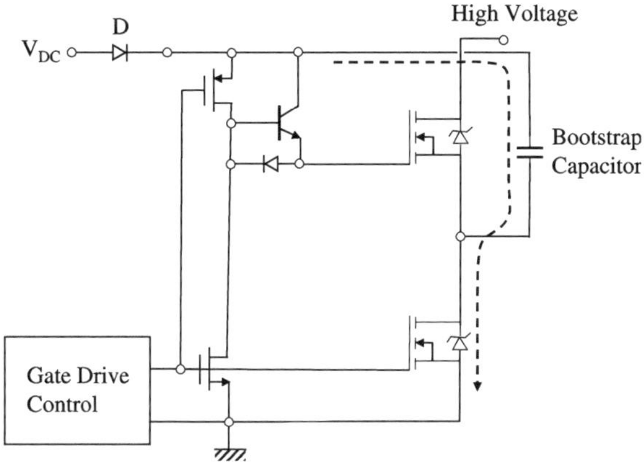

Fig. 4.27. Bootstrap technique used in the gate drive circuit for NMOSFET half-bridge configuration.

supply. The charge stored in the bootstrap capacitor is then used as the power supply for the high-side gate drive unit. For such a bootstrap configuration, the lower MOSFET (or together with its fly-wheeling diode) should have sufficient period of conduction time to complete charging the bootstrap capacitor. Thus, there may be certain limit on the duty cycle of operation.

Many commercial gate driver integrated circuits, such as the Motorola MDC1000A, International Rectifier IR2110, offer a simple and cost-effective solution for driving power MOSFETs. The MDC1000A MOSFET gate driver (shown in Fig. 4.28) has a SCR active gate turn-off (Prentice *et al.*, 1992). A diode provides the charging current required to switch on the MOSFET. The turn-on time of the power MOSFET is limited by the forward recovery time of this diode. The forward voltage of the diode clamps the SCR off as long as the input control signal remains high. When the gate control signal is low, the SCR turns on rapidly and discharges the gate charge of the MOSFET. The gate-to-source voltage is less than 1 V, and well below the threshold voltage of the power MOSFET. The IR2110 (shown in Fig. 4.29) has two drivers: a floating high-side driver and a ground-referenced driver (Giandomenico *et al.*, 1985, Clemente and Dubhashi, 1990). Although the IR2110 has good noise immunity, gate undervoltage lockout limits the protection features. Power for the floating high-side circuit is obtained through a bootstrap technique. When the low-side switch is on, the output is low and bootstrap is charged from the supply via a charging diode. When the high-side output is enabled, the load voltage goes high and the charging diode blocks the bootstrap capacitor from the supply. All the current used by the floating driver is supplied from the

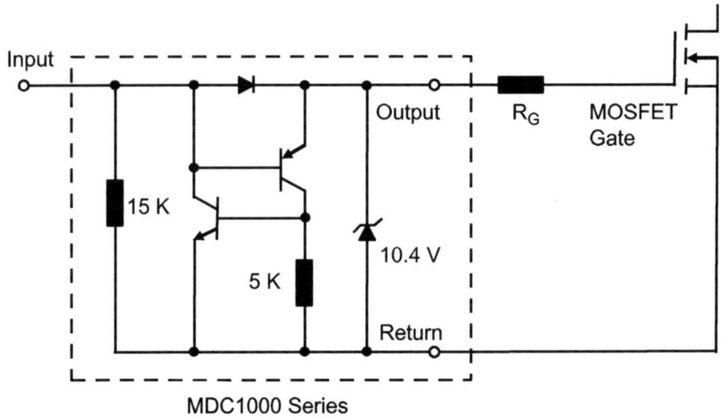

Fig. 4.28. MDC1000 turn-off device for power MOSFET (*source*: Motorola Datasheet MDC1000A/D 1991).

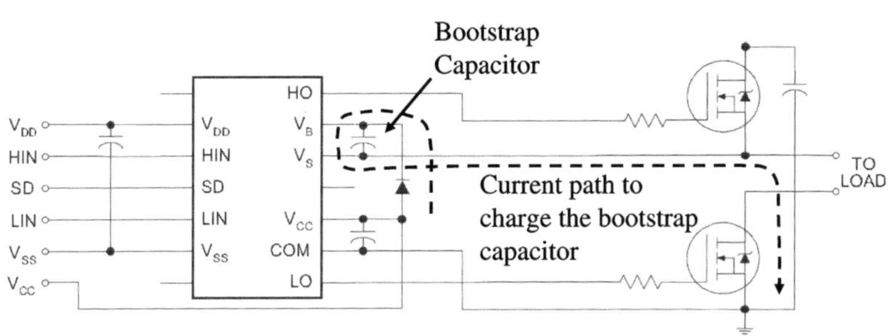

Fig. 4.29. IR2110 high- and low-side driver in bootstrap operation (*source*: International Rectifier Datasheet IR2110(-1-2)(S)PbF/IR2113(-1-2)(S)PbF 2005).

bootstrap capacitor, which must be refreshed at regular intervals so that its voltage does not discharge below the undervoltage shutdown level. The refresh rate depends on the amount of internal power required to operate the upper floating power IC and the external driver energy required by the power switch. The IR2110 gate driver can operate in most applications from frequencies in the tens of Hz to hundreds of kHz (Bliss *et al.*, 1992).

References

S. S. Ang, *Power-Switching Converters* (Marcel Dekker, Inc., New York, 1995).

B. J. Baliga, *Modern Power Devices* (John Wiley and Sons, New York, 1987), pp. 263–343.

R. A. Blanchard, Power MOS transistor with a plurality of longitudinal grooves to increase channel conducting area, US Patent 4,393,391 (1983).

V. Barkhordarian, MOSFETs, in *Power Electronics Handbook*, ed. T. L. Skvarenina (CRC Press, 2002).

J. Bliss, J. Prentice and I. Cohen, MOSFET gate drivers reach new performance goals-Part I: Low-voltage MGD's, *PCIM Europe*, Mar 14–20 (1992).

U. Bussmann, A. K. Robinson, P. L. F. Hemment and G. J. Campisi, Silicon-on-insulator device islands formed by oxygen implantation through patterned masking layers, *J. Appl. Phys.* **70**(8), 4584–4592 (1991).

S. Clemente and B. R. Pelly, Understanding the power MOSFET switching performance, *Proc. IEEE Ind. Appl. Soc. Meeting*, 763–776 (1981).

S. Clemente and A. Dubhashi, HV floating MOS-gate driver IC, in *International Rectifier Application Notes*, AN-978A (International Rectifier, El Segundo, California, 1990).

M. N. Darwish and K. Board, Optimization of breakdown voltage and on-resistance of VDMOS transistors, *IEEE Trans. Electron Devices* **31**, 1769–1773 (1984).

J. M. Dilhac, D. Zerrouk, C. Ganibal, P. Rossel and M. Bafleur, Fabrication of SOI structures by uniform zone melting recrystallization for high voltage ICs, *IEEE Int. Symp. Power Semiconductor Devices ICs*, 215–218 (1996).

T. Efland, P. Mei, D. Mosher and B. Todd, Self-aligned RESURF to LOCOS region LDMOS characterization shows excellent R_{sp} vs BV performance, *IEEE Int. Symp. Power Semiconductor Devices ICs*, 147–150 (1996).

D. Giandomenico, D.-S. Kuo and C. Hu, Oscillations in multichip power MOSFETs, *Power Conversions Intell. Motion Mag.*, 74–78 (August, 1985).

F. E. Holmes and C. A. T. Salama, V-groove MOS transistor technology, *Electron. Lett.* **9**, 457–458 (1973).

J. Kassakian and D. Lau, An analysis and experimental verification of parasitic oscillation in paralleled power MOSFETs, *IEEE Trans. Electron Devices* **31**, 959–963 (1984).

E. Kobori, N. Izumi, N. Kumamoto, Y. Hamazawa, M. Yamamoto, K. Yamamoto and A. Kamisawa, Efficiency of power devices using full Cu metallization technologies, *IEEE Int. Symp. Power Semiconductor Devices ICs*, 67–70 (1999).

Y. C. Liang, S. Xu, C. Ren and P.-D. Foo, New partial SOI LDMOS device for high power-added efficiency in 2 GHz RF power amplifier applications, *IEEE Ind. Electron. Control Instr. Conf. Proc.*, 1001–1006 (2000).

D.-G. Lin, S. L. Tu, Y. C. See and P. Tam, A novel LDMOS structure with a step gate oxide, *IEEE Int. Electron Devices Meeting Digest*, 963–966 (1995).

S. Merchant, R. Baird, P. Hui, R. Thoma and J. Victory, High-performance 20–30 V LDMOS transistors in a 0.65 μm-based BiCMOS compatible process, *IEEE Bipolar/BiCMOS Circuits Technol. Meeting*, 202–205 (1997).

S. Merchant, R. Baird, S. Chang, P. Hui, V. Macary and M. G. Neaves, High-performance 13–65 V rated LDMOS transistors in an advanced smart power technology, *IEEE Int. Symp. Power Semiconductor Devices ICs*, 225–228 (1999).

R. S. Muller and T. I. Kamins, *Device Electronics for Integrated Circuits*, Chaps. 8 and 9 (John Wiley and Sons, New York, 1986).

B. Murari, F. Bertotti and G. A. Vignola, (eds.), *Smart Power ICs: Technologies and Applications* (Springer, Berlin, 1996).

C. T. Nguyen, S. C. Kuehne, P. Renteln and S. S. Wong, Quasi-SOI MOSFETs using selective epitaxy and polishing, *IEEE Int. Electron Devices Meeting Digest*, 341–344 (1992).

J. Prentice, I. Cohen, D. Everett, A. Alderman and S. Clemente, MOSFET gate drivers reach new performance goals — Part II: Low-voltage and high-voltage MGDs, *Power Conversions Intell. Motion Mag.*, 24 (April, 1992).

R. S. Ramshow, *Power Electronics Semiconductor Switches*, Chap. 6 (Chapman & Hall, 1993).

C. Ren, Y. C. Liang, and S. Xu, New RF LDMOS structure with improved power added efficiency for 2 GHz power amplifier, *IEEE Region 10 Conf. Proc.*, 29–34 (2002).

C. Ren, J. Cai, Y. C. Liang, P. H. Ong, N. Balasubramanian and J. K. O. Sin, The partial silicon-on-insulator technology for RF power LDMOSFET devices and on-chip microinductors, *IEEE Trans. Electron Devices* **49**(12), 2271–2278 (2002).

K. Shenai, Optimally scaled low-voltage vertical power MOSFETs for high-frequency power conversion, *IEEE Trans. Electron Devices* **37**(4), 1141–1153 (1990).

Siliconix Incorporated, *MOSPOWER Applications Handbook*, ed. R. Severns and J. Armijos (Siliconix Inc., Santa Clara, California, USA, 1984), pp. 5.15–5.32

M. Trivedi, P. Khandelwal and K. Shenai, Performance modelling of RF power MOSFETs, *IEEE Trans. Electron Devices* **46**, 1794–1802 (1999).

S. Xu, K. P. Gan, P. D. Foo, Y. Su and Y. Liu, Graded gate VDMOSFET, *IEEE Electron Device Lett.* **21**(4), 176–178 (2000).

S. Xu, C. H. Ren, Y. C. Liang, P.-D. Foo and J. K. O. Sin, Theoretical analysis and experimental characterization of the dummy gated VDMOSFET, *IEEE Trans. Electron Devices* **48**(9), 2168–2176 (2001).

H. J. M. Veendrick, *MOS ICs: From Basics to ASICs* (VCH Publishers, Inc. New York, 1992), pp. 28–29, 58–69.

Y. Zhu, Y. C. Liang, S. Xu, P.-D. Foo and J. K. O. Sin, Folded gate LDMOS transistor with low on-resistance and high transconductance, *IEEE Trans. Electron Devices* **48**(12), 2917–2928 (2001).

5

Insulated-Gate Bipolar Transistor

5.1. Introduction

MOSFET devices have been widely used since 1970s for various electronic signal and power applications. For power switching applications, there exists the requirement for the device to sustain a high operating voltage, e.g. a few hundred volts. This is made possible by placing a thick and lightly doped drift region as an extension of the MOSFET drain emitter to accommodate the field depletion. By doing so, the penalty on the device will be having a higher on-state conduction resistance as the MOSFET current conduction is made of majority carrier transport, and the conductivity modulation occurred in the drift region by the majority carriers is normally insufficient. In 1980s, a new type of device named insulated-gate bipolar transistors (IGBT) was invented to improve the situation. By its operational principle, the device is a MOS-driven power bipolar device which has both types of carriers flowing in the drift region to lower the conduction resistance. The device has been known earlier by many different names such as, the insulated-gate rectifier (IGR) (Baliga *et al.*, 1982), insulated-gate transistor (IGT) (Chang *et al.*, 1983, 1984), conductivity-modulated field-effect transistor (COMFET) (Russell *et al.*, 1983; Goodman *et al.*, 1983), and bipolar-mode MOSFET (IGBT) (Nakagawa *et al.*, 1984). For such a MOS-driven power bipolar transistor, it combines the advantages of both the MOSFET gate control and the nature of bipolar transistor into a single device. Due to the high-impedance characteristic of its MOSFET input gate, the driver circuitry for the IGBT is essentially similar to that for the power MOSFET. With the key feature of having bipolar conductivity modulation in the lightly doped drift region, it lowers the on-state power dissipation for devices of high-voltage rating. However, the bipolar excess carriers needed for low on-state conduction voltage do make the device turn-off speed much slower than that of the power MOSFET of similar current rating. The IGBT device exhibits a current tailing, e.g. in sub-microseconds range, during turn-off which restricts the device to operate at high switching frequency. Carrier lifetime control is then needed to improve the situation.

The IGBT device offers a good solution to medium-range frequency (e.g. <100 kHz), high-voltage high-current switching applications. Under special switching techniques, such as zero current switching resonant converters, the IGBT has been shown to operate in the hundreds of kilohertz range (Rangan *et al.*, 1987). It has been shown theoretically and experimentally that the p-channel IGBT can have performance characteristics comparable to those of n-channel IGBT (Chang *et al.*, 1984), thus allowing the use of complementary devices in power electronics applications. Also, the IGBT device can handle a larger current density compared to the power bipolar junction transistor (BJT) and the power MOSFET. For example, at a breakdown voltage of 600 V, the IGBT device carries a current density at about 20 times that of the power MOSFET and at about five times that of the power BJT. However, the presence of a diode knee voltage in its current–voltage characteristic prevents the IGBT device to be used in applications that require a very low forward voltage drop. Since 1990, IGBT devices of voltage ratings between 600 V and 1200 V have been commonly used in power conversion applications. Voltage ratings of 2500 V and above are also recently available in the market.

In power electronic circuits, it is possible to directly replace a power MOSFET by an IGBT to improve the conduction efficiency (but be careful on the long turn-off time). In general, an IGBT has a smaller die size than a similarly rated power MOSFET. Since the cost of a device is partially related to the silicon die area, the smaller die size of the IGBT makes the device a lower cost solution compared to the power MOSFET at similar current rating. Also, the IGBT device is ideally suited for inductive switching owing to the uniform current distribution in the device during turn-off. However, the device has two disadvantages in comparison to the MOSFET counterpart. First, is its slower switching speed due to the bipolar transistor structure, and the second one is prone to parasitic thyristor latch-up when operating at high current or high *dv/dt* power switching conditions.

In this chapter, the fundamental structure of the IGBT device is described along with its current–voltage characteristics. The device characteristics on switching and temperature effect are also described. This is followed by the introduction of lateral IGBT structure, integrated current sensor, and over-current protection. Other related types of MOS-controlled bipolar devices are generally described at the end.

5.2. Device Structure and Current–Voltage Characteristics

The device structure for IGBT is similar to that of a double-diffused MOSFET (DMOS) with the exception of having the p^+ substrate (anode) for the IGBT device as shown in Fig. 5.1. As such, it is a four-layer structure that resembles that of a thyristor. Unlike the thyristor where the device latches, an IGBT is designed to turn on without any regenerative action and the MOS-gate remains in control. A wide-base p–n–p structure is formed at the bottom three layers with the p^+ substrate as the emitter, the n^--drift layer as the base, and the p-body of the MOSFET as the collector region. As such, the thickness and doping density of the n^--drift layer determines the breakdown voltage of the device. To follow the convention of the four-layer structure, the top n^+ source is named as the cathode while the p^+ substrate is the anode for the IGBT device in this chapter.

Except the leakage current, there is no visible current-flow when a negative voltage is applied to the anode with respect to the cathode because the junction between the p^+-body and the n^--drift layer is reverse-biased. The IGBT is now operating in its reverse blocking mode with the I–V curve as shown in Fig. 5.2. Most of the depletion region is extended into the lightly doped n^--drift layer. The reverse blocking voltage is essentially the BV_{CBO} of the p^+ substrate/n^--drift/p-body transistor. As such, the doping and thickness of the n^- layer are chosen to yield the desired blocking voltage. It should be noted that a proper junction edge termination and passivation technique

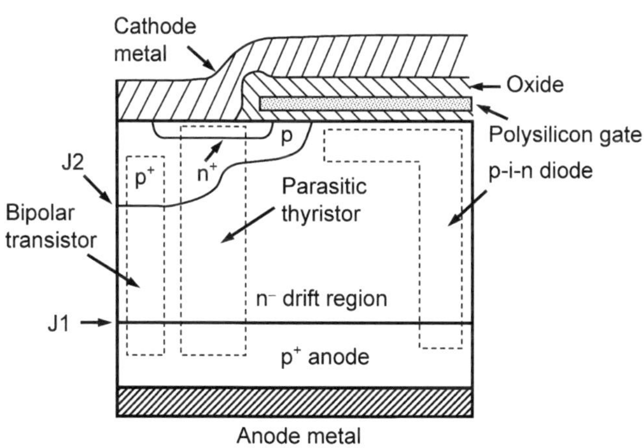

Fig. 5.1. Device structure of an IGBT device.

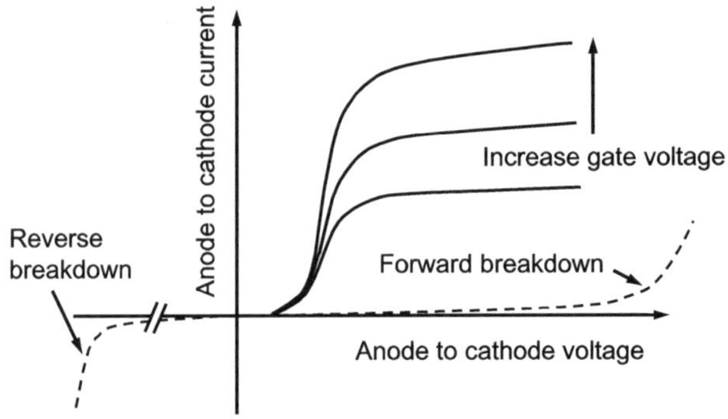

Fig. 5.2. Forward and reverse blockings (dash lines), and forward conduction *I–V* curves (solid lines) of an IGBT device.

must be employed to achieve the optimum reverse blocking voltage. When a positive voltage is applied to the anode terminal, with the gate shorted to the cathode (ground) terminal, the IGBT is operating in its forward blocking mode since the junction between the p-body and n^--drift region is reverse-biased. The IGBT device is said to operate in its forward conduction state if a gate voltage of greater than the threshold voltage is applied under positive anode-to-cathode bias condition. Similar to the MOSFET device, a conductive channel is induced underneath the oxide gate in the p-body region. Electrons flow from the n^+-cathode to the n^--drift region while the p^+ substrate injects holes into the n^--drift layer region to form the bipolar conductivity modulation in the drift region. The *I–V* curves for various gate voltages are shown in Fig. 5.2. The injected hole concentration increases as the anode-to-cathode voltage increases. Thus, the forward current of the IGBT increases similarly to that of a p–i–n diode. The forward current starts to saturate when a significant voltage drop develops across the MOSFET conducting channel in the p-body region. These current–voltage characteristics are similar to those of a power MOSFET, except for the presence of the diode knee voltage at the starting point of current conduction.

In applications where the IGBT device is not required to block a reverse voltage, an asymmetrical IGBT structure is formed with an n-buffer layer placed between the p^+ substrate and the much lightly doped n^--drift layer as shown in the lower part of Fig. 5.3. In the symmetrical structure as shown in the same figure, the doping density and thickness of the n^- layer are chosen to prevent punchthrough to the p^+-anode of the IGBT. In the case of the asymmetrical IGBT structure which is a classic punchthrough structure, the electric field distribution changes from the triangular shape to the rectangular-alike shape. Thus, the forward blocking capability of the asymmetrical device is

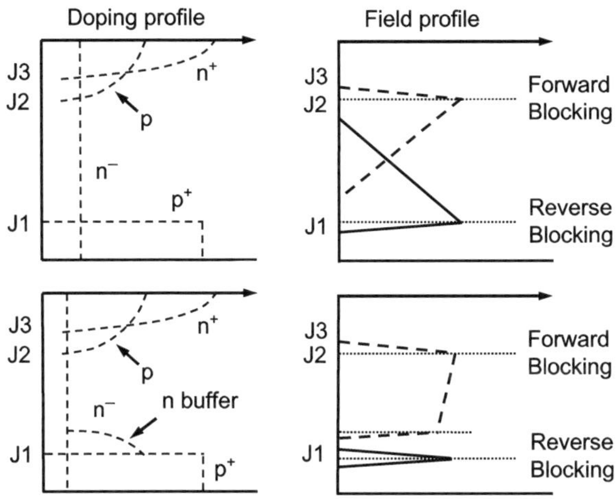

Fig. 5.3. Symmetrical (upper) and asymmetrical (lower) IGBT device doping profiles and the field distributions.

increased approximately by a factor of 2 more than that of the symmetrical structure if a similar n^--drift layer thickness was used. Therefore, with a shorter drift region length it can be used to enhance the forward conduction characteristics. And, due to less amount of excess-charge storage, this asymmetrical IGBT structure has a lower turn-off time compared to that of the symmetrical IGBT structure.

5.2.1. Forward Conduction Characteristics

Forward conduction occurs when a positive bias is applied to the anode with respect to the cathode and with a positive gate voltage greater than the applied threshold voltage. Current flow from anode to cathode must pass through a p–n junction formed by the p^+ substrate and n^--drift region, and the MOS channel. Thus, a diode knee is present in its initial forward current–voltage characteristic as shown in Fig. 5.2. Based on the structure, the IGBT can be represented as the equivalent circuit shown in Fig. 5.4, which is a p–i–n diode in series with a MOSFET device, named as the p–i–n/MOSFET model. Or, because the IGBT device is also a MOS-driven BJT structure, it can alternatively be modeled by the BJT/MOSFET equivalent circuit as shown in Fig. 5.5 (Baliga, 1987).

5.2.1.1. *p–i–n/MOSFET Model*

The p–i–n/MOSFET model, as shown in Fig. 5.4, fully explains the presence of the diode knee of the IGBT current–voltage characteristics. At low forward

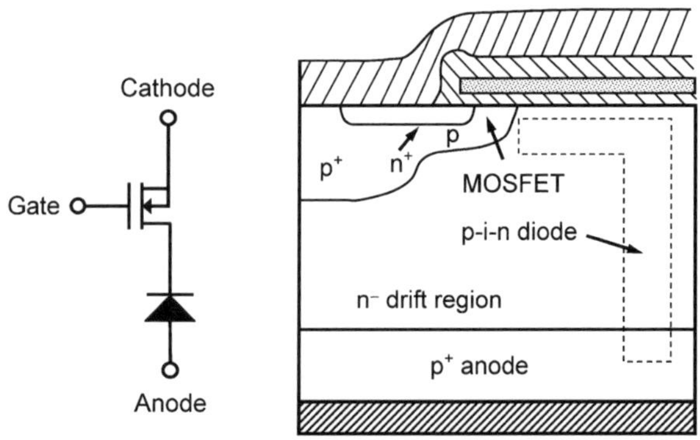

Fig. 5.4. The diode–MOSFET equivalent device model.

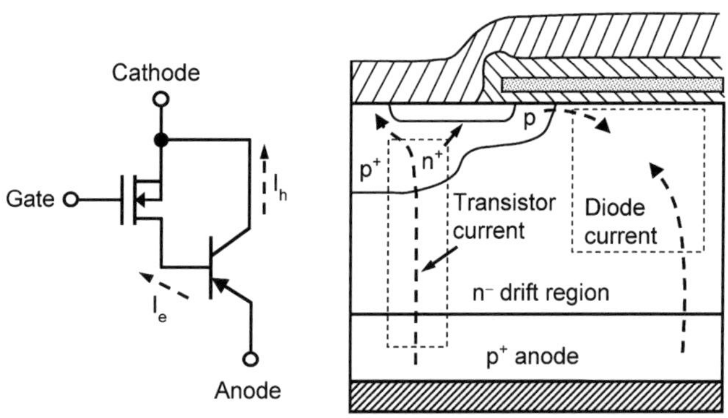

Fig. 5.5. The BJT–MOSFET equivalent device model.

biases, the current density is related to the voltage drop across the p–i–n diode by Eq. (3.56) and repeated here under medium to high forward bias condition,

$$J_{\mathrm{f,pin}} = \frac{2qD_{\mathrm{a}}n_{\mathrm{i}}}{d}F\left(\frac{d}{L_{\mathrm{a}}}\right)e^{\frac{qV_{\mathrm{f}}}{2kT}}, \tag{5.1a}$$

$$\mathrm{F}\left(\frac{d}{L_{\mathrm{a}}}\right) = \frac{\frac{d}{L_{\mathrm{a}}}\tanh\left(\frac{d}{L_{\mathrm{a}}}\right)}{\sqrt{1 - B^2\tanh^4\left(\frac{d}{L_{\mathrm{a}}}\right)}}e^{\frac{-qV_{\mathrm{n}^-}}{2kT}}, \tag{5.1b}$$

$$B = \frac{\frac{\mu_{\mathrm{n}}}{\mu_{\mathrm{p}}} - 1}{\frac{\mu_{\mathrm{n}}}{\mu_{\mathrm{p}}} + 1}, \tag{5.1c}$$

where D_a is the amipolar diffusion coefficient, L_a is the amipolar diffusion length, μ_n and μ_p are the electron and hole mobilities in the drift region, and d is half of the drift region length. Thus, the voltage drop across the p–i–n diode can be expressed as

$$V_{f,pin} = \frac{2kT}{q} \ln\left[\frac{J_{f,pin}d}{2qD_a n_i F\left(\frac{d}{L_a}\right)}\right]. \tag{5.2}$$

For diode conduction under low bias conduction, the term *kT* shall be used instead of *2kT* in Eqs. (5.1) and (5.2), as described in Chapter 3. As the p–i–n diode is connected in series with the MOSFET, and if the upper transistor current is ignored, the same diode current will flow through the MOSFET, i.e.

$$I_{D,MOSFET} = I_{f,pin}. \tag{5.3}$$

For the MOSFET operating in its linear region, the current flowing through the device can be expressed as a function of its drain-to-source voltage drop as in Eq. (5.4):

$$I_{D,MOSFET} = \frac{\mu_n C_{ox} W}{L}(V_G - V_T)V_{DS,MOSFET}, \tag{5.4}$$

where μ_n is the effective channel mobility, $V_{DS,MOSFET}$ is the drain-to-source voltage of the MOSFET, V_T is the channel threshold voltage, L is the channel length, W is the channel width and C_{ox} is the gate oxide capacitance. From this equation, the voltage drop across the MOSFET can be expressed as

$$V_{DS,MOS} = \frac{I_{D,MOSFET} L}{\mu_n C_{ox} W(V_G - V_T)}. \tag{5.5}$$

Hence, the voltage drop across the IGBT is the sum of the voltage drops across the p–i–n diode and the MOSFET potions (Baliga, 1987):

$$V_f = \frac{2kT}{q} \ln\left[\frac{I_f d}{2qWW_d D_a n_i F\left(\frac{d}{L_a}\right)}\right] + \frac{I_f L}{\mu_n C_{ox} W(V_G - V_T)}, \tag{5.6}$$

where W_d is the effective drift region width contributed to the diode conduction. From the above equation, the p–i–n/MOSFET model characterizes the IGBT current–voltage relationship and it shows a diode knee at low forward current region. Just above the diode knee voltage, the current starts to increase exponentially as shown in Fig. 5.2. As the anode-to-cathode voltage of the IGBT continues to increase, the MOSFET channel eventually pinches off near the side of n^--drift region. As such, the IGBT anode current is now limited by the MOSFET channel in saturation. The collector current is now saturated and is given by

$$I_{C,Sat} = \frac{\mu_n C_{ox} W}{2L} (V_G - V_T)^2. \tag{5.7}$$

The major shortcoming of this p–i–n/MOSFET model is the omission of the hole current component flowing in the p-base region of the upper n–p–n transistor.

5.2.1.2. *BJT/MOSFET Model*

In the more appropriate BJT/MOSFET model as shown in Fig. 5.5, the MOSFET provides the base drive for the p-body/n^--drift layer/p^+ substrate BJT. The electron current component flowing through the MOSFET channel is labeled as I_e in Fig. 5.5, while the hole current component injected from the p^+ substrate into the transistor n^--drift region, of which, received by the upper p^+/cathode contact is labeled as I_h. The electron and hole current components are related by the p–n–p common-base current gain α_{PNP} as

$$I_h = I_e \left(\frac{\alpha_{PNP}}{1 - \alpha_{PNP}} \right). \tag{5.8}$$

Understand that, the current of IGBT is the sum of both hole current and electron current components. Therefore,

$$I_f = I_h + I_e = \frac{I_e}{(1 - \alpha_{PNP})} = \frac{I_h}{\alpha_{PNP}}. \tag{5.9}$$

It can be seen that the MOSFET channel current is only a part of the anode current of the IGBT. Unlike the case in p–i–n/MOSFET model, the entire anode current is assumed to flow through the MOSFET channel. The common-base current gain α_{PNP} of the wide-base lightly doped p–n–p transistor is mainly determined by the base transport factor and can be approximately given by

$$\alpha = \frac{1}{\cosh\left(\frac{L_B}{L_a}\right)}, \tag{5.10}$$

where L_B is the undepleted base region of the p–n–p transistor, or essentially the thickness of the n^--drift layer, which equal to *2d* of the diode drift region length, under on-state conduction. L_a is the amipolar diffusion length at the medium and high carrier injection level. In a typical IGBT structure, the p–n–p transistor gain is about 0.5. The current–voltage relationship across the IGBT in this BJT/MOSFET model is similar to Eq. (5.6) except that the current flowing through the MOSFET is replaced by the electron current component (Baliga, 1987). Thus,

$$V_{\mathrm{f}} = \frac{2kT}{q}\ln\left[\frac{I_{\mathrm{f}}d}{2qWW_{\mathrm{d}}D_{\mathrm{a}}n_{\mathrm{i}}F\left(\frac{d}{L_{\mathrm{a}}}\right)}\right] + \frac{(1-\alpha_{\mathrm{PNP}})I_{\mathrm{f}}L}{\mu_{\mathrm{n}}C_{\mathrm{ox}}W(V_{\mathrm{G}}-V_{\mathrm{T}})}. \tag{5.11}$$

It is noted that the on-state voltage drop is smaller according to the BJT/MOSFET model as compared to that of the p–i–n/MOSFET model since the current flowing through the MOSFET channel is lower in the BJT/MOSFET model by a factor of $1/(1-\alpha_{\mathrm{PNP}})$. Similarly, the saturated current is also modified accordingly

$$I_{\mathrm{C,Sat}} = \frac{1}{(1-\alpha_{\mathrm{PNP}})}\frac{\mu_{\mathrm{n}}C_{\mathrm{ox}}W}{2L}(V_{\mathrm{G}}-V_{\mathrm{T}})^2. \tag{5.12}$$

From the above equation, the small-signal transconductance of the IGBT in the active region of operation can be derived as

$$g_{\mathrm{m}} = \frac{1}{(1-\alpha_{\mathrm{PNP}})}\frac{\mu_{\mathrm{n}}C_{\mathrm{ox}}W}{L}(V_{\mathrm{G}}-V_{\mathrm{T}}). \tag{5.13}$$

5.2.2. Output Resistance

In practice, an IGBT exhibits a finite anode output resistance due to (a) the reduction in effective channel length with higher anode voltage, and (b) the increase in current gain of the p–n–p transistor as its undepleted base width is reduced. Especially, a significant reduction in anode output resistance at higher anode voltage occurs in the symmetrical structure (Baliga, 1986). In the asymmetrical structure, the depletion width in the n^--layer does not change much with further increase in anode voltage because of the relatively high doping density in the n-buffer layer. Therefore, the current gain of the p–n–p transistor remains almost constant for all anode voltages. As such, the output resistance is expected to be more stable for the asymmetrical IGBT structure. The lower current gain of the p–n–p transistor in the asymmetrical IGBT structure due to the presence of the n-buffer layer also helps to keep the output resistance higher. For symmetrical structure, electron irradiation can be employed to increase the anode output resistance by lowering the carrier lifetime and subsequentially, the minority carrier diffusion length and the current gain of the p–n–p transistor can be lowered.

5.3. Switching Characteristics

To switch an IGBT from its forward conduction state to the off-state, it is necessary to reduce the gate charge to zero in order to remove the inversion

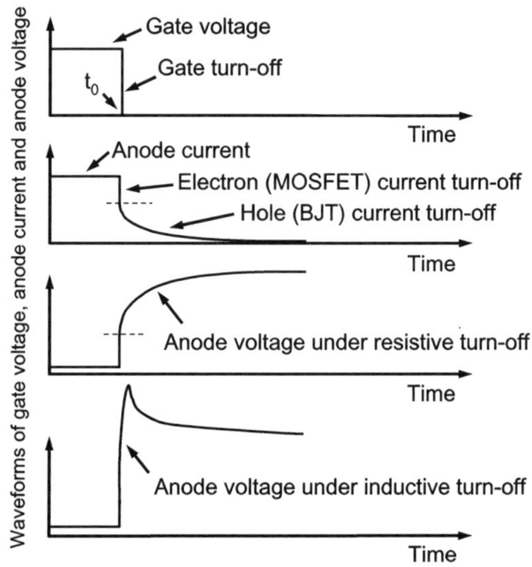

Fig. 5.6. Waveforms during IGBT turn off, namely (from top) gate voltage, anode current, anode voltage at resistive switching, and anode voltage at inductive switching.

channel in the p-base region, so as to cut off the flow of electrons from the n^+-cathode to the n-drift region. After MOS-channel is turned off, the excess carriers confined in the long drift region will then slowly die down through recombination process similar to that of an open-base transistor. A current tail results as shown in Fig. 5.6. As such, the turn-off process occurs in two stages, namely immediate turn-off of the MOSFET-controlled electron current and the slow die-down of the open-base transistor (mainly) hole current.

The turn-off characteristic of an IGBT is governed not only by its internal device properties, but also influenced by the external circuitry which the device is connected to. When the gate voltage drops below the threshold voltage, the anode current reduces abruptly due to the disappearance of the MOS-channel electron current, I_e, as shown in Fig. 5.6. Meanwhile the anode-to-cathode voltage rises suddenly to compensate the reduction in the load voltage by a lower current. Therefore, the device, in particular the drift region, is now the blocking part of the supply voltage. The sudden change in the effective undepleted base width will bring the current gain of the p–n–p transistor to a higher level and this will have a significant influence on the IGBT dynamic latch-up characteristics. Dynamic latch-up phenomenon will be described later. If the device is connected to an inductive load, then the voltage across the IGBT device may be much higher due to the inductive effect caused by the sudden drop of conduction current. Now, the anode current continues to flow through the device during carrier recombination process made by the high concentration minority carriers stored in the n^--drift region during on-state. Due to the relatively long minority carrier lifetime in the n^--drift region, which is necessary to achieve a low on-state conduction voltage, the current tail normally lasts for a long time which is comparable to that of bipolar devices. For this, similar methods of carrier lifetime control, e.g. proton irradiation (Mogro-Campero *et al.*, 1985) can be employed in the drift region near the p^+-anode to reduce the tail time. For the asymmetrical structure, the added n-buffer layer does provide a suitable and quicker recombination site for minority holes near the buffer region and this structure greatly reduces the turn-off tail time.

The magnitude of the abrupt drop in the anode current is determined by the amount of MOSFET electron current and is related to the current gain of the p–n–p transistor, i.e.

$$\Delta I_f = I_e = (1 - \alpha_{PNP}) I_f. \tag{5.14}$$

At the time point t_0, the hole-current flow remains the same (although the p–n–p transistor gain becomes higher at this point) as that during the on-state conduction just before turn-off. The magnitude of the transient corrector current is

$$i_f(t_0) = I_h(t_0) = I_f - \Delta I_f = \alpha_{PNP} I_f. \tag{5.15}$$

After t_0, the anode current decreases exponentially at a rate determined by the high-level minority carrier lifetime. The current waveform can be approximately expressed as

$$i_f(t) \approx I_h(t_0) e^{-\frac{t}{\tau_h}} = \alpha_{PNP} I_f e^{-\frac{t}{\tau_h}}. \tag{5.16}$$

Usually, the turn-off time is defined as the time taken for the anode current to decay to 10% of its on-state value, i.e.

$$t_{off} = \tau_h \ln(10\,\alpha_{PNP}). \tag{5.17}$$

Therefore, the turn-off time increases with the p–n–p transistor current gain and the minority carrier lifetime in the drift region.

5.4. Latch-up

The IGBT device has an inherent four-layer thyristor structure by looking along from the p^+ anode to the n^+ cathode. Once the inherent thyristor latches up, the device will remain in its conduction state only till shut-down of the supply voltage. The MOS-gate control does not have any influence to interrupt

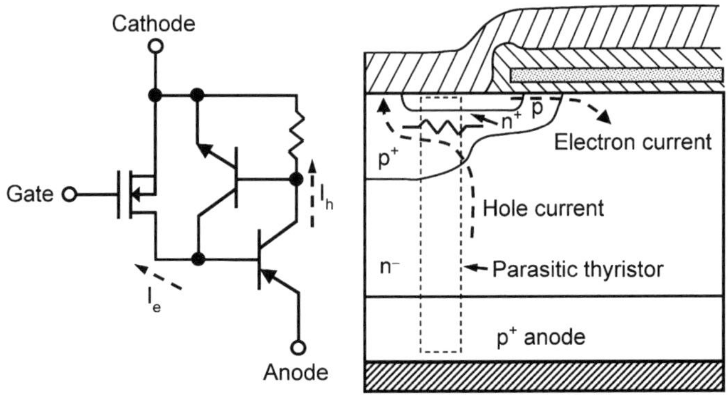

Fig. 5.7. The parasitic thyristor in IGBT structure.

the current conduction in the latched thyristor. Figure 5.7 shows the schematic IGBT model with the parasitic thyristor.

It is understood that, the thyristor latch-up phenomenon occurs when the sum of the upper n–p–n transistor gain and the lower p–n–p transistor gain approaches unity. We have so far studied the p–n–p transistor gain during the device on-state conduction and during turn-off operations. Here, together with n–p–n transistor gain, they will determine the operational limit of maximum on-state current conduction and the voltage–current ranges in turn-off dynamics. The latch-up phenomena can be identified in two types, namely the static latch-up and the dynamic latch-up, according to the operational status when it occurs. This is so because the transistor gain is not a constant value but a function of biasing conditions. During normal operations, the p-base/n^+-emitter junction in the upper n–p–n transistor is moderately forward-biased. This is caused by the hole current passing through the p-base to arrive at the top of cathode contact. The voltage across the junction is normally not sufficiently large to boost up the n–p–n transistor current gain of α_{npn} to a high level in order to trigger the transistor in action. In a sense, during normal IGBT operation, the top n–p–n transistor is in the dormant state, while the bottom transistor is in the active (conduction) state.

When the conduction current increases, both electron and the hole components increase with a similar trend, although not the same proportion due to the moderate variation of α_{PNP} current gain as a function of the current. The voltage across the p-base/n^+-emitter junction is approximately equal to the voltage drop at the p-base under the n^+-emitter layer, and can be expressed as

$$V_{p/n+} = R_p I_h = R_p \alpha_{PNP} I_f. \tag{5.18}$$

To activate the n–p–n transistor, the voltage across the top junction needs to be effectively forward-biased, i.e. $V_{p/n+} \geq 0.7\,\text{V}$. Therefore, the upper limit of steady-state current before latch-up point can be calculated by using Eq. (5.18) as

$$I_{f,SL} = \frac{0.7}{R_p \alpha_{PNP}} = \frac{0.7}{\rho_p \frac{L_E}{W d_p} \alpha_{PNP}}, \tag{5.19}$$

where ρ_p is the resistivity of the p-base as a function of the doping concentration of the region, L_E is the n^+-emitter length, W is the device width in the paper direction, and d_p is the p-base depth under the n^+-emitter. Figure 5.8 gives a detailed schematic to highlight these parameters. It is understood that Eq. (5.19) is an approximate equation with the assumption that the whole hole current will flow through the p-base region underneath the n^+-emitter. In reality, there is certain percentage of the hole current which does not flow underneath the n^+-emitter.

During the initial stage of turn-off process, the electron current diminishes to zero after the MOS channel is removed. However, the hole current remains unchanged, and still flowing through the p-base region to bias the upper junction. Then, it will be a reasonable assumption to say that the n–p–n transistor gain remains similar at the beginning of the turn-off process. But, this shall not be treated as an absolute assurance of safe device operation from latch-up. The reason is, because the p–n–p transistor gain changes from its on-state value to a higher value during the turn-off transition. Equation (5.10) indicates that the p–n–p transistor gain is a function of the undepleted base width, L_B. During on-state, the base width, L_B, is almost the same as the n^--drift region length $2d$, and it stays at the same value for various current levels. During turn-off, the IGBT device sustains part of the supply voltage at and after the initial stage of the period. The high voltage depletes the drift region and shortens the effective base width of the p–n–p transistor. Considering the change in n-drift region

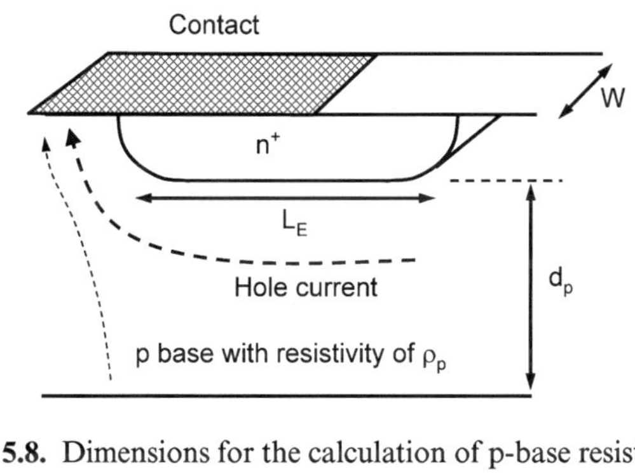

Fig. 5.8. Dimensions for the calculation of p-base resistance.

width for the symmetrical IGBT device, Eq. (5.10) can now be re-written as

$$\alpha_{\mathrm{PNP,DY}} = \frac{1}{\cosh\left(\frac{2d-\sqrt{\frac{2\varepsilon_{\mathrm{S}} V_{\mathrm{DY}}}{qN_{\mathrm{d}}}}}{L_{\mathrm{a}}}\right)}, \tag{5.20}$$

where V_{DY} is the voltage across the IGBT during the turn-off dynamic, N_{d} is the drift region doping, and ε_{S} is the silicon dielectric constant. For the case of asymmetrical structure, the undepleted base width is approximately equal to the n-buffer width. The transient voltage across the device raises the p–n–p transistor gain and this makes the latch-up limit lower and the device more liable to be latched up. In a sense, when an IGBT device operates below the latch-up current limit during steady-state does not imply that it can be turned off successfully. The ratio of static latch-up current limit versus the dynamic one can be derived as

$$\frac{I_{\mathrm{f,SL}}}{I_{\mathrm{f,DL}}} = \frac{1-\alpha_{\mathrm{PNP,SL}}}{1-\alpha_{\mathrm{PNP,DL}}} = \frac{1-\frac{1}{\cosh\left(\frac{2d}{L_{\mathrm{a}}}\right)}}{1-\frac{1}{\cosh\left(\frac{2d-\sqrt{\frac{2\varepsilon_{\mathrm{S}} V_{\mathrm{DL}}}{qN_{\mathrm{d}}}}}{L_{\mathrm{a}}}\right)}} \approx \frac{1-2e^{-\frac{2d}{L_{\mathrm{a}}}}}{1-2e^{-\frac{2d-\sqrt{\frac{2\varepsilon_{\mathrm{S}} V_{\mathrm{DL}}}{qN_{\mathrm{d}}}}}{L_{\mathrm{a}}}}}, \tag{5.21}$$

where V_{DL} is the voltage across the IGBT device at the dynamic latch-up during turn-off. For the inductive load, the voltage overshoot during turn-off is larger than that of the resistive load. It is then expected to have a lower dynamic latch-up current limit. To reduce the inductive overshoot, a gate resistor can be used to adjust the rate of MOS channel turn-off and this will vary the rate of current change at the initial stage of turn-off, i.e. the electron current component. This has a greater influence on the voltage overshoot and therefore on the dynamic latch-up current limit.

Various effects were made to bring up the latch-up current limit, such as introduction of deep p^+ diffusion under the n^+-emitter, adjusting the n^+-emitter length (Baliga *et al.*, 1984), lifetime reduction in the n-base region (Chang *et al.*, 1983; Goodman *et al.*, 1983), retrograde p-well and dual p-well implantation for lateral IGBT structure (Disney and Plummer, 1993), self-aligned trenched cathode contact (Nezar and Mok, 1993), and formation of metal sinker to the substrate (Liang *et al.*, 1999). Among them, the p^+ sinker layer on vertical structure, as shown in Fig. 5.9(a), is commonly used to reduce the p-base resistance under the n^+-emitter to lower the upper n–p–n transistor gain. On SOI substrate, the trenched metal sinker, as shown in Fig. 5.9(b), is a good candidate to raise the latch-up limit and to relieve the thermal stress caused by the buried oxide layer.

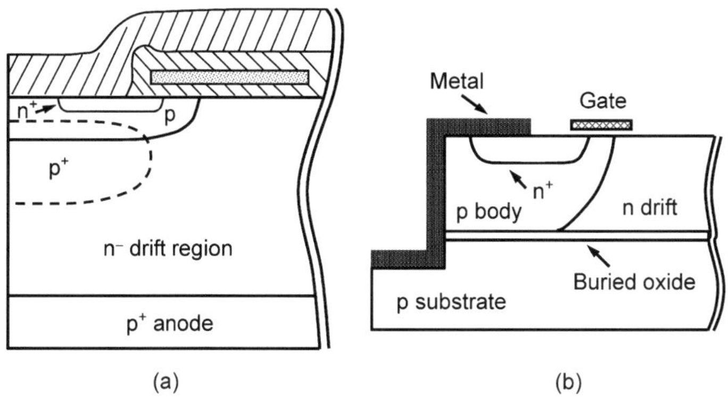

Fig. 5.9. (a) The p^+ implantation under the emitter on episubstrate and (b) the trench metal structure on SOI substrate.

5.5. Temperature Effects

Temperature variations pose an important role in the IGBT performance. Key features, such as the forward conduction voltage, turn-off time, and latch-up current level are all affected (Baliga, 1987). The MOS channel region has a positive temperature coefficient, but the bipolar junction has a negative temperature coefficient. This makes the on-state voltage coefficient of an IGBT device, $V_{ce(on)}$, at low-current region to be negative, e.g. more like a p–i–n diode. And at medium current region, the conduction voltage varies less than that of the MOSFET device as the device temperature increases. At high-current region, the channel resistance will dominate the on-state behavior, and the temperature coefficient on forward voltage becomes more positive.

The IGBT turn-off time is dominant by the open-base transistor tail time made by the recombination process. When temperature goes higher, the minority carrier lifetime is found to be higher and the recombination process takes a longer time to complete. Higher temperature also makes the p–n–p transistor current gain to be higher and subsequently the hole/electron current composition becomes higher as well. A higher percentage of hole current compounds the lifetime effect and makes the bipolar current tail time even longer.

High temperature creates the negative effect on the latch-up current level as well. Basically, both transistor gains increase with temperature and so the p-base resistance. The p-base resistance is influenced by the variations of the intrinsic carrier concentration and the carrier mobility when temperature varies. A higher temperature will raise the intrinsic concentration and lower the carrier mobility. The amount of variation is a function of the background doping concentration. For a medium-high doping region, the increment on the intrinsic concentration plays less dominant role compared to the effect on reduction in carrier mobility. Overall, the p-base resistance becomes higher at a higher temperature. All three variations make the structure more liable to be latched up.

5.6. Series and Parallel Operations

Connecting IGBT devices in series allows high-power/high-voltage semiconductor switches to be realized. In series operation, the current flowing through each IGBT is the same but the anode-to-cathode voltage of each of the IGBTs may be different due to inherent differences in the device parameters. Because of the differences in both device parameters and gate drive circuits, a control for voltage balancing needed for each device to endure an equal magnitude of anode-to-cathode voltage in the transient and steady-state operations, e.g. an active gate drive balancing technique for series-connected IGBTs was reported (Gerster, 1994). For this technique, a closed-loop control is integrated in each gate drive circuit to adjust the anode-to-cathode voltage of each of the series-connected IGBTs to an equal value.

Paralleling of IGBT devices helps to reduce conduction losses and thermal stress (Dapkus, 1994). It should be noted that parallel configuration to take advantage of the lower price of smaller devices should not be attempted without due consideration of technical and gate drive complexity. In general, all IGBTs are operating at the linear region, i.e. at low-voltage conduction. Hence, the two key device parameters, *I–V* slope and starting knee voltage, decide the current sharing under steady-state conduction. However, during transient operations, the current balancing becomes more complicated than simply matching these two key parameters. Other parameters, such as the gate threshold voltage, the input and output capacitances, the stray inductance in each leg, and even the amount of excess charge storage in the drift region become equally important during transient operations.

When two or more IGBT devices are parallel-connected, the anode-to-cathode voltage across each device is hard-wired to be the same during steady state. Thus, for a given load current, one IGBT will carry a higher or lower proportion of current than other mismatched ones if the gate voltage is not properly adjusted. As long as the device current remains below the maximum specified limit on the data sheet, the current imbalance is initially not critically important. However, the device that carries more and more current may exceed the rated junction temperature of about 150°C if the heat sink design does not take the current imbalance into consideration. One factor that can reduce current imbalance is by identifying and matching the temperature coefficients (Baliga, 1985) of IGBT devices used in parallel. Experimental results (Yang and Liang, 1996) show that incremental current variation, $\Delta I_{A,IGBT}/\Delta I_{LOAD}$ (the change in IGBT current versus the change of total load current) of each IGBT depends on the operating current level. This means, an IGBT shares the least portion of current at low-current level may eventually share the highest portion of current at the high-current level. This makes the uncontrolled current matching more difficult. That is, by matching the current sharing at one current level does not guarantee the equal sharing at another current level. A good gate control is thus needed to ensure equal current sharing at all current levels (e.g. Tabata *et al.*, 1998) by *di/dt* control and with closed-loop balancing controller (Hofer *et al.*, 1996).

5.7. Device Operations under Soft Switching

The use of soft-switching inverters in industry has good potential since it allows the effective utilization of faster device switching speed for higher switching frequency at lower loss, lower *dv/dt* for low EMI (electromagnetic interference), and handling of higher power density. Soft-switching operations are formed in zero voltage or zero current conditions by adding additional components in the circuit. Figure 5.10 shows the circuit and the current–voltage waveforms associated with the zero-voltage (ZV) conditions. Under soft-switched conditions, the IGBT produces a voltage spike and a current bump during switching transitions. The phenomena are mainly due to the lag of conductivity modulation in the drift region during turn-on, and the ineffective removal of stored charge

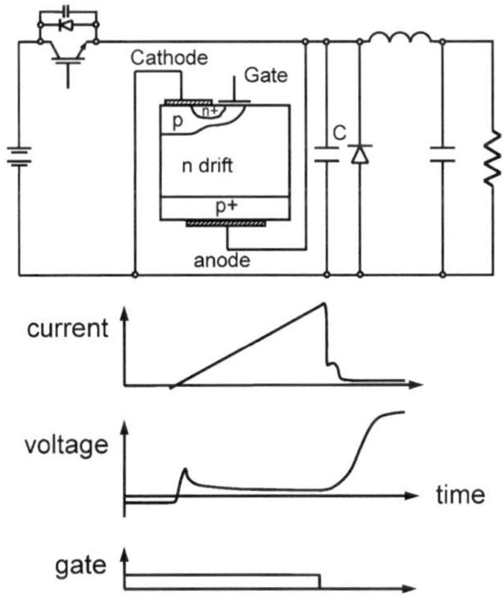

Fig. 5.10. IGBT device under zero-voltage switching and the voltage and current waveforms.

in the same region during turn-off. The phenomena have been observed and analyzed as mentioned in various publications (Kurnia *et al.*, 1992; Widjaja *et al.*, 1994, 1995).

During zero-voltage transition, prior to the current flow into the IGBT anode, a negative voltage of about −1 V clamped by the reverse-conducting diode is applied across the device. The gate signal is applied prior to the current to become forward conduction. Although technically the IGBT device has been turn-on, there is so far no current flowing through the device. When the in-rush current flows through the IGBT device, similar to a p–i–n diode, the forward recovery phenomenon occurs in the lightly doped drift region. As such, the dynamic voltage spike occurs at the initial period of the zero-voltage switching. The spike disappears when the voltage across the n-drift region drops to its normal steady-state value at full conductivity modulation. During turn-off period, the anode voltage increases rather slowly due to the snubber capacitor C in parallel. Without a strong depletion field, the excess carriers in the transistor base region cannot be removed effectively at the initial period. They were swept away by the increased electric field later on when the depletion junction gradually builds up. The final outcome gives a current bump at the tail of the current waveform. Figure 5.11 shows the experimental waveforms of a commercial IGBT under zero-voltage soft-switching conditions.

5.7.1. Dual-Gate IGBT for ZV Soft Switching

The dual-gate bidirectional IGBT as shown in Fig. 5.12 can improve the situation by applying suitable gate signal to the second gate (Yuan, 1999). Prior to the current flow into the IGBT for zero-voltage turn-on transition, the auxiliary gate is turned on for some period of time. A small portion of the diode current will be diverted to flow through the IGBT in the reverse direction. Thus, the hole carriers will be injected from the upper p^+-emitter into the bulk drift region for recombination with the electron carriers flowing through the channel along the auxiliary gate. Then, the resistance of drift region will be dramatically reduced by the existence of hole–electron carriers prior to forward conduction. The relief on the "conductivity modulation lag" reduces the formation of forward voltage spike during turn-on transition. Figure 5.13 gives the simulation data for different *dJ/dt* conditions for both ordinary IGBT (forward current density of 225 A/cm^2) and the dual-gate IGBT (for the same forward current density and the reverse current density of 16 A/cm^2). The results show that, under fast transient condition, the forward voltage spike can be reduced by up to 60%.

Under the zero-voltage turn-off transition, the anode voltage increases rather slowly due to the snubber capacitor across the IGBT device. By switching on the auxiliary gate, the inversion layer is formed to short the junction J1

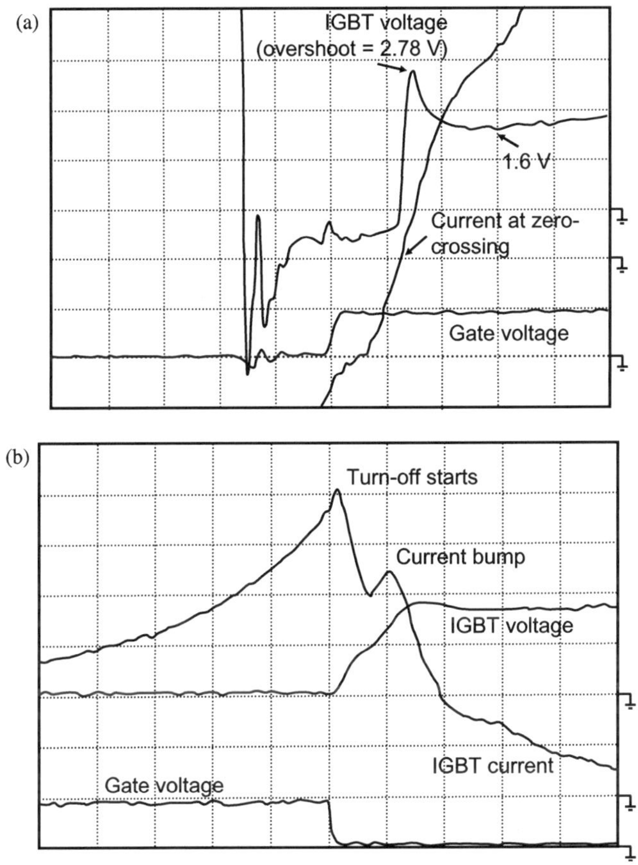

Fig. 5.11. (a) Measured forward voltage spike during turn on: time base (1 μs/div), gate voltage (20 V/div), IGBT current (2 A/div), and IGBT voltage (1 V/div). (b) Measured current bump during turn off: time base (1 μs/div), gate voltage (20 V/div), IGBT current (1 A/div), and IGBT voltage (20 V/div).

so as to reduce the hole injection into the drift region during the turn-off period. The accumulation of excess hole carriers can be greatly relieved. As a result, a shorter turn-off time can be achieved. Simulation results are shown in Fig. 5.14. When the ordinary IGBT turns off with snubber capacitor, a current bump occurs as expected after the MOS-channel was cut off. It is found that, without lifetime control ($\tau_{HL} = 2.18\,\mu$s), the dual-gate structure has a better performance than that of the ordinary structure with lifetime control ($\tau_{HL} = 0.95\,\mu$s).

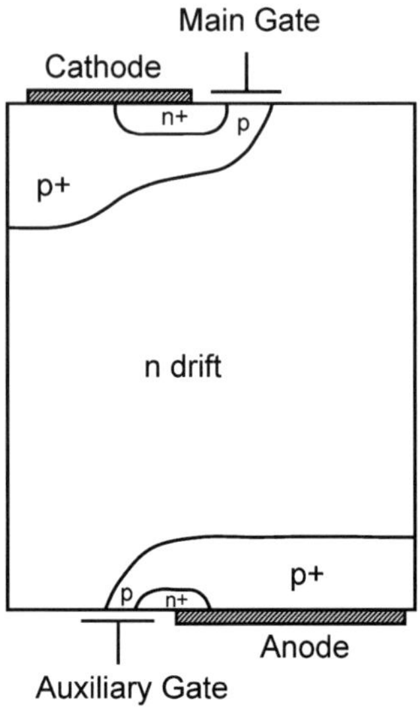

Fig. 5.12. Dual-gate IGBT device for soft-switching applications.

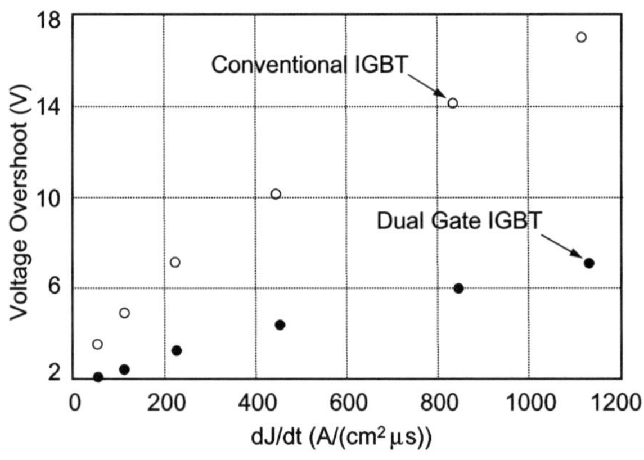

Fig. 5.13. Simulated forward voltage overshoots for the conventional and dual-gate IGBTs of the same drift region.

5.8. Lateral IGBT Structure

Much interest in development is focused on the integrated power modules using (Bi)CMOS and DMOS process technologies. The integration feature requires the power IGBT structure to be formed in the lateral way, so that other circuit devices can also be fabricated on the same substrate. The basic lateral

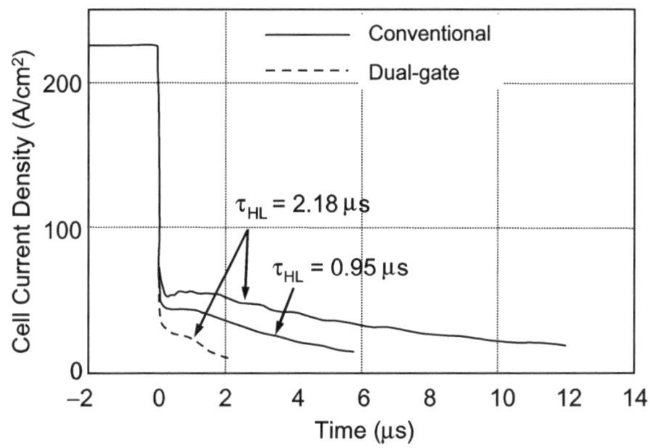

Fig. 5.14. Simulated turn off times for conventional and dual-gate IGBTs of the same drift region.

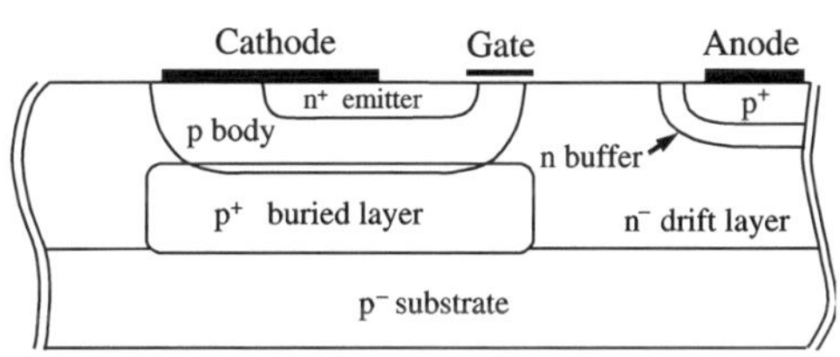

Fig. 5.15. Lateral IGBT structure on p-substrate.

IGBT structure on p bulk substrate is shown in Fig. 5.15, which consists of a gate channel region, p–n–p vertical bipolar transistor, and a p–n–p lateral bipolar transistor. The majority of the anode current flows laterally through the n-drift region in parallel with the p-substrate to arrive at the channel/p-body region then to cathode contact. Some of the anode current flows in the substrate and may be collected by the substrate contact or other part of the circuit elements. The n-buffer layer next to the anode p^+-emitter is placed to raise the breakdown voltage as similar to that of the punchthrough vertical structure. The p^+ buried layer creates a low-resistance path between the p-body and the p-substrate. This increases the current conductivity from the substrate to the body region and thus increases the lateral current ratio.

The lateral current ratio is defined as the lateral current collected by the cathode versus the anode current. When the IGBT is fabricated on the SOI wafer, the lateral current ratio is then 100%. A good lateral IGBT device is said to achieve a high lateral current ratio, proper forward and reverse breakdown voltages, a low on-state conduction voltage, and the turn-off time below a few hundred nanoseconds. Liang and Hor (1995) showed the variations of parameters on drift region length, drift region doping, buffer layer doping, and inclusion of buried layer to the influence on device performance. On bulk wafer, the lateral current ratio can be increased by including the buried layer and shortening the drift region length. The forward breakdown voltage is affected by the drift region concentration as similar to that of the vertical structure. The reverse breakdown voltage occurs at the buffer region and affected by its concentration and length. The turn-off time is a function of drift region concentration, drift region length, and buffer layer concentration. By including a large buried layer, higher drift layer doping, shorter drift layer length, and higher buffer layer doping will shorten the turn-off time. However, a higher buffer layer concentration will cause a higher on-state voltage drop. Similar to the vertical IGBT structure, lifetime control can also be made in the n^--drift region near the n-buffer to speed up the turn-off transient.

5.9. Integrated Current Sensor

A current sensor in power electronic circuits is usually needed in two aspects, namely to provide the precise current information for feedback control (Chow *et al.*, 1992; Manduteanu, 1993; Shen *et al.*, 1994; Liang and Hor 1995; Lang *et al.*, 1998), and otherwise to provide critical overcurrent information for protection purpose (Seki *et al.*, 1994; Robb *et al.*, 1994; Shimizu *et al.*, 1994). For the latter, the sensing ratio on overcurrent protection can be much coarse compared to the former one which needs to be kept almost constant at all current levels. In some cases, due to MOS-channel saturation at the high-current region, the IGBT anode voltage can be used to represent the overcurrent information for protection purpose (Luo *et al.*, 2000). The advantage of having an integrated current sensor on the same die is to provide the direct sensing information with less parasitic involved in comparison to the usage of external current sensors, not to mention that the external sensing circuits can be complex and bulky. Preferably, the integrated sensor is fabricated together during the process of IGBT fabrication.

The accurate current sensor suitable for smart power integration in the lateral insulated-gate bipolar transistor (LIGBT) structure is described in this section. The same sensor structure can also be applied for other types of MOS-controlled bipolar structures. For MOS-bipolar devices, the electron and hole current components need to be sensed individually in order to precisely predict the amount of device current at all operating current levels. The rationale, in brief here, is that the electron/hole current composite ratio, i.e. electron current versus hole current, does not keep to a constant level when the device current level varies. Other more stringent requirements for a constant sensing ratio are on the temperature variations and in transient operation.

The integrated sensor should not affect the normal operation of power devices. This requires the current taken by the sensor to be very small compared to the main device current. Besides, the sensing current should be maintained possibly at a constant ratio with minimum variation in respect to the main current under large load variations, e.g. during both the steady-state and transient operations, gate voltage variations, and over the range of operating temperature.

The basic principle in the design of a sensor is simple and it requires a parallel current path. Using the principle of the parallel resistors shown in Fig. 5.16, the sensor current can be made to be a small and fixed ratio to the main current by properly controlling the resistive ratio between R_s and R_m. Therefore, the sensing ratio will be $I_s/I_o = R_m/(R_s + R_m)$. By putting the value of R_s to be much greater than that of R_m, the sensing ratio can be made to be very small.

A sensor structure was designed (Liang *et al.*, 1998) to enable sensing of the electron current together with the hole current in the correct proportion. In this structure, two additional regions and contacts are added to the standard LIGBT structure at the cathode area. The device has the cathode split into two contacts with one sensor contact inserted in between. The device structure is shown in Fig. 5.17 and its peak doping levels, junction/profile depths, and contact positions are given in Table 5.1. The anode and n-buffer layers are not shown in the figure since the main focus is on the sensor structure in the cathode area.

The electron current sensor layer (sensor-1 contact) and the hole current sensor layer (sensor-2 contact) are routed and electrically connected together. The doping concentration of the n sensor layer can be determined through process and device simulations, such that the electron current sensing ratio is of

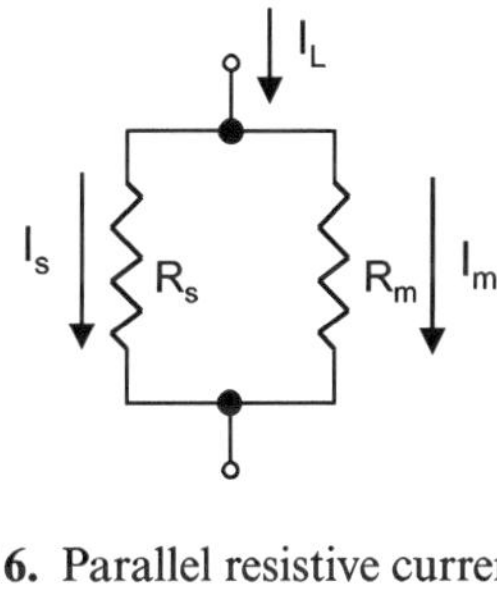

Fig. 5.16. Parallel resistive current paths.

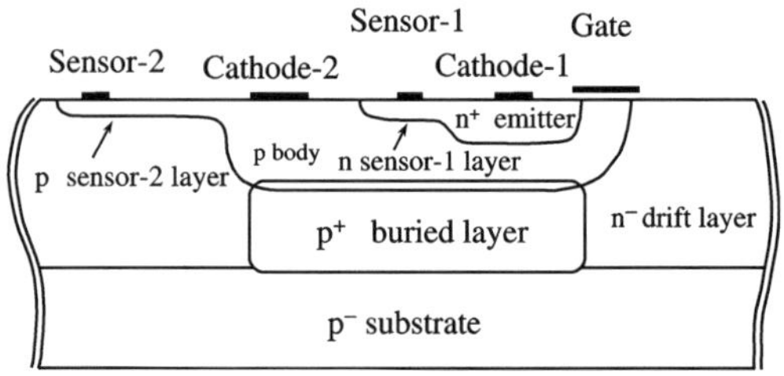

Fig. 5.17. Electron and hole current sensors integrated in the lateral IGBT structure (anode and n-buffer are not shown).

Table 5.1. LIGBT (with current sensor) layer dimensions and doping profiles.

Layer	Lateral dimension (μm)	Peak doping concentration (cm^{-3})	Depth (μm)
n^+-emitter	6.0	1×10^{20}	1.0
p-body	15.2	1×10^{18}	5.0
p^+ buried	12.0	2×10^{18}	Between 8.0 and 12.0
n^--drift	45.0	1×10^{15}	10.0
p^+ anode	5.0	2×10^{19}	2.0
n-buffer	7.0	4×10^{16}	5.0
p^- substrate	—	2×10^{14}	40.0
p sensor	7.3	2×10^{15}	1.0
n sensor	3.5	5×10^{18}	1.5

the same proportion as the hole sensing ratio. This criterion is a very important and is the prerequisite to ensure the constant sensing ratio over the whole range of current and gate voltage variations. The two cathode contacts are also routed together. Both the cathode and the sensor contacts are grounded externally. Referring to Fig. 5.17, it must be noted that the position and length of cathode-1 contact are important factors in ensuring sufficient potential difference within the n^+ layer between cathode-1 and sensor-1 contacts. This potential difference, typically about a few hundred millivolts, would result in some electron current flowing into the sensor-1 contact.

The design gives a good performance in sensing ratio when the cathode current and gate voltage are varied. However, it suffers by a large variation in sensing ratio when the operating temperature varies. For the doping levels in Table 5.1, the electron sensing ratio varies in a range of as large as ±40%

between 250 K and 450 K. For the hole sensing ratio, it varies in a smaller scale of ±9%. This makes the variation on the overall sensing current to be rather large.

Here, the physics behind the variation and the solution to remedy the problem will be described. Intuitively, the variation in sensing ratio is caused by the variation in (parallel) resistance ratio. And, there are two main factors that cause the resistance ratio variation with temperature, namely the modulation of carrier mobility and the change of intrinsic carrier concentration. These two factors jointly, but not equally, affect the semiconductor resistivity of the conduction paths taken by the electron and hole currents. Thus, the sensor ratio varies. These two factors have opposite temperature dependency, interacting to minimize the spread of the sensor ratio variation. The hole current sensing ratio was found to be relatively more constant as the counteracting becomes more effective. However, the situation was not so for electron current sensing ratio.

The carrier mobility in the cathode region is a function of doping concentration, types of dopants, and temperature as described in (Masetti *et al.*, 1983). Although the magnitude of variation is a function of the dopant concentration, the general trend is that the mobility drops when the temperature increases. Therefore, the higher the temperature is, the higher the resistivity will be. The second factor is on the intrinsic carrier concentration. When the temperature rises, the intrinsic carrier concentration will go up. This phenomenon will counteract the mobility degradation to lower the resistivity. However, this phenomenon is less obvious in the n^+-emitter/sensor region due to the high background doping concentration. It is more obvious in the p-body/sensor region for the reason of lower doping concentration. This explains the situation well on why the hole current sensing ratio has less variation with temperature compared to the electron current sensing ratio.

An innovative approach to solve this problem of sensing ratio variation was found through analyzing the current flow pattern of electron carriers. And, the approach of having the "flat-top" doping profile in the n^+-emitter region is introduced to minimize the difference in doping concentration between the paths taken by the sensor and cathode currents. When looking at the distribution of doping concentration along vertical cut from wafer surface to the substrate, as shown in Fig. 5.18, the "flat-top" profile gives a constant doping concentration till certain depth and then starts to fall off. This approach replaces the original Gaussian doping profile associated with a typical diffusion process. In comparison, the profile now has a more uniform doping concentration for the entire cathode region and a steep transition in doping concentration at the internal junction. Therefore, the cathode current and the sensor current would flow over a region of the same doping concentration.

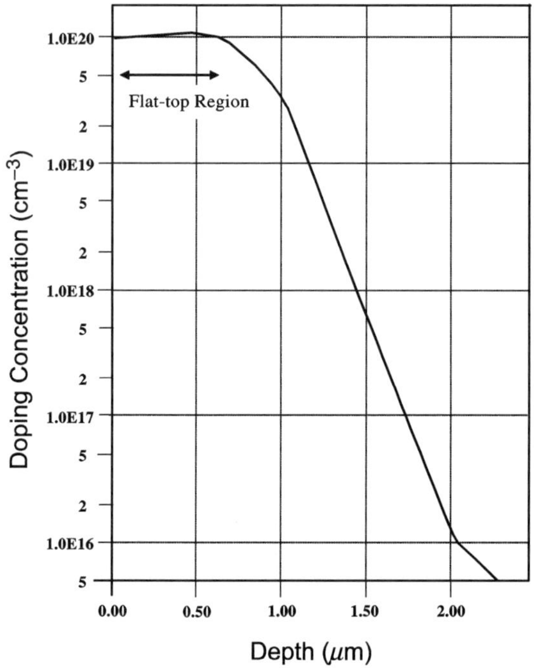

Fig. 5.18. The doping profile along the vertical cut showing the flat-top portion.

The result on applying this technique gives a significant improvement in the variation of the electron current ratio against temperature variation as the resistivity of the main current path will vary at the same manner as that of the sensor current path. With this approach, the variation of the total sensing ratio against temperature can be reduced to about ±9% between 250 K and 450 K, as obtained by the analysis results.

5.9.1. Fabrication Aspects

The numerical results indicate a clear need for a flat-top profile at n^+-cathode region to reduce the variation of the electron current ratio against temperature. To create the flat-top profile, a double-implantation, single-anneal method is needed. For both the first and second implantation, the dosage of phosphorus impurity implanted into the silicon is similar. They differ only in the acceleration energy with which the impurity ions are injected into the silicon wafer. Adjustments should be made to produce a set of implantation energies that will create a flat-top doping profile.

However, after the double-implantation single-anneal process, it was found to have the persistent dip of concentration profile at the surface of the n^+-emitter region although the rest of the region whose concentration beneath the surface is flat. Therefore, an oxidation step was added after the annealing and drive-in step to rectify this problem. The doping impurity (initially present in the silicon) redistributes at the interface due to the segregation effect (Sze, 1985) to lift n^+ concentration up and remove the dip near the surface. The flat-top profile of the n^+-emitter region is shown in Fig. 5.18 from the process simulator by applying the above steps at cathode region. As to the sensor region, by choosing a shallower doping profile, process simulation shows that it was possible to obtain a good flat top profile using just a single-implantation process.

Detailed fabrication process steps are described here. The process started with p-type substrate with ⟨100⟩ wafer of uniform doping of $2 \times 10^{14}\,\text{cm}^{-3}$. The p-type buried layer was formed by Boron implant with energy 200 keV and dose $8 \times 10^{14}\,\text{cm}^{-2}$ followed by a 10 minutes post-implant anneal in nitrogen ambient at 1000°C. Then a 5 μm epitaxial-layer was grown with phosphorus doping of $1 \times 10^{15}\,\text{cm}^{-3}$. This layer basically forms the low-doped n-drift layer. The p-body region was then formed by boron implant with energy 200 keV and dose $7 \times 10^{14}\,\text{cm}^{-2}$ followed by a 300 minutes post-implant anneal/drive-in in nitrogen ambient at 1050°C. The n-buffer region was then formed by phosphorus implant with energy 100 keV and dose $2 \times 10^{14}\,\text{cm}^{-2}$ followed by a 220 minutes post-implant anneal/drive-in in nitrogen ambient at 1050°C. The p-anode region was then formed by boron implant with energy 50 keV and dose $3 \times 10^{15}\,\text{cm}^{-2}$ followed by a 10 minutes post-implant anneal in nitrogen ambient at 1000°C. The p-sensor region was then formed by boron implant with energy 120 keV and dose $2 \times 10^{11}\,\text{cm}^{-2}$ followed by a 120 minutes post-implant anneal/drive-in in nitrogen ambient at 1000°C. Afterwards the gate oxide is grown in dry oxygen ambient for 20 minutes at 1000°C. Finally, a double implanted n^+-emitter and shallow n-sensor regions were formed followed by metallization and passivation. Figure 5.19 shows the simulated process profiles which were created by the series of process steps discussed here.

5.9.2. Performances

The wafer micrograph in Fig. 5.20 shows the fabricated lateral IGBT and the integrated sensor contact positions. The dc performance was tested with various anode bias, gate bias, and device temperature. A special heating platform was set up to maintain the wafer at a predefined elevated temperature. Measurements were made for the linear region (function as a power switch) as well as the saturation region (function as an amplifier).

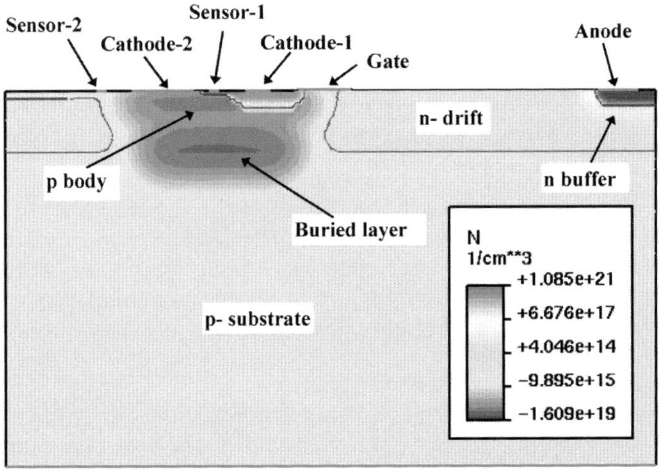

Fig. 5.19. Device profile obtained by process simulation.

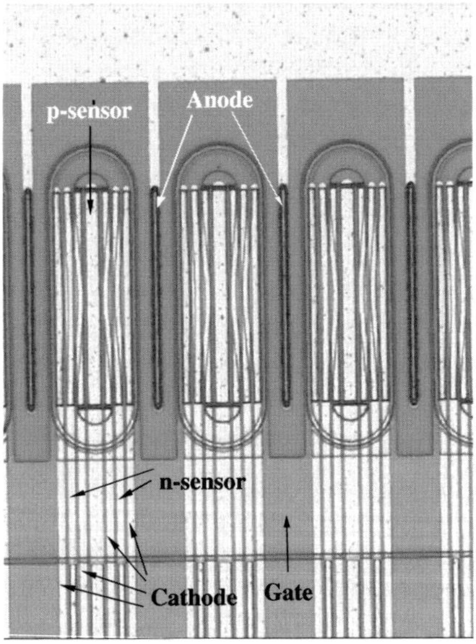

Fig. 5.20. The micrography of a fabricated lateral IGBT with integrated current sensor.

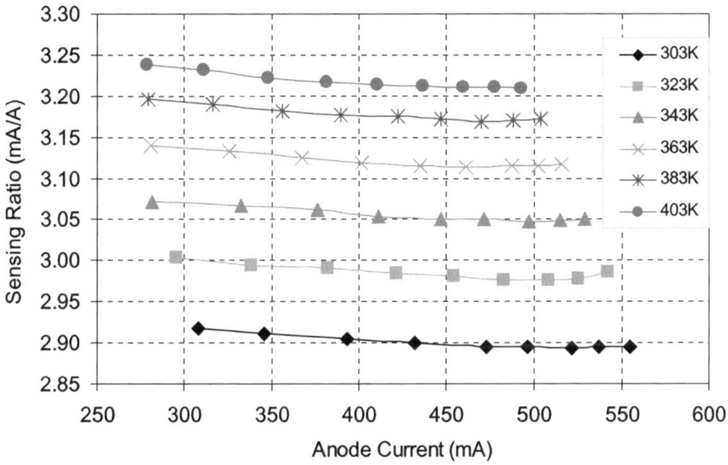

Fig. 5.21. Variations of current sensing ratio against anode current at various temperatures (under linear operating mode, as a switch) with a constant gate bias.

5.9.2.1. *DC Measurement in the Linear Operating Region (Switch Operation)*

Figure 5.21 shows the measured values of sensing ratio over a range of temperatures with anode current values between 250 mA and 600 mA per cell group. The curves show that the sensor current ratio remains fairly constant with the anode current density and has a stronger dependence on the temperature variations. It is observed that the sensing ratio becomes less dependent on temperature variation at a higher temperature. A total variation of around ±5.2% from the average sensing ratio is observed when the device is at temperature range between 303 K and 403 K.

The gate bias sensitivity on the current sensing ratio is shown in Fig. 5.22. The values of sensor ratio change in a peak-to-peak range from 1.96% to 2.41% (or within ±1.21%) with the gate voltage. The sensing ratio changes by ±5.22% for temperature from 303 K to 403 K as seen. A similar characteristic is observed that the temperature variation dominates the current ratio variation and the effect becomes weaker toward the higher temperature region.

5.9.2.2. *DC Measurement in the Saturation Operating Region*

Similar dc tests were conducted with the device operating in the saturation region (as an amplifier device). The device operating in saturation region conducts a higher current and at a higher anode voltage compared to the linear region. The device then has a higher loss, and therefore, a higher thermal stress.

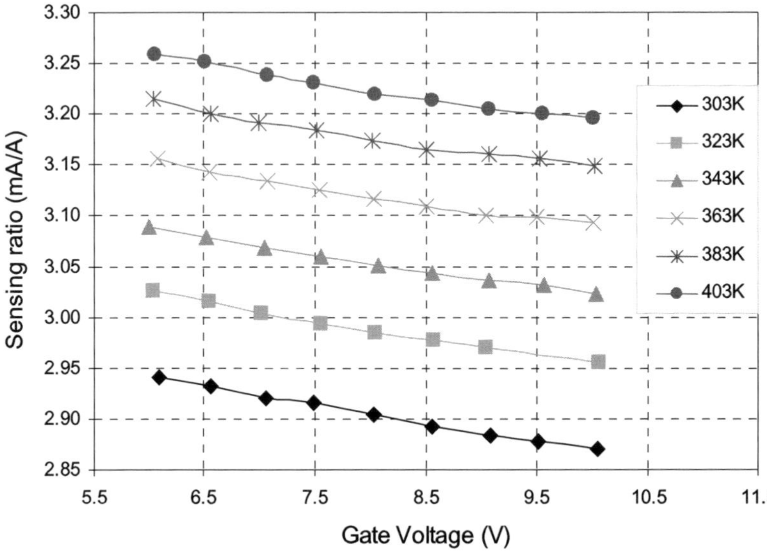

Fig. 5.22. Variations of current sensing ratio against gate voltage at various temperatures (linear operating region, as a switch) with a constant anode current.

Furthermore, the current increases with the high temperature under the same electrical biasing condition. When the compound current was sufficiently large, the thyristor latch-up effect was triggered.

The measurement results for ramping anode voltage are plotted in Fig. 5.23, while the results for ramping gate voltage measurement are plotted in Fig. 5.24. Similar to the case of linear region, the temperature effect dominates the current sensing ratio variation. The total sensing ratio varies within $\pm 0.85\%$ with respect to the anode current variation, and within $\pm 1.73\%$ with respect to the gate voltage variation.

5.9.2.3. *Transient Response*

Transient response was observed during the switching of the lateral device from on-state to off-state. The waveforms are shown in Fig. 5.25. The turn-off time, measured from 100% to 10% of the load current is around 20 μs. The sensor ratio tracks the transient responses quite well.

5.10. Safe Operating Area

The safe operating area (SOA) is an important aspect for operations in steady-state conduction, blocking, and switching. The shape of SOA for an

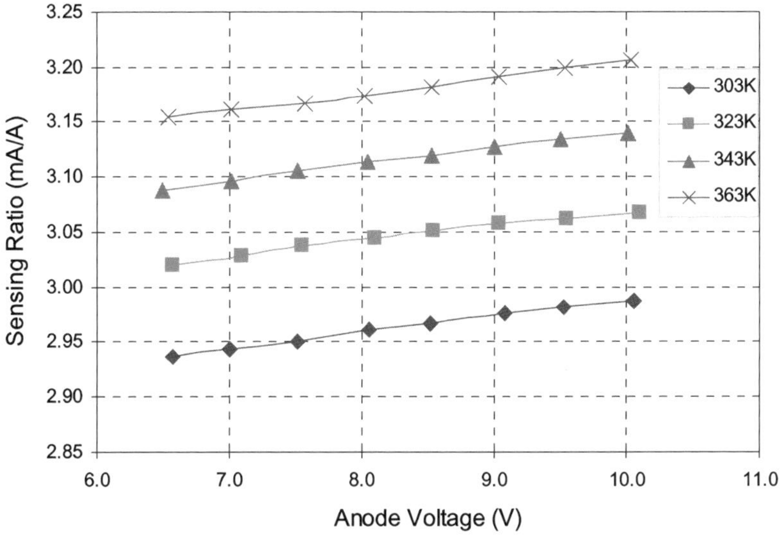

Fig. 5.23. Variations of current sensing ratio against anode voltage at various temperatures (saturation operating region, as an amplifier device).

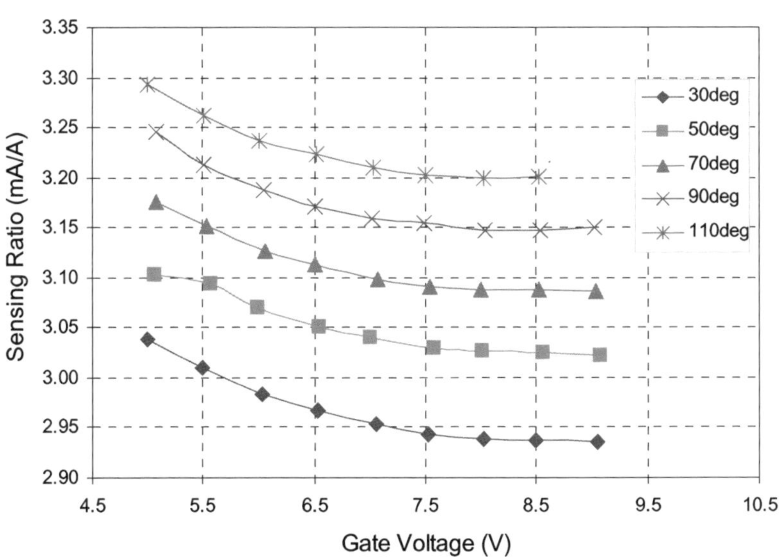

Fig. 5.24. Variations of current sensing ratio against gate voltage at various temperatures (saturation operating region, as amplifier device).

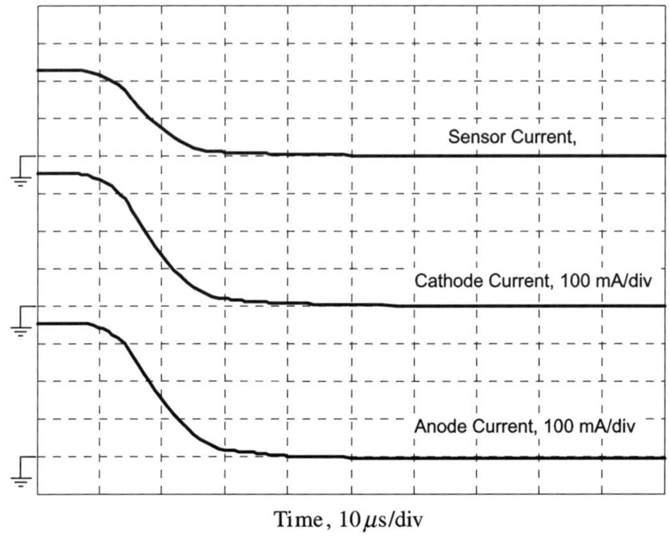

Fig. 5.25. Transient response waveforms during device turn off.

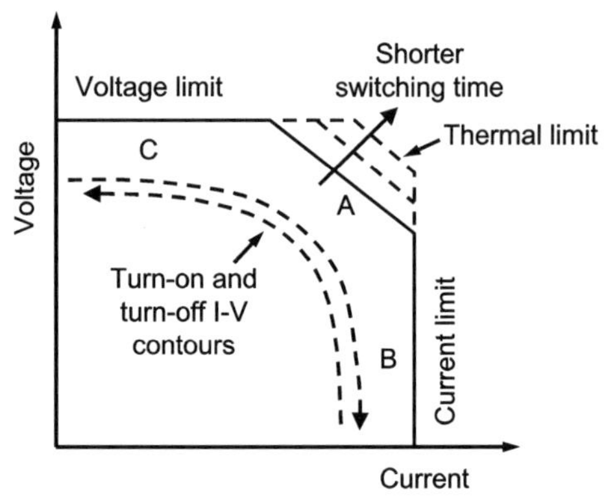

Fig. 5.26. The device SOA region.

IGBT is similar to that of a typical power MOSFET and is shown in Fig. 5.26. The corner shape of the safe operating area depends on device switching speed. For different switching speeds, the forward operating area is thermally limited and the SOA varies as shown by the boundary line A. This is the boundary where both the anode current and voltage are simultaneously large, i.e. the high thermal stress region. Boundary line B is limited by the maximum allowable anode current of the IGBT which is below the latch-up limit. This limit is usually reached at higher gate voltage and especially at high operating temperature. Boundary line C is set by the maximum anode-to-cathode breakdown voltage of the IGBT. The voltage is determined by the open-base p–n–p transistor, i.e. BV_{CBO}. Practically, the SOA of an IGBT is more robust than that of the power MOSFET.

5.11. Overcurrent Protection

The overcurrent protection scheme is usually necessary to built a part of the function in power integrated circuits. The protection scheme needs to distinguish different types of fault conditions, e.g. moderate over-load or sever short-circuit, and to react accordingly based on the device SOA limit. At the same time, the protection circuit should be concise and suitable for integration. The behaviors of the IGBT device under various fault conditions were studied in the literature and useful protection schemes were proposed (Biswas *et al.*, 1991; Chokhawala *et al.*, 1995; Valentine, 1995). The protection in practice should cover a wide range of overcurrent condition. Under moderate overload, the device current is higher than its continuous current rating (I_{CCR}) but remains below the maximum pulse current level (I_{MPC}). The failure mechanism for IGBT device under the moderate overload is thermal runaway. Under the short-circuit condition, the current is much larger and usually exceeds the I_{MPC} rating. Under such a high current, the IGBT device needs to be turned off immediately or less than a few microseconds to avoid fatal destruction.

Usually, a time-delay control is implemented for overcurrent protection to avoid unnecessary shutdown caused by transient current surges. In practice, it ranges from a few microseconds for the catastrophic fault to a few milliseconds for the moderate overload condition. The variation of delay time is a nonlinear function of load current as shown in Fig. 5.27. The delay-time limit is derived from the device SOA. In general, it covers four major zones, namely the thermal limitation, the maximum rating on pulsed current, the short-circuit withstanding capability, and the maximum current level limited by the transistor gain (I_{MAX}). At room temperature, the boundary on thermal limitation at certain current level is determined by the time taken to reach 450 K junction temperature under single-pulse and normal-gate voltage condition. Depends on the gate voltage, the maximum short-circuit current level varies (Shen, 1996).

A resistor–capacitor first-order circuit can be used and to be charged by the anode voltage to serve as a simple timer for turn-off delay time. The voltage across the capacitor can be expressed as

$$V_C = KI_A(1 - e^{-\frac{t}{RC}}), \tag{5.22}$$

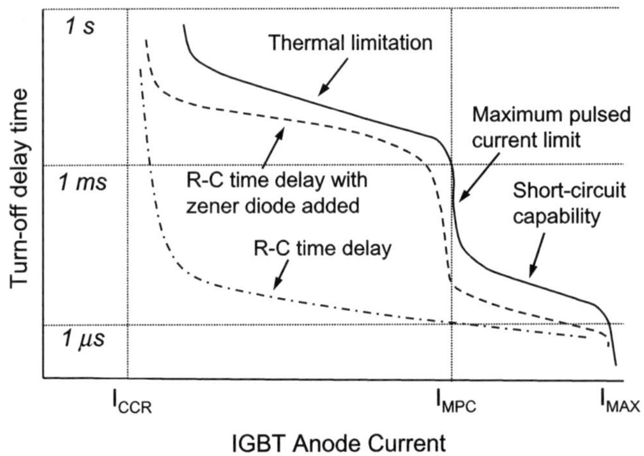

Fig. 5.27. The turn-off delay time under different overcurrent conditions.

where R and C are the resistance and capacitance and K is the conversion gain factor between the anode voltage and anode current, i.e. $KI_A = V_A$. Under overload condition, the IGBT device works in its saturation region, e.g. at high voltage and high current. The value K can approximately be assumed as a constant. If a voltage V_p is set as the reference voltage for the shutdown threshold, that is, when V_C is equal or higher than V_p, then the IGBT gate drive will be removed. The delay time can be found as

$$t_{\text{to-turn-off}} = RC \ln \frac{KI_A}{KI_A - V_p}. \tag{5.23}$$

The curve of time delay using simple R–C timer is shown in Fig. 5.27. It can be seen that, although the delay time is controlled within the SOA transient limit, it has a far distance away from the actual thermal limitation boundary. This is because the time constant needs to be kept small to meet the critical limit on short-circuit withstanding limit. In this case, the allowable overload capability of the device is not fully utilized. One way to make the full use of the SOA transient limit is to insert a zener diode in parallel with the resistor to provide a voltage-dependent two-stage time constant. This is shown by (Z // R2) and C1 in the circuit of Fig. 5.28. When the anode voltage, which is approximately proportional to the anode current, goes up and the voltage across R2 goes above the zener diode breakdown voltage, V_Z, the zener diode breaks down and the capacitor C1 will be charged up at a much smaller time constant. This is because the zener diode

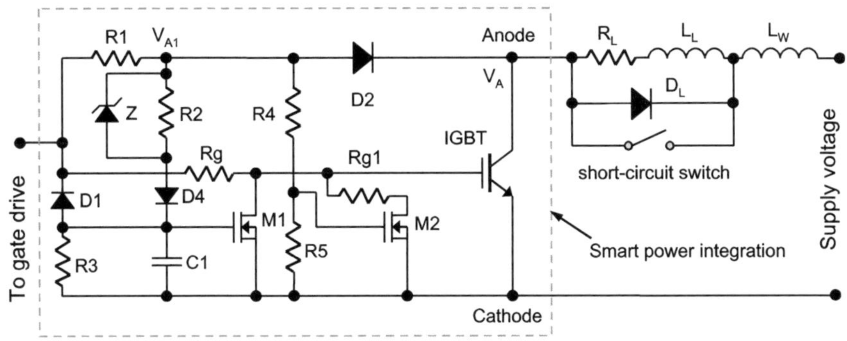

Fig. 5.28. Overcurrent protection circuit.

has a smaller dynamic resistance after breakdown. The delay time can now be expressed as:

$$t_{\text{to-turn-off}} = \begin{cases} (R_Z//R_2//R_3)C_1 \ln\left(\dfrac{KI_A - V_Z}{KI_A - V_Z - V_p}\right) & \text{for } I_A > I_{MPC}; \\ (R_2//R_3)C_1 \ln\left(\dfrac{KI_A}{KI_A - V_p}\right) & \text{for } I_{MPC} > I_A > I_{CCR}; \\ \text{Infinitive} & \text{for } I_{CCR} > I_A. \end{cases} \quad (5.24)$$

The curve with zener diode added is also shown in Fig. 5.27 which has a closer match to the profile of SOA transient boundary.

For new generation IGBTs, they have a higher I_A/V_g gain for high conductivity and low anode voltage drop. But, they also have a lower short-circuit withstanding time (Otsuki *et al.*, 1993; Seki *et al.*, 1994; Iwamuro *et al.*, 1995). The short-circuit withstanding time is typically below 5 μs. Although the delay time can be made very short by reducing the value of capacitance C1 to a minimum, in practice, it may suffer from parasitics and noise interference. Therefore, it would not be a reliable configuration for the R–C network to provide a delay time in the sub-microsecond zone. Also, the delay time should be relatively longer than the switching transition time to ensure its normal operation. One solution to prolong the allowable time before turn-off is to lower the gate voltage as the short-circuit withstanding capability is also related to the gate voltage.

The protection circuit in Fig. 5.28 has incorporate the feature to lower the gate voltage when a direct short-circuit occurs. When this happens, the voltage across the IGBT goes nearly the same level as the supply voltage. The diode

D2 provides the isolation and keeps voltage V_{A1} to a constant level which is related to the magnitude of gate drive voltage. Resistors R4 and R5 divide the voltage and turn the MOSFET M2 on. When M2 is turned on, it pulls Rgl to ground and lowers the voltage to the IGBT gate contact. A lower gate bias will then prolong the short-circuit withstanding capability. At the same time, capacitor Cl keeps being charged up via zener diode and R2. When the capacitor voltage reaches the threshold value V_p, MOSFET M1 turns on and the IGBT gate contact is grounded. The device is now turned off.

The operation can be quantitatively analyzed. Under the normal operating condition, the anode voltage is low and diode D2 is at its on-state. The voltage V_{A1} is

$$V_{A1} = V_A + V_{D2}, \tag{5.25}$$

$$V_{C1} = \frac{R_3}{R_2 + R_3}(V_{A1} - V_{D4}) < V_{th,M1}, \tag{5.26}$$

where V_{D2} is the voltage across diode D2, V_{D4} is the voltage across diode D4, $V_{th,M1}$ is the threshold voltage to turn on MOSFET M1. When overcurrent occurs, V_A will increase and this brings V_{A1} to be higher as well. As a result, capacitor Cl is charged to a higher voltage. The time constant is

$$\tau_1 = (R_2//R_3)C_1 = \frac{R_2 R_3}{R_2 + R_3} C_1. \tag{5.27}$$

When V_{C1} reaches the threshold voltage $V_{th,M1}$, M1 turns on to pull down the voltage to the gate contact. The IGBT is then turned off.

A zener diode is used to distinguish between the moderate overload and the catastrophic conditions. When V_A rises above $V_Z + V_{th,M1}$, the device will be regarded as in the short-circuit operation, and the turn-off delay time must be shortened within its withstanding capability. The time constant is now

$$\tau_2 = (R_z//R_2//R_3)C_1 \approx R_z C_1, \tag{5.28}$$

which is much lower than τ_1. As mentioned, M2, R4, R5, and Rgl are added to provide the current-limiting feature by reducing the voltage to the gate contact to prolong the short-circuit withstanding time. When the short-circuit occurs, diode D2 is turned off, and the voltage to the gate contact of MOSFET M2 is

$$V_{gate,M2} = \frac{R_5}{R_1 + R_4 + R_5} V_{gate-drive-input} \geq V_{th,M2}, \tag{5.29}$$

where $V_{th,M2}$ is the threshold voltage of MOSFET M2. As the IGBT gate voltage is reduced, the fault current can be limited to a lower value. The response time for the current-limiting effect to appear is determined by the time constant of

$$\tau_3 = \frac{(R_1 + R_4)R_5}{R_1 + R_4 + R_5} C_{g,M2}, \tag{5.30}$$

where $C_{g,M2}$ is the gate input capacitance of MOSFET M2. This time constant is normally very short and below 1 μs. While the short-circuit withstanding capability is extended, the time delay circuit (with zener diode breakdown) functions concurrently to turn off the gate voltage to IGBT in the similar manner as described earlier, but with a more comfortable delay-time period. The turn-off delay-time is described as shown in Eq. (5.24). Figure 5.29 shows the measured circuit performance to protect an IGBT rated around 15 A. Based on the SOA limit at gate voltage of 15 V, the time constant of τ_1 is set to be 7ms and τ_2 is set to be 16 μs. The triangular data point is measured at a reduced gate voltage when the protection circuit lowers it from 15 V to 10 V to prolong the short-circuit withstanding capability.

The protection circuit can be fabricated together with the IGBT device to form the smart-power integration on the same silicon die, e.g. Fig. 5.30 showing the integration of key components and Fig. 5.31 showing part of the silicon die. The chip was fabricated with a total of eight masks on a single-side n^- (5×10^{14} cm^{-3})$\langle 100 \rangle$ epi-wafer with zener diode as an external component. Resistors are not shown and they can be formed by polysilicon deposition. Short-circuit faults have been applied to verify the effectiveness of the proposed protection scheme. Two types of short-circuit faults, namely the "hard switch fault" and the "fault under load", can occur during the operation of power electronic circuits. The "hard switch fault" condition is described as that the short-circuit fault exists before the IGBT is turned on, that is, the IGBT will be turned on

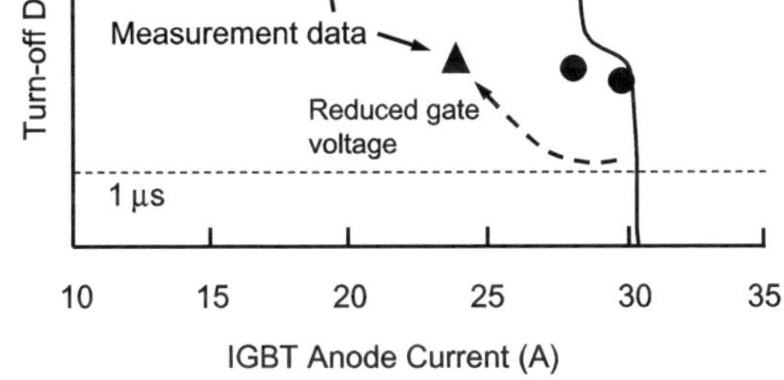

Fig. 5.29. Measured turn-off delay time of the protection circuit of Fig. 5.25.

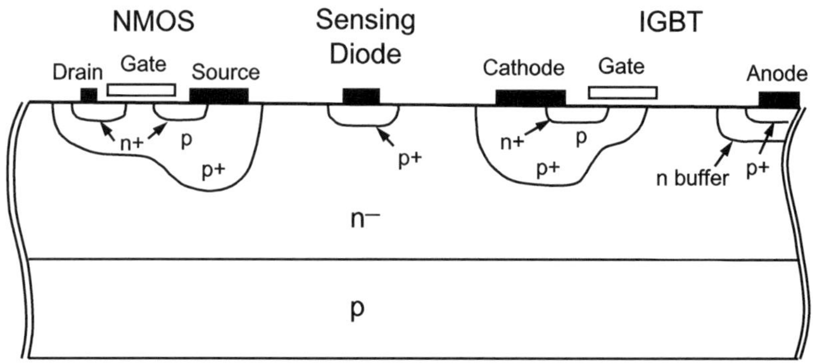

Fig. 5.30. Integration of key components for the overcurrent protection circuit on bulk substrate.

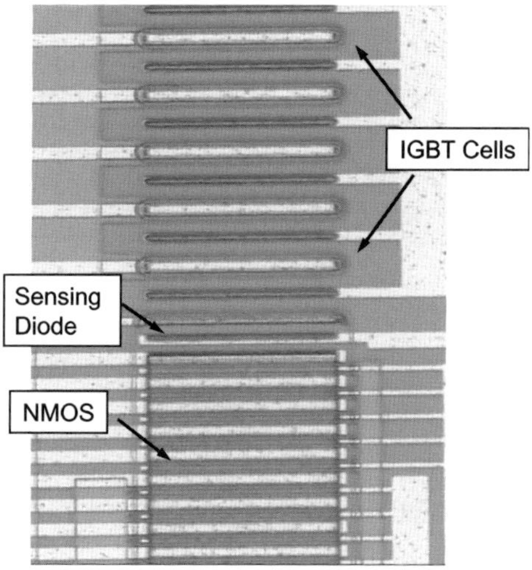

Fig. 5.31. Micrography of key components of the overcurrent protection integrated circuit.

under short-circuit condition. The "fault under load" condition is described as that the short-circuit fault occurs during the normal IGBT conduction state.

Figure 5.32 shows the measured waveforms of anode voltage and current of hard switched fault (at 800 A/cm^2) at supply voltage of 105 V and gate voltage of 10 V. The device recovers from the fault by shutting down the gate drive after about 10 μs. Another short-circuit fault under load (at 1060 A/cm^2), as shown in Fig. 5.33, is applied to the device, and similar recovery is made after about 5 μs.

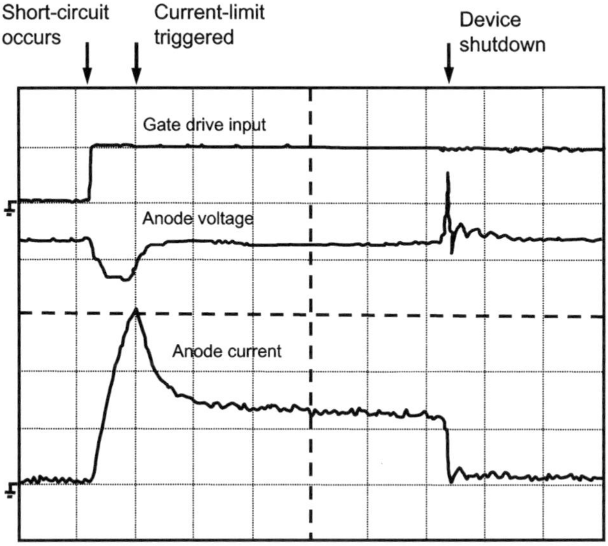

Fig. 5.32. Measured hard-switch fault (at 105 V supply voltage, 800 A/cm^2 anode current density); time scale: 2 μs/div; anode voltage scale: 25 V/div; anode current scale: 2 A/div.

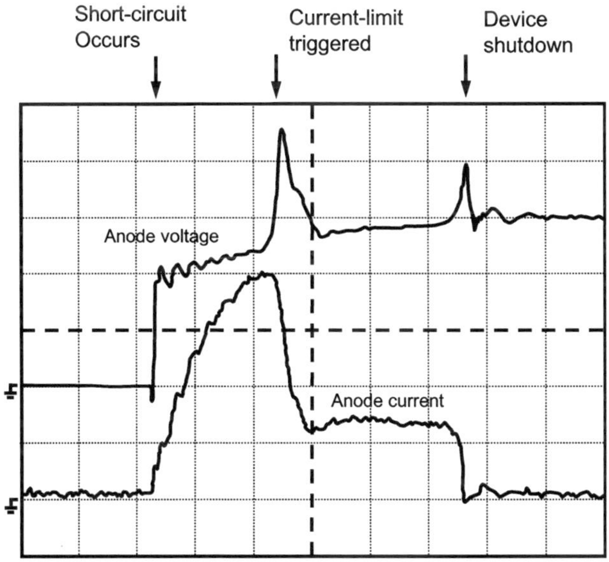

Fig. 5.33. Measured fault under load (at 105 V supply voltage, 1068 A/cm^2 anode current density); time scale: 1 μs/div; anode voltage scale 35 V/div; anode current scale: 2 A/div.

5.12. Vertical IGBT Fabrication Process

To fabricate a vertical IGBT device, it follows the following basic steps (Chang, 1995; Luo *et al.*, 2000) as shown in the table below.

Step	Process	Recipe
1	Starting wafer: n^-/p^+ epi-wafer ⟨100⟩ drift region concentration and thickness determined by the breakdown voltage specification	Wafer cleaning
2	Field oxidation for 4000 Å	Dry–wet–dry oxidation 1100°C, 10 min (dry), 25 min (wet), 5 min (dry)
3	Photolithography, p-well mask	
4	Field oxide etching	Buffered oxide etching (BOE)
5	p-well implant: BF_2	80–100 keV, 5E15/cm^2
6	Photoresist strip	
7	p-well drive-in and oxidation	1050°C, 60–70 min (wet)
8	Photolithography, active area mask	
9	Oxide etching	BOE
10	Photoresist strip	
11	Gate oxidation	900°C, 30–60 min (dry)
12	Polysilicon deposition (LPCVD)	650°C, 5000 Å
13	Oxide deposition (APCVD)	
14	Photolithography, polygate mask	
15	APCVD oxide etching	
16	Polysilicon etching	HNO_3: CH_3COOH: HF
17	Gate oxide and APCVD oxide etching	BOE
18	p-body (channel region) implant: B	35–40 keV, 1E14/cm^2, 7°
19	p-body drive-in and oxidation	900°C, 20 min (dry O_2)1100°C, 180 min, N_2
20	Oxide removal	
21	n^+ cathode implant: As	80 keV, 5E15/cm^2

Step	Process	Recipe
22	n^+ cathode drive-in	900°C, 30 min (dry O_2)950°C, 30 min, N_2
23	Oxide removal	
24	PCVD oxide and BPSG densification	1.5 K SiO_2–7.5 K BPSG 900°C, 30 min
25	Photolithography, contact mask	
26	Contact opening	BOE
27	Photoresist strip	
28	Al deposition, 1.5–2 μm	400°C, 30 min
29	Photolithography, metal mask	
30	Al etching	
31	Sintering	

The corresponding device profiles are shown in Fig. 5.34 by process simulation.

5.13. Related MOS-Bipolar Structures

5.13.1. Emitter Switched Thyristor (EST)

The emitter switched thyristor was designed to have a lower on-state voltage drop and a higher voltage–current saturation capability. The device structure evolved from the conventional EST (Shekar *et al.*, 1991a; Bhalla and Chow, 1994), the dual-channel EST (Shekar *et al.*, 1991b), and to the recent dual-gate EST (Sridhar and Baliga, 1996). The device operation is initially similar to that of the IGBT, however, due to the latch-up of the main parasitic thyristor and dual-gate operation, it has a good on-state voltage–current capability and higher saturation capability. The device structure is shown in Fig. 5.35. In the structure, two MOS gates, namely Gates 1 and 2, are used to form the inversion channels for conduction. During the operation, Gate 2 can be turned off to switch the operation mode from dual-gate operation to dual-channel conduction (Sridhar and Baliga, 1996), if desired.

When both gates are applied a positive voltage, the device functions like an IGBT, however with multiple conduction paths due to two cathode contacts. From cathodes, electron carriers flow through the MOS channels to arrive at the n^--drift region. And from anode, hole carriers flow into the drift region for conductivity modulation and also partially through the p-base region forming the transistor current. The main thyristor latch-up occurs when the current goes to a higher level and the p-base/n^+-junction becomes forward-biased.

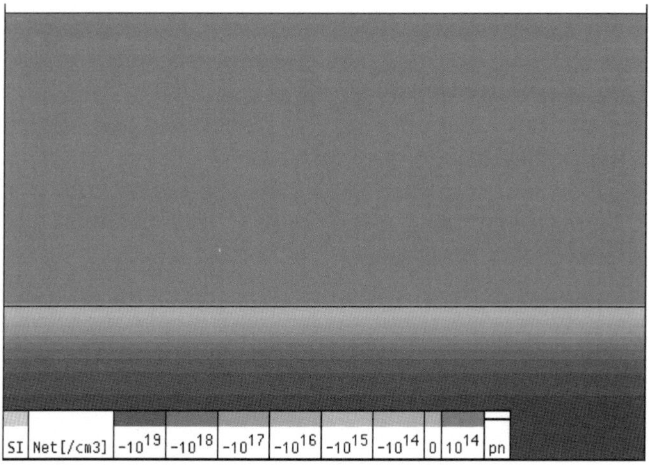

(a) Starting wafer.

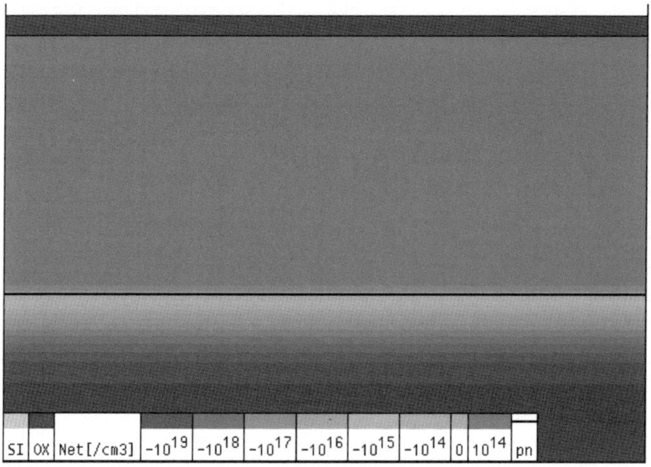

(b) Wafer cleaning and field oxidation.

Fig. 5.34. IGBT process details by simulation.

Now, the device enters the EST region with a large part of current supply through the main thyristor. The on-state voltage reduces to be as low as about 1.1 V at high current conduction. The thyristor current is partitioned and flows through both MOS channels. The voltage rating of the forward safe operation area is now limited by the breakdown voltage of the short-channel lateral MOSFET under Gate 2. If the bias at Gate 2 is turned off, the device can

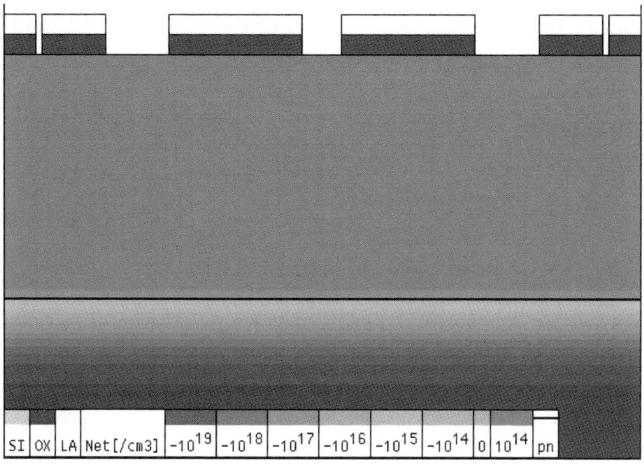

(c) P.R. coating, p⁺ well mask, and field oxide etch.

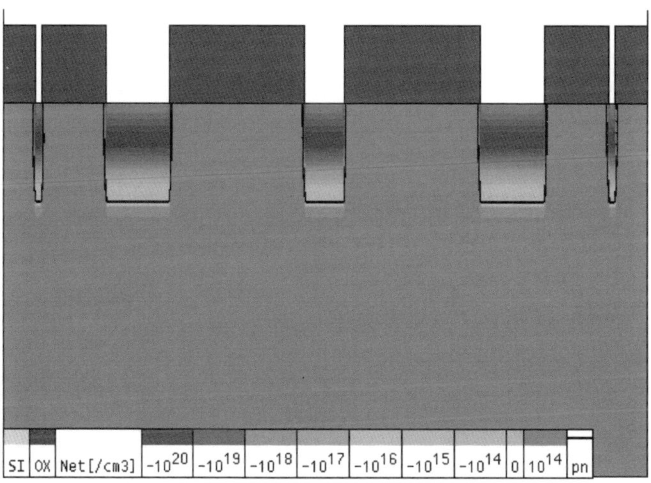

(d) p⁺ well implant and P.R. strip.

Fig. 5.34. (*Continued*)

operate at a higher saturation voltage as that of the dual-channel EST. The structure has two p^+ regions to divert the hole current to two cathodes. At a very high current density, the gate control is lost due to the latch-up of the parasitic thyristors. As reported in Sridhar and Baliga (1996), the normal operation region is around 800 A/cm^2 with the parasitic thyristor latch-up current

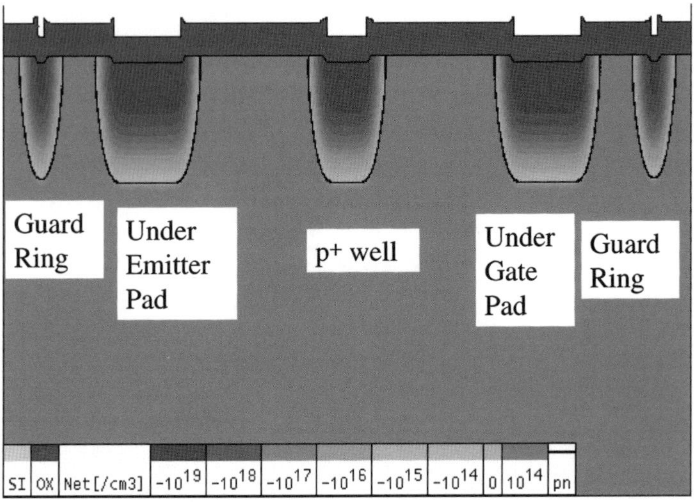

(e) p^+ well drive-in.

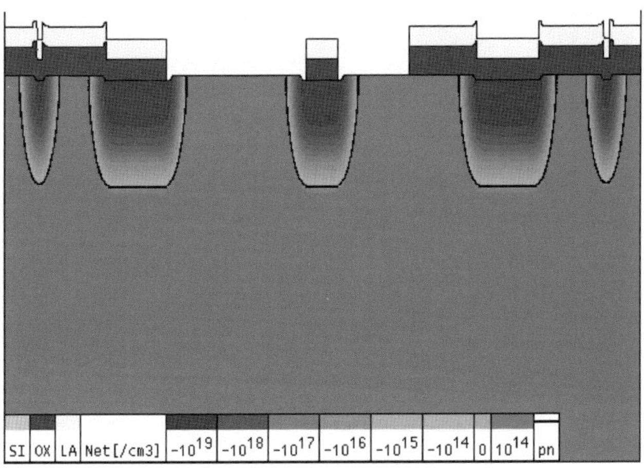

(f) P.R. coating, active area mask, and oxide etch.

Fig. 5.34. (*Continued*)

density at above 1000 A/cm^2. Figure 5.36 gives the *I–V* curves of high voltage, e.g. 600 V, MOSFET, IGBT, and the dual-gate EST devices for a brief on-state comparison. For the modern third generation IGBT, devices, the forward voltage drop at 100 A/cm^2 ranges from 1.3 V to 1.7 V, which is still higher than that of EST at around 1.1 V drop.

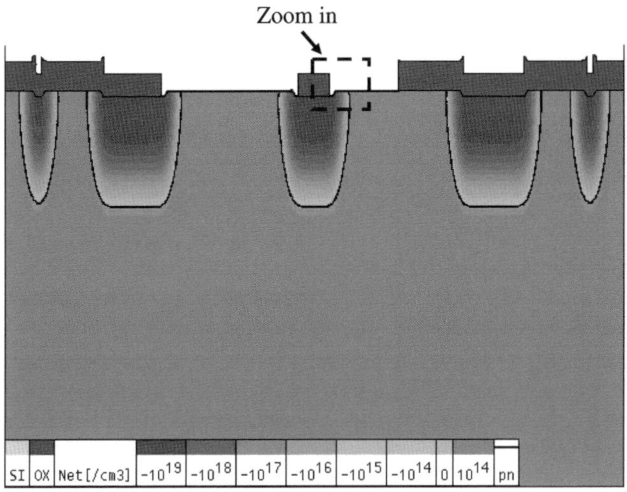

(g) P.R. strip and gate oxide growth.

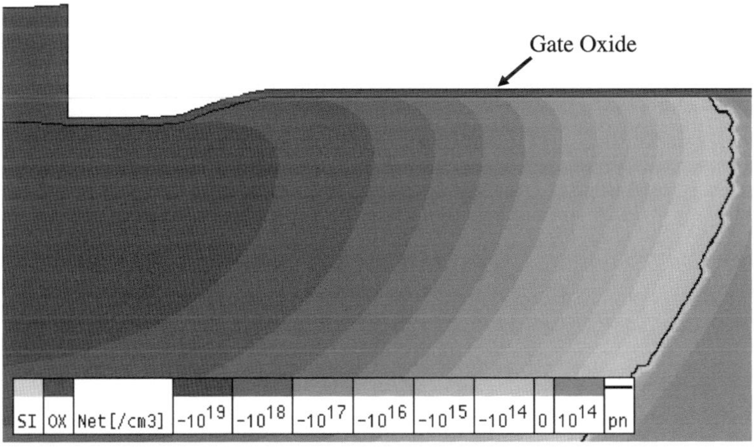

(h) P.R. strip and gate oxide growth.

Fig. 5.34. (*Continued*)

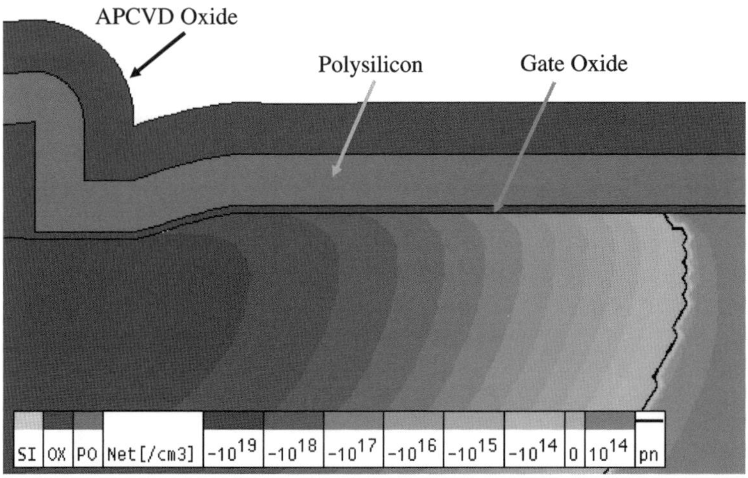

(i) Polysilicon deposition and APCVD oxide deposition.

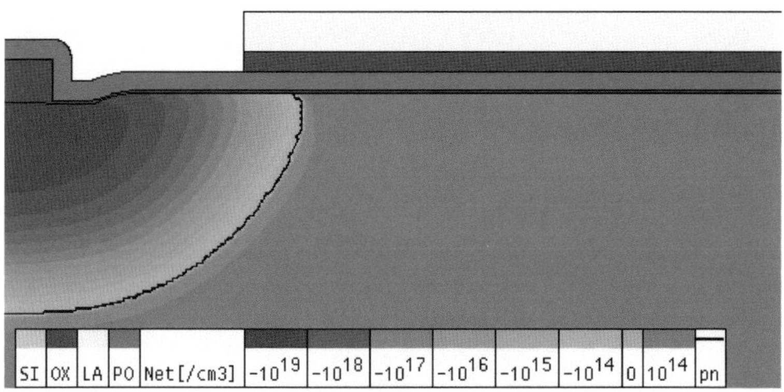

(j) P.R. coating, poly mask, and APCVD oxide etch.

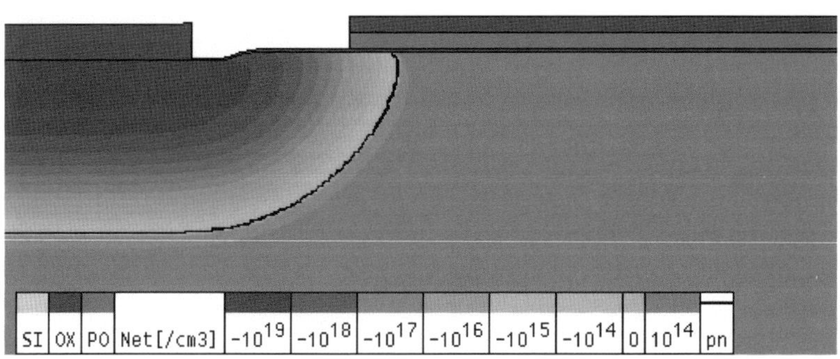

(k) P.R. strip and poly etch.

Fig. 5.34. (*Continued*)

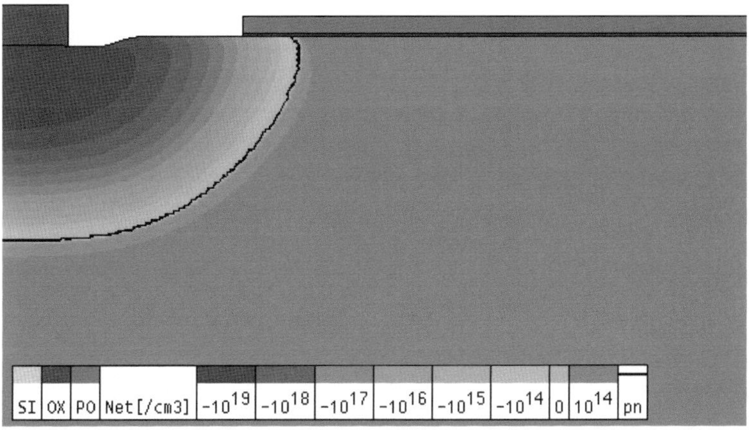

(l) Gate oxide and APCVD oxide etch.

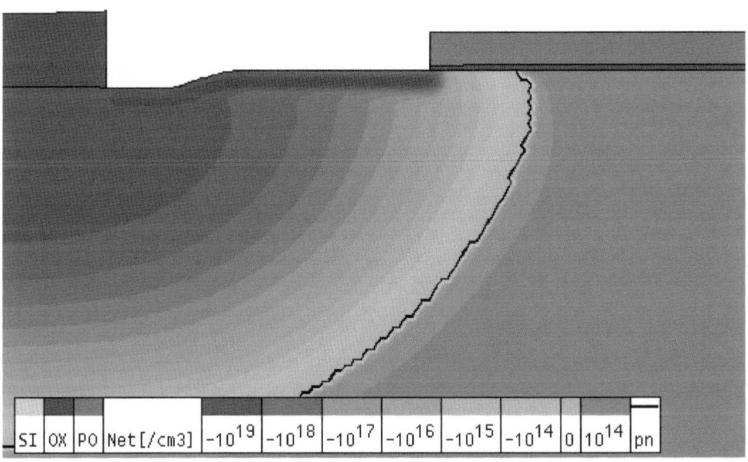

(m) p-Base implantation.

Fig. 5.34. (*Continued*)

5.13.2. Base-Resistance-Controlled Thyristor (BRT)

The structure of the asymmetric dual-gate BRT is shown in Fig. 5.37 (Kurlagunda and Baliga, 1995) which evolved from the earlier structure of BRT by Nandakumar *et al.* (1991). The structure utilizes the MOS gate control to divert the hole current in the transistor p-base to switch the device operation modes between a thyristor and an IGBT. When Gate 1 is applied a

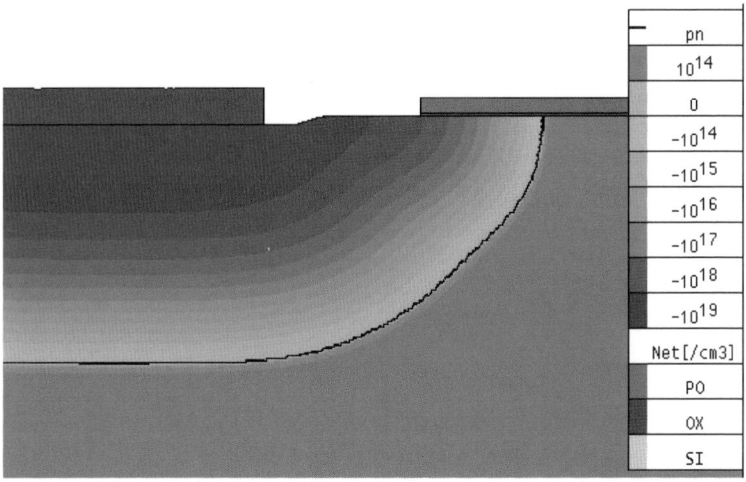

(n) p-Base drive-in.

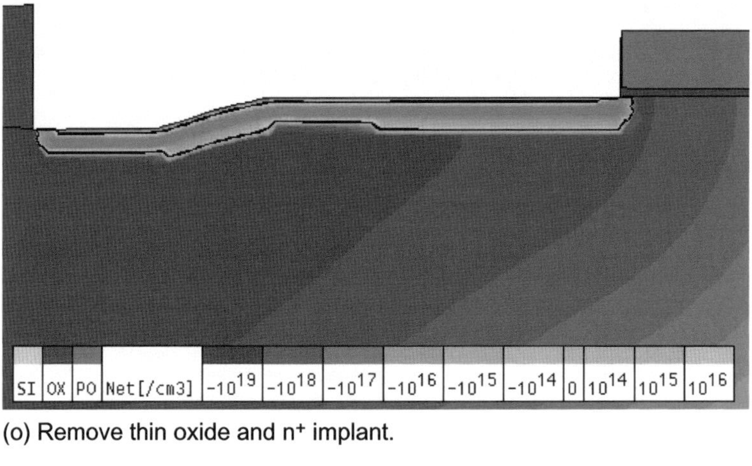

(o) Remove thin oxide and n⁺ implant.

Fig. 5.34. (*Continued*)

positive voltage and Gate 2 is either applied a positive voltage or is grounded, the device enters the IGBT mode of operation. When the current is sufficiently large to bias the upper cathode junction, the intended thyristor latch-up occurs. And, the on-state voltage reduces to a lower level, e.g. around 1.1 V at 100 A/cm^2. Published data show that the device can operate at 1.24 V, which is 0.5 V below that of the IGBT device at the current density of 300 A/cm^2

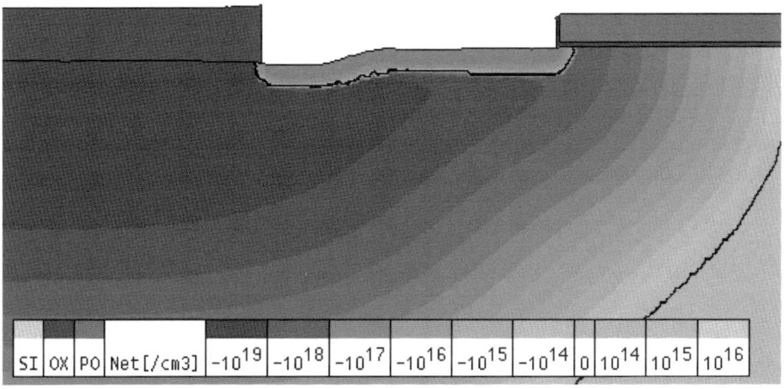

(p) n⁺ drive-in.

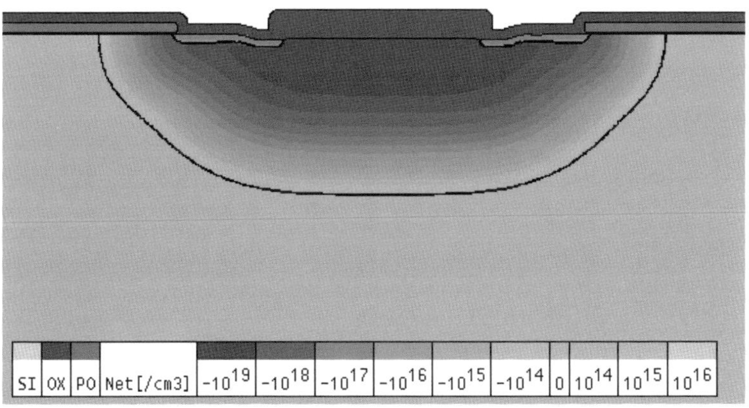

(q) Remove thin oxide and APCVD oxide deposition.

Fig. 5.34. (*Continued*)

(CSIHPDPIC, 1996). When the device turn-off needs to be made, the operation mode needs to be switched back to the IGBT mode prior to the actual turn-off. When Gate 1 remains at the positive, Gate 2 is applied a negative voltage to turn on the p-MOS channel to divert the transistor base current. This diversion, if strong enough, will lower the upper transistor gain and break the regenerative latch-up. The magnitude of the negative gate voltage applied to Gate 2, ranging from −5 to −20 V, needs to be approximately linearly proportional to the amount of thyristor current to be controlled. When the device operates in the IGBT mode, turn-off can be made by switching off the voltage at Gate 1, as that of IGBT turn-off.

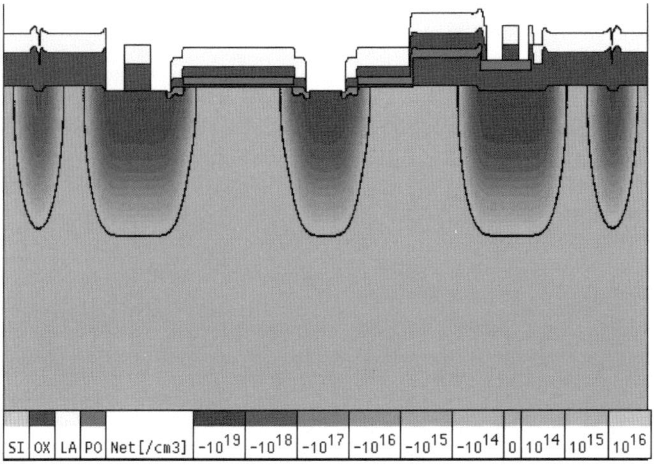

(r) P.R. coating, contact mask for contact open.

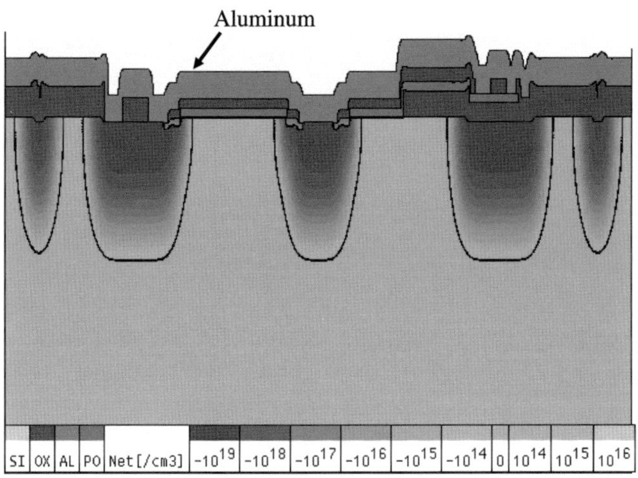

(s) P.R. strip and Al deposition.

Fig. 5.34. (*Continued*)

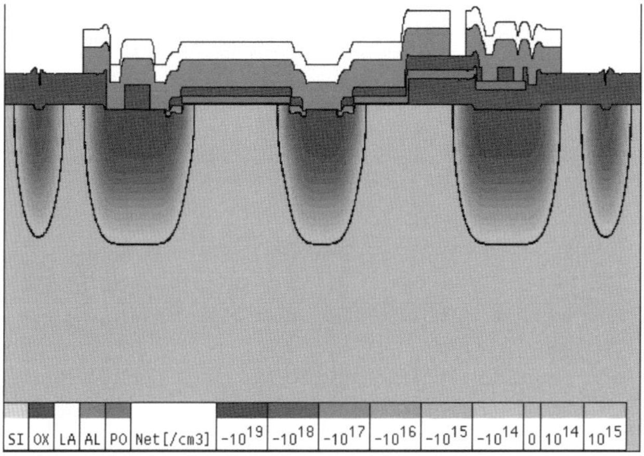

(t) P.R. coating, metal mask, and Al etch.

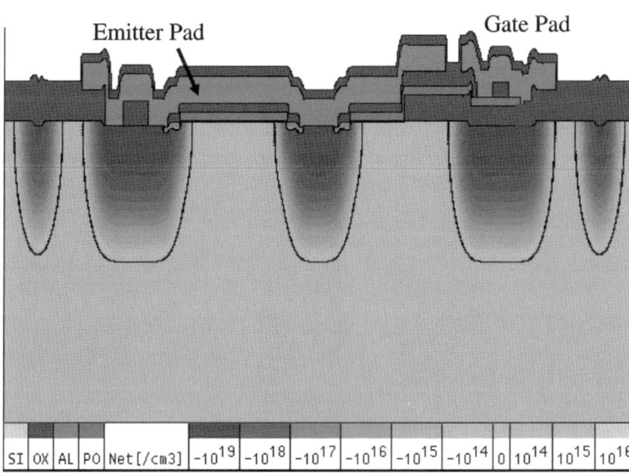

(u) P.R. strip, alloy sintering, and passivation.

Fig. 5.34. (*Continued*)

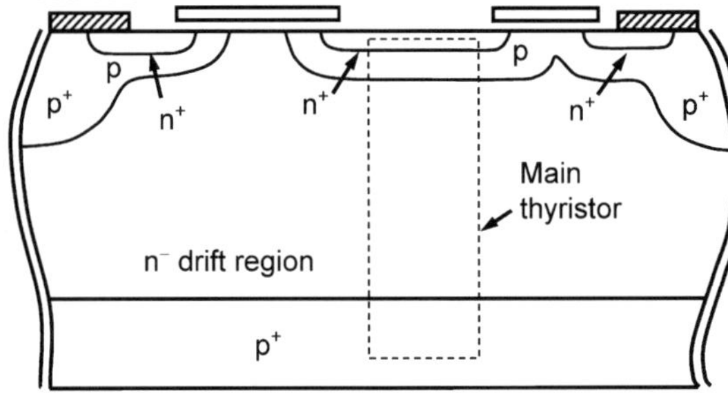

Fig. 5.35. The dual-gate emitter switched thyristor.

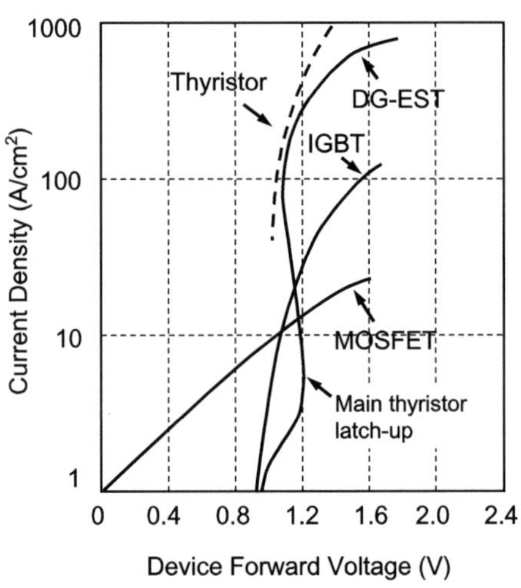

Fig. 5.36. Brief comparison on the on-state *I–V* characteristics of MOSFET, IGBT, EST, and thyristor devices (at 600 V rating).

5.13.3. Injection-Enhanced Insulated-Gate Bipolar Transistor (IEGT)

The IEGT device was developed to lower the on-state voltage drop (Kitagawa *et al.*, 1993). Of the development, the device rated at 2500 V has a voltage drop of around 4 V at room temperature. The device structure is shown in Fig. 5.38. It has a deep trench with oxide sidewall to inject the electron carriers via the accumulation layer formed by the electric field along the extended vertical gate. The higher amount of electron injection, as shown in Fig. 5.38, raises the

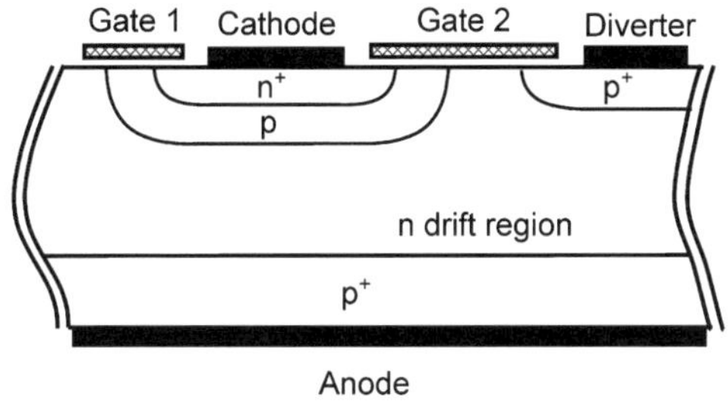

Fig. 5.37. Dual-gate base-resistance-controlled thyristor.

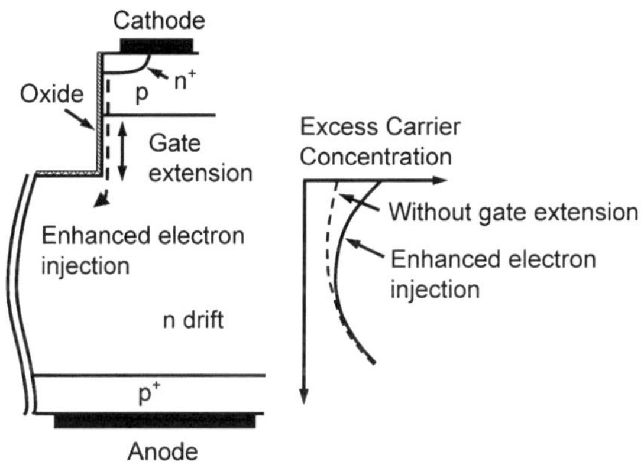

Fig. 5.38. The injection-enhanced IGBT and the excess carrier concentration in the bulk drift region.

electron concentration in the drift region near the p-body then in term it helps to lower the on-state conduction voltage.

5.13.4. MOS-Controlled Thyristor (MCT)

The MCT was developed (Temple, 1984; Bauer *et al.*, 1991) for the similar objectives to have a lower on-state conduction voltage than IGBT devices. The basic device structure is shown in Fig. 5.39 which has both n-channel and p-channel to turn on and to turn off the device. Published data show that a 600 V rated MCT device can be able to carry about 10 times higher current than that of the IGBT device, and about 100 times higher than the MOSFET device (CSIHPDPIC, 1996).

When the Gate 2 is applied a positive voltage, the device functions like an IGBT and due to the prolonged n-emitter, it is very susceptible to latch-up,

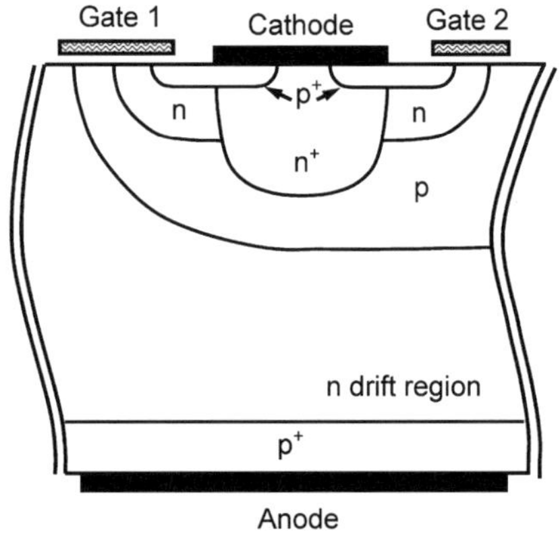

Fig. 5.39. MOS-controlled thyristor.

as an intended characteristic. To turn off the device, a negative gate voltage is needed to be applied to both Gates 1 and 2 to short the upper n^+-emitter/p-base junction, so as to reduce the current gain of n–p–n transistor in order to break the regenerative latch-up. Similar to BRT device, the magnitude of the negative gate voltage should be proportional to the amount of thyristor current to be turned off. CSIHPDPIC (1996) gives the relationship of turn-off current densities and gate voltages for single-cell and array-group of MCT devices. The negative gate voltage varies according to different cell pitches and, for a fixed cell pitch it is almost linearly proportional to the thyristor current density.

References

B. J. Baliga, M. S. Adler, P. V. Gray, R. Love and N. Zommer, The insulated gate rectifier (IGR): A new power switching device, *IEEE Int. Electron Devices Meeting Digest*, 264–267 (1982).

B. J. Baliga, M. S. Adler, P. V. Gray and R. P. Love, Suppressing latch-up in insulated gate transistors, *IEEE Electron Device Lett.* **5**, 323–325 (1984).

B. J. Baliga, Temperature behavior of insulated gate transistor characteristics, *Solid-State Electron.* **28**, 289–297 (1985).

B. J. Baliga, Analysis of the output conductance of insulated gate transistors, *IEEE Electron Device Lett.* **7**(12), 686–688 (1986).

B. J. Baliga, *Modern Power Devices*, Chap. 7, (John Wiley and Sons, New York, 1987).

A. Bhalla and T. P. Chow, 550V, n-channel emitter switched thyristors with an atomic lattice layout geometry, *IEEE Electron Device Lett.* **15**, 452–454 (1994).

F. Bauer, E. Halder, K. Haddon, P. Roggwiller, T. Stockmeier, J. Burgler, X. Fichtner, S. Muller, M. Westermann, J.-M. Moret and R. Vuilleumier, Design aspects of MOS-controlled thyristor elements: Technology, simulation and experimental results, *IEEE Trans. Electron Devices* **38**(7), 1605–1611 (1991).

S. Biswas, B. Basak and K. Rajashekara, A modular gate drive circuit for insulated gate bipolar transistors, *Proc. IEEE Ind. Appl. Soc. Ann. Meeting*, 1490–1496 (1991).

J. R. Chang, The study of IGBT latchup characteristics, M. Eng Thesis, National Tsing Hua University, Taiwan (1995).

M. F. Chang, G. C. Pifer, B. J. Baliga, M. S. Adler and P. V. Gray, 25 amp, 500 volt insulated gate transistors, *IEEE Int. Electron Devices Meeting Digest*, 83–86 (1983).

M. F. Chang, G. C. Pifer, H. Yilmaz, R. F. Dyer, B. J. Baliga, T. P. Chow and M. S. Alder, Comparison of N and P channel IGTs, *IEEE Int. Electron Devices Meeting Digest*, 278–281 (1984).

R. Chokhawala, J. Catt and L. Kiraly, A discussion on IGBT short-circuit behavior and fault protection schemes, *IEEE Trans. Ind. Appl.* **31**, 256–263 (1995).

T. P. Chow, Z. Shen, D. N. Pattanayak, E. J. Wildi, M. S. Adler and B. J. Baliga, Modelling and analysis of current sensors for n-channel vertical IGBT's, *IEEE Int. Electron Devices Meeting Digest*, 253–256 (1992).

Committee of Specialist Investigators on High performance Power Devices and Power IC (CSIHPDPIC), *Power Semiconductor Device and Power IC Handbook* (Corona Publishing Co. Ltd. Japan, 1996), pp. 386–388.

D. A. Dapkus, Equalized junction temperature is the key in paralleled IGBT reliability, *Power Conversion and Intell. Motion Mag.* **20**(11), 10–18 (1994).

D. R. Disney and J. D. Plummer, SOI LIGBT devices with a dual p-well implant for improved latching characteristics, *Proc. Int. Symp. Power Semiconductor Devices ICs*, 254–258 (1993).

C. Gerster, Fast high-power/high-voltage switch using series-connected IGBTs with active gate-controlled voltage-balancing, *Proc. IEEE Power Electron. Specialist Conf.*, 469–472 (1994).

A. M. Goodman, J. P. Russell, L. A. Goodman, J. C. Neuse and J. M. Neilson, Improved COMFETs with fast switching speed and high current capability, *IEEE Int. Electron Devices Meeting Digest*, 79–82 (1983).

P. Hofer, N. Karrer and C. Gerster, Paralleling intelligent IGBT power modules with active gate-controlled current balancing, *Proc. IEEE Power Electron. Specialist Conf.*, 1312–1316 (1996).

N. Iwamuro, Y. Harada, T. Yamazaki, N. Kumagai and Y. Seki, A new vertical IGBT structure with a monolithic overcurrent, over-voltage, and over-temperature sensing and protecting circuit, *IEEE Electron Device Lett.* **16**, 399–401 (1995).

M. Kitagawa, I. Omura, S. Hasegawa, T. Inoue and A. Nakagawa, A 4500 V injection enhanced gate bipolar transistor (IEGT) operating in a mode similar to a thyristor, *IEEE Int. Electron Devices Symp. Digest*, 679–682 (1993).

R. Kurlagunda and B. J. Baliga, The dual-gate BRT, *Proc. IEEE Int. Symp. Power Semiconductor Devices ICs*, 29–33 (1995).

A. Kurnia, O. H. Stielau, G. Venkataramanan and D. M. Divan, Loss mechanisms in IGBT's under zero voltage switching (ZVS), *Proc. IEEE Power Electron. Specialist Conf.*, 1011–1018 (1992).

Y. C. Liang and V. Hor, Design of lateral IGBT power devices with current sensor, *Proc. IEEE Ind. Appl. Soc. Meeting*, 1010–1015 (1995).

Y. C. Liang, G. S. Samudra and V. S. S. Hor, Design of integrated current sensor for lateral IGBT power devices, *IEEE Trans. Electron Devices* **45**(7), 1614–1616 (1998).

Y. C. Liang, S. Xu, C. Ren and J. Luo, New SOI structure for LIGBT with improved thermal and latch-up characteristics, *Proc. IEEE Power Electron. Drive Sys. Conf.*, 258–261 (1999).

J. Luo, Y. C. Liang and B. J. Cho, Design of LIGBT protection circuit for smart power integration, *IEEE Trans. Ind. Electron.* **47**(4), 744–750 (2000).

G. V. Manduteanu, A power semiconductor diode with integrated forward-current sensor, *Solid-State Electron.* **36**, 1335–1338 (1993).

G. Masetti, M. Severi and S. Solmi, Modelling of carrier mobilities against carrier concentration in arsenic-, phosphorus- and boron-doped silicon, *IEEE Trans. Electron Devices* **30**, 764–769 (1983).

A. Mogro-Campero, R. P. Love, M. F. Chang and R. F. Dyer, Shorter turn-off times in insulated gate transistors by proton implantation, *IEEE Electron Device Lett.* **6**, 224–226 (1985).

A. Nakagawa, H. Ohashi, M. Kurata, H. Yamaguchi and K. Watanabe, Non-latch-up 1200 V bipolar mode MOSFET with large SOA, *IEEE Int. Electron Devices Meeting Digest*, 860–861 (1984).

M. Nandakumar, B. J. Baliga and M. S. Shekar, A new MOS-gated power thyristor structure with turn-off achieved by controlling the base resistance, *IEEE Electron Device Lett.* **12**, 227–229 (1991).

A. Nezar, and P. K. T. Mok, Latch-up prevention in insulated gate bipolar transistors, *Proc. Int. Symp. Power Semiconductor Devices ICs*, 236–239 (1993).

M. Otsuki, S. Momota, A. Nishiura and K. Sakurai, The 3rd generation IGBT towards a limitation of IGBT performance, *Proc. IEEE Int. Symp. Power Semiconductor Devices ICs*, 41–45 (1993).

R. Rangan, D. Y. Chen, J. Yang and J. Lee, Application of the IGT/COMFET to zero-current switching resonant converters, *IEEE Power Electron. Specialists Conf.*, 55–60 (1987).

S. P. Robb, A. A. Taomoto and S. L. Tu, Current sensing in IGBTs for short-circuit protection, *Proc. IEEE Int. Symp. Power Semiconductor Devices ICs*, 81–85 (1994).

J. P. Russell, A. M. Goodman, L. A. Goodman and J. M. Nielson, The COMFET: A new high conductance MOS gated device, *IEEE Electron Device Lett.* **4**, 63–65 (1983).

Y. Seki, Y. Harada, N. Iwamuro and N. Kumagi, A new IGBT with monolithic overcurrent protection circuit, *Proc. IEEE Int. Symp. Power Semiconductor Devices ICs*, 31–35 (1994).

M. S. Shekar, B. J. Baliga, M. Nandakumar, S. Tandon and A. Reisman, Characteristics of the emitter switched thyristor, *IEEE Trans. Electron Devices* **38**, 1619–1623 (1991a).

M. S. Shekar, B. J. Baliga, M. Nandakumar, S. Tandon and A. Reisman, High-voltage current saturation in emitter switched thyristors, *IEEE Electron Device Lett.* **120**, 227–229 (1991b).

Z. Shen, K. C. So and T. P. Chow, Comparative study of integrated current sensors in n-channel IGBT's, *Proc. IEEE Int. Symp. Power Semiconductor Devices ICs*, 75–80 (1994).

Z. Shen and S. Robb, Monolithic integration of vertical IGBT and intelligent protection circuits, *Proc. IEEE Int. Symp. Power Semiconductor Devices ICs*, 295–298 (1996).

Y. Shimizu, Y. Nakano, Y. Kono, N. Sakurai, Y. Sugawara and S. Otaka, A high performance intelligent IGBT with overcurrent protection, *Proc. IEEE Int. Symp. Power Semiconductor Devices ICs*, 37–41 (1994).

S. Sridhar and B. J. Baliga, The dual gate emitter switched thyristor (DG-EST), *IEEE Electron Device Lett.* **17**, 25–27 (1996).

S. M. Sze, *Semiconductor Devices: Physics and Technology* (John Wiley and Sons, New York, 1985).

M. Tabata, S. Igarashi and K. Kuroki, Control methods of current balancing for parallel connected IGBTs, *Proc. IEEE Int. Symp. Power Semiconductor Devices ICs*, 101–104 (1998).

P. Tan and Y. C. Liang, Performance analysis of MOS-controlled bipolar power devices, *Proc. IEEE Power Electron. Drive Syst. Conf.*, 87–92 (1995).

V. A. K. Temple, MOS controlled thyristors (MCTs), *IEEE Int. Electron Devices Meeting Digest*, 282–285 (1984).

R. Valentine, Power module control design, *Proc. IEEE Ind. Appl. Soc. Ann. Meeting*, 904–910 (1995).

I. Widjaja, A. Kurnia, D Divan and K. Shenai, Computer simulation and design optimization of IGBT's in soft-switching converters, *Proc. Int. Symp. Power Semiconductor Devices ICs*, 105–109 (1994).

I. Widjaja, A. Kurnia, K. Shenai and D. M. Divan, Switching dynamics of IGBT's in soft-switching converters, *IEEE Trans. Electron Devices* **42**, 445–454 (1995).

C. H. Yang and Y. C. Liang, Investigation on parallel operations of IGBTs, *Proc. IEEE Ind. Electron. Control (IECON) Conf.*, 1005–1010 (1996).

X. Yuan, Design of IGBT devices for soft-switching applications, M. Eng Thesis, National University of Singapore (1999).

6

Superjunction Structures

6.1. Introduction

For power devices, the off-state blocking capability is one of the most important parameters in the performance besides its on-state voltage drop and switching speed. In most cases, these three important parameters of a single device are always interlinked and having certain trade-off relationships adopted by the device designers. For example, when applications requiring fast switching, the MOSFET device of unipolar charge transport (the majority carrier) for minimum excess charge storage is utilized. For such a power device, the blocking capability is a function of the doping concentration of its epitaxial drift layer and the minimum thickness of the layer, called the drift length. The former determines the maximum allowable field occurring at the junction of p-body/drift layer before avalanche breakdown, and the later determines the amount of breakdown voltage sustainable by the depletion region. For one device to have a higher blocking capability, it shall have a lightly doped drift layer and a sufficiently long drift distance between p-body and drain. However, due to the nature of unipolar carrier transport, the degree of conductivity modulation within the epitaxial drift region is not as prominent as that of the bipolar junction transistor where both electron and hole carriers are injected into the drift region to lower the on-state voltage drop across the region. The on-state voltage of the power MOSFET will be higher if the background doping of the epitaxial drift region is low, and also if it is long for higher blocking capability. Therefore, there exists a trade-off relationship which prescribes the technical limit between the on-state resistance and the blocking voltage of the MOSFET devices for them to be achievable in device design. That is, for a MOSFET device of a particular voltage rating, one can achieve only a certain minimum (the best) on-state resistance and can never be lower than that limit. In principle, the minimum achievable on-state resistance for each voltage rating can be calculated. These calculated data form the boundary line of so-called "ideal silicon limit" for MOSFET power devices. The relationship indicates that the achievable on-state resistance (R_{on}) is approximately proportional to the power of 2.5 times of the device breakdown rating (i.e. $V_{\text{BR}}^{2.5}$).

The reason why the superjunction structure receives great attention in the late 1990s and more nowadays is that, the structure actually breaks the "ideal silicon limit" of the conventional power MOSFET and itself forms a new limit which is almost a linear relationship ($R_{\text{on}} \propto V_{\text{BR}}^{1.17}$) for the achievable device performance. The impact is very obvious when dealing with devices of 100 V or higher ratings.

This chapter starts with the formation of ideal silicon limit equation followed by the detailed description of the basic superjunction structure and the derivation of the superjunction silicon limit equation. Process technology, the limitation and the difficulty on device formation are also discussed. Several new device structures, namely the poly-flanked MOSFET, the oxide-bypassed MOSFET, and the graded oxide-bypassed MOSFET based on the superjunction charge compensation and field diversion principles are discussed in this chapter. Finally, the lateral superjunction device on the partial SOI platform is described.

6.2. The Unipolar Ideal Silicon Limit

Basically, the off-state voltage across a device is sustained by the reverse-biased p–n junction. If for multiple grounded structures (especially for lateral structure), there exist more than one reverse-biased junction. Except for the rare dielectric breakdown, the device voltage rating is normally capped by the junction breakdown.

When an external voltage is applied, electric field exists within the depletion zone at both sides of the p–n junction. The normal distribution of the electric field profile is linearly going down when moving away from the junction. The rate of change of electric field is a function of the doping profile for each side of the junction. For a lightly doped region, the field profile tends to be flatter because of a relatively wider depletion region. Comparing to the region having higher doping concentration, for the same maximum field at the junction, the lightly doped region is able to sustain a higher voltage. The phenomenon can be fully understood when we look at a simple abrupt p–i–n junction under reverse voltage bias V_{R} as shown in Fig. 6.1.

The doping concentration on the left side of the junction is much higher than that of the right side. Assuming, the doping profile is homogeneous for each side and abrupt transition at the junction, the distributions of charge, electric field, and voltage can be found by solving the Poisson equation.

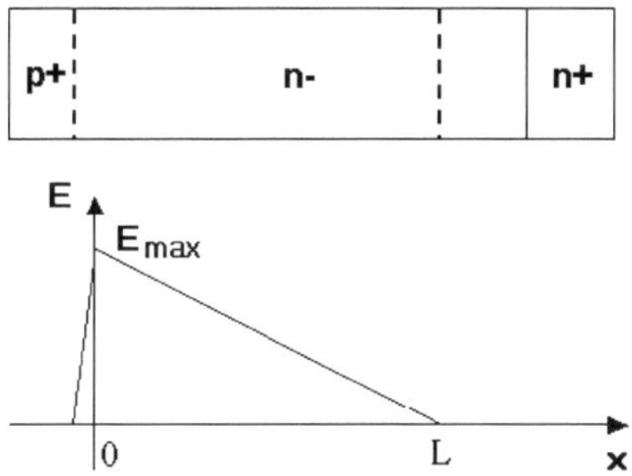

Fig. 6.1. Reverse-biased p–n junction showing the distributions of charge, electric field, and voltage.

The junction critical field and breakdown phenomena have been discussed in Chapter 2 in detail. Here is just to briefly go through with it for the purpose of derivation of the unipolar silicon limit. For a reverse-biased p^+–n^- abrupt junction, the Poisson equation for the right side of depletion region (positive ion density N_D in atom/cm^3) can be written as

$$\frac{d^2V}{dx^2} = -\frac{qN_D}{\varepsilon_s} = -\frac{d\hat{E}}{dx}, \tag{6.1}$$

where ε_s is the dielectric constant for silicon. Solving the electric field with boundary condition of $\hat{E} = 0$ at $x = L$,

$$\hat{E}(x) = \frac{qN_D}{\varepsilon_s}(L - x) \quad \text{and} \quad \hat{E}_{max} = \frac{qN_D}{\varepsilon_s}L, \tag{6.2}$$

where L is the depletion width on the right side.

The maximum electric field is located at the junction ($x = 0$) and the slope of the field line is a function of the doping concentration N_D. The voltage distribution along the depletion region can also be found as

$$V(x) = \frac{qN_D}{\varepsilon_s}\left(Lx - \frac{x^2}{2}\right) \quad \text{and} \quad V_R = \frac{qN_D L^2}{2\varepsilon_s}. \tag{6.3}$$

The equation indicates that most of the voltage drop is near the junction. For the area away from the junction, it has much lower electric field and shares a small amount of voltage. When a fixed voltage, V_R is applied, the depletion width L can be found as

$$L = \sqrt{\frac{2\varepsilon_s V_R}{qN_D}} = \frac{\hat{E}_{max}\varepsilon_s}{qN_D}. \tag{6.4}$$

The breakdown voltage V_{br} is defined as the voltage when the maximum electric field $\hat{E}_{max}$ reaches the critical electric field $\hat{E}_c$. Therefore, substituting Eq. (6.4) to Eq. (6.3), the blocking voltage of a normal p–i–n diode is given by

$$V_{br} = \frac{qN_D\left(\frac{\hat{E}_c\varepsilon_s}{qN_D}\right)^2}{2\varepsilon_s} = \frac{\varepsilon_s\hat{E}_c^2}{2qN_D}. \tag{6.5}$$

In silicon, it is known that the impact ionization coefficient is approximately

$$\alpha = 1.8 \times 10^{-35}\hat{E}^7. \tag{6.6}$$

And, the avalanche integral for breakdown condition is

$$\int_0^L \alpha \cdot dx = 1. \tag{6.7}$$

By combining Eqs. (6.2), (6.6), and (6.7), we may get the depletion region length L_{br} at breakdown

$$L_{br} = 2.6 \times 10^{10} N_D^{-7/8}. \tag{6.8}$$

The critical electric field $\hat{E}_c$ can be found as

$$\hat{E}_c = 4023 N_D^{1/8}. \tag{6.9}$$

Thus, at a certain doping concentration N_D, the breakdown voltage can be obtained by

$$V_{br} = 5.24 \times 10^{13} N_D^{-3/4}. \tag{6.10}$$

Therefore, the specific on-state resistance of the drift region is

$$R_{d,sp} = \frac{L_{br}}{q\mu N_D} = 6.21 \times 10^{-9} V_{br}^{2.5} \quad (\Omega \cdot \text{cm}^2) \tag{6.11}$$

which is proportional to the power of 2.5 of the device breakdown rating. In the literature, the specific on-state resistance of a conventional vertical double-diffusion MOSFET (VDMOS) is given by (Hu, 1979)

$$R_{on,sp} = \frac{27V_{br}^2}{8\mu\varepsilon_s\hat{E}_c^3}. \tag{6.12}$$

Substitute $\hat{E}_c = 8.2 \times 10^5 V_{br}^{-0.2}$ (V$\cdot$ cm^{-2}) and $\mu = 710V_{br}^{0.1}$(cm$^2 \cdot$ V$^{-1}\cdot$ s^{-1}) into the equation, it yields (Fujihira, 1997)

$$R_{on,sp} = 8.3 \times 10^{-9} V_{br}^{2.5} \quad (\Omega \cdot \text{cm}^2). \tag{6.13}$$

Equations (6.11) and (6.13) can be used to represent the unipolar ideal silicon limit on MOSFET performance merit.

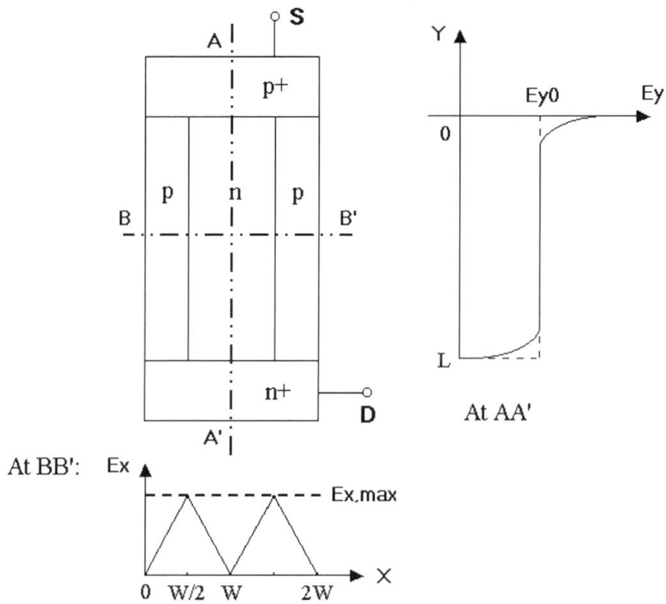

Fig. 6.2. Superjunction structure and the approximate electric field profile at $V_{DS} > 0$.

6.3. The Superjunction Structure

The typical SJ structure is shown in Fig. 6.2 (Chen, 1993). Different from the conventional structure as shown in Fig. 6.1, SJ devices provide the superior performance on blocking voltage and on-state resistance by replacing the monotonic drift region with the alternative heavily doped p–n semiconductor layers. Figure 6.2 also gives the approximate electric field profiles within the SJ device under the bias of $V_{DS} > 0$.

Optimum Doping Concentration

For the SJ device of a given breakdown voltage, there exists an optimum doping concentration $N_{d,op}$ that results in the best performance. Equation (6.14) gives the depletion width of p–n junction under certain reverse bias voltage V.

$$W_{dep} = \sqrt{\frac{2\varepsilon_s V(N_A + N_D)}{qN_A N_D}}, \tag{6.14}$$

where N_A and N_D are the doping concentration of p- and n-column, respectively. In ideal SJ structure above-mentioned, it is required that

$$N_A = N_D = N_{d,op} \quad \text{and} \quad W_p = W_n \tag{6.15}$$

to have the best performance. Assume such a condition that, when the device is at breakdown, the depletion region of SJ structure is just pinched-off horizontally. That is, the p–n column width $W_p = W_n = W$ is equal to the lateral depletion width of W_{dep}. By combining Eqs. (6.10), (6.14) and (6.15), the relationship between $N_{d,op}$ and W is approximated as

$$N_{d,op} = 1.2 \times 10^{12} W^{-8/7}. \tag{6.16}$$

SJ $R_{on,sp}$ Calculation

From the device simulation, one knows that the breakdown of SJ structure always happens at the interface between p- and n-columns. In SJ structure, due to the influence from the sidewall p-column, there provides an additional horizontal electric field component ($\hat{E}_x$) in comparison to that of the conventional p–i–n diode where the electric field is only in the vertical direction. Along the central vertical line of n or p-column, horizontal electric fields generated from neighboring p–n junction are in opposite direction and counteract each other. Thus, only the vertical electric field manifests. Where at the interface of p–n columns, $\hat{E}_x$ reaches the maximum value $\hat{E}_{x,\max}0$ and vertical electric field ($\hat{E}_y$) maintains nearly constant, the total electric field ($\hat{E}$) is the sum of both and becomes the highest. The profile of vertical field $\hat{E}_y$ is shown in Fig. 6.2. To simplify the derivation of blocking voltage, we assume that the p–n column length L is large enough, $\hat{E}_y$ has a constant (average) value of $\hat{E}_{y0}$ and breakdown occurs when the total electric field reaches the critical electric field of silicon ($\hat{E}_c$), which is expressed by

$$\hat{E}_c = \sqrt{\hat{E}_{x,\max}^2 + \hat{E}_{y0}^2}. \tag{6.17}$$

We define

$$\hat{E}_{x,\max} = \alpha \hat{E}_c \tag{6.18}$$

then

$$\hat{E}_{y0} = \sqrt{1-\alpha^2}\,\hat{E}_c, \tag{6.19}$$

where α is the coefficient which has the value between 0 and 1.

From the p–i–n junction structure, $\hat{E}_{x,\max}$ is given by

$$\hat{E}_{x,\max} = \frac{qN_A L}{2\varepsilon_s} = \frac{qN_D L}{2\varepsilon_s} < \hat{E}_c. \tag{6.20}$$

Combining Eqs. (6.18) and (6.20), it yields

$$q\mu N_D = (2\mu\alpha\varepsilon_s\hat{E}_c)/L. \tag{6.21}$$

Thus, the $R_{\mathrm{on,sp}}$ in the region of $(0 < x < W)$ is

$$R_{\mathrm{on,sp}} = \frac{2L}{q\mu N_{\mathrm{D}}} = \frac{WL}{\mu\alpha\varepsilon_{\mathrm{s}}\hat{E}_{\mathrm{c}}}. \tag{6.22}$$

Therefore if $L \gg W$, V_{br} is given by

$$V_{\mathrm{br}} \approx \hat{E}_{y0}L = \sqrt{1-\alpha^2}\hat{E}_{\mathrm{c}}L. \tag{6.23}$$

Combining Eqs. (6.22) and (6.23), it yields

$$R_{\mathrm{on,sp}} = \frac{WV_{\mathrm{br}}}{\alpha\sqrt{1-\alpha^2}\mu\varepsilon_{\mathrm{s}}\hat{E}_{\mathrm{c}}^2}. \tag{6.24}$$

Numerical simulations by MEDICI were carried out in order to find out the relationship between the electric field components of SJ structure and the critical electric field at various doping concentrations and different column width and length. Table 6.1 gives the results between $\hat{E}_{x,\mathrm{max}}$ and $\hat{E}_{\mathrm{c}}$ according to different dimensions and doping profiles of SJ structures. To guarantee the data accuracy, $\hat{E}_{x,\mathrm{max}}$ is extracted from the mesh point at the center of the vertical line along the p–n interface. $\hat{E}_{\mathrm{c}}$ is calculated from Eq. (6.9).

It was observed that, $\alpha = \hat{E}_{x,\mathrm{max}}/\hat{E}_{\mathrm{c}}$ has a nearly constant value when $L \gg W$ is satisfied. As shown in Fig. 6.3, when W/L is less than 1/4, all the simulation results of $\hat{E}_{x,\mathrm{max}}/\hat{E}_{\mathrm{c}}$ fall into the range of 0.669–0.673. Therefore, the average value of 0.672 is an acceptable value for α.

The minimum $R_{\mathrm{on,sp}}$ is then calculated by $R_{\mathrm{on,sp}} \approx \frac{2WV_{\mathrm{br}}}{\mu\varepsilon_{\mathrm{s}}\hat{E}_{\mathrm{c}}^2}$ with $\alpha = 0.672$. By using the average constant mobility and Eqs. (6.9) and (6.10), the $R_{\mathrm{on,sp}}$ for the vertical structure can be calculated as

$$R_{\mathrm{on,sp}} \approx 2.18 \times 10^{-3} WV_{\mathrm{br}}^{1.17}. \tag{6.25}$$

Using Eq. (6.25), the ideal SJ limits of specific on-resistance versus breakdown voltage at different column widths are plotted together with the ideal unipolar silicon limit of conventional MOSFET device, as shown in Fig. 6.4. Other detailed derivation by considering the mobility variation and lower critical electrical field within the SJ structure was made (Chen, 1999). Overall, the $R_{\mathrm{on,sp}}$ of an SJ structure can be considered to be proportional to the breakdown voltage in a range between $V_{\mathrm{br}}^{1.17}$ and $V_{\mathrm{br}}^{1.3}$.

6.3.1. SJ Electric Field Profiles

Figure 6.5 shows the simulation plots of SJ structure with $W = 5\,\mu$m and $L = 15\,\mu$m at different positive V_{DS} bias when $V_{\mathrm{GS}} = 0$ V produced by MEDICI.

Table 6.1. α values between $\hat{E}_{x,\max}$ and $\hat{E}_{\text{crit}}$ at different ratios of W/L.

N (cm^{-3})	**9.8×10^{16}**	**4.47×10^{16}**	**1.27×10^{16}**	**7.1×10^{15}**	**3.2×10^{15}**	**1.5×10^{15}**
W (μm)	0.5	1	3	5	10	19.5
$\hat{E}_c$ (V/cm)	5.33×10^{15}	4.84×10^{15}	4.13×10^{15}	3.84×10^{15}	3.48×10^{15}	3.16×10^{15}
$L = 15\mu$m						
$\hat{E}_{x,\max}$ (V/cm)	3.569×10^5	3.256×10^5	2.773×10^5	2.546×10^5	1.956×10^5	1.126×10^5
W/L	0.033	0.067	0.2	0.333	0.667	1.3
$\alpha A \hat{E}_{x,\max}/\hat{E}_c$	0.670	0.673	0.671	0.663	0.562	0.356
$L = 20\ \mu$m						
$\hat{E}_{x,\max}$ (V/cm)	3.569×10^5	3.256×10^5	2.775×10^5	2.577×10^5	2.158×10^5	1.430×10^5
W/L	0.025	0.05	0.15	0.25	0.5	0.975
$\alpha = \hat{E}_{x,\max}/\hat{E}_c$	0.670	0.673	0.672	0.671	0.620	0.453
$L = 50\ \mu$m						
$\hat{E}_{x,\max}$ (V/cm)	3.569×10^5	3.256×10^5	2.775×10^5	2.585×10^5	2.329×10^5	2.065×10^5
W/L	0.01	0.02	0.06	0.1	0.2	0.39
$\alpha = \hat{E}_{x,\max}/\hat{E}_c$	0.670	0.673	0.672	0.673	0.669	0.654

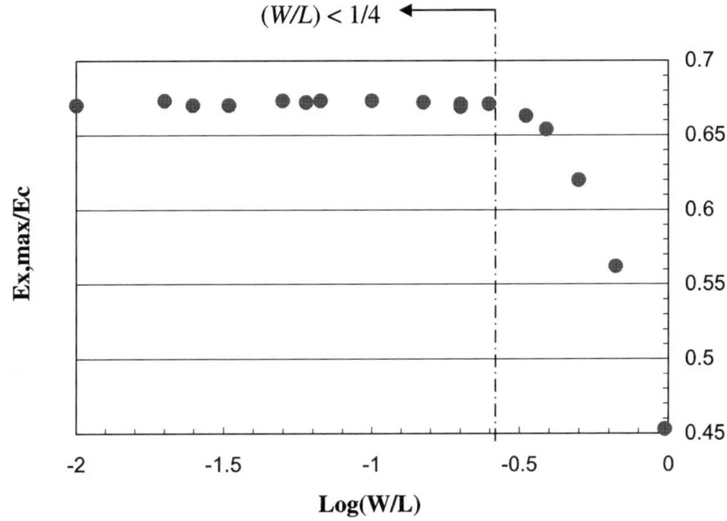

Fig. 6.3. Relationship between $\hat{E}_{x,\max}/\hat{E}_{\mathrm{c}}$ and W/L for SJ structure.

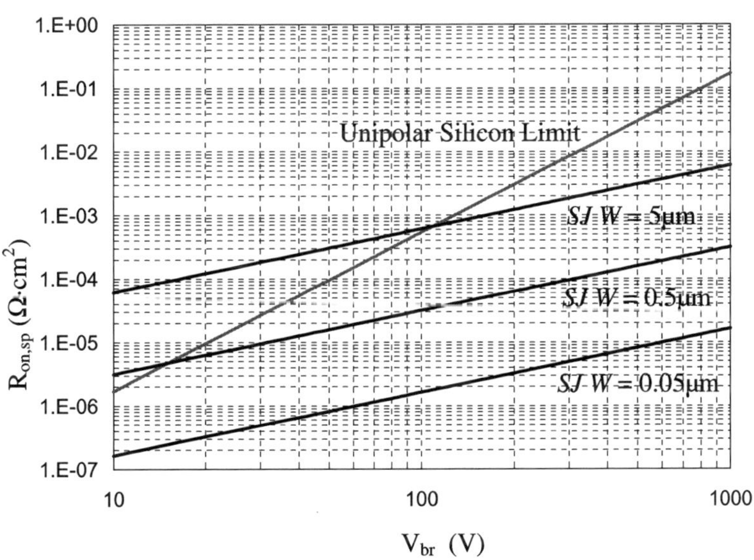

Fig. 6.4. $R_{\mathrm{on,sp}}$ vs V_{br} performance merit of ideal unipolar silicon limit and SJ limits at $W = 5\,\mu\mathrm{m}$, $0.5\,\mu\mathrm{m}$, and $0.05\,\mu\mathrm{m}$.

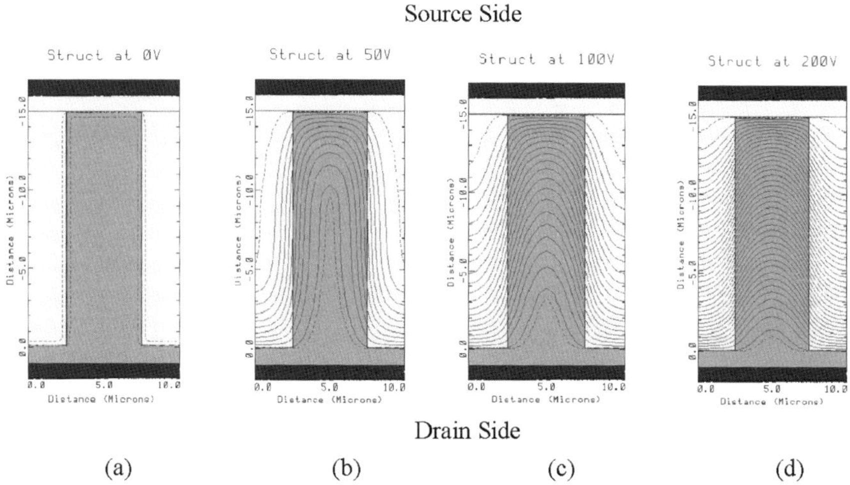

Fig. 6.5. Equipotential lines within SJ structure at different V_{DS} bias before breakdown.

V_{br} is simulated to be 261 V for this structure. The doping concentrations N_A and N_D for p- and n-column in the structure are exactly equal to get the ideal result. However, in the practical fabrication, perfect match of $N_A = N_D$ is difficult to achieve. The analysis of charge imbalance will be discussed in the later part of this chapter. When $V_{DS} = 0$ (see Fig. 6.5(a)), the depletion region that results from built-in potential between p–n columns is very small. The built-in potential (V_{bi}) is approximated by

$$V_{bi} \approx \frac{kT}{q} \ln \left(\frac{N_A N_D}{n_i^2} \right). \tag{6.26}$$

From Eq. (6.14) we know that in the case of $N_A = N_D = 7 \times 10^{15}\,\text{cm}^{-3}$, the built-in depletion width (W_{bi}) is about 0.5 μm. After $V_{DS} > 0$, the depletion region starts to extend. It is observed in Figs. 6.5(b)–(d) that the depletion region first merges on the top part of the drift region at a small bias, then moves down with the increase of the bias.

At the breakdown, the entire drift region is fully depleted as shown in Fig. 6.6. The distribution of equipotential lines (the parallel curves shown in Fig. 6.6(a)) is nearly uniform within the drift region. The E-field vector plots in Fig. 6.6(b) represents that electric field crowds obviously at the p–n column interface, especially at the top and bottom regions. Along the vertical direction at the center of the n-drift region, only vertical electric field is found. The above simulation result verifies the electric field profile as shown in Fig. 6.2.

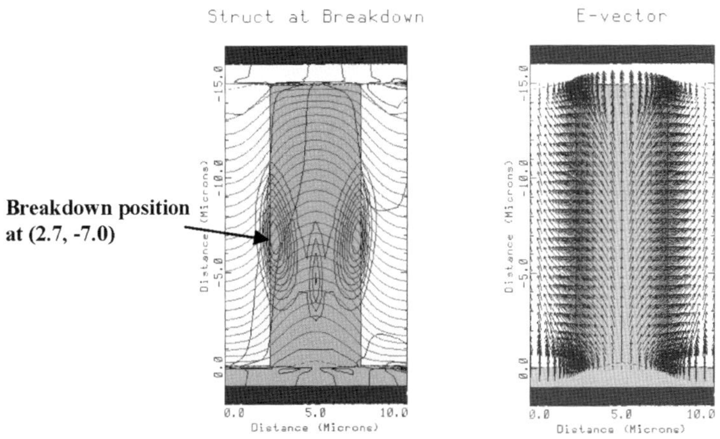

(a) (b)

Fig. 6.6. (a) Equipotential lines at 10 V interval, impact ionization representation and (b) E-field vector plots for SJ structure at breakdown.

As shown in Fig. 6.7(a), when increasing $V_{\rm DS}$, the vertical electric field profiles of SJ structure at the center of n-column ($x = 5\,\mu$m) start to transform from a triangle shape as that in the conventional case to a trapezium-like distribution. The electric field at p–n interface ($x = 2.5\,\mu$m) as shown in Fig. 6.7(b) is almost in the same shape at any bias but different in magnitude as determined by the applied bias.

6.3.2. Charge Imbalance

To utilize the benefit of SJ structures to the maximum, the principle of charge compensation must be satisfied as

$$Q_{\rm n} = Q_{\rm p} \quad \text{and} \quad W_{\rm n} \times N_{\rm D} = W_{\rm p} \times N_{\rm A}. \tag{6.27}$$

And, the doping integral over a layer perpendicular to the current flow direction remains smaller than the specific breakthrough charge for silicon of about $2 \times 10^{12}\,{\rm cm}^{-2}$ (Deboy *et al.*, 1998). Thus, the enhanced doping level of the current carrying n region results in a significant drop in resistivity.

Experimental data from Shenoy *et al.* (1999) reveal that the change of $V_{\rm br}$ in a SJ structure is dependent on the value of the charge imbalance (i.e. $\Delta Q = |Q_{\rm p} - Q_{\rm n}|$). It becomes lower when ΔQ increases. Other effects on charge imbalance to the device transient performance include higher electric field, peak recovery current, and higher turn-off losses. Numerical simulations prove

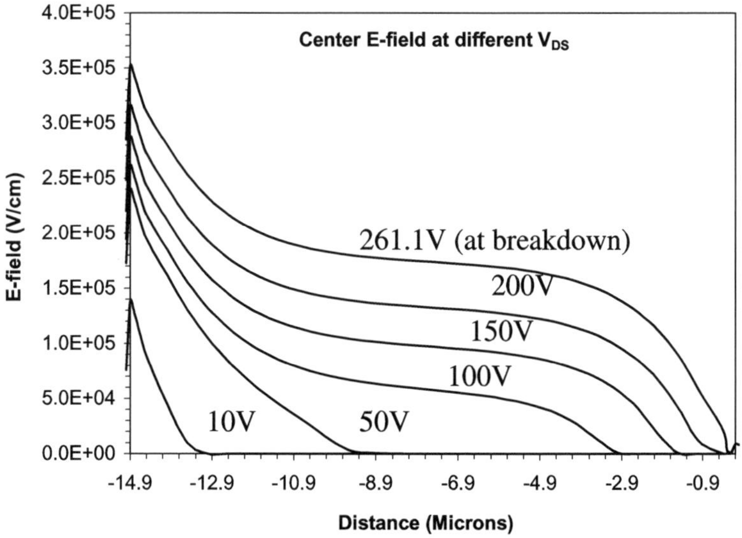

(a)

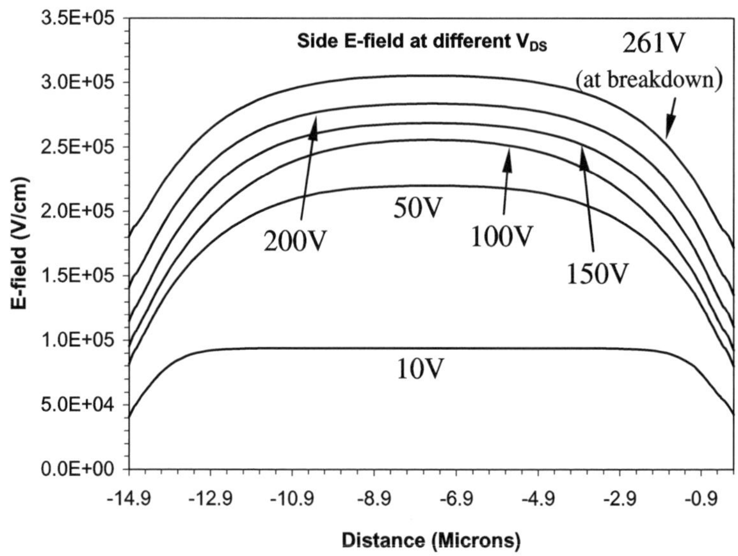

(b)

Fig. 6.7. Electric field plots along (a) $x = 5\,\mu\text{m}$ and (b) $x = 2.5\,\mu\text{m}$ at different V_{DS}.

that the SJ device is highly sensitive to charge imbalance especially for devices designed for low on-state resistance.

6.3.3. Fabrication Technologies

The COOLMOS above 600 V by Siemens was the first commercially available silicon device exploiting the SJ concept (see Fig. 6.8). It is able to reduce the resistivity by a factor of 5, which leads to the reduction of chip size in comparison to that of the conventional MOSFET of the same voltage rating. Also, the fall time of COOLMOS is short due to quick removal of carriers of charge storage. It comes to compete with IGBT devices in medium-high voltage, high frequency applications where switching losses of IGBT become more predominant.

The distinguished feature in the process technology of COOLMOS is that multi-epitaxy technology is used in fabrication to achieve the near charge balance between p–n columns. However, there are some drawbacks (Minato *et al.*, 2000):

- Repetitive masking, epitaxy growth, and implantation steps are required to realize the p–n columns especially when the column is long, which leads to a higher cost and process complexity.

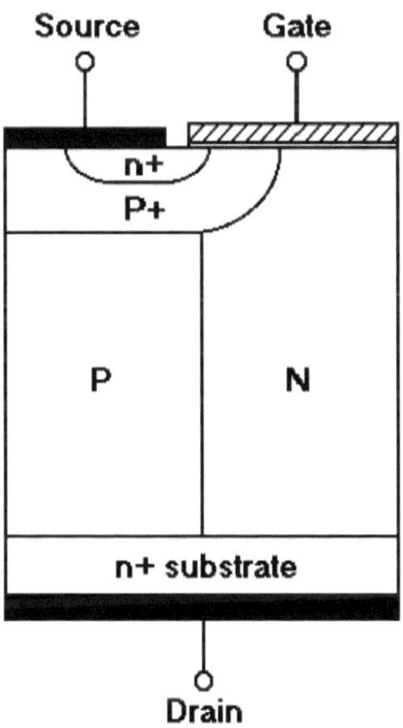

Fig. 6.8. Typical COOLMOS structure (mirrored cross-section).

- A long thermal budget in drive-in treatment is needed to anneal the deep and vertically buried impurity regions formed by the repetitive epitaxy growth and implantation. This causes dopant interdiffusion between p–n columns resulting in a nonabrupt transition junction and lower effective impurity concentration. Consequently, the performance merit of both V_{br} and $R_{on,sp}$ will be affected.

Super Trench MOSFET (STM) (Nitta *et al.*, 2000) has its vertical p- and n-columns formed within the mesa region between adjacent trenches. The trench is then filled with insulator as in Fig. 6.9(a). The p–n layers are made by boron and phosphorus tilted implant at the opposite sidewalls of the deep silicon trenches. The advantage of having this device structure is that, compared to the multi-epitaxy technology in COOLMOS, it simplifies the fabrication process by having only one additional mask over the conventional DMOS process to achieve the near SJ performance.

The trench width, which takes up the area in the drift region, must be wide enough to guarantee the depth of the trench and the suitability of tilted implantation. This will reduce the conduction area especially when p–n column dimensions become small. Therefore, it limits the fabrication of STM devices for low voltage application, e.g. below 200 V, where the device has small column width.

The structure of vertical deep trench RESURF DMOS (VTR-DMOS) is shown in Fig. 6.9(b) (Glenn and Siekkinen, 2000). It can be formed through

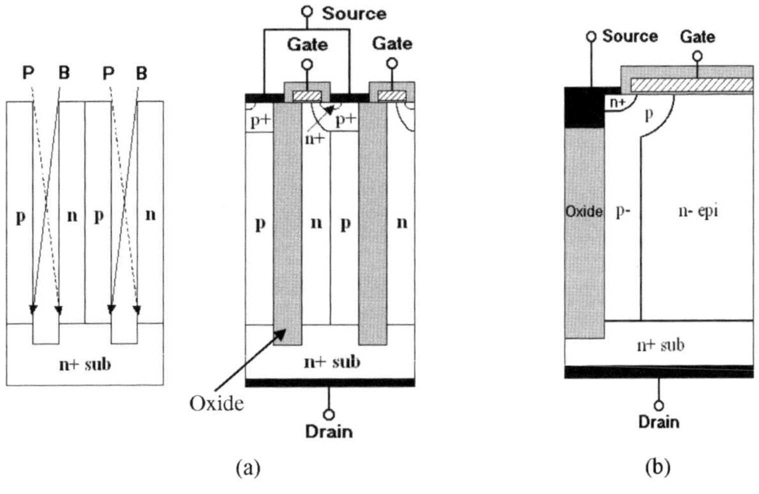

Fig. 6.9. (a) STM structure an (b) VTR-DMOS structure.

the following steps in the n-epitaxy region.

- p-well formation.
- Silicon deep trench etching in the n-epitaxy layer.
- Thin *in situ* boron-doped LPCVD poly-Si deposition and boron diffusion drive-in at 1000°C.
- Dry oxidation to convert poly-Si into SiO_2 (the sidewall SiO_2 is to passivate the surface along the trench sidewall).
- n+ implantation.
- TEOS filling in the trench.

For a design case study, simulation result of the VTR-DMOS shows that the $R_{on,sp}$ and V_{br} are in the range between the best case of $25.7\,m\Omega \cdot cm^2$ and $V_{br} = 880$ V, and the worse case of $26.1\,m\Omega \cdot cm^2$ and $V_{br} = 700$ V. While for the conventional DMOS, it is of $143\,m\Omega \cdot cm^2$ and $V_{br} = 710$ V.

6.4. Practical SJ Performance

For the constraints existing in device structure or fabrication technology, p–n column interdiffussion is one of the problems causing significant performance degradation. In this section, the phenomenon and the study of practical superjunction device performance under given thermal cycles are discussed.

Most of the fabricated SJ devices are in vertical structure. To achieve the best performance in superjunction structure, precisely charge-balanced p- and n-voltage-sustaining columns must be formed to have exactly the same doping concentration and the same column width. However, to achieve an ideal matching condition in practice is rather difficult, if not impossible. In addition, even if the doping of both p–n columns could be made equal, the quality of sidewall junction formed by the neighboring p–n columns would play an another important role to influence the breakdown voltage. This constraint becomes more dominant when p- and n-column widths get smaller for devices rated in low voltage region, e.g. below 200 V. For a column width less than $3\,\mu m$, the approach of using multiple-layer epitaxial growth and implantation process (Deboy *et al.*, 1998) to form vertical p–n columns becomes unsuitable due to the constraints of multiple misalignments of masks, auto-doping in the epitaxial process and thermally induced dopant interdiffusion in annealing process. Instead, the silicon trench and selective epitaxial growth processes can be used at a lower temperature to overcome the above-mentioned constraints (Yamauchi *et al.*, 2002; Chang and Sze, 1996; Regolini *et al.*, 1989).

However, other follow-on fabrication processes involving thermal cycles for MOSFET gate and source formation are still needed. These high temperature thermal cycles promote dopant interdiffusion between nonideal p–n columns and give rise to the formation of a wider dopant transition region at the junction, resulting in more nonuniform dopant distribution. The sidewall junction quality is thus worsened, and in turn, this lowers the achievable breakdown voltage of superjunction devices having narrow column width.

To study the realistic SJ performance under given process thermal cycles, two vertical SJ MOSFET structures, namely DMOS and UMOS were compared in process and device simulations. The schematic view of DMOS and UMOS structures are shown in Figs. 6.10(a) and (b).

The conducting n-column is set initially to have the equal doping and width as the nonconducting p-column at the selective epitaxy growth stage then followed by all needed thermal processes in order to find the SJ device performance in practice. In process simulation, a silicon wafer of n-epitaxial layer grown on n^+ substrate is used. A trench is etched and p-column is

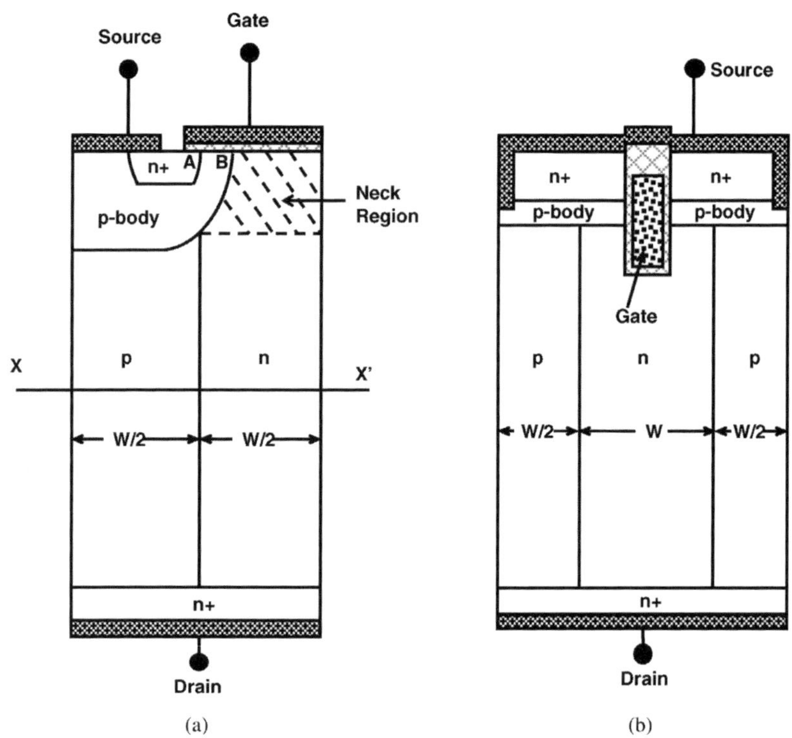

Fig. 6.10. The vertical superjunction power MOSFET structures: (a) DMOS structure and (b) UMOS structure.

grown by low-temperature selective epitaxy. The epitaxial layer now containing interdigitated p- and n-columns is further processed to form p-body, gate and source regions so as to complete the MOSFET structure. The p-body and source regions are self-aligned to the gate edge to ensure fixed channel length for structural consistency for all cases in the investigation. The channel length is fixed at 0.3 μm. The right edge of source region, i.e. point A in Fig. 6.10, is always aligned with the junction of p–n columns. The distance between p-body right edge, i.e. point B, and the p–n column junction is maintained at a reasonable distance, so that the neck region in DMOS structure will not be too small. The device geometries are summarized as following. The minimum source length is 0.5 μm, of which the overlap with gate oxide region is 0.1 μm. The column width varies from minimum of 1.3 to 9 μm at 0.1 μm step. No obvious degradation on breakdown voltage by the intercolumn dopant diffusion in SJ structure was observed when the width is more than 9 μm in the simulated breakdown voltage range from 80 to 1000 V. The neck length for DMOS is equal to half of the column width minus the channel length, i.e. it varies from 0.35 μm to 4.2 μm. The epitaxial thickness is adjusted to the minimum for each specific breakdown voltage so that smallest $R_{\mathrm{on,sp}}$ can be achieved. All essential thermal cycles such as gate oxide growth, n^+ source–drain anneals and p-body drive-in are included in TSUPREM process simulation. The main thermal processes in MOSFET device fabrication are listed in Table 6.2. The thermal budget listed is taken from the actual fabrication condition in the run sheet, which is intended for high voltage device.

To find the realistic and optimum SJ performance curve, the approach is to find the minimum $R_{\mathrm{on,sp}}$ at various column width, doping concentration, and drift length at each breakdown voltage. For each breakdown voltage, the column width (W) is set at the maximum value and the initial concentration, (N) is set at 30% of the theoretical value of $N = 1.2 \times 10^{12} \times W^{-8/7}$ to start with the numerical iterations. The drift region depth is set initially at a value by assuming uniform field of 10 V/μm over the drift length. The structure formed by the process simulation (TSUPREM) is imported in device simulation (MEDICI) to simulate device performance. The iterations are to find the locally optimal doping concentration first at each column width, and for

Table 6.2. Main thermal cycles for MOSFET device fabrication.

Thermal process	Time (min)	Temperature (°C)
Gate oxide growth	76	1000
P-body drive-in	20	1100
Source/drain anneal	30	900

all different column widths to determine the globally optimal column width and doping concentration to yield the minimum $R_{on,sp}$ for a given breakdown voltage.

For example, let us consider the case of column width $W = 1.7\,\mu$m for device breakdown rating of 300 V. The calculated optimal concentration based on the ideal SJ theory for $W = 1.7\,\mu$m is 2.44×10^{16} cm^{-3}. In practice, the optimal concentration is likely to be lower; hence, it starts from the concentration of 8×10^{15} cm^{-3} (about 30% of ideal value). The concentration is gradually increased while the width W is fixed. For each concentration, the drift length to sustain the expected breakdown voltage, i.e. 300 V, is adjusted accordingly. Then, $R_{on,sp}$ is found with that concentration and epitaxial thickness. As can be seen in Fig. 6.11, under realistic process conditions, the increase in the doping concentration enables a lower $R_{on,sp}$ and the performance point moves toward the point Q where the optimal value is reached. This occurs at the concentration of 1.3×10^{16} cm^{-3}, a value below the theoretically calculated one mainly due to influence of dopant interdiffusion. During thermal process, the substrate dopant may also out-diffuse into the epitaxial drift region. The out-diffusion phenomenon has been considered and included in the drift length adjustment, so that the drift length after thermal process is at the correct quantity to sustain the target voltage, i.e. 300 V in the current example.

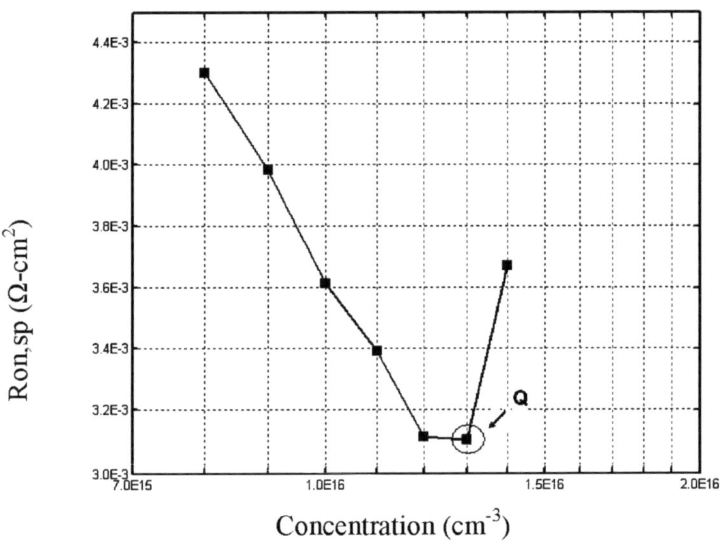

Fig. 6.11. The relationship between $R_{on,sp}$ and column doping concentration at given breakdown voltage (300 V) and column width (1.7 μm). The optimum concentration for column width of 1.7 μm is 1.3×10^{16} cm^{-3}.

For each column width, the optimal concentration and $R_{on,sp}$ to achieve a specific breakdown voltage can be found using the methodology described above. By observing $R_{on,sp}$ varying with changing W, the globally optimal $R_{on,sp}$ can be found for a given breakdown voltage. For the breakdown voltage of 300 V, $R_{on,sp}$ versus W curve is shown in Fig. 6.12. The optimal $R_{on,sp}$ at 300 V rating is $2.24 \times 10^{-3} \Omega \cdot cm^2$ at the column width of 2.3 μm. Hence, by following the same approach, it is now possible to find the globally optimal $R_{on,sp}$ for all different breakdown voltages. This has been done for breakdown voltages of 80, 100, 200, 300, 500, 800, and 1000 V, respectively.

The realistic optimal SJ performance curves under intercolumn diffusion influence for both DMOS and UMOS of minimum $R_{on,sp}$ versus V_{br} are shown in Fig. 6.13 together with the ideal silicon limit lines of conventional structure and ideal SJ performance lines for different column width W (Fujihira, 1997) and different Δ/L where Δ is the hexagonal column width and L is the drift region depth as defined in Chen *et al.* (2002). Experimental data extracted from the published work (Deboy *et al.*, 1998; Nitta *et al.*, 2000; Minato *et al.*, 2000; Glenn and Siekkinen, 2000; Gan *et al.*, 2002) are also labeled in Fig. 6.13. It is

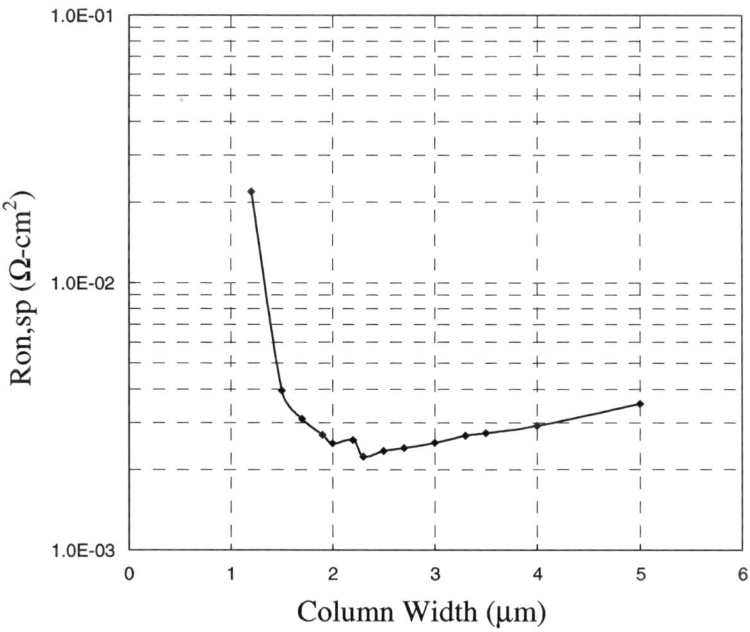

Fig. 6.12. The relationship between $R_{on,sp}$ and the column width at given breakdown voltage (300 V). The optimal $R_{on,sp}$ at 300 V is $2.24 \times 10^{-3} \Omega \cdot cm^2$ at column width $W = 2.3\,\mu$m.

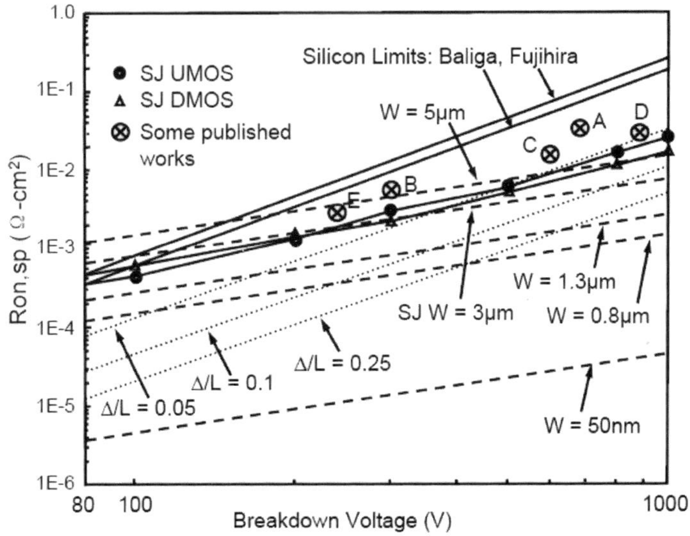

Fig. 6.13. The simulated practical optimum SJ performance curves (SJ DMOS and SJ UMOS) with dopant interdiffusion included compared with the ideal SJ performance lines, the ideal silicon limit curves and some published data (A: Deboy *et al.*, 1998; B: Nitta *et al.*, 2000; C: Minato *et al.*, 2000; D: Glenn and Siekkinen, 2000; E: Gan *et al.*, 2002).

shown that all the experimental data have a higher specific on-state resistance above the realistic optimal SJ performance curves. With dopant interdiffusion included in the practical process, the realistic optimal SJ performance curves are not linear on the log–log plane. Compared with the ideal silicon limit, when the breakdown voltage is below 80 V, the realistic $R_{\text{on,sp}}$ of SJ structure is higher than that of the ideal unipolar silicon limit.

The interdiffusion influence on breakdown voltage is studied. To show the transition between ideal and realistic situation, the doping profiles changing from ideal case without thermal budget to the realistic case with full thermal budget are shown in Fig. 6.14. The simulated SJ VDMOS has a column width of 2 μm and ideal doping concentration of 2×10^{16} cm^{-3}. The doping profiles are made by cutting across the p–n columns. The dopant interdiffusion creates a broader dopant transition zone at p–n junction and it also alters the junction position away from its original metallurgical position of equal p- and n-column widths.

By integrating the doping concentration in p- and n-columns, respectively, the difference between p-column doping integral Q_{P} and n-column doping

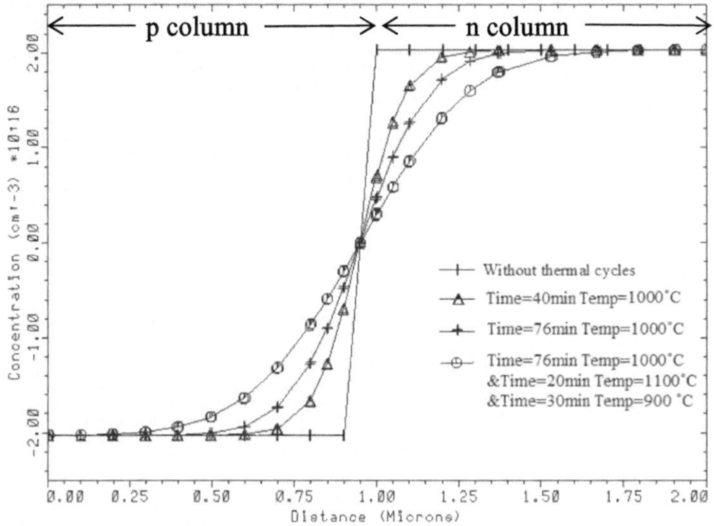

Fig. 6.14. The doping profiles change from the ideal profile to the realistic case with full thermal budget. The thermal budget for each doping profile is labeled in the figure. The SJ VDMOS has the column width of 2 μm and ideal concentration of $2.03 \times 10^{16}\text{cm}^{-3}$. The doping profiles are cut along line X–X′ as in Fig. 6.10(a).

integral Q_n can be found and is defined as the one-dimensional (1D) p–n doping integral imbalance. The two-dimensional (2D) p–n doping imbalance can be obtained by numerically integrating the 1D p–n imbalance along the entire length of SJ column in the drift region. The results show that, for higher doping concentration the imbalance of doping integral is more severe for a given thermal cycle. In turn, the sustainable voltage becomes lower. Hence lower column concentration is preferred to maintain the high breakdown voltage.

The realistic relationship between the optimal concentration and column width for breakdown voltages of 300, 500, and 1000 V together with the theoretical concentration curve are shown in Fig. 6.15. All the three practical curves follow the same trend and deviate from the ideal curve when the column width becomes small.

To illustrate the phenomenon caused by doping integral imbalance, Fig. 6.16 shows the potential contours and depletion boundaries of two similar structures at breakdown voltage. They have the same column width of 1.5 μm and epitaxial depth of 22 μm but only the column dopant concentrations are different. The concentration in the structure of Fig. 6.16(a) is at the

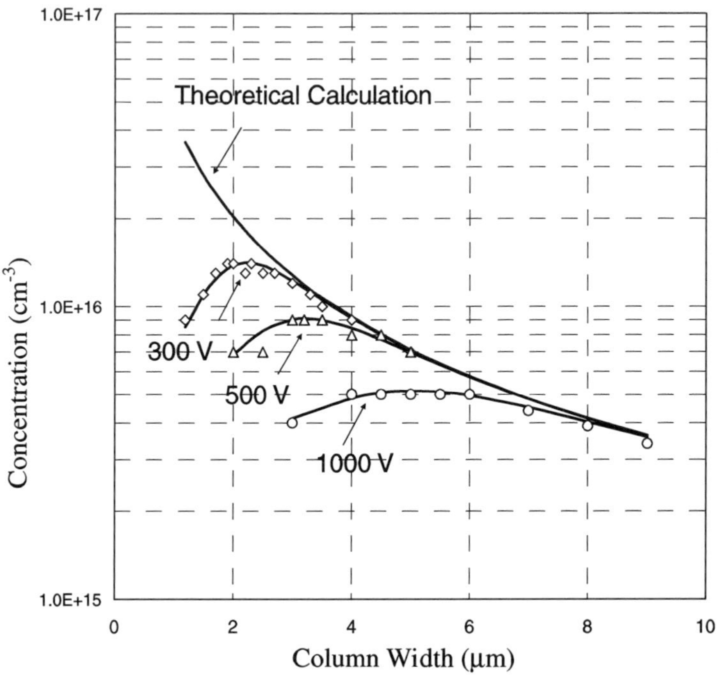

Fig. 6.15. The relationship between the simulated optimal concentration and column width at 300, 500, and 1000 V along with the ideal concentration curve which is calculated using the equation $N = 1.2 \times 10^{12} \times W^{-8/7}$ are shown.

SJ theoretical value of $2.82 \times 10^{16}\text{cm}^{-3}$. The best acceptable concentration of $1.1 \times 10^{16}\text{cm}^{-3}$ under practical conditions is used in the second structure of Fig. 6.16(b). Simulated results indicate the breakdown voltages of 149.5 V for structure (a) and 301.1 V for structure (b).

By observing the depletion boundaries (dashed lines in Fig. 6.16) in both structures, the lower breakdown voltage is due to incomplete depletion of n-column. In Fig. 6.16(a), the n-column is not fully depleted at breakdown voltage indicating that the SJ field action did not occur effectively. To study the relationship between p–n doping integral imbalance and incomplete column depletion, the depletion status of n-column, while p-column is just fully depleted was investigated. It was confirmed by the simulation that the doping integral of the undepleted n-column region is approximately equal to the 2D p–n doping integral imbalance.

The influence of p–n doping integral imbalance can also be observed in the slope of electrical field (E-field) profiles. The E-field profiles in

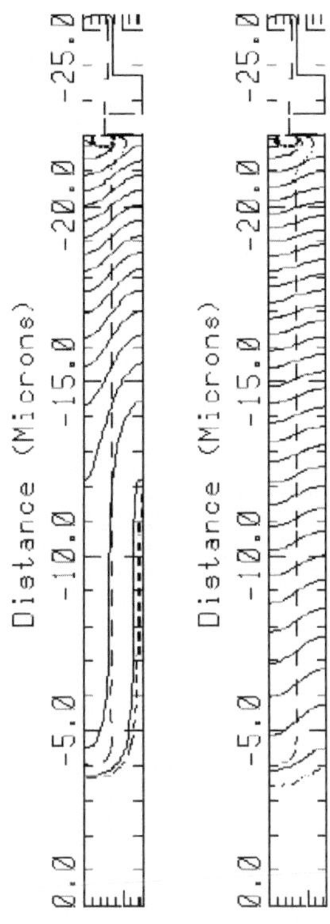

(a) (b)

Fig. 6.16. Solid lines are the potential contours and dashed lines are the depletion boundaries. Both structures have the same column width of 1.5 μm and epitaxial depth of 22 μm. The column concentrations are different, namely (a) the theoretical concentration of 2.8×10^{16}cm^{-3} having the breakdown voltage of 149 V, and (b) the optimal concentration of 1.1×10^{16}cm^{-3} having the breakdown voltage of 301 V.

Figs. 6.17(a) and (b) are corresponding to the structures of Figs. 6.16(a) and (b). The E-field profiles are plotted along the cut at the center of n-column along the drift region. The applied reverse voltage varies from 5 V to the respective breakdown voltage in a step of 5 V.

The particular E-field profile at the voltage when p-column is just fully depleted is marked as a dark line. The triangle-like shape of E-field in Fig. 6.17(a) indicates the incomplete depletion of n-column addressed by the doping integral imbalance. The slope of E-field profile maintains approximately the same value after p-column is fully depleted. Knowing the critical E-field is similar when the column concentration varies in a limited range, the

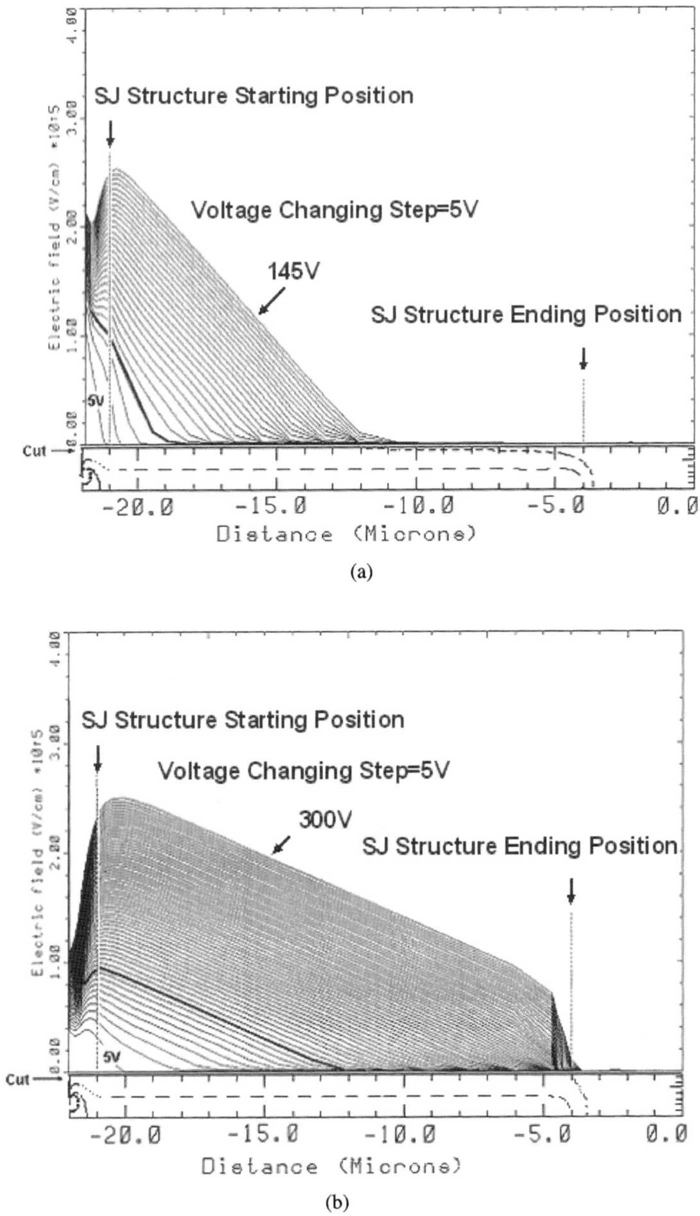

Fig. 6.17. The electric field profiles within n-column, (a) and (b) are corresponding to the structures in Figs. 6.16(a) and (b). The cuts are at the center of n-column with the voltage step of 5 V to the respective breakdown voltage.

slope of the E-field profile after p-column is fully depleted can be used as an index for the breakdown voltage. The lower E-field slope indicates a better depletion of n-column and as a consequence of higher breakdown voltage.

Simulation results indicate that the slope of the E-field profile is approximately proportional to the 2D p–n doping integral imbalance at fixed column width. Figure 6.18 shows the relationship between 2D doping integral imbalance and the E-field slopes at 1.5 μm at different concentrations.

With the above observations, the three realistic curves in Fig. 6.15 can be better explained. Take the 300 V curve as an example, the prospective slope of E-field at the breakdown voltage can be identified at a certain value. At the same time, it is known to be proportional to the 2D p–n doping integral imbalance at a fixed column width. When the theoretical concentration corresponding to a smaller column width gets higher, the charge imbalance becomes worse. When the imbalance gets larger, the E-field slope will get steeper. Thus, to maintain the E-field slope for a specific breakdown voltage, the doping concentration needs to be reduced to lower imbalance in order to maintain the full depletion in p–n columns.

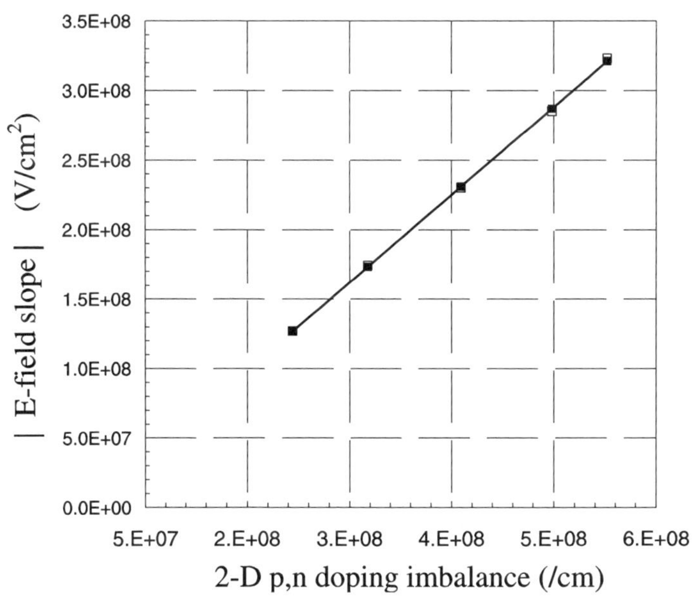

Fig. 6.18. The relationship between 2D p–n doping integral imbalance and the E-field slopes.

6.4.1. The Practical Concentration Equation

Figure 6.15 shows the practical concentration deviates from the theoretical concentration when the column width becomes small. As the column width increases, the practical concentration curve merges with the theoretical curve. The merger of the two curves shows that as column concentration decreases to a certain value and the column width is large enough, the doping integral imbalance caused by the interdiffusion becomes unimportant and plays a very little role. The merge point of the practical curve and the theoretical curve is dependent on the breakdown voltage rating. As the rating becomes higher, the merger point occurs at larger column width, i.e. lower concentration. Higher breakdown voltage corresponds to longer drift region depth which will make 2D doping integral imbalance larger. Hence, column doping concentration needs to be reduced in order to keep the 2D doping integral imbalance low. The merger point can be empirically identified by the product of the breakdown voltage and the theoretical concentration. The product has a constant value of approximately 4×10^{18} V/cm^3 for ratings between 300 V and 1000 V as observed from the simulation results of a given thermal processes. It is recommended that this constant in product is used to identify the merger point on the concentration–column width graph at a given breakdown voltage. The concentration in columns should be lower than the one predicted by the theoretical equation once the column width falls below the merger point. Hence a simple method to gauge onset of deviation from ideal concentration at different column width is possible. A simple empirical equation to fit all the curves in Fig. 6.15 was found. In this equation, realistic doping concentration N for a given breakdown rating based on actual given process is

$$N \approx \frac{1.2 \times 10^{12} \times W^{-\frac{8}{7}}}{1 + \frac{3}{V_{\mathrm{br}}^{0.7}} e^{\frac{7(W_{\mathrm{m}}-W)}{W_{\mathrm{m}}}}}, \tag{6.28}$$

where W_{m} is found by $1.2 \times 10^{12} \times W_{\mathrm{m}}^{-\frac{8}{7}} \times V_{\mathrm{br}} = 4 \times 10^{18}$ as indicated earlier. The numerator of Eq. (6.28) is the expression of theoretical concentration, and the denominator is introduced due to the interdiffusion influence.

This proposed equation can be used to predict the optimum practical concentration for the given thermal cycle in processing. This equation is also useful for device design under known thermal cycles. For example, if a 400 V breakdown rating power MOSFET device is required to have $R_{\mathrm{on,sp}}$ as small as possible, the optimum practical concentration at different column width to sustain the breakdown voltage can be calculated from the equation. Through which, a set of column widths is chosen and the corresponding column concentrations are calculated at these column widths to find the most suitable ones

Table 6.3. The calculated practical concentration and the theoretical concentration for the device breakdown rating of 400 V.

Column width (μm)	Practical conc. (cm^{-3})	Theoretical conc. (cm^{-3})
1.5	7.22E+15	2.81E+16
2	9.51E+15	2.02E+16
2.5	1.09E+16	1.57E+16
3	1.08E+16	1.27E+16
3.5	1.00E+16	1.07E+16
4	8.94E+15	9.17E+15
4.5	7.94E+15	8.02E+15

with lowest on-state resistance. The chosen column widths are 1.5, 2, 2.5, 3.5, 4, and 4.5 μm. The calculated practical column concentrations and theoretical column concentrations are listed in Table 6.3. From there, the optimum column width is found to be 2.5 μm with the concentration of 1.09×10^{16}cm^{-3}. The table also shows that for column width larger than 3.7 μm, which is at the merger point (W_m) for 400 V rating, the practical concentration and theoretical concentration are approximately equal.

Two groups of devices were simulated by TSUPREM and MEDICI with given thermal cycles. They have the same drift region depth of 27 μm, the column widths, and concentration data obtained from Table 6.3. One group uses the practical concentration and the other group uses the theoretical concentration. After the same thermal cycles given by Table 6.2, the simulation results are shown in Fig. 6.19. Breakdown voltages for the group with practical concentrations are always above 400 V. As for the other group, it increases when column width increases. When the column width is larger than W_m, the breakdown voltages can reach over 400 V. Figure 6.19 has in effect proven the suitability and usefulness of Eq. (6.28).

6.4.2. Practical SJ Performance Equation

Based on the practical concentration equation given in Eq. (6.28), the practical SJ performance equation is derived. The specific on-state resistance of the drift region is given by

$$R_{\text{on,sp}} = \frac{L_{\text{drift}}}{q\mu(N/2)}. \tag{6.29}$$

Due to the existence of neck region in SJ VDMOS, the mismatch between the simulated resistance and Eq. (6.29) is visible. On the other hand, there is no neck region in SJ UMOS; the derived resistance is close to the simulated one.

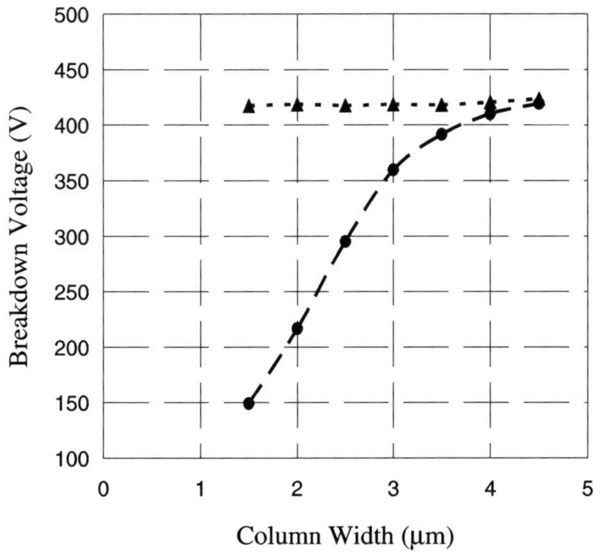

Fig. 6.19. The relationships between breakdown voltage and column widths for two groups of devices of the same drift region length of 27 μm and the same set of column widths. One group uses the theoretical concentration whose breakdown voltages are shown by long dashed lines and dots. The other group uses the practical concentration whose breakdown voltages are shown by short dash line and triangles.

Equation (6.29) can be derived further to become

$$R_{\mathrm{on,sp}} \approx 1.8 \times 10^{-2} W^{8/7} \left\{ 1 + \frac{3}{V_{\mathrm{br}}^{0.7}} e^{\frac{7(W_{\mathrm{m}}-W)}{W_{\mathrm{m}}}} \right\} V_{\mathrm{br}}^{7/6}, \tag{6.30}$$

where $R_{\mathrm{on,sp}}$ is in $\Omega \cdot \mathrm{cm}^2$; W in cm; and V_{br} in V.

Using Eq. (6.30), the practical SJ performance curves in the breakdown range of 300–1000 V for different column widths are plotted in Fig. 6.20. These curves show that, in practice, the $R_{\mathrm{on,sp}}$ does not necessarily decrease with a smaller column width. There exists an optimum column width for each breakdown voltage. The column width corresponding to the lowest on-state specific resistance for a breakdown rating among the practical superjunction curves is the optimum column width.

6.5. Polysilicon Flanked VDMOS (PF VDMOS)

To lower the fabrication cost and simplify the fabrication process of superjunction VDMOS device, as well as to overcome the interdiffusion problem

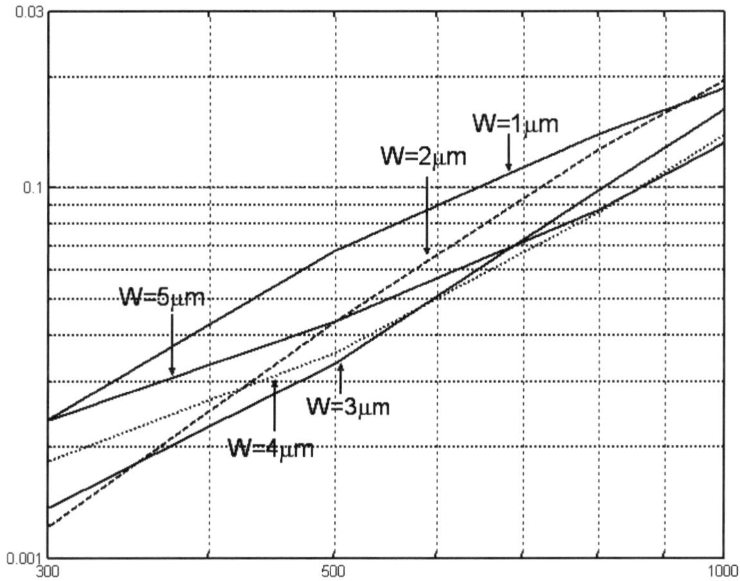

Fig. 6.20. Practical SJ performance curves under the nominal thermal cycles at column widths of 1, 3 and 5 μm.

between p–n columns, a technology named PF VDMOS was introduced. The fabrication of PF VDMOS were reported in Gan *et al.* (2002) and it could break the conventional MOSFET silicon limit. The full PF structure is shown in Fig. 6.21 with the edge termination.

To minimize the interdiffusion problem in SJ devices, a thin oxide film is grown between p- and n-columns. This is achieved by etching a trench in n-epi layer and followed by a quick dry oxidation. The oxide on the top and bottom surfaces is removed and the thin oxide is left on the trench sidewall only. To achieve uniform doping for p polysilicon column, boron tilted implantation into the conformal polysilicon layer, polysilicon refill and drive-in are carried out subsequently in that order. Finally, the polysilicon is etched back with slight over-etch to ensure that all polysilicon is removed from the active n-epi region. The subsequent fabrication steps are the same as those of conventional VDMOS devices in the n-epi columns. Thus, both high doping concentration and smaller column width are now achieved in alternating p–n layers to improve on-resistance while maintaining the high blocking voltage. Instead of complex multi-epitaxial growth, this structure can be fabricated by relatively simple steps of etching, deposition, implantation, and polysilicon refill. This brings forth the new milestone that SJ MOS-devices

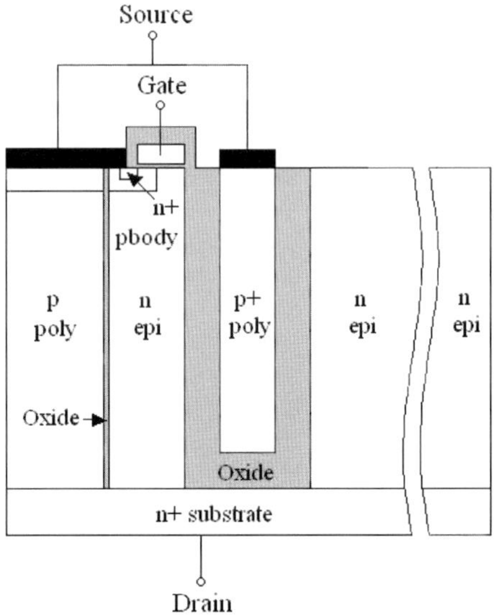

Fig. 6.21. The PF SJ VDMOS structure with edge termination.

can now be fabricated by the standard cleanroom facilities without epitaxy growth.

Device simulation by TSUPREM, MEDICI and measurement of fabricated PF VDMOS devices were carried out. At the same blocking voltage of 240 V, the doping concentration of $N_D = 7 \times 10^{15}\,cm^{-3}$ in the PF VDMOS is much higher than that in the conventional VDMOS of $N_D = 1.2 \times 10^{15}\,cm^{-3}$.

The SEM photograph of PF VDMOS is shown in Fig. 6.22. To avoid premature breakdown occurs at the edge of the device, an edge termination with thick trench oxide and doped polysilicon is formed. The doped polysilicon is connected to the source electrode. The pads are placed on thick oxide next to the active region of the device. The thick oxide is needed to isolate the high voltage applied on the pads. Figure 6.23 shows the SEM picture of 15 μm in depth, and 2.5 μm in width of multiple trenches. The distance between two neighboring trenches is 1.2 μm. Wet oxidation and CVD oxide deposition are followed to make a whole cubic oxide region.

Measured breakdown voltage (V_{br}) of one fabricated PF SJ VDMOS sample is 187 V, which is 2.4 times higher than that of conventional VDMOS (breakdown at 55 V) at the same epi doping concentration of $7 \times 10^{15}\,cm^{-3}$. At $V_{GS} = 10$ V, measured $R_{on,sp}$ of the PF VDMOS at the current density of 100 A/cm^2 is 3.68 mΩ-cm^2 which is lower than the silicon limit of 3.97 mΩ-cm^2.

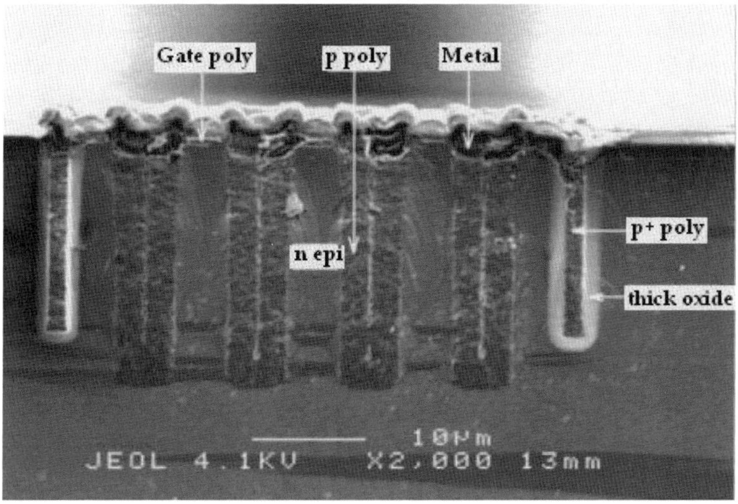

Fig. 6.22. SEM photograph of PF SJ VDMOS with MTO termination.

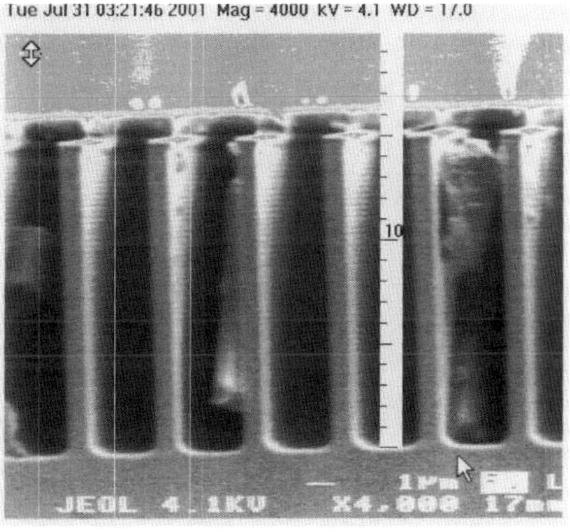

Fig. 6.23. SEM photograph of silicon trenches for the thick oxide formation.

The process sensitivity for the result of charge imbalance is an important concern. It is observed that, for the designed PF SJ VDMOS structure, the measured V_{br} is reduced when the mismatch ratio goes up. Charge imbalance of $\pm 10\%$ results in 22% reduction of V_{br}. For device fabrication, this mismatch

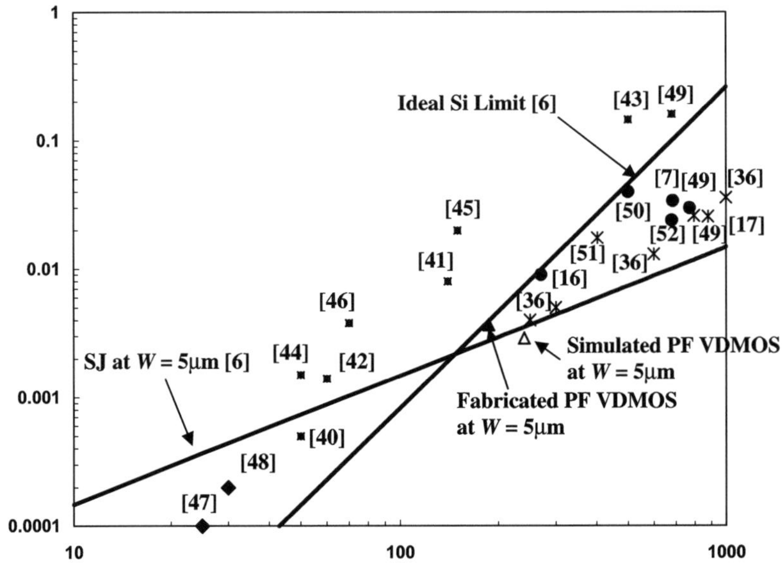

Fig. 6.24. $R_{on,sp}$ vs V_{br} performance of PF SJ VDMOS compared to ideal silicon limit, superjunction structure at p–n column width $W = 5\,\mu m$ (Fujihira, 1997), and the other works on SJ VDMOS devices extracted from (7: Deboy *et al.*, 1998; 16: Nitta *et al.*, 2000; 17: Glenn and Siekkinen, 2000; 36: Minato *et al.*, 2000; 40: Shenai *et al.*, 1998; 41: Berta *et al.*, 1991; 42: Wong *et al.*, 1991; 43: Ajit *et al.*, 1991; 44: Weyers and Vogt, 1992; 45: Ifstrom *et al.*, 1992; 46: Kobayashi *et al.*, 1995; 47: Baliga *et al.*, 1992; 48: Fujishima *et al.*, 2002; 49: Huang *et al.*, 2000; 50: Chen *et al.*, 2000; 51: Kondekar *et al.*, 2002; 52: Onishi *et al.*, 2002).

phenomenon does happen because the doping profile in p-poly is sensitive to the angle of tilted implantation and the thermal diffusion.

Figure 6.24 gives the $R_{on,sp}$ vs V_{br} performance of some published VDMOS and SJ devices. Most of the SJ devices are designed for high blocking voltage, the PF VDMOS at 200 V range becomes more attractive for applications of medium voltage rating. The fabricated PF VDMOS does break the ideal silicon limit.

6.6. Oxide Bypassed (OB) SJ MOSFET

It was recognized that the SJ structure is attractive for its high blocking ability. However, the performance of fabricated SJ devices is handicapped in reality by the realization of charge balance. So far the ideal charge balance in SJ p–n columns cannot be precisely realized due to the limitation of fabrication technologies. The introduction of oxide bypassed (OB) MOSFETs is an alternative to overcome the difficulty mentioned above. It was found that if the p-column of SJ structure was replaced by a thick oxide layer next to a polysilicon layer, the new structure will maintain a high breakdown voltage. In this section, basic theory and research efforts on OB structure will be described in detail.

The possible OB structures of OBVDMOS and OBUMOS are shown in Fig. 6.25. Compared to the SJ MOSFET, the OB structure is accomplished by making deep trenches next to the n-drift region. Along the trench sidewall, there is a grown thick oxide layer with the cavity filled with highly doped polysilicon which is connected to the source electrode of the MOSFET, i.e. at the ground potential. This OB structure is meant to deplete the n-column (drift region) by mimicking the superjunction lateral field insertion, and thus it modulates the electric field profile and enhances the breakdown voltage.

Under the blocking state, the depletion zone starts to form at the oxide/silicon interface. With the increasing voltage bias, the depletion zone expands laterally to the center of n-drift region, and at a certain voltage, the entire drift region becomes fully depleted. The OB structure moderates the electric field at drift region and results in a higher breakdown voltage. Apparently, OB structure provides a similar function like the p-column of SJ devices. Experimental measurement shows that the breakdown voltage (V_{br})

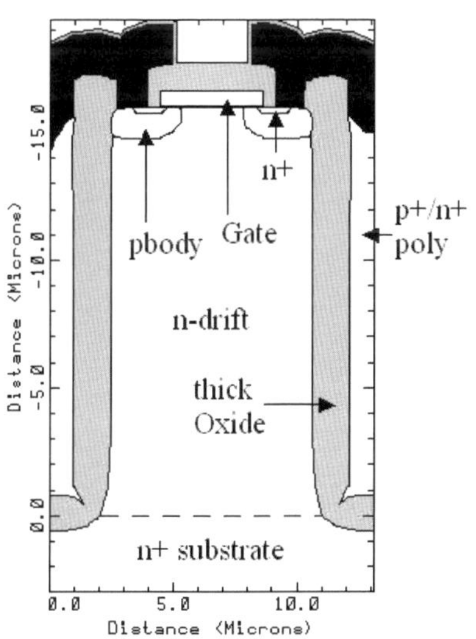

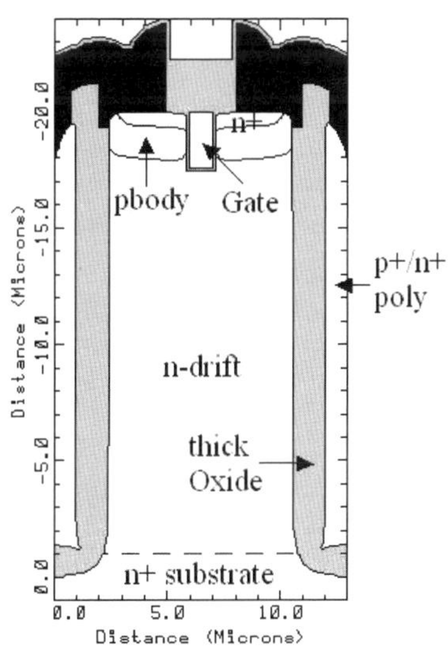

Fig. 6.25. Schematics of OBVDMOS (left) and OBUMOS (right).

is a function of the oxide thickness (t_{ox}), the impurity concentration (N_D), and the width (W) of the drift region between the two neighboring thick oxide sidewalls.

The Si/oxide/polySi structure can be treated as a capacitor. That is, the voltage drop on both sides of the thick oxide can be obtained by:

$$V = \frac{Q}{C} = \frac{Q \cdot t_{ox}}{A \cdot \varepsilon_{ox}} = \frac{qN_D \cdot W \cdot t_{ox}}{2\varepsilon_{ox}}, \tag{6.31}$$

where Q is the charge of the capacitor, C is the capacitance, and ε_{ox} is the permittivity of oxide. To prevent premature breakdown and achieve optimum performance, the sidewall field must deplete the n-drift region completely before the main p–n junction (between p-body and n-drift region) breaks down, i.e. $W_{dep} = W/2$, when $V \leq V_{br}$ should be satisfied. Hence, we can derive the maximum drift region concentration as (Liang *et al.*, 2001):

$$N_D = 2.90 \times 10^{11} \cdot \left(t_{ox} \cdot \frac{W}{2}\right)^{-\frac{4}{7}}. \tag{6.32}$$

The SEM photo of OBUMOS is shown in Fig. 6.26. The 1.5 μm thick oxide structure provides the field effect to be exerted to deplete the n-column laterally. Owing to this additional lateral depletion, the doping in the n-region

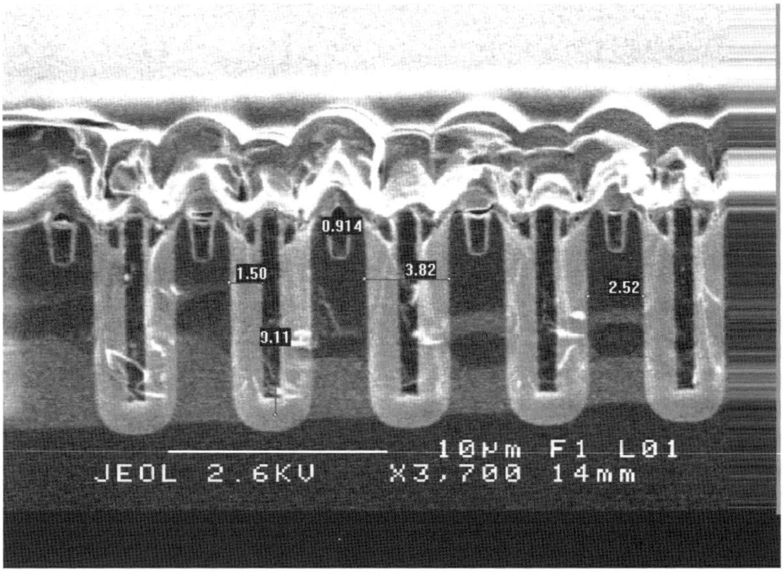

Fig. 6.26. SEM picture of fabricated OBUMOS with $t_{ox} = 1.5\,\mu$m.

can be raised higher to a value comparable to that in the SJ devices and thus overcoming the conventional unipolar silicon limit. There is no requirement in precisely doping matching as p-column does not exist in this structure.

In fully depleted drift region, there are two electric-field peaks along the silicon neighboring oxide sidewall region, namely one appears at the p-body/n-drift junction (around $y = -14.5\,\mu$m in Fig. 6.25) and the other is at the bottom of the OB sidewall (around $y = -1\,\mu$m) because the maximum horizontal electric field $\hat{E}_{\text{Si},x\,\text{max},\text{OB}}$ inserted into the drift region appears there. The maximum horizontal electric field is a fraction of the critical electric field at breakdown, defined as

$$\hat{E}_{\text{Si},x\,\text{max},\text{OB}} = \alpha_{\text{OB}}\hat{E}_{\text{c}}, \tag{6.33}$$

where $0 < \alpha_{\text{OB}} < 1$, and

$$\hat{E}_{\text{Si},x\,\text{max},\text{OB}} = \frac{\varepsilon_{\text{ox}}}{\varepsilon_{\text{Si}}}\hat{E}_{\text{ox},x\,\text{max},\text{OB}} = \frac{\varepsilon_{\text{ox}}}{\varepsilon_{\text{Si}}}\frac{V_{\text{br}}}{t_{\text{ox}}}. \tag{6.34}$$

Thus, the sidewall oxide thickness can be related to the breakdown voltage as

$$t_{\text{ox}} = \frac{\varepsilon_{\text{ox}}V_{\text{br}}}{\varepsilon_{\text{Si}}\alpha_{\text{OB}}\hat{E}_{\text{c}}}. \tag{6.35}$$

For OB structure, the electric field in the drift region is less uniformly distributed as compared to that in p–n superjunction counterpart. Here, we define that $\hat{E}_{\text{ave,OB}}$ is the average electric field in y-direction which is always smaller than the critical electric field,

$$\hat{E}_{\text{ave,OB}} = \beta_{\text{OB}}\hat{E}_{\text{c}}, \tag{6.36}$$

where β_{OB} is a coefficient which depends on the structure dimensions and $0 < \beta_{\text{OB}} < 1$. Then the drift region length can be obtained as

$$L = \frac{V_{\text{br}}}{\hat{E}_{\text{ave}}} = \frac{V_{\text{br}}}{\beta_{\text{OB}}\hat{E}_{\text{c}}}. \tag{6.37}$$

The on-state resistance is approximately equal to the resistance of drift region, i.e. $R_{\text{on,sp}} \approx R_{\text{drift}}$ as the drift region is long for a power device. The unit width of the OB device is $(W + 2t_{\text{ox}} + C)$, in which C is the polysilicon contact width, W is the drift region width, and t_{ox} is the sidewall oxide thickness. Thus, we have the specific on-state resistance calculated as

$$R_{\text{on,sp}} \approx \frac{L}{q\mu N_{\text{D}}} \cdot \frac{W + 2t_{\text{ox}} + C}{W}. \tag{6.38}$$

The carrier mobility μ for n-type doping with dependency on doping level at 300 K is

$$\mu_n = \frac{5.10 \times 10^{18} + 92N_D^{0.91}}{3.75 \times 10^{15} + N_D^{0.91}}. \tag{6.39}$$

By using the relationships derived earlier, the expression of specific on-state resistance (for very small contact width C) is given by

$$R_{on,sp} \approx 1.87 \times 10^{-2} \beta_{OB}^{-1} V_{br} H^{\frac{9}{13}} \frac{3.75 \times 10^{15} + 4.87 \times 10^{13} H^{-0.56}}{5.10 \times 10^{18} + 4.48 \times 10^{15} H^{-0.56}}$$

$$\times \left[1 + 2.16 \times 10^{-6} \left(\alpha_{OB}^{-1} W^{-1} V_{br}\right) H^{\frac{1}{13}}\right],$$

where

$$H = \alpha_{OB}^{-1} W V_{br}. \tag{6.40}$$

By numerical simulation using MEDICI, we can obtain α_{OB} vs w/L (shown in Fig. 6.27(a)) and β_{OB} vs W/L relationships (shown in Fig. 6.27(b)). From the results, we identify the following coefficients:

$$\left.\begin{aligned} \alpha_{OB} &\cong 0.9 \\ \beta_{OB} &= 0.9W/L \end{aligned}\right\} \quad \text{if } W/L \le 0.4. \tag{6.41}$$

Consequently, the breakdown voltage is derived as

$$V_{br} = \hat{E}_{ave,OB} L = \beta_{OB} \hat{E}_c L = 0.9 \hat{E}_c W. \tag{6.42}$$

Equation (6.42) shows that, when $W/L \le 0.4$, the breakdown voltage depends on drift region width W only and it is not sensitive to the drift region length L. Placing (6.42) into (6.35), one gets

$$t_{ox} \approx 0.33\, W. \tag{6.43}$$

By defining $L = \gamma W$, where γ is the aspect ratio, the specific on-state resistance is then given by

$$R_{on,sp} \approx 9.43 \times 10^3 \gamma W H^{\frac{8}{13}} \frac{3.75 \times 10^{15} + 4.87 \times 10^{13} H^{-0.56}}{5.10 \times 10^{18} + 4.48 \times 10^{15} H^{-0.56}}. \tag{6.44}$$

At $W/L \le 0.4$, for a given breakdown voltage, the specific on-state resistance is a function of the aspect ratio (γ) of the drift region. Simulated vertical component of electric field along the silicon neighboring oxide sidewall region is shown in Fig. 6.28. There are two field peaks in the drift region where breakdown point may locate: one occurs at the junction between p-body and n-drift,

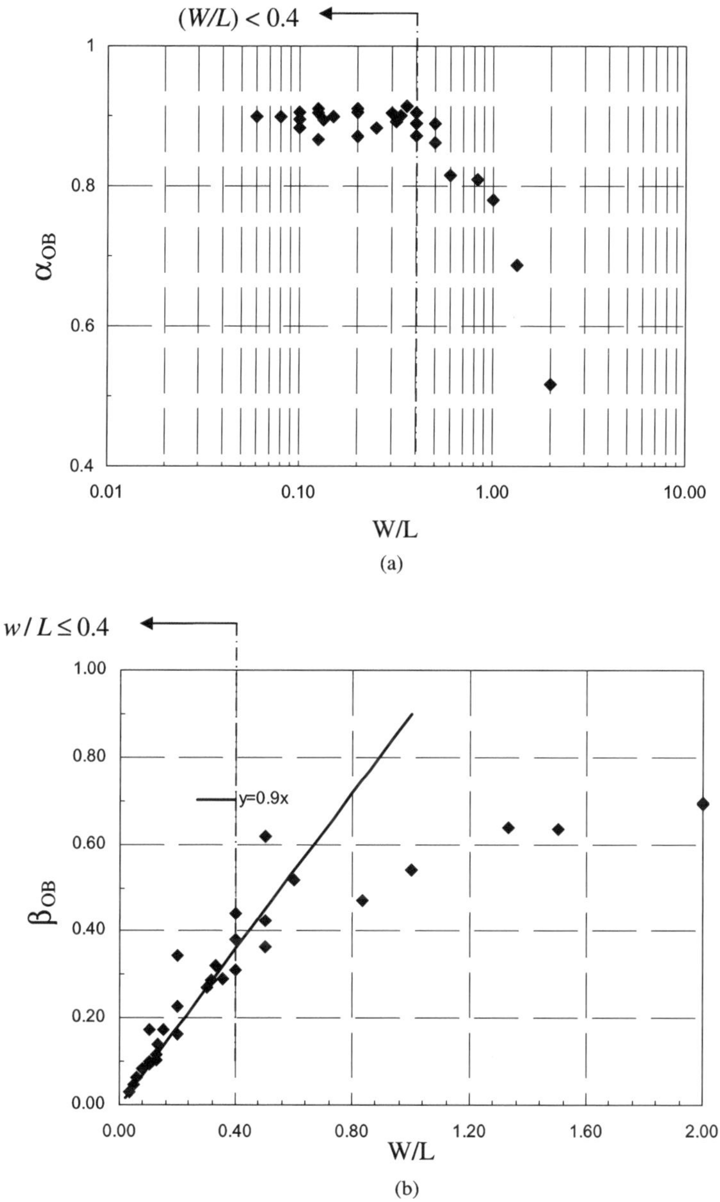

Fig. 6.27. α_{OB} and β_{OB} coefficients of OB device as a function of *W/L*.

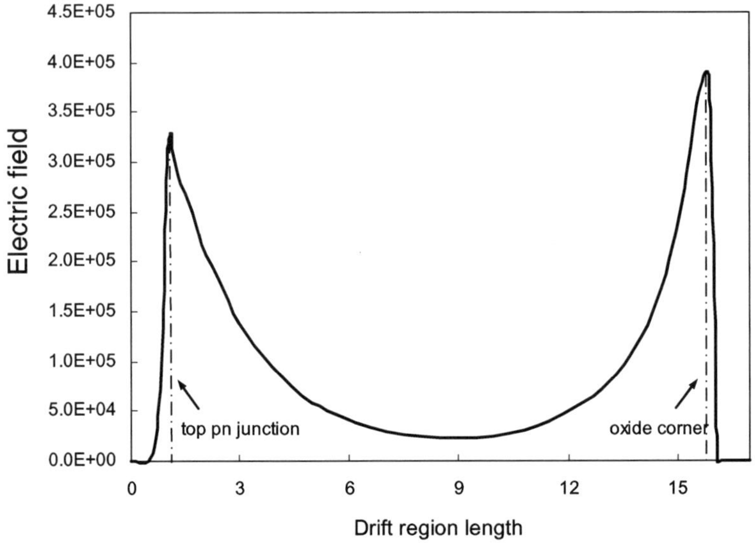

Fig. 6.28. Vertical electric field distribution near the Si–SiO_2 interface in silicon region with $L = 15\,\mu$m, $W = 3\,\mu$m, and $N_d = 1.3 \times 10^{16}\,\text{cm}^{-3}$.

the other occurs at the bottom of drift region near the oxide corner. At a given W with increased L, the vertical electric field in the middle of the drift region will become lower. When L becomes several times larger than W, the breakdown voltage is no longer improved by making L further larger, which means the drain–source voltage supported by OB structure becomes saturated and thus independent of the drift region length. Hence, there is an optimum drift region length to achieve the maximum breakdown voltage for a given drift region width. Further increasing the length beyond that will not help to enhance the breakdown voltage, and not to degrade the specific on-state resistance. This is a very different phenomenon in comparison to p–n SJ structure, in which breakdown voltage follows the drift region length in proportion.

Figure 6.29 shows the relationship on breakdown voltage and drift region length with drift region width W equals to 0.5, 1, 6, 10, and 20 μm by simulations. The result verifies that there exists a saturated breakdown voltage for a given drift region width at certain aspect ratio. Figure 6.30 shows the comparison of theoretical prediction from Eq. (6.44) with the simulation results, in which L/W equals to 2 with W varies from 0.5 to 20 μm. The simulation data have a little higher on-state resistance than the derivation. This is because only the drift region resistance was considered in the derivation, while the simulated resistance was the total device on-state resistance.

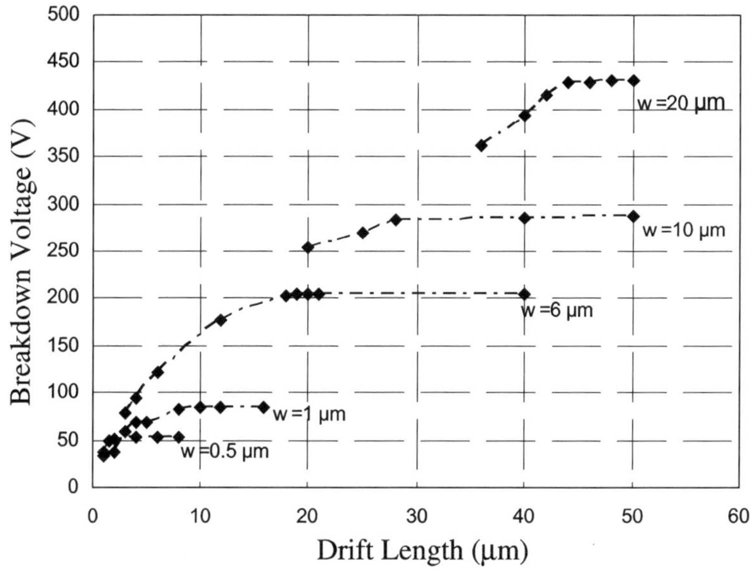

Fig. 6.29. Simulated breakdown voltage and drift region length relationship with drift region width $W = 0.5$, 1, 6, 10, and 20 μm in OBUMOS structure.

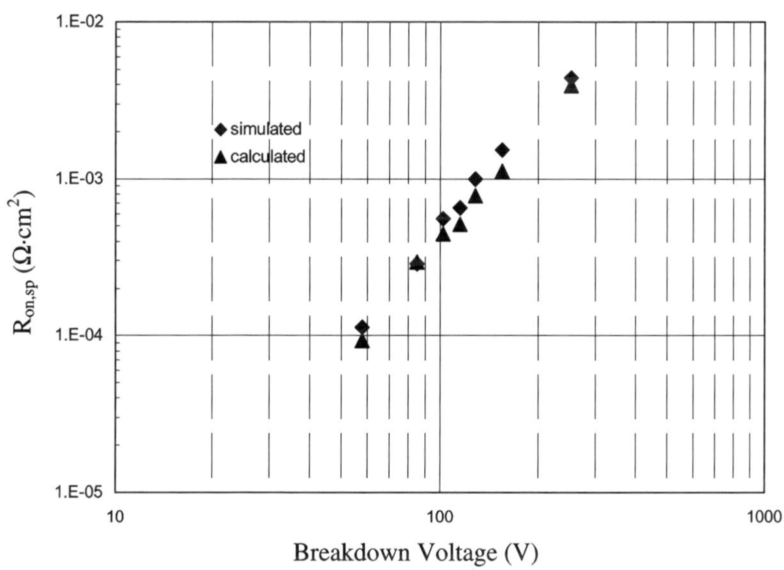

Fig. 6.30. Theoretical and the analytical data on the performance merit of OBUMOS devices at various breakdown ratings.

In summary, the OB structure has two obvious drawbacks, namely (a) the breakdown voltage saturation issue and (b) the nonuniform electric field distribution in the drift region compared to that in p–n SJ devices. These two drawbacks limit the application scope of OB structure for high voltage devices. To improve this, a gradient oxide bypassed (GOB) structure was proposed afterward to provide a uniform electric field distribution like that in p–n SJ devices to achieve a comparable performance.

6.7. Graded Doping in Drift Region

In the conventional MOSFET structure, one may prefer to have a nonuniform doping profile in the drift region to alter the field profile for a higher breakdown and, at the same time, to reduce the on-state resistance. The optimum doping profile in the drift region of the conventional MOSFETs $N_d(y)$ can be found to be (Baliga, 1995)

$$N_D(y) = \frac{\varepsilon_s \hat{E}^2_{crit}}{3qV_{br}\sqrt{1-(2\hat{E}_{crit}y/3V_{br})}}. \tag{6.45}$$

According to the equation, the doping concentration will increase from the surface (source side) to the end of the drift layer (drain side). Using this doping profile, the specific on-resistance is given by

$$R_{on,min} = \frac{3V^2_{br}}{\varepsilon_s \mu_n \hat{E}^3_c}, \tag{6.46}$$

where $\hat{E}_c$ is the critical electric field for breakdown. This specific on-state resistance is lower than that of the uniformly doped profile (Hu, 1979).

For the OB structure, the advantage of using graded doping is similarly to adjust the E-field for a more uniform distribution in the drift region. If the doping profile is gradually increased from the top to bottom as described in Eq. (6.47), a uniform E-field along the drift region can be achieved. And, a better breakdown voltage can be obtained compared to the ordinary OB devices with uniformly doped drift region. For the OB structure, $N_D(y)$ is found to be (Baliga, 1992)

$$N_D(y) = \left[\frac{\varepsilon_s}{\frac{qm}{2}\left(\frac{\varepsilon_s}{\varepsilon_{ox}}t_{ox}+\frac{m}{4}\right)}\right]\cdot\left(\frac{V_{br}}{d-t_{ox}}\right)\cdot y. \tag{6.47}$$

One way to implement the above is to grow a number of epitaxial layers with gradient dopings. In such a way, the fabrication complexity will be proportional to the epi-thickness required for the device. Fortunately, by having the partial graded doping at the lower part of the drift region, the on-resistance and blocking capability can also be improved to a large extend. The process becomes simple by adding one doped epi-layer in between the drift region and the substrate, i.e. to become a double epi-wafer, followed by the thermal diffusion. When undergoing the thermal process, the dopant from the heavily doped substrate diffuses into the doped epi-region and from the doped epi-region to the lightly doped drift region. As a result, a graded doping profile can be formed by dopant redistribution. Simulations were done on OBUMOS at three different graded profiles as shown in Fig. 6.31 for different breakdown capabilities.

6.8. Tunable Oxide Bypassed MOSFETs

In the standard OBUMOS structure, poly contact at the oxide trench, p-body, and n^+ source are all connected to the ground potential. However, if the poly contact is separated, it will potentially provide an additional control to influence the device performance.

In Fig. 6.32, both OB-VDMOS and TOB-VDMOS structures are shown. For the TOB structure, we discover that the blocking voltage, on-state resistance, and transconductance (G_m) can be tunable if an external voltage is applied at the poly contact, named the control electrode, to adjust the electric field at the thick oxide sidewall. This discovery enables the OB device to have the $V_{br} \sim R_{on,sp}$ position moving on the performance merit plane to be further

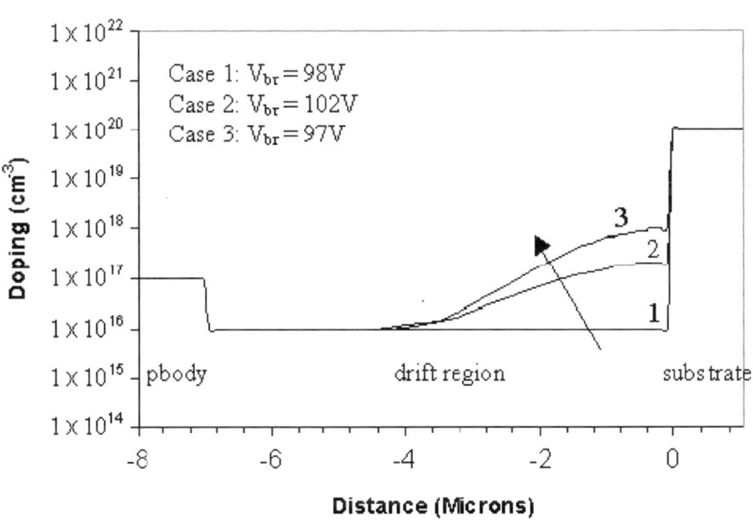

Fig. 6.31. Doping profiles along the n-drift region of OBUMOS.

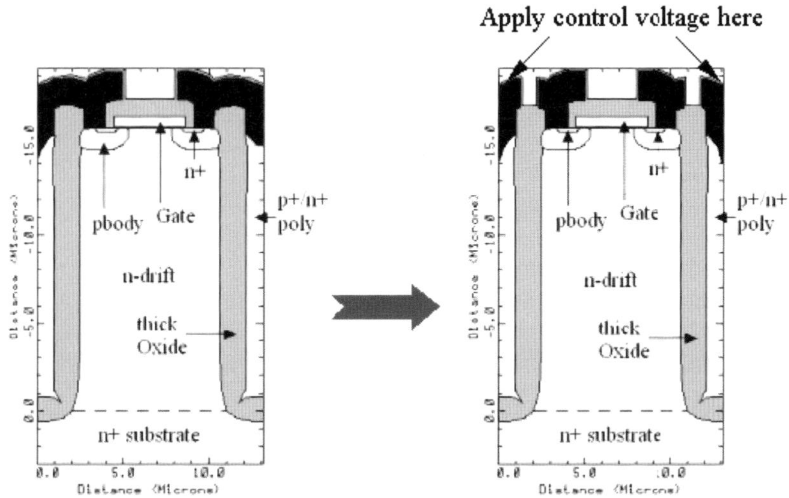

Fig. 6.32. OB-VDMOS (left) and TOB-VDMOS structures (right).

Table 6.4. V_{br} and the breakdown point at different control voltages from 0 to 100 V.

Control voltage (V)	V_{br}(V)	Breakdown point (μm)
0	199.3	(10.8, −0.76)
10	208.8	(10.8, −0.76)
20	218.2	(10.8, −0.76)
30	227.7	(10.8, −0.76)
50	246.6	(10.8, −0.76)
60	256	(10.2, −14.3)
80	131.5	(10.2, −14.3)
100	81.0	(10.4, −14.3)

away from the ideal silicon limit line, and at the same time to have a higher G_m if for small signal applications.

Off-state

Table 6.4 gives the TOB VDMOS V_{br} and the breakdown point position (in (x, y) coordinates) at different control voltages from 0 to 100 V with $W = 8\,\mu$m, $N_D = 2 \times 10^{15}\,\text{cm}^{-3}$ and $t_{ox} = 1.5\,\mu$m, where W is the width, N_D is the doping concentration of the drift region, and t_{ox} is the sidewall oxide thickness.

It is clear that, with the increase of control voltage, V_{br} increases at first to the maximum and then decreases. The positive control voltage provides an additional electric field in a direction opposite to the original one to partially counteract the influence from the drain bias. However, the control voltage also pushes the depletion upwards to quickly extend to the upper part of the n-drift region. This increases the electric field near the p-body/n-drift junction. Therefore, the breakdown point changes place when the electric field of the p-body/n-drift junction reaches the critical value.

Figure 6.33 shows the structures at breakdown for TOB VDMOS at different control voltage of 0, 50, 60, and 100 V. In each simulated structure, the parallel equipotential lines are plotted at 10 V interval, and solid curves gathering at certain places in drift region represent the impact ionization at breakdown. It is seen that for OB device without control voltage, breakdown happens at the silicon/oxide interface near the bottom corner of OB region. For TOB devices, with control voltage, there are more potential lines gathering around p-body region for a higher breakdown voltage. Final breakdown occurs by the impact ionization at the location right below p-body.

On-state

When the device is at on-state, a vertical accumulation layer is formed at the interface of MTO and n-drift region due to the MOS effect produced by the positive control voltage at the poly-electrode. This accumulation layer provides a lower resistance path for current flow in the drift region and the results in the reduction of on-state resistance. Simulations show that, for TOB VDMOS with $W = 8\,\mu$m, $N_D = 2 \times 10^{15}\,\text{cm}^{-3}$, and $t_{ox} = 1.5\,\mu$m, the specific on-state resistance is 5.26 m$\Omega \cdot$cm^2 without control voltage and 4.58 mΩ-cm^2 at 60 V control voltage.

Investigation of 200V TOBUMOS

The TOBUMOS structure using trench gate is shown in Fig. 6.34. We define an additional electrode called "control" electrode in the polysilicon region, which is separated from source electrode in device simulation with $N_D = 3 \times 10^{15}\,\text{cm}^{-3}$. The result in Fig. 6.35 shows that there is a linear relationship between the control bias and V_{br} when the positive control bias is less than 60 V. Beyond 60 V, there is a drastic decrease in V_{br}.

The results also indicate that during off state the avalanche breakdown at 0 V control bias occurs at position A (see Fig. 6.34). At positive control bias, the electric field in the top part of the structure becomes higher and so does the breakdown voltage. When the bias is below 60 V, the increase in V_{br} is approximately equal to the amount of the control bias. When the bias becomes

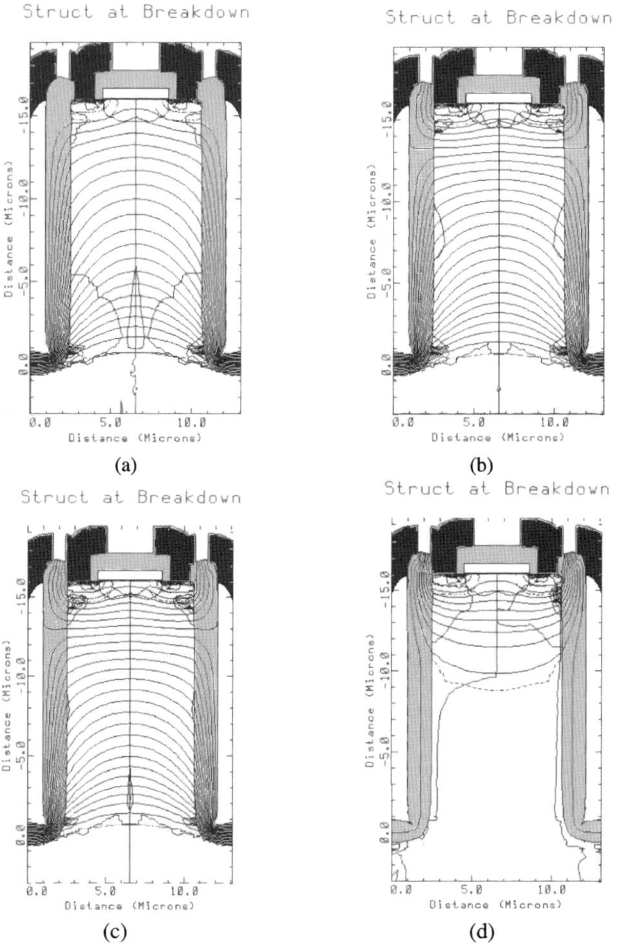

Fig. 6.33. Structures at breakdown for TOB VDMOS at different control voltages of (a) 0 V, (b) 50 V, (c) 60 V, and (d) 100 V.

higher, the breakdown point moves from positions A to B or C. Table 6.5 gives the relationship of V_{br}, $R_{on,sp}$, and breakdown location under different control bias for TOBUMOS with $N_D = 3 \times 10^{15}\,cm^{-3}$.

During on-state, the lateral electric field produced by the positive bias at the control electrode acts on the device and leads to the formation of vertical accumulation layer at the interface between the oxide and n-drift region. Figure 6.36 shows the current flow lines and potential lines for TOBUMOS structure with $N_D = 3 \times 10^{15}\,cm^{-3}$ at $V_{GS} = 10\,V$, and $V_{DS} = 30\,V$ under

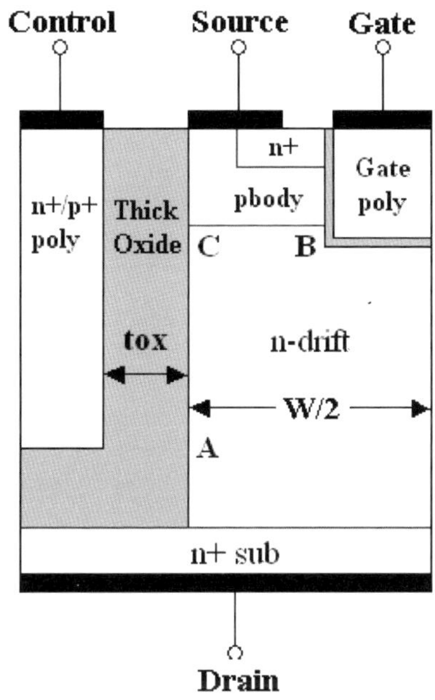

Fig. 6.34. Schematic structure of half TOBUMOS with trench gate.

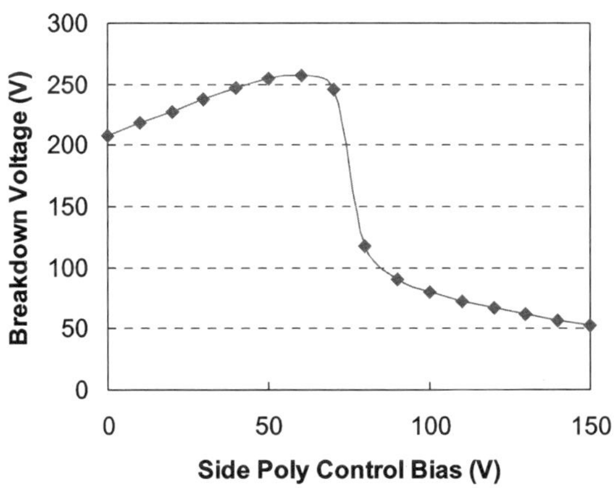

Fig. 6.35. V_{br} vs control bias of the high-voltage TOBUMOS ($N_D = 3 \times 10^{15}$ cm^{-3}) at $t_{ox} = 1.5\,\mu$m and $W/2 = 1.5\,\mu$m.

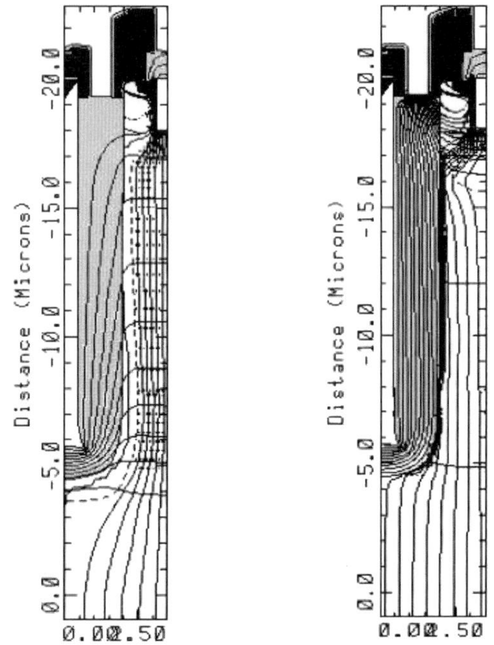

Fig. 6.36. TOBUMOS with $N_D = 3 \times 10^{15}\,\text{cm}^{-3}$ at $V_{GS} = 10\,\text{V}$ and $V_{DS} = 30\,\text{V}$ at control bias of 0 V (left) and 60 V (right).

different control bias of 0 and 60 V. It is obvious that, due to the existence of the accumulation layer, conductivity at on-state can be increased. This results in the reduction of on-state resistance.

In Table 6.5, for the bias varies from 0 V to 60 V, V_{br} will increase by 48.8 V and $R_{on,sp}$ decrease by 1.6 mΩ-cm^2. The results are depicted in Fig. 6.37 on the performance merit plane together with the unipolar silicon limit and superjunction limit. For each N_D, the control bias increases in a step of 10 V from 0. All the points with increasing control bias are shown in the diagram. The number of points in each of the cases is different as the lines terminate at the point where V_{br} is just about to drop. The minimum $R_{on,sp}$ obtained under 20 V control bias with $N_D = 6 \times 10^{15}\,\text{cm}^{-3}$ is much lower than the ideal silicon limit and superjunction limit ($W = 5\,\mu\text{m}$).

The simulation indicates that the maximum control voltage ($V_{con,max}$), which makes the highest V_{br}, mainly depends on N_D if the device structure is fixed. Based on one optimized OBUMOS structure with $W/2 = 1.5\,\mu\text{m}$, $t_{ox} = 1.5\,\mu\text{m}$ at $N_D = 6 \times 10^{15}\,\text{cm}^{-3}$, $V_{con,max} = 20\,\text{V}$. A higher $V_{con,max}$ of 60 V can be obtained if N_D decreases to $3 \times 10^{15}\,\text{cm}^{-3}$.

Table 6.5. Relationship of V_{br}, $R_{on,sp}$, and breakdown location listed under different control bias for TOBUMOS with $N_D = 3 \times 10^{15}\,cm^{-3}$.

Control bias (V)	V_{br}(V)	$R_{on,sp}$(mΩ-cm^2)	Breakdown position
0	208.2	4.9	A
10	218.0	4.3	A
20	227.9	4.0	A
30	237.7	3.8	A
40	247.2	3.6	A
50	255.3	3.4	A
60	257.0	3.3	A
70	245.9	3.2	B
80	118.3	—	B
90	90.3	—	B
100	79.7	—	B
110	72.7	—	B
120	67.0	—	B
130	61.8	—	C

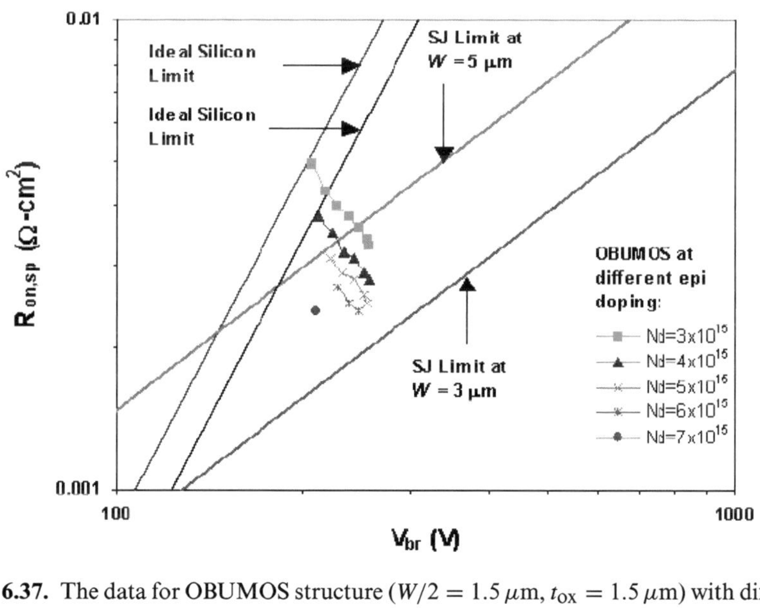

Fig. 6.37. The data for OBUMOS structure ($W/2 = 1.5\,\mu$m, $t_{ox} = 1.5\,\mu$m) with different epi doping are plotted for different control bias varying from 0 to 60 V for $N_D = 3 \times 10^{15}\,cm^{-3}$; 0 to 50 V for $N_D = 4 \times 10^{15}\,cm^{-3}$; 0 to 40 V for $N_D = 5 \times 10^{15}\,cm^{-3}$; 0 to 20 V for $N_D = 6 \times 10^{15}\,cm^{-3}$ and 0 V for $N_D = 7 \times 10^{15}\,cm^{-3}$, respectively.

Table 6.6. Comparison of V_{br} with the same $R_{on,sp} = 2.4\,\text{m}\Omega\text{-cm}^2$ for SJ device at p–n column width of 5 and 3 μm, ideal silicon limit, OBUMOS with $N_D = 7 \times 10^{15}\,\text{cm}^{-3}$ and TOBUMOS with $N_D = 6 \times 10^{15}\,\text{cm}^{-3}$ at 20 V control bias.

	SJ at $W = 5\,\mu$m	SJ at $W = 3\,\mu$m	Ideal silicon limit	Original OBU-MOS	Tunable OBUMOS
V_{br}	162.1 V	307 V	152.9 V	210 V	246.8 V

Table 6.6 gives the comparison of V_{br} with the same $R_{on,sp}$ of 2.4 mΩ-cm^2. With the same given $R_{on,sp}$, the SJ devices at $W = 5\,\mu$m and $W = 3\,\mu$m can support V_{br} of 162.1 and 307 V, respectively; while the ideal silicon limit is 152.9 V. In contrast, V_{br} of OBUMOS device is 37.3% beyond the ideal silicon limit and for TOBUMOS, it is increased by 61.4%. Therefore, the important advantage of the TOBUMOS device is that its performances, such as the on-state resistance, breakdown voltage, small signal transconductance gain, and cut-off frequency, can be tuned by applying suitable control voltage although within a limited range to the optimum value.

6.9. Gradient Oxide Bypassed (GOB) Structure

Gradient oxide bypassed (GOB) structure provides improvement over OB structure to enhance the uniformity of electric field profile in the drift region. The structure has a slanted oxide sidewall at a specific slope as shown in Fig. 6.38. The design with uneven oxide thickness, i.e. thinner at the top and thicker at the bottom of the trench sidewall relieves the electric field stress at the bottom of the drift region compared to that in the OB structure. The drift region with a more uniform electric field distribution can support a higher breakdown voltage. Consequently, GOB will have a better specific on-state resistance vs breakdown voltage performance comparable to that of p–n SJ structure.

Derivations are made here on $R_{on,sp}$ vs V_{br} relationship for GOB structure. From Fig. 6.39 we can assume that the electric field along y-direction in the fully depleted GOB drift region is uniform except for very small parts near the p-body and the n$^+$ substrate. For $L/W \gg 1$, we can then assume $\hat{E}_y$ as constant and the potential at any location in drift region can be expressed as a linear function of the vertical distance from the p-body/n-drift region junction as

$$V(y) = \hat{E}_y \cdot y. \tag{6.48}$$

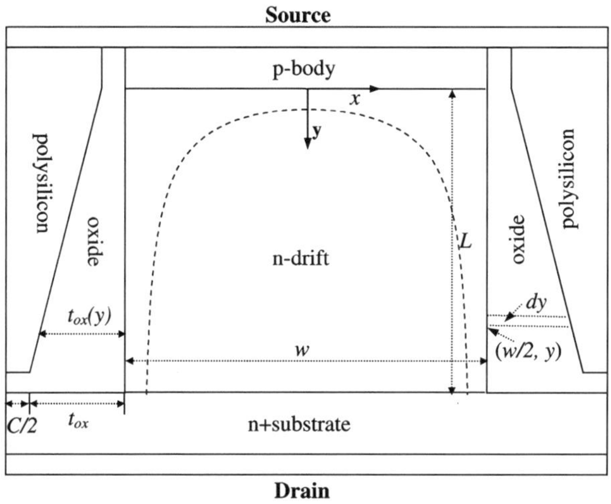

Fig. 6.38. The schematic of GOB structure.

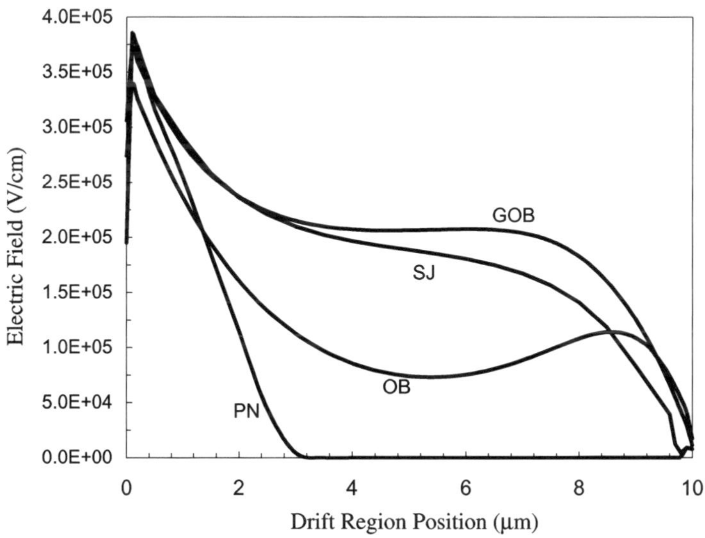

Fig. 6.39. Electric field distributions of PN junction, superjunction (SJ), OB and GOB structures for the same drift region doping concentration.

The device voltage can be expressed in linearly proportional to the drift region length

$$V_{\rm DS} \cong \hat{E}_y L = V_{\rm br}. \tag{6.49}$$

Consequently,

$$\hat{E}_y = \beta_{\rm GOB}\hat{E}_{\rm c}, \tag{6.50}$$

where $\beta_{\rm GOB}$ is a field coefficient, $0 < \beta_{\rm GOB} < 1$. Using Expressions (6.49) and (6.50), we get

$$L = \frac{V_{\rm br}}{\beta_{\rm GOB}\hat{E}_{\rm c}}. \tag{6.51}$$

At an arbitrary point ($W/2$, y), for a small vertical distance dy, Si–SiO_2–polySi structure can be approximately considered as a capacitor

$$dQ = V_{\rm ox}(y)dC_{\rm ox} = V_{\rm ox}(y)\frac{\varepsilon_{\rm ox}}{t_{\rm ox}(y)} \cdot dy, \tag{6.52}$$

where dQ is the charge across this incremental capacitor, $V_{\rm ox}(y)$ is the voltage drop across the oxide at y. Lateral electric field produced by voltage drop across the oxide helps to deplete the drift region. When the drift region is fully depleted

$$V_{\rm ox}(y) \approx \hat{E}_y \cdot y = \beta_{\rm GOB}\hat{E}_{\rm c} \cdot y. \tag{6.53}$$

Integrating both sides of Eq. (6.52)

$$\begin{aligned}\int_0^{Q_{\rm total}} dQ &= \int_0^{C_{\rm ox}} V_{\rm ox}(y)dC_{\rm ox} = \int_0^{L} \left(\beta_{\rm GOB}\hat{E}_{\rm c}y\right) \cdot \frac{\varepsilon_{\rm ox}}{t_{\rm ox}(y)}dy \\ &= \int_0^{L} \beta_{\rm GOB}\hat{E}_{\rm c}\varepsilon_{\rm ox}K \cdot dy,\end{aligned} \tag{6.54}$$

where $\frac{y}{t_{\rm ox}(y)} = K$, and K is the sidewall oxide slope. From Eq. (6.54) that

$$qN_{\rm D}\frac{W}{2}L = \beta_{\rm GOB}\hat{E}_{\rm c}\varepsilon_{\rm ox}KL. \tag{6.55}$$

Thus, the optimum oxide slope K can be derived as

$$K = \frac{qWN_{\rm D}}{2\beta_{\rm GOB}\varepsilon_{\rm ox}\hat{E}_{\rm c}}. \tag{6.56}$$

For the lateral electric field depleting the drift region and uniform electric field distribution in the drift region, GOB and SJ structures are in similar manner. Therefore, the optimum doping concentration in GOB drift region can be approximated as that in the superjunction structure. The oxide slope can then determined to achieve the optimum relationship between the breakdown voltage and specific on-state resistance.

The specific on-state resistance of GOB structure can be derived as

$$\begin{aligned} R_{\mathrm{on,sp}} \approx R_{\mathrm{drift}} &= \frac{L}{q\mu N_{\mathrm{D}}}\left(\frac{W + 2t_{\mathrm{ox}} + C}{W}\right) \\ &= 4.01 \times 10\beta_{\mathrm{GOB}}^{-1} W^{9/7} V_{\mathrm{br}} \frac{1}{\mu}\left(1 + \frac{C}{W} + \frac{2t_{\mathrm{ox}}}{W}\right), \end{aligned} \tag{6.57}$$

where t_{ox} is the bottom oxide thickness, $(w + 2t_{\mathrm{ox}} + C)$ is the total width of the GOB device, C is the polysilicon contact area at the bottom, t_{ox} is related to the breakdown voltage and the doping concentration of the drift region as

$$t_{\mathrm{ox}} = \frac{L}{K} = \frac{\dfrac{V_{\mathrm{br}}}{\beta_{\mathrm{GOB}}\hat{E}_{\mathrm{c}}}}{\dfrac{qWN_{\mathrm{d}}}{2\beta_{\mathrm{GOB}}\varepsilon_{\mathrm{ox}}\hat{E}_{\mathrm{c}}}} = \frac{2\varepsilon_{\mathrm{ox}} V_{\mathrm{br}}}{qWN_{\mathrm{D}}}. \tag{6.58}$$

Placing Eq. (6.58) into Eq. (6.57)

$$R_{\mathrm{on,sp}} = 4.01 \times 10\beta_{\mathrm{GOB}}^{-1} W^{9/7} V_{\mathrm{br}} \frac{1}{\mu}\left(1 + \frac{C}{W} + 7.1 \times 10^{-6} W^{-\frac{6}{7}} V_{\mathrm{br}}\right). \tag{6.59}$$

As β_{GOB} is similar to β_{SJ}, thus $\beta_{\mathrm{GOB}} \approx 0.4$. For a small polysilicon contact area, $C \approx 0$, thus

$$\begin{aligned} R_{\mathrm{on,sp}} &\approx 1.0 \times 10^{2} W^{9/7} V_{\mathrm{br}} \frac{3.75 \times 10^{15} + 9.82 \times 10^{10} W^{-1.04}}{5.10 \times 10^{18} + 9.03 \times 10^{12} W^{-1.04}} \\ &\quad \times (1 + 7.1 \times 10^{-6} W^{-\frac{6}{7}} V_{\mathrm{br}}). \end{aligned} \tag{6.60}$$

From Eq. (6.58), we can see that t_{ox} is proportional to V_{br}. The specific on-state resistance in Eq. (6.60) placed in two different V_{br} regions described as follows:

(1) For low V_{br}, $R_{\mathrm{on,sp}} \propto V_{\mathrm{br}}$

For low V_{br}, t_{ox} is comparable to $W/2$, which makes the GOB $R_{\mathrm{on,sp}}$ and V_{br} relationship linear. This dependency is very similar to that of SJ. When t_{ox} is small, GOB can have a better performance than SJ.

(2) For high V_{br}, $R_{\mathrm{on,sp}} \propto V_{\mathrm{br}}^{2}$

When V_{br} becomes higher, t_{ox} is clearly larger than $W/2$, the conduction area becomes smaller compared with the total cell area (ratio $= (W/(W + 2t_{\mathrm{ox}} + C))$. Therefore, $R_{\mathrm{on,sp}}$ increases almost quadratically with V_{br} and hence the structure is no longer beneficial.

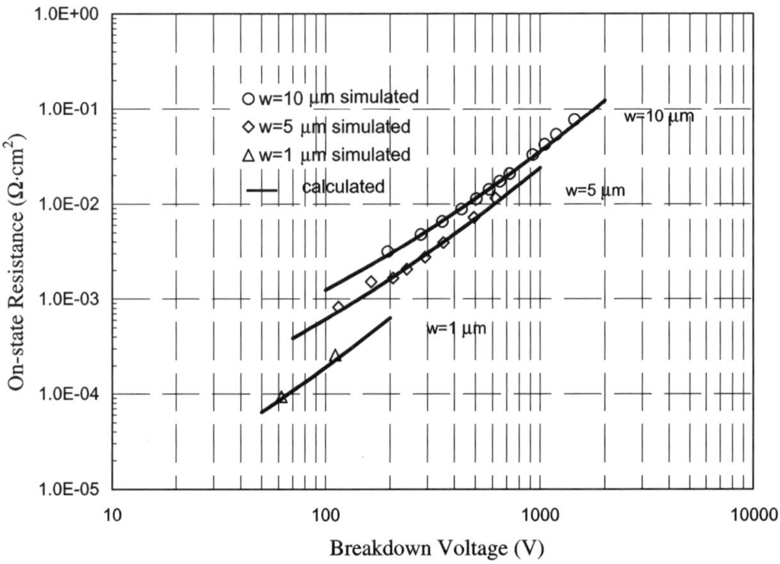

Fig. 6.40. Theoretical and analytical performance merit data for GOB structure.

Simulations by MEDICI were carried out to verify the theoretical predictions. Figure 6.40 shows the comparison of theoretical prediction from Eq. (6.60) with the simulation results. Simulation data and theoretical prediction match very well.

6.10. Lateral Superjunction Power MOSFET

Superjunction power LDMOS device if implemented on the bulk Si substrate suffers from the substrate-assisted depletion (SAD) effect, which causes the charge imbalance and thus limits the performance of the SJ LDMOS device. By adjusting the p–n column width of SJ structure (Lin *et al.*, 2003) or combining both SJ and RESURF concepts (Nassif-Khalil *et al.*, 2004), or adding an n-buffer layer under the SJ structure (Park and Salama, 2005), the performance of the SJ LDMOS device on the bulk Si substrate can be improved. However, the SAD effect still cannot be fully eliminated. To completely eliminate the SAD for the best performance of SJ LDMOS, an insulated substrate should be utilized. SJ LDMOS device on the sapphire substrate was demonstrated with good experimental results (Nassif-Khalil and Salama, 2003). Nevertheless, the sapphire substrate is not the mainstream for silicon wafer process. Simulation results have been reported on the SJ LDMOS devices on SOI substrate (Xu *et al.*, 2000; Ng *et al.*, 2001; Amberetu *et al.*, 2002; Park *et al.*, 2001). However,

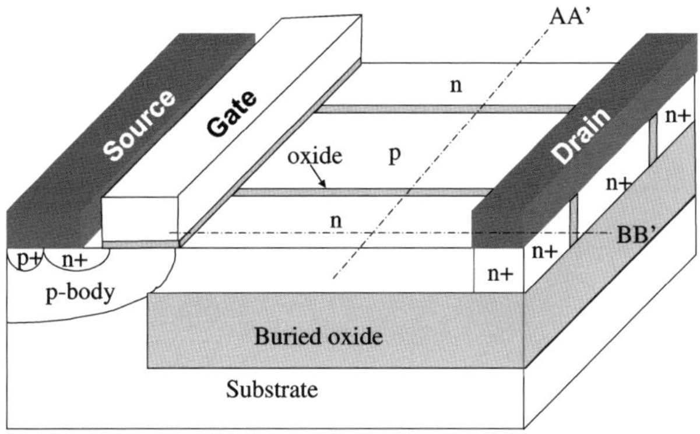

Fig. 6.41. The lateral superjunction MOSFET structure on the partial SOI platform.

for SOI SJ LDMOS, the fabrication is handicapped due to the bottleneck difficulty on formation of thick buried oxide in order to completely block the substrate depletion.

Partial SOI (PSOI) structure formed using trench etch and direct thermal oxide pinch off technique can tailor the thick buried oxide easily (Ren *et al.*, 2002). Together with the trench etch and poly refill processes, lateral SJ structure can be integrated on the PSOI platform (Fig. 6.41). Such PSOI SJ LDMOS device and process technology enables the making of the device on bulk silicon substrate without sacrificing its electrical and thermal performance. The technology provides all the benefits of PSOI, such as good substrate isolation, low leakage current, fast switching speed, good heat dissipation as well as the advantages of SJ structure, such as high breakdown voltage and low specific on-state resistance.

The alternative p–n columns are placed on the thick buried oxide. The buried oxide is terminated near the p-body and source, so as to reserve a thermal dissipation window. Unlike the conventional SJ structure, a thin oxide layer is formed in between the p–n columns to avoid the p–n column interdiffusion (Gan *et al.*, 2002). The p–n column widths W_p and W_n are predetermined by the thermal oxidation ratio to obtain the continuous buried oxide as shown in Fig. 6.42. The p–n column doping concentrations need to be adjusted accordingly by looking at the p–n column widths for good charge balance. One set of optimum p–n column widths and doping concentration using 2D process simulator Tsuprem and 3D device simulator Davinci is listed in Table 6.7 for reference.

Figure 6.43 shows the field distribution at the onset of breakdown. The solid lines are the equipotential lines and the dashed lines are the depletion

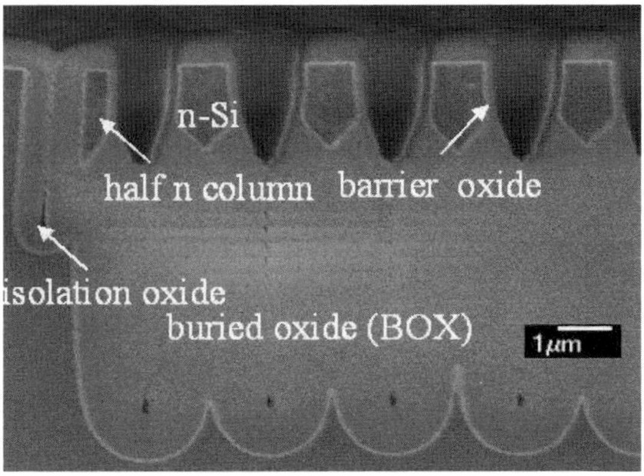

Fig. 6.42. SEM photo of the PSOI platform.

Table 6.7. p–n column width and doping concentration for PSOI-LDMOS.

p conc. (N_A)	1.87×10^{16}	2.72×10^{16}	3.50×10^{16}
n conc. (N_D)	4.20×10^{16}	5.16×10^{16}	7.73×10^{16}
p col. $W_p(\mu m)$	1.50	1.29	1.02
n col. $W_n(\mu m)$	0.98	0.79	0.61

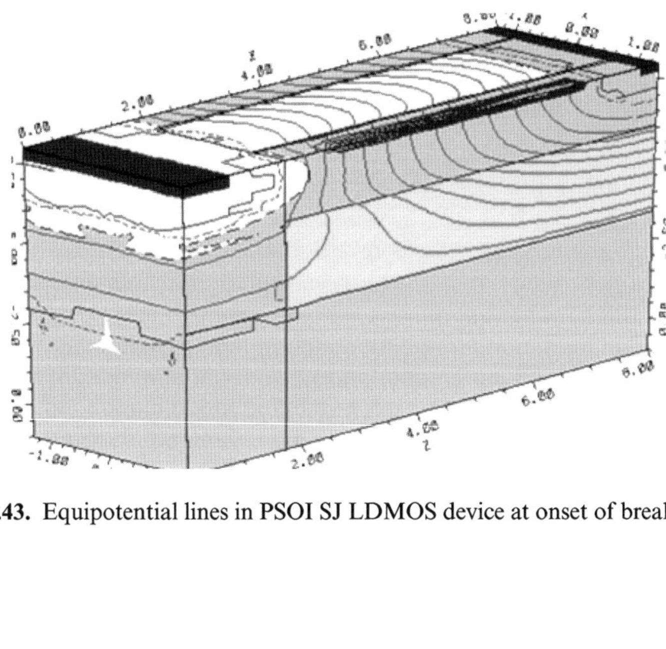

Fig. 6.43. Equipotential lines in PSOI SJ LDMOS device at onset of breakdown.

region boundary. We can see that the drift region is fully depleted and the potential is equally distributed at the onset of the breakdown.

The designed device has a breakdown voltage of 126 V for the drift region n-column width of 0.98 μm, length of 5.5 μm, depth of 1 um, pinch size of 2.5 μm and n-column doping concentration of $4.2 \times 10^{16}\,cm^{-3}$. The specific on-state resistance of this device with a planar gate is 0.439 mΩ-cm^2 by simulation. This outstanding performance is attributed to the elimination of the substrate-assisted depletion and the suppression of the interdiffusion between p–n columns.

6.10.1. Device Process Technology

The process started with the n^-–p^+ epi-wafer with the epi-layer thickness of 14 μm and the resistivity of 12$\Omega \cdot$cm. The key process steps are illustrated in Fig. 6.44. The oxide/SiN/oxide triple hard mask was used to etch the first 1.8 μm Si trench; SiN with the pad oxide was then deposited and anisotropically etched, after that the second 4.2 μm Si trench was etched (a), followed by the thermal oxidation to form the partially buried oxide (b), the buried oxide thickness is determined by the second trench depth; after that, SiN hard mask was removed and n-column phosphorus tilt implantation with dose of $2.2 \times 10^{14}\,cm^{-2}$ and the energy of 120 keV was done, then 500 Å oxide was grown on the trench sidewalls as a diffusion barrier (c); an additional mask was used to wet etch the sidewall oxide for a contact window between the p-column and p-body (d); 3 kÅ poly was then deposited in the trench, followed by $2 \times 10^{14} cm^{-2}$ dose and 60 keV energy boron tilted implantation, 6 kÅ poly-refill and poly-etch back (e); finally, gate/source/drain were formed as those in the conventional LDMOS process (f). In this process, both SJ and PSOI structure were formed simultaneously using the same Si trench, therefore no additional mask is required to form the lateral SJ structure on the PSOI platform except one more is needed to open the contact window between the p-body and p-poly as shown in (d).

The fabricated devices have a channel length of 1 μm and gate oxide thickness of 375 Å. The device pitch size is 1.6 μm and the depth of Si on PSOI is about 1 μm. The cross section (AA$'$ in Fig. 6.41) SEM picture of the alternative SJ p–n columns with 500 Å oxide diffusion barrier formed on the PSOI platform is shown in Fig. 6.42. For the measurement results, the specific on-state resistance at 15 V gate voltage is 2.8 m$\Omega \cdot$ cm^2. The breakdown voltage of this device is 74.5 V which is 3.5 times of that of the conventional LDMOS fabricated on the same PSOI platform (Chen, 2007) (Fig. 6.45).

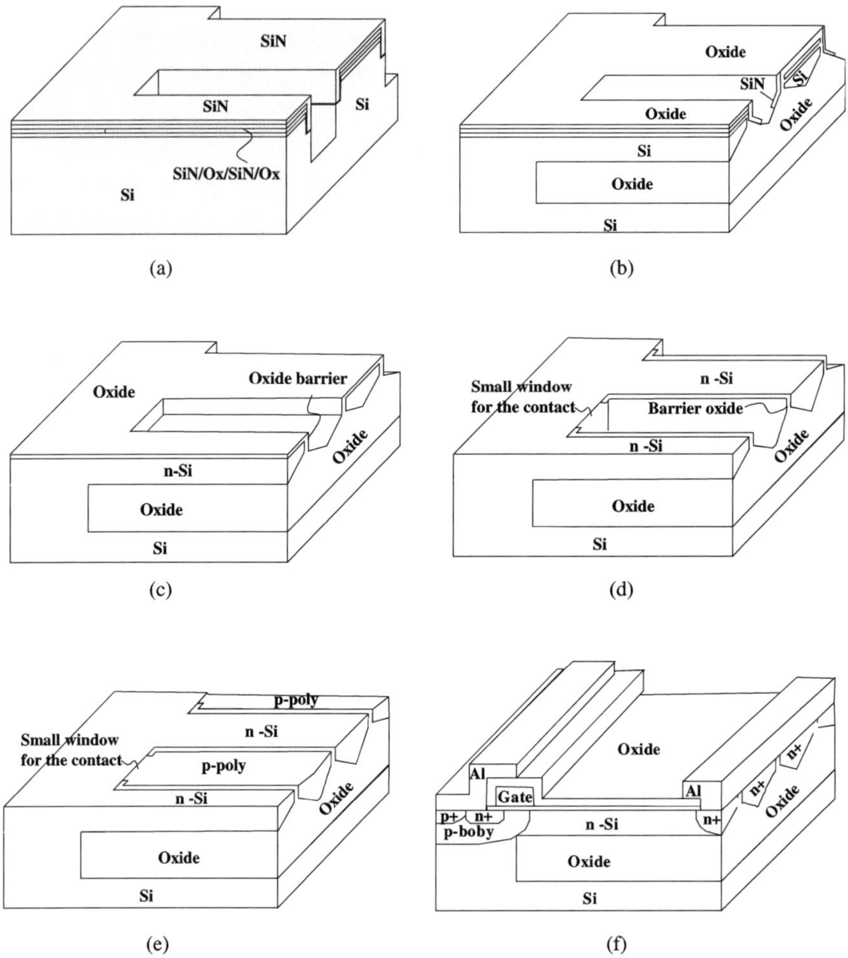

Fig. 6.44. Key process steps for the PSOI SJ-LDMOS: (a) two-step trench etching, (b) thermal oxide growth, (c) n-column phosphorus implantation and oxide barrier layer growth, (d) oxide etch using the additional mask, (e) poly deposition followed by the boron tilted implantation, poly refill and planarization, (f) gate/source/drain formation.

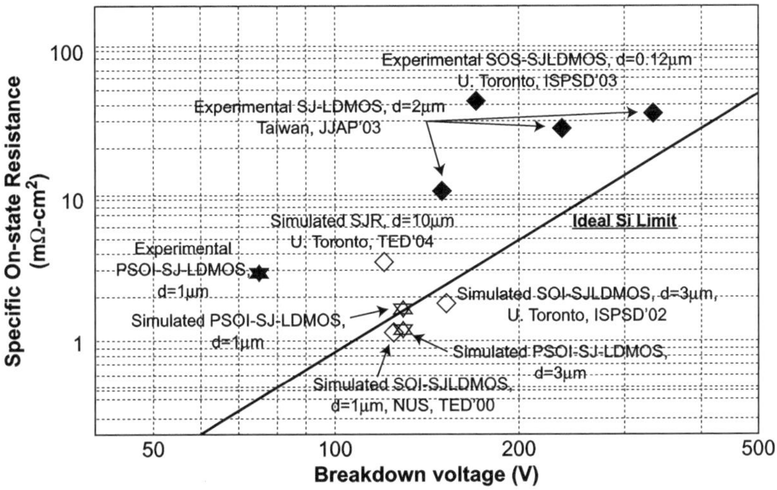

Fig. 6.45. Measured performance merit of SJ LDMOS in comparison with other simulation and measurement data published earlier.

References

T. Fujihira, Theory of semiconductor superjunction devices, *Japanese J. Appl. Phys.* **36** (10, Part 1), 6254–6262 (1997).

G. Deboy, M. Marz, J. Stengl, J. Tihani and H. Weber, A new generation of high voltage MOSFETs breaks the limit line of silicon, *Int. Electron Devices Meeting*, 683–685 (1998).

X.-B. Chen, Semiconductor power devices with alternating conductivity type high-voltage breakdown regions, US Patent 5216275 (1993).

X.-B. Chen, M. S. Towers, P. Mawby and K. Board, High-voltage structure with embedded oppositely dope regions, *IEEE Proc. Circuits. Device Syst.* **146**(2), 90–94 (1999).

Y. Chen, K. D. Buddharaju, Y. C. Liang, G. S. Samudra and H. H. Feng, Superjunction power LDMOS on partial SOI platform, in *IEEE Int. Symp. Power Semiconductor Devices and ICs*, Korea, May 2007, pp 177–180.

T. Nitta, T. Minato, M. Yano, A. Uenisi, M. Harada and S. Hine, Experimental results and simulation analysis of 250V super trench power MOSFET (STM), *Proc. Int. Symp. Power Semiconductor Devices ICs*, 77–80 (2000).

J. Glenn and J. Siekkinen, A novel vertical deep trench RESURF DMOS (VTR-DMOS), *Proc. Int. Symp. Power Semiconductor Devices ICs*, 197–200 (2000).

Y. C. Liang, K. P. Gan and G. S. Samudra, Oxide-bypassed VDMOS (OBVDMOS): An alternative to superjunction high voltage MOS power devices, *IEEE Electron Devices Lett.* **22**(8), 407–409 (2001).

J. M. Park, R. Klima and S. Selberherr, Lateral trench gate super-junction SOI-LDMOSFETs with low on-resistance, *Eur. Solid-State Devices Res. Conf.*, 283–286 (2002).

B. J. Baliga, *Power Semiconductor Devices* (PWS Publishing Company, Boston 1995).

C. Hu, Optimum doping profile for minimum ohmic resistance and high-breakdown voltage, *IEEE Tran. Electron. Devices* **26**(3), 243–245 (1979).

P. M. Shenoy, A. Bhalla and G. M. Dolny, Analysis of the effect of charge imbalance on the static and dynamic characteristics of the super junction MOSFET, *Proc. Int. Symp. Power Semiconductor Devices ICs*, 99–102 (1999).

T. Minato, T. Nitta, A. Uenisi, M. Yano, M. Harada and S. Hine, Which is cooler, trench or multi-epitaxy?, *Proc. Int. Symp. Power Semiconductor Devices ICs*, 73–76 (2000).

K. P. Gan, X. Yang, Y. C. Liang, G. S. Samudra and Y. Liu, A simple technology for superjunction device fabrication: Polyflanked VDMOSFET, *IEEE Electron Devices Lett.* **23**(10), 627–629 (2002).

K. Shenai, P. A. Piacente, R. Saia, C. S. Korman, W. Tantraporn and B. J. Baliga, Ultralow resistance, selectively silicided VDMOSFET's for high-frequency power switching applications fabricated using sidewall oxide spacer technology, *IEEE Trans. Electron Devices* **35**(12), 2459 (1988).

F. Berta, J. Fernàndez, S. Hidalgo, P. Godignon, J. Rebollo and J. Millán, A simplified low-voltage smart power technology, *IEEE Electron Device Lett.* **12**(9), 465–467 (1991).

S. L. Wong, M. J. Kim, J. C. Young and S. Mukherjee, A scaled CMOS-compatible smart power IC technology, *Proc. Int. Symp. Power Semiconductor Devices ICs*, 51–55 (1991).

J. S. Ajit, B. J. Baliga, S. Tandon and A. Reisman, The minority-carrier injection-controlled field-effect transistor (MICFET), *IEEE Tran. Electron Devices* **38**(12), 2694–2695 (1991).

J. Weyers and H. Vogt, A 50 V smart power process with dielectric isolation by SIMOX, *IEEE Int. Electron Devices Meeting*, 225–228 (1992).

T. Ifström, U. Apel, H.-G. Graf, C. Harendt and B. Höfflinger, A 150-V multiple up-drain VDMOS, CMOS, and bipolar process in direct-bonded silicon on insulator on silicon, *IEEE Electron Devices Lett.* **13**(9), 460–461 (1992).

K. Kobayashi, T. Hamajima, H. Kikuchi, M. Takahashi and K. Arai, An intelligent power device using poly-Si sandwiched wafer bonding technique, *Proc. Int. Symp. Power Semiconductor Devices ICs*, 58–62 (1995).
B. J. Baliga, T. Syau and P. Venkatraman, The accumulation-mode field-effect transistor: A new ultralow on-resistance MOSFET, *IEEE Electron Devices Lett.* **13**(8), 427–429 (1992).
N. Fujishima, A. Sugi, S. Kajiwara, K. Matsubara, Y. Nagayasu and C. A. T. Salama, A high-density low on-resistance trench lateral power MOSFET with a trench bottom source contact, *IEEE Tran. Electron Devices* **49**(8), 1462–1468 (2002).
S. Huang, G. A. J. Amaratunga and F. Udrea, Analysis of SEB and SEGR in super-junction MOSFETs, *IEEE Tran. Nucl. Sci.* **47**(6), 2640–2647 (2000).
X.-B. Chen, X. Wang and J. K. O. Sin, A novel high-voltage sustaining structure with buried oppositely doped regions, *IEEE Trans. Electron Devices* **47**(6), 1280–1285 (2000).
P. N. Kondekar, M. B. Patil and C. D. Parikh, Analysis and design of superjunction power MOSFET: CoolMOS^{TM} for improved on resistance and breakdown voltage using theory of novel voltage sustaining layer, *Proc. Int. Conf. Microelectron* **1**, 209–212 (2002).
Y. Onishi, S. Iwamoto, T. Sato, T. Nagaoka, K. Ueno and T. Fujihira, 24 $m\Omega cm^2$ 680 V silicon superjunction MOSFET, *Proc. Int. Symp. Power Semiconductor Devices ICs*, 241–244 (2002).
S. Yamauchi, Y. Urakami, N. Tuji and H. Yamaguchi, Defect-less trench filling of epitaxial Si growth by H_2 annealing, *Proc. Int. Symp. Power Semiconductor Devices ICs*, 133–136 (2002).
C. Y. Chang and S. M. Sze, *ULSI Technology* (McGraw Hill, 1996), pp. 131–134.
J. L. Regolini, D. Bensahel, E. Scheid and J. Mercier, Selective epitaxial silicon growth in 650°C–1100°C range in a reduced pressure chemical vapor deposition reactor using dichlorosilane, *Appl. Phys. Lett.* **54**, 658–659 (1989).
X.-B, Chen, H.-Q. Yang and M. Cheng, New silicon limit of power devices, *Solid-State Electron* **46**, 1185–1192 (2002).
M.-J. Lin, T.-H. Lee, F.-L. Chang, C.-W Liaw and H.-C. Cheng, Lateral superjunction reduced surface field structure for the optimization of breakdown and conduction characteristics in a high-voltage lateral double diffused metal oxide field effect transistor, *Japanese J. Appl. Phys.* **42**(12, Part 1), 7227–7231 (2003).
S. G. Nassif-Khalil, L. Z. Hou and C. A. T. Salama, SJ/RESURF LDMOST, *IEEE Trans. Electron Devices* **51**(7), 1185–1191 (2004).

Y. Park and C. A. T. Salama, CMOS compatible super junction LDMOST with N-buffer layer, *Proc. ISPSD*, 163–16 (2005).

S. G. Nassif-Khalil and C. A. T. Salama, 170 V super junction-LDMOST in a 0.5 μm commercial CMOS/SOS technology, *ISPSD 2003*, 228–231 (April 2003).

S. Xu, K. P. Gan, G. S. Samudra, Y. C. Liang and J. K. O. Sin, 120 V interdigitated-drain LDMOS (IDLDMOS) on SOI substrate breaking power LDMOS limit, *IEEE Trans. Electron Devices* **47**(10), 1980–1985, (2000).

R. Ng, F. Udrea, K. Sheng, K. Ueno, G. A. J. Amaratunga and M. Nishiura, Lateral unbalanced super junction (USJ)/3D-RESURF for high breakdown voltage on SOI, *Proc. Int. Symp. Power Semiconductor Devices ICs*, 395–398 (2001).

M. A. Amberetu, C. A. T. Salama and E. S. Rogers, 150-V class superjunction power LDMOS transistor switch on SOI, *Proc. Int. Symp. Power Semiconductor Devices ICs*, 101–104 (2002).

C. Ren, J. Cai, Y. C. Liang, P. H. Ong, N. Balasubramanian and J. K. O. Sin, The partial silicon-on-insulator technology for RF power LDMOSFET devices and on-chip microinductors, *IEEE Trans. Electron Devices* **49**(12), 2271–2278 (2002).

J. M. Park, T. Grasser, H. Kosina and S. Selberherr, Numerical study of partial-SOI LDMOSFET power devices, *Int. Semiconductor Devices Res. Symp.*, 114–117 (2001).

7

Silicon Carbide Power Devices

7.1. Introduction

With more than 50 years of development, silicon (Si) power devices and process technologies have become mature, with slight improvements often gained only through complicated device structures or tight control of the process steps and material properties, which increases the manufacturing cost significantly. Some of these advancements and innovations have been introduced in previous chapters. It is generally accepted that the potential for improvement of Si power devices is approaching its limits. Because of this, advancements in silicon are becoming more challenging, so attempts to identify an alternative semiconductor device structure or material that enables significant breakthroughs in performance are continuing. A figure of merit $\epsilon\mu E_{\mathrm{g}}^{3}$ was proposed that could be used to evaluate the theoretical performance of unipolar devices made from different semiconductors based on the breakdown voltage and the specific on-resistance (Baliga, 1982), which later evolved into $\epsilon\mu\hat{\mathrm{E}}_{\mathrm{C}}^{3}$, often referred to as Baliga's Figure of Merit, shortened as BFOM. To explain the importance of BFOM, a simple derivation is provided. Combining Eqs. (2.130) and (2.131) with Eq. (3.93), and assuming that all the resistance in the device is dominated by the drift region resistance in which conductivity modulation does not occur, and noting that the definition of the specific on-resistance of a unipolar device can be expressed as,

$$R_{\mathrm{on,sp}} = R_{\mathrm{on}} \times \mathit{Area} = \frac{x_{\mathrm{d}}}{q\mu_{\mathrm{n}}N_{\mathrm{d}}}, \tag{7.1}$$

one can establish the following relationship and clearly observe how BFOM relates V_{BR} and $R_{\mathrm{on,sp}}$, the two most important static features of a power device.

$$\frac{V_{\mathrm{BR}}^{2}}{R_{\mathrm{on,sp}}} = \frac{\varepsilon_{\mathrm{s}}\,\mu_{\mathrm{n}}\,\hat{E}_{\mathrm{C}}^{3}}{4} = \frac{\mathrm{BFOM}}{4}, \tag{7.2}$$

where $\hat{E}_C$ is the critical electric field. Conceptually, with a wider bandgap, it would be more difficult to energize an electron in the valence band and free it to the conduction band. Therefore, the impact ionization rate is smaller in a wide bandgap semiconductor as compared with Si, which translates into a larger critical electric field $\hat{E}_C$. Empirically, it is found that (Hudgins *et al.*, 2003)

$$\hat{E}_C = 1.75 \times 10^5 (E_g)^{2.36} . \tag{7.3}$$

Therefore, from Eq. (7.2) it is expected that devices made from wide bandgap semiconductors will have a better BFOM and provide a superior performance in terms of lower $R_{on,sp}$ and larger V_{BR}. It is for this reason that wide bandgap semiconductors, such as silicon carbide (SiC), gallium nitride, and diamond, have been receiving increased attention in recent years. For SiC, the initial studies into its potential use for power devices began in the early 1990s. After years of fundamental research in both academia and industry, SiC devices were commercialized for power applications in 2001, with the first device being a Schottky diode. Now, both SiC diodes and switches ranging from 600 V up to 1.7 kV are commercially available, and are now manufactured by half a dozen companies worldwide. A few companies even manufacture devices for specific applications that provide voltages of 3.3 kV, 6.5 kV, or even higher. It is generally accepted that SiC devices will occupy a significant portion of the power semiconductor device market in the years to come.

This chapter intends to provide readers with a basic understanding of SiC material and device technology. The first part of this chapter provides an introduction to the material properties which define the advantages of SiC, before moving on to the process challenges that are sometimes the origin of the limits on current SiC devices. The basic principles and performance of SiC rectifiers, which are the most successful SiC products to date, will be discussed in the second part of the chapter, along with an analysis of some of the design issues. In the third part of the chapter, the focus then transfers to SiC unipolar switching devices, which are the emerging and most anticipated SiC products, followed in the fourth part by a detailed description of SiC bipolar devices for future ultrahigh voltage and ultrahigh power applications. In the final part, SiC lateral devices for possible power integrated circuits will be briefly discussed.

7.2. SiC Material Properties and Processing Technologies

Typically, the potential and the performance of a semiconductor device are limited either by its material properties in nature, or by the processing technology currently available. Therefore, it is essential to gain an understanding of these basic concepts before investigating details of the device structures and their characteristics. More importantly, we need to be aware of what to expect from a new material like SiC and any electronic devices manufactured from it.

7.2.1. Material Properties

SiC is a group IV–IV compound semiconductor with a carbon-to-silicon atomic ratio of 1:1. More than 170 polytypes can be found in the literature, each being differentiated by the stacking sequence of the SiC bilayers. Each polytype has its own unique properties. Table 7.1 lists the key physical properties at room temperature for some of the most common SiC polytypes, together with those for pure Si. These include 3C-SiC, where C refers to a cubic crystal unit cell, also called β-SiC, 6H- and 4H-SiC, where H refers to a hexagonal crystal unit cell, also called α-SiC. These properties are closely related to the performance of power semiconductor devices.

The first thing to be noticed in this table is the increased critical electric field where the bandgap is larger, which was discussed in the previous section. In a practical situation, the critical electric field in 4H-SiC is related to the background doping concentration, as described in an empirical equation given by Konstantinov *et al.* (1997)

$$\hat{E}_C = \frac{2.49 \times 10^6}{1 - 0.25 \log\left(\frac{N_d}{10^{16}}\right)}. \tag{7.4}$$

Table 7.1. Material properties for 3C-, 6H-, 4H-SiC, and Si.

	Si	3C-SiC	6H-SiC	4H-SiC
E_g (eV)	1.12	2.36	3.05	3.26
n_i (cm^{-3})	1.5×10^{10}	6.9	3×10^{-6}	1.6×10^{-8}
E_c (MV/cm)	0.3	2	3	3
μ_n (cm^2/Vs)	1500	800	400 (a)* 85 (c)**	900 (a)* 1050 (c)**
μ_p (cm^2/Vs)	450	40	95	120
v_{sat} (cm/s)	1×10^7	2×10^7	2×10^7	2×10^7
k (W/cmK)	1.5	3.3	3.3	3.3
ε_s	$11.8\varepsilon_0$	$9.7\varepsilon_0$	$9.7\varepsilon_0$	$9.7\varepsilon_0$
Melting point (°C)	1414	2700	2700	2700
BFOM (MW/cm^2) normalized to Si	1	130	47***	575***

*parallel to the a-axis.
**parallel to the c-axis.
***electron mobilities of (c) are used.

The next feature that can be noticed is the extremely low intrinsic carrier density n_i for SiC, which is a direct consequence of the large bandgap, as can be seen from Eq. (2.14). Because of this, the junction leakage current of a SiC device is expected to be several orders of magnitude smaller than that of a Si device, as discussed in Sec. 3.3.2. The capability of SiC devices to withstand high temperatures is also as a result of this feature. The electron concentration at different temperatures can be calculated through Eq. (2.21) by considering the contribution from donors and the valence band, and the results for both Si and 3C-SiC are shown in Fig. 7.1. Here, the doping concentration is assumed to be $N_\mathrm{d} = 10^{16}/\mathrm{cm}^3$, and the donor level is assumed to be 50 meV below the conduction band. The ionized dopant concentration N_d^+ will be reduced in the freeze-out region, where the electron concentration is smaller than the doping concentration. N_d^+ can be derived as

$$N_\mathrm{d}^+ = \frac{N_\mathrm{d}}{1 + g_\mathrm{D} e^{(E_\mathrm{F} - E_\mathrm{D})/kT}}, \tag{7.5}$$

where g_D is the donor-site degeneracy factor, which is typically assumed to be 2. The dashed line in the figure represents n_i at different temperatures.

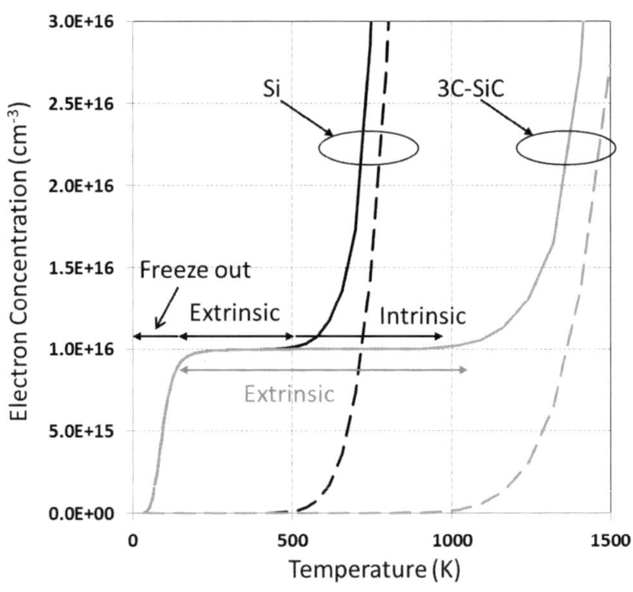

Fig. 7.1. Electron concentrations for $1 \times 10^{16}\,\mathrm{cm}^{-3}$ doped n-type Si and 3C-SiC at different temperatures.

From this figure, it can be seen that the conductivity of a semiconductor in the extrinsic temperature range is controlled by the doping concentration, and is the range where the device maintains its normal functions. At temperatures above the extrinsic temperature range, the electron concentration increases drastically when it is not being controlled by dopants. Because 3C-SiC has a wider bandgap, the temperature at which n_i begins to dominate the electron concentration is around 1000 K, extending the upper limit of the extrinsic temperature range to a much higher temperature than that of Si, which is around 500 K. Similar conclusions can be drawn for both 4H-SiC and 6H-SiC, as well as for other wide bandgap semiconductors.

Because of the different crystal structures, each of the three most common SiC polytypes listed in Table 7.1 has its own unique electron and hole transport characteristics. 3C-SiC has a smaller bandgap and a smaller critical electric field, and is therefore less appealing in terms of BFOM. However, the biggest obstacle for the commercialization of 3C-SiC is the difficulty in producing the bulk material, as will be explained in the next section. For 4H- and 6H-SiC, the electron mobility is different along, and perpendicular, to the c-axis, as shown in the table. By comparing the physical property values, it can be concluded that 4H-SiC is superior because of the larger and more isotropic electron mobilities in different directions. Moreover, currents are conducted along the c-axis in the state-of-the-art vertical SiC power devices, making 4H-SiC the prevailing choice among all the polytypes. The electron saturation velocity is also higher for SiC, but that aspect is more important for RF devices and less critical for power devices. The poor hole mobilities can also be observed in all of the SiC polytypes presented in Table 7.1, making the p-type SiC very resistive. This will impose some limitations on the performance of SiC power devices. For bipolar devices in particular, p-type regions are required in order to provide hole injection into the drift region when the devices are in the on-state, but significant parasitic resistance will be introduced when there is a p-type region in the current path.

Another benefit of SiC is its high thermal conductivity, which is comparable to copper, and is more than twice that of Si, making heat dissipation in the device more efficient. This is particularly important for a power device since the junction temperature is always elevated from room temperature because of significant self-heating. The very high melting point comes from the strong atomic bonding strength of SiC, which also implies stable chemical inertness. This property can be either beneficial or unfavorable, depending on the perspective. From one viewpoint, one might expect greater ruggedness and reliability for SiC devices. On the other hand, however, the process requirements for SiC would be more challenging as it is difficult to manipulate the material through typical chemical reactions, as will be discussed in the next section.

7.2.2. Processing Technologies

Since the melting temperature of SiC is very high, it is not easy to melt and grow SiC boules at a standard pressure in the same manner as the Czochralski process for Si. Rather, the majority of single crystal SiC materials available today are obtained through gas phase reactions. A typical example is growing SiC boules by providing Si and C atoms through the vaporization of SiC powders, followed by condensation on a seed in a graphite crucible, which is a process called the modified Lely method (Tairov and Tsvetkov, 1978). The typical process temperature is around 2200°C, with a temperature gradient of about 30°C/cm, decreasing in temperature towards the seed to achieve a growth rate in the order of several hundreds of μm per hour, with the dominant growth direction being the c-axis. With the process being performed at such a high temperature, large thermal stress can be expected in the material, and it is not easy to create large diameter, high-quality wafers that have very low defect densities. In the early 1990s, the largest available size for a SiC wafer was less than 1 inch (Powell and Rowland, 2002). As of 2016, the largest wafer sizes that are commercially available are 6 inch, or 8 inch even with the best R&D efforts. The most critical defect that exists in SiC wafers is the open core screw dislocation, or often called micropipe, with a diameter in the μm range. This kind of defect propagates along the c-axis and appears on every wafer diced from the same boule, creating a significant yield loss in the number of functional devices. In the early 1990s, the typical micropipe density reported for SiC wafers was around 100/cm^2. Today, this has improved to less than 1/cm^2, with the best quality wafers being free of micropipes. An example of this is the device area as large as 1 cm^2 reported in the literature (Zhang *et al.*, 2012). Other types of defects also exist in SiC wafers, such as basal plane dislocations, and threading edge and threading screw dislocations, with densities typically in the range of tens of thousands per square centimeter (Ha *et al.*, 2002), as well as point defects such as interstitials and vacancies. The existence of these defects greatly reduces the carrier lifetime of SiC because some will certainly act as recombination centers. As a result, the minority carrier lifetime observed for SiC is typically less than 1 μs (Hassan and Bergman, 2009), as compared with more than 1 ms that is possible in Si (Zhao, 2004b). SiC wafers can be n-type doped to a resistivity of around 0.01 $\Omega \cdot$ cm, or p-type doped to a resistivity of more than 2 $\Omega \cdot$ cm. Additionally, semi-insulating wafers with a resistivity greater than $10^9 \Omega \cdot$ cm can be created using vanadium doping, or can be produced simply by controlling the impurities to a very low level, typically below 10^{15} cm^{-3}. Semi-insulating wafers using vanadium doping, or high-purity semi-insulating wafers, are mainly used as carrier wafers for RF devices in order to reduce the substrate loss, and are not suitable for vertical power devices. On the other hand, the high resistivity of p-type SiC, together with a substrate thickness of 350 μm for 4 inch wafers, will add too much substrate resistance for vertical devices. As a result, the wafers used for 4H-SiC power devices are predominantly n-type.

One of the requisites for building semiconductor devices is the ability to dope the material using a designated dopant at a controllable doping concentration. Unfortunately in SiC, the diffusivities of most useful dopants, such as aluminum, phosphorus or nitrogen, are very small, even at a temperature of 1750°C (Seshadri *et al.*, 1998). Boron shows an appreciably better diffusion (Bracht, 2000), but is still not applicable for the active areas of a device due to its very high resistivity. Feasible options for introducing dopants in the practical fabrication of SiC devices include *in situ* doping during epitaxy and ion implantation. During the early development of SiC technologies, homogeneous epitaxy on SiC wafers was difficult to control because different polytypes may co-exist in the epilayer, with 3C-SiC inclusions most commonly observed, since 3C is the most stable polytype among all others at lower growth temperatures. Epitaxy on 4–8-degree off-axis SiC wafers was proposed in order to solve this problem (Itoh *et al.*, 1994; Kimoto *et al.*, 1997a; Wada *et al.*, 2006). With the exposure of the crystal structure from the off-cut surface of the substrate, the epilayer is able to be grown following the stacking sequence of the substrate, while maintaining a single polytype with a much improved quality at a reduced growth temperature of 300°C. A schematic diagram showing the concept of step-controlled growth is shown in Fig. 7.2. Nowadays, high quality SiC homogenous epitaxy can be performed using commercialized CVD tools at temperatures above 1500°C, using hot/warm wall designs to increase the growth rate to above 30 μm/h.

There have been some attempts to grow 3C-SiC heterogeneously on large Si wafers, but with limited success (Nagasawa *et al.*, 2006). Even today, the material quality and the defect density still do not meet the requirements for large current, high voltage power devices because the large lattice mismatch of around 20% between the 3C-SiC and the Si crystals is difficult to overcome. Power devices require a thick and lightly doped drift region to support a high reverse voltage, but, in SiC, the possibility of forming a drift region using deep diffusion, as in some Si power devices, has been ruled out. Therefore, the drift region in a SiC power device is, almost without exception, grown epitaxially.

To control the doping concentration to an acceptably low level, the concept of site-competition has been introduced (Larkin *et al.*, 1994). By adjusting the C/Si ratio in the ambient during growth, the incorporation efficiency of the dopants to the targeted lattice sites in the crystal can be tuned and, therefore, the doping concentration can be adjusted. For example, aluminum, which is widely used as an acceptor in SiC, tends to occupy Si sites, and nitrogen, which is a popular donor in SiC, tends to occupy C sites. Therefore, by decreasing the C/Si ratio to within an appropriate range, while maintaining other growth

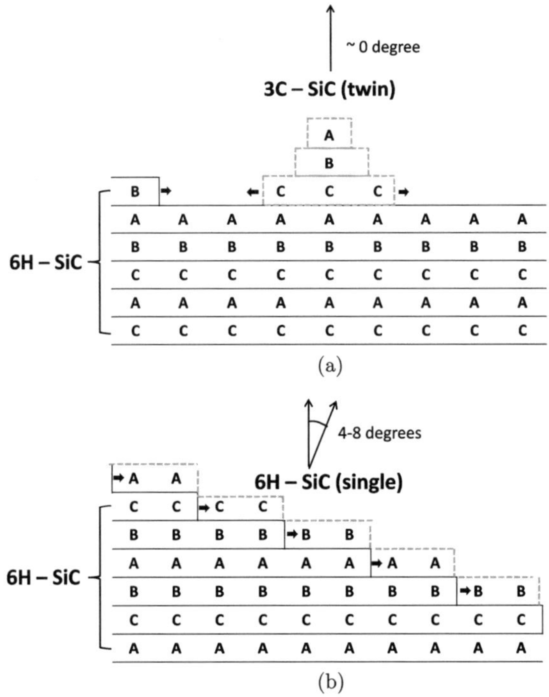

Fig. 7.2. Illustrations of (a) 3C inclusion formation during epitaxy on a nearly zero degree off c-axis surface, (b) step controlled homogenous epitaxy on a 4–8-degree off c-axis surface. Each A, B, and C represents different possible arrangements of SiC bilayers, which defines the polytype.

parameters at the same level, the nitrogen doping concentration in the layer can be controlled in a wide range because the completion rate from carbon to the same atom sites is less. The surface of a SiC wafer is either terminated with Si atoms, which is called the Si-face or the (0001) face in Miller indices, or with C atoms, which is called the C-face or the (000-1) face. It was found that the epitaxy growth of SiC on the Si-face will yield a much lower background doping concentration under the same growth conditions. Therefore, the Si-face is more favorable for the lightly doped drift layer than the C-face. As a consequence, the majority of SiC power devices are created on the Si-face.

Another way to dope SiC is by using ion implantation, which is very often used to create heavily doped regions in the device structure, such as the source and drain in a MOSFET. Aluminum and boron are used to create p-type regions in SiC devices using ion implantation, while nitrogen and phosphorus are used to create n-type regions. The implantation profile and junction depth can be predicted with a reasonable accuracy by using Monte Carlo simulation (Tian, 2008). After dopant ions are introduced in the material, a high temperature anneal is required in order to recover the crystal quality and to incorporate the species into appropriate substitutional sites. It was reported that, in order for boron to be activated effectively, the annealing temperature must be greater than 1650°C to achieve an activation percentage of 50% (Capano *et al.*, 1999). For aluminum, annealing at 1300°C for 12 h is enough to achieve an activation percentage of more than 40% (Saks *et al.*, 2001), but, practically, annealing is performed at temperatures greater than 1600°C for less than 1 h so as to achieve a reasonable production throughout.

Typically, all implantations in device fabrication are activated together using a single annealing process. A temperature of above 1600°C is sufficient to activate both nitrogen and phosphorus implantations, and it was found that phosphorus implantation can achieve sheet resistances for the n+ regions that are an order of magnitude lower than nitrogen (Kimoto *et al.*, 1997c; Capano *et al.*, 2000). Sheet resistance values of less than 100 Ω/square were frequently achieved using phosphorus implantation, which is suitable for device purposes. An explanation for the lower sheet resistance of phosphorus is its nearly 100% complete activation, up to the solubility limit, while this is not the case for nitrogen (Bockstedte *et al.*, 2004). Ion implantation performed at 500°C or above was found to be beneficial in restoring the crystal quality and in activating the dopant better, which is explained by the lower extent of damage introduced when using higher temperature implantations (Kimoto *et al.*, 1997c). One of the concerns regarding activation annealing at 1600°C and above is the roughening of the surface, which is sometimes highly critical for device performance, such as when the surface serves as the MOS interface or the Schottky metal interface, for example. It was reported that Si and C atoms will escape at such high temperatures, and the edges of the bilayer start to be exposed on the surface, forming terraces on off-axis cut wafers, which is a phenomenon known as step bunching (Kimoto *et al.*, 1997b; Capano *et al.*, 1998). Several methods have been proposed that are intended to preserve the surface morphology during high temperature annealing, including covering the surface with a capping layer, such as graphite and AlN, and annealing the wafer in a Si-rich ambient, such as providing an excessive pressure of silane gas higher than typical conditions. The rms surface roughness can be as low as 1 nm, even after annealing at 1700°C for 30 min with a graphite cap, and the sheet resistance is virtually unaffected thanks to the capping process (Negoro *et al.*, 2004a).

Even though dopants can be incorporated into SiC with a reasonable efficiency, another important issue to be considered in the device operation is the dopant level in the wide energy bandgap. The donor levels for n-type dopants in 4H-SiC, such as phosphorus and nitrogen, are 40–60 meV below the conduction band, which is similar to popular n-type dopants used for Si (Schmid *et al.*, 2002; Capano *et al.*, 2000), and are shallow enough to achieve an activation level of close to 100% at room temperature. As a result, a sheet resistance of less than 100 Ω/square at room temperature can be routinely produced if the implantation doses are sufficient. However, this is not the case for a p-type dopant such as aluminum or boron. The shallowest reported acceptor level for aluminum is around 190 meV above the valence band, and is 285 meV for boron (Troffer *et al.*, 1997; Sridhara *et al.*, 1998). As a consequence of the very deep levels in the energy bandgap, the ionization percentage for p-type dopants in SiC is very low at room temperature, which has a significant impact on the device characteristics, in particular, for bipolar devices. The ionization percentage at different temperatures can be estimated by employing the same equations used to obtain Fig. 7.1, except that we are now considering the dopant ionization in the free-out temperature range. The degeneracy factor for the acceptors g_A is assumed to be 4. Figure 7.3 illustrates the calculated ionization percentage versus the doping concentrations at 300 K and 600 K for boron and aluminum. At a doping level of 1×10^{18} cm^{-3}, the ionization percentage at room temperature is about 6% for aluminum, and only about 1% for boron. The percentage becomes even worse at higher doping concentrations. As a result of this, the best p-type sheet resistance that has been achieved is 2.3 kΩ/square for aluminum implantation and activation at 1800°C (Negoro

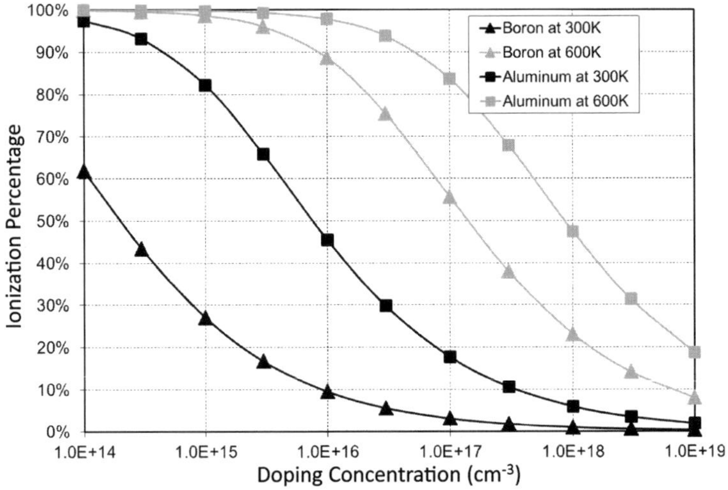

Fig. 7.3. Calculated ionization percentage for different doping concentrations of boron and aluminum at 300 K and 600 K.

et al., 2004b), which is at least one order of magnitude greater than the typical n-type sheet resistance.

Another issue that arises with the low ionization percentage and the high resistance of p-type SiC is the large p-type contact resistivity. Ohmic contacts to n-type SiC are achieved using a Ni-based metal scheme, and the reported contact resistivities are typically around $1 \times 10^{-6}\ \Omega \cdot cm^2$ (Nikitina *et al.*, 2005). For p-type SiC contacts, Ti/Al-based metal schemes are frequently used, with special care taken in order to achieve contact resistivities below $1 \times 10^{-4}\ \Omega \cdot cm^2$ (Crofton *et al.*, 2002).

Other typical fabrication processes for Si devices, such as lithography, thin film deposition, dry etching, and thermal oxidation, can also be applied to SiC, but with some modifications. One exception to this is wet etching of single crystal SiC, which is almost impossible at reasonable temperatures. It is also difficult to determine an appropriate material that has a good selectivity to serve as the mask for selective wet etching. Consequently, selective etching of SiC is performed exclusively using plasma chemistry, typically with flourine-based gases such as NF_3, CF_4, and SF_6. Etch rates of more than 500 nm/min can be accomplished using inductively coupled plasma etching (Jiang *et al.*, 2003). Thermal oxidation can be performed on SiC surfaces, with SiO_2 being the major reaction product. The majority of the carbon atoms react with oxygen to form CO and escape from the surface, while some remain in the SiO_2 layer in the form of carbon interstitials and clusters, concentrated at the interface of the SiO_2 and the SiC (Afanasev *et al.*, 1997). It is generally accepted that those remaining carbon atoms are the reason for the inferior SiO_2 layers grown on SiC compared to those grown on Si. A detailed discussion of this phenomenon will be covered later in the chapter related to SiC MOSFETs.

7.3. High Voltage Designs for SiC Devices

To achieve high voltages in SiC devices, care must be taken when selecting the doping concentration and the thickness of the drift region so as to achieve the optimal V_{BR}, in the same manner used for Si. Furthermore, the edge termination must be properly implemented in order to realize a reasonable V_{BR} that is close to the ideal value, while remaining within the limits and constraints imposed by SiC material and processing technologies.

7.3.1. Drift Region and Ideal Breakdown Voltage

Based on Eq. (2.150), Fig. 7.4 illustrates that the ideal V_{BR} for different doping concentrations and drift region thicknesses for 4H-SiC can be generated as a quick reference. This figure serves a starting point for all SiC power device designs. According to the voltage ratings defined in the specifications based

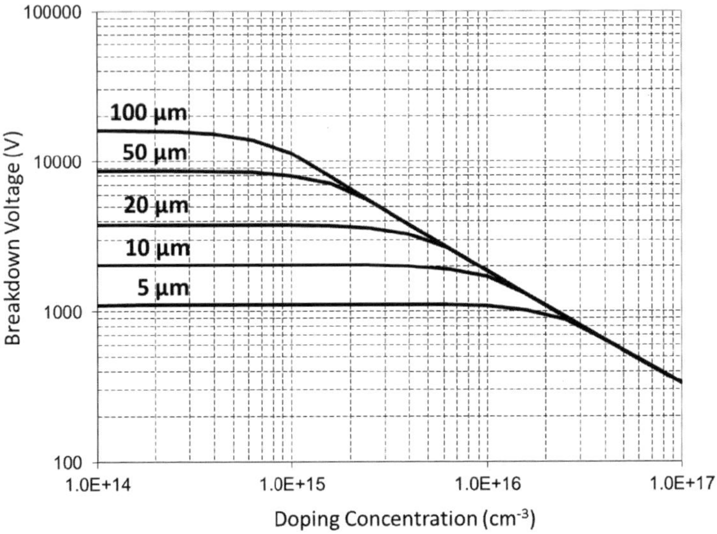

Fig. 7.4. Ideal breakdown voltage for the 4H-SiC drift region for different doping concentrations and thicknesses.

on the targeted applications, the ideal V_{BR} that has a reasonable margin and safety factor can be selected.

Once a suitable V_{BR} has been selected, the next step is to determine the ideal doping concentration and the drift region thickness for that enables the minimum $R_{on,sp}$ or V_F to be achieved. For unipolar devices, this can be accomplished by expressing $R_{on,sp}$ as a function of the drift region thickness W with the assistance of Eq. (2.150), and then letting $dR_{on,sp}/dW = 0$. The corresponding optimized drift region thicknesses and doping concentrations are:

$$W_{opt} = \frac{3\,V_{BR}}{2\,\hat{E}_C}, \tag{7.6}$$

$$N_{opt} = \frac{4\,\varepsilon_s\,\hat{E}_C^2}{9q\,V_{BR}}. \tag{7.7}$$

For example, for an ideal V_{BR} for 4H-SiC of 1200 V, W_{opt} is 6 μm, and N_{opt} is $1.79 \times 10^{16}\,\text{cm}^{-3}$, assuming a constant $\hat{E}_C$ of 3 MV/cm, while for Si, W_{opt} is 60 μm and N_{opt} is $2.20 \times 10^{14}\,\text{cm}^{-3}$, assuming an $\hat{E}_C$ of 0.3 MV/cm. From this simple calculation, we know that in the drift region for 4H-SiC, the thickness is reduced by a factor of 10, and the doping concentration is increased by a factor of 100, yielding a much reduced $R_{on,\,sp}$. It is also worth mentioning that the optimal design for unipolar devices is always in the punchthrough region. The same methodology can be applied for bipolar devices, but V_F should be considered instead of $R_{on,sp}$. However, simple form expressions for W_{opt} and N_{opt} are not possible, since, taking the PIN diodes discussed in Sec. 3.3.1 as an example, the expression for V_F is rather complicated. However, another key parameter, the ambipolar lifetime τ_a, should also be provided in the analysis. Furthermore, in a practical application, the turn-off loss related to the storage charge should also be included in the overall analysis. Consequently, at this point, we will leave the discussion of V_F of bipolar devices for later sections.

The Superjunction principle was first proposed in the 1990s, and was intended to overcome the limits to the V_{BR}–$R_{on,sp}$ trade-offs in conventional Si vertical devices (Chen, 1993). In recent years, attempts have been made to realize the superjunction concept in SiC in order to further improve the $R_{on,sp}$. In 2014, Kosugi *et al.* first demonstrated the idea of implementing a superjunction in SiC through a combination of MeV implantation and the multi-epitaxial growth method. A V_{BR} of 1545 V and a $R_{on,sp}$ of $1.06\,\text{m}\Omega \cdot \text{cm}^2$ was achieved on a 5.5 μm thick drift layer, clearly showing the advantage of superjunctions in SiC.

7.3.2. Edge Termination for SiC Power Devices

In general, the discussions in Sec. 2.11 can be directly extended to SiC, but a few important points must be noted. First, because diffusion in SiC is not practical, the junction depth and junction curvature formed by implantation at the edges of the active area in SiC devices is typically less than 1 μm in the vertical direction, resulting in significant crowding of the electric field. As a consequence, edge termination is indispensable in SiC devices in order to realize a breakdown voltage that is close to its ideal value. Second, due to the difficulties encountered when attempting wet etching of SiC, the etched contour termination has never been applied to SiC devices. Third, there was a concern about the deep levels of the acceptors in the depleted p-type region in SiC, either in the active area or in the edge termination. The question is whether the acceptors are fully or partially ionized in the depletion region, and whether they are ionized quickly enough to avoid any adverse effects when switching, resulting in premature breakdown. In 1999, Lades *et al.* conducted a numerical study and showed that, at room temperature, the time constant for aluminum acceptors to be fully ionized in the depletion region is only 300 ps, which is less than the practical switching times for power devices, and, therefore, it would not affect normal operation. For boron, however, the time constant is 100 ns, which might raise concerns for fast switching power devices. At lower temperatures, the time constant begins to increase for both dopants, and attention must be paid to the ionization transients in the device design, even for aluminum. The fourth point relates to the junction termination extension (JTE), which was introduced in Sec. 2.11.7, and is a common edge termination method used for ultra-high voltage SiC devices. One of the most important design parameters in JTE is the implantation dose Q_{JTE}, which is defined as the dopant density per unit area. The optimal Q_{JTE} is roughly

$$Q_{\mathrm{JTE}} = \frac{\varepsilon_{\mathrm{S}} \hat{E}_{\mathrm{C}}}{q}. \tag{7.8}$$

Incorporating the values in the equation, the Q_{JTE} for 4H-SiC is $1.6 \times 10^{13}/\mathrm{cm}^2$, which is about 10 times larger than that for Si, mainly due to the fact that $\hat{E}_{\mathrm{C}}$ is 10 times larger. In addition, it must be kept in mind that, in practice, Q_{JTE} should vary in different device structures, since $\hat{E}_{\mathrm{C}}$ is a function of the doping concentration, as described in Sec. 2.9.1, whereas we assume it is a constant in the discussion here.

Several edge termination structures for SiC devices have been demonstrated. Floating guard rings, as discussed in Sec. 2.11.3, are also effective for SiC devices, particularly for 600–1000 V. Floating guard rings have the advantage that they can be incorporated in the same p+ process for the active area, thereby saving costs. For example, they can be incorporated with the p+ implantation for the anode in PIN diodes. However, at higher voltages, this approach is less attractive since spacing and width of the guard rings must be fine-tuned in order to achieve a suitable optimization, meaning that its use is limited by the capabilities of the lithography. JTE, on the other hand, can be realized without needing to consider this limit and is therefore more favorable for higher voltage devices (Perez *et al.*, 2005). However, a major issue faced by JTE is its high sensitivity to Q_{JTE} and, particularly for SiC, it is difficult to precisely control the activation percentage of the implanted JTE dose because of the very high annealing temperature required. Several advanced JTE structures have been proposed that are intended to improve the sensitivity of V_{BR} against the variation of Q_{JTE}. As a rule of thumb, multizone JTE (MZ-JTE) using a gradual doping approach will improve the tolerance of V_{BR} to Q_{JTE}. However, defining multiple zones requires multiple lithography steps and multiple implantations which, consequently, will increase the cost. Several methods have been proposed that are aimed at creating multiple zones with a reduced number of lithography and process steps. Examples of these methods include the stepped oxide mask JTE (Losee *et al.*, 2004), the etched MZ-JTE (Ghandi *et al.*, 2009), the space-modulated JTE (SM-JTE) (Feng *et al.*, 2012), and the counter-doped JTE (CD-JTE) (Huang *et al.*, 2015). Figure 7.5 shows the structures of single-zone JTE, MZ-JTE, SM-JTE, and CD-JTE, using PIN diodes as an example. The p-type regions in the edge termination for an n-type epilayer can be formed using either ion implantations or selective etching. Figure 7.6 shows simulated breakdown voltage versus JTE dose for different

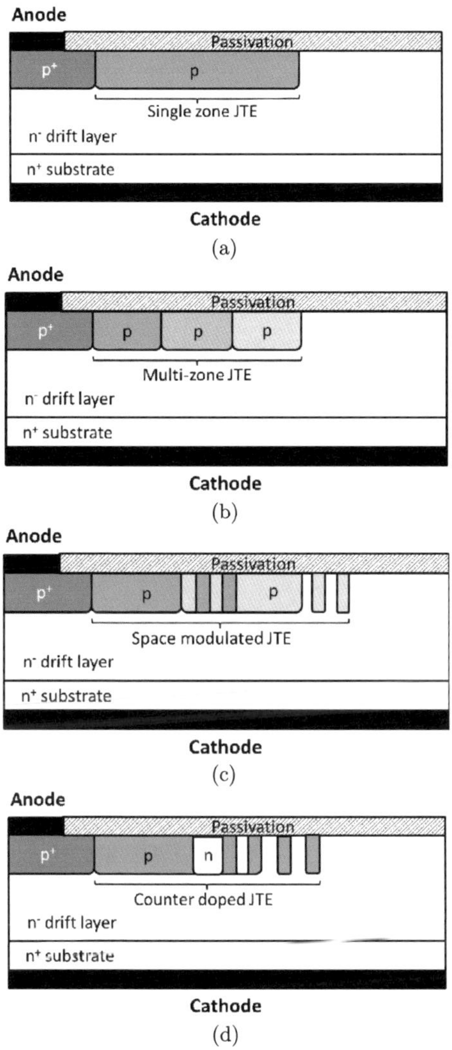

Fig. 7.5. Examples of JTE structures (a) Single-zone JTE, (b) MZ-JTE, (c) SM-JTE, (d) CD-JTE.

JTE structures with a 100 μm width for a 30 μm thick, 2×10^{15} cm^{-3} doped 4H-SiC drift layer. For a single-zone JTE, the breakdown voltage is very sensitive to JTE dose and the optimal value is about 1×10^{13} cm^{-2}. With the advanced structures, the sensitivity is significantly improved and reasonable breakdown voltages can be achieved with a wide range of JTE dose.

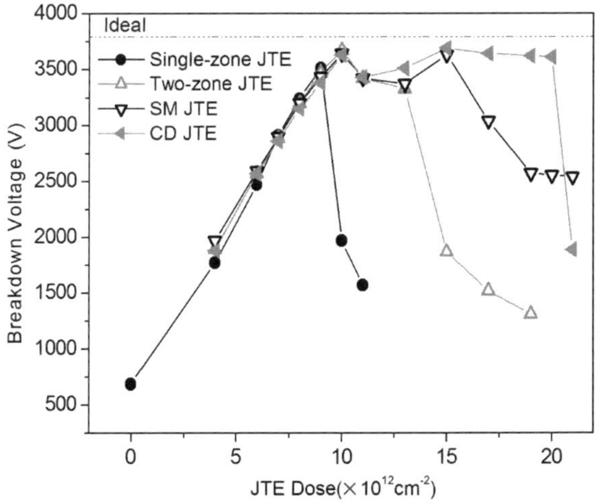

Fig. 7.6. Simulated breakdown voltage versus JTE dose for different JTE structures with a 100 μm width for a 30 μm thick 4H-SiC drift layer.

Furthermore, since the surface charges will compensate for the effects generated from the ionized dopants in the JTE region, a JTE structure that has a greater tolerance to $Q_{\rm JTE}$ will also have a greater tolerance to the surface charges introduced by the passivation processes. Breakdown voltages of more than 20 kV have been experimentally demonstrated using SM-JTE on a 186 μm, $2.3 \times 10^{14}\,\text{cm}^{-3}$ n-type drift layer (Miyake *et al.*, 2012b). Bevelled edge termination, which is typically used for wafer-sized Si devices, has also been applied to SiC, where a breakdown voltage of more than 1000 V was demonstrated using a p-type drift region that was 15.8 μm thick and doped at $6.1 \times 10^{15}\,\text{cm}^{-3}$ (Huang *et al.*, 2012).

7.4. SiC Rectifiers

After gaining a general understanding of SiC material and processing technologies, together with $V_{\rm BR}$ design and edge termination for SiC power devices, we are in a good position to discuss their performance. This section is devoted to the performance and important design issues related to some of the most common device structures for SiC rectifiers, such as Schottky barrier diodes (SBDs), junction barrier Schottky (JBS) diodes, and PIN diodes.

7.4.1. SiC Schottky Barrier Diodes

SBDs are simple semiconductor devices in terms of structural complexity, yet have a broad range of applications. The most important design issue for SBDs is the control of the barrier height ψ_b, since this determines the trade-off between the on- and off-state characteristics, as discussed in Sec. 3.4. In 1997, Itoh and Matsunami (1997) showed that the barrier heights for different metals on the Si-face of an n-type 4H-SiC material are based on an empirical relationship, i.e.

$$\psi_b = 0.70\phi_M - 1.95(\text{V}) \quad \text{for n-type 4H-SiC}, \tag{7.9}$$

which shows the effectiveness of controlling the barrier height ψ_b by selecting metals that have a suitable work function. This also implies that there is very little Fermi level pinning at the metal–semiconductor interface because of the superior chemical inertness of SiC. For comparison, an empirical relationship for n-type Si is also shown (Sze, 1981).

$$\psi_b = 0.27\phi_M - 0.55(\text{V}) \quad \text{for n-type Si}. \tag{7.10}$$

In the previous literature, titanium, nickel, and gold have commonly been used for studying the characteristics of SiC SBDs, while molybdenum and tungsten have also been tested for high temperature applications on account of their excellent thermal stability (Nakamura *et al.*, 2005). A variety of thermal treatments have also been attempted on SiC SBDs, and it was discovered that the barrier height can be varied and tuned by forming metal silicide and metal carbide at the interface (Kinoshita *et al.*, 2010). Because the bandgap for SiC is larger than Si, both the barrier height and the turn-on voltage for SiC SBDs are typically larger, but this can be compensated for thanks to their superior reverse and switching characteristics. For some applications, the rectifiers in the circuits are required to have a large surge current capability. In this case, JBS diodes are more favorable than SBDs (Brosselard *et al.*, 2008). The typical structures for SBDs, JBSs, and PIN diodes are shown in Fig. 7.7 for comparison.

With the p+ islands underneath the anode to form parasitic PIN structures that are parallel with the Schottky diode, the forward voltage drop will be reduced at a sufficiently large surge current because the parasitic PIN will turn on and inject holes in the drift region, enabling conductivity modulation. As a result, the JBS diodes are able to dissipate less joule heat and are less prone to damage at a high current density than Schottky diodes. The forward I–V curves for SBD, PIN, and JBS diodes are compared in Fig. 7.8. Another benefit of JBS diodes is that the Schottky metal will be protected by the p+ islands at reverse biases, lowering the electric field at the interface, and reducing the reverse leakage current.

Another approach that can be implemented to further increase the surge current capabilities of SiC JBS diodes is to use a thin substrate. Experimental results showed that the surge current capabilities can be increased by 50% using

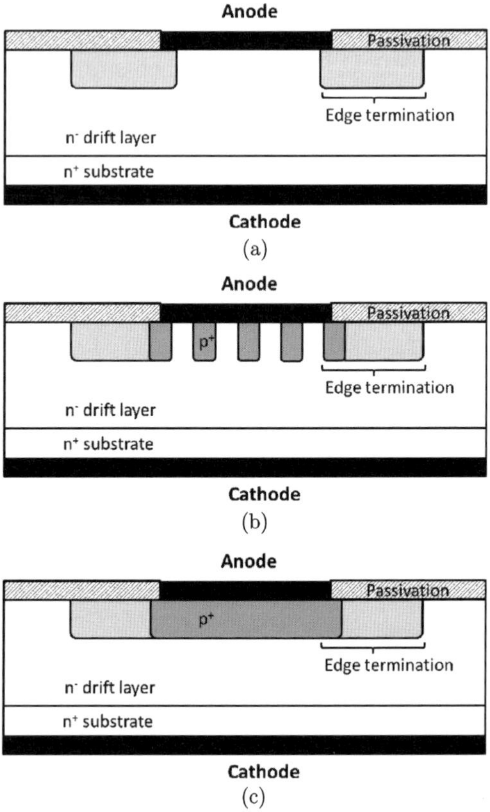

Fig. 7.7. Typical structures for (a) SBDs, (b) JBS diodes, (c) PIN diodes.

a 90 μm substrate and an Ag-based backside solder in comparison to a 300 μm substrate and a standard Pb-rich Sn based solder (Nakanishi *et al.*, 2014). The specific on-resistance was also improved by $0.4\,\mathrm{m}\Omega \cdot \mathrm{cm}^2$, which is about 20% of the total resistance. By taking advantage of the very small reverse recovery charges, the aim is to replace Si PIN diodes with SiC SBDs and JBSs of the same voltage rating. The switching performance has been tested for SiC JBSs, and, as a result of the reduced switching loss, the overall system efficiency was improved. For example, replacing 600 V Si PIN diodes with 600 V SiC SBDs in a 300 W boost power factor corrector increased the efficiency by 2% (Spiazzi *et al.*, 2003), and the EMI performance was also improved. To date, a breakdown voltage as high as 10 kV has been reported in the literature for SiC SBDs and JBSs that have a drift layer that is more than 100 μm thick (Zhao *et al.*, 2003b). JBS diodes rated at 50 A are commercially available in TO 247 and similar packages, or in bare dies.

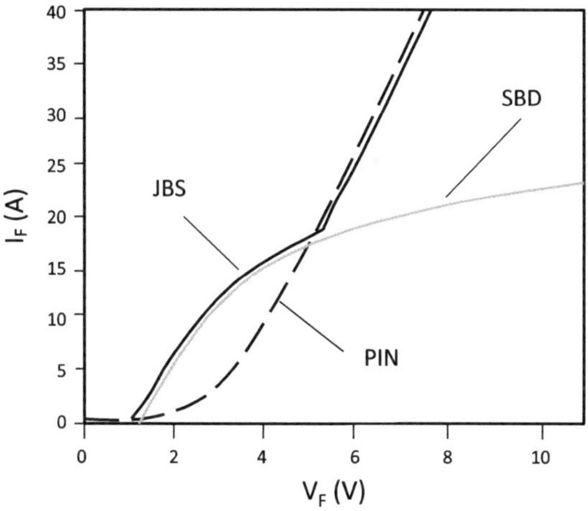

Fig. 7.8. Comparison of forward I–V curves for SBD, PIN, and JBS diodes.

7.4.2. SiC PIN Diodes

Early SiC PN diodes were first demonstrated for blue LEDs in the 1970s (Matsunami *et al.*, 1977). Later, the attention was shifted to power electronics applications as the material production became more mature. Since the built-in voltage for a 4H-SiC PN junction is around 3 V, the forward voltage drop of PIN diodes at 100 A/cm^2 is typically 3.5–4 V, making them only attractive for ultra-high voltage and low frequency applications (Morisette and Cooper, 2002). One of the critical design issues associated with PIN diodes is the carrier lifetime. It was found that better forward characteristics could be achieved when using an epitaxial p+ layer as the anode rather than an implanted p+ layer (Losee *et al.*, 2009), which can be attributed to the low carrier lifetime of the junction in the implanted region due to the difficulties related to the activation of the implanted species and in the recovery of the crystal quality. Moreover, the drift region must be conductivity modulated in order to achieve a reasonable forward voltage drop. Figure 7.9 shows the estimated forward voltage drops where different thicknesses are used for the drift region, together with the ambipolar lifetimes, based on Eq. (3.53), assuming an electron mobility of 800 cm^2/Vs, a hole mobility of 120 cm^2/Vs, a p+ anode doping concentration of 10^{19} cm^{-3}, and a p+ cathode doping concentration of 10^{19} cm^{-3}. If, for example, we take PIN diodes rated at 10 kV, the thickness of the drift region should be around 100 μm thick to allow for a safe design margin. From Fig. 7.8, it can be seen that the ambipolar lifetime should be larger than 10 μs in order to create a forward voltage less than 4 V. As mentioned in the previous section, the carrier lifetime for the SiC epilayer is typically several hundreds of nanoseconds.

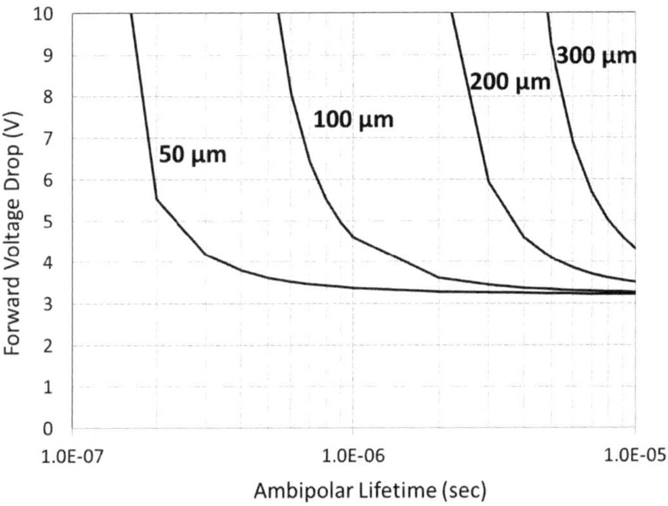

Fig. 7.9. Estimated forward voltage drop for a 4H-SiC PIN diode based on different drift layer thicknesses and ambipolar lifetimes.

Other than optimizing the growth conditions to gain a marginal improvement, several methods have been shown to be effective in greatly enhancing the carrier lifetime. In 2009, Hiyoshi and Kimoto showed that one of the impediments to the lifetime of 4H-SiC, is $Z_{1/2}$, which is suspected to be a deep level carbon vacancy in the energy bandgap, and can be annihilated by the injection of carbon interstitials through the thermal oxidation of the SiC surface. As a result, the carrier lifetime, which is measured via the decay in photoconductance, is increased from 0.73 μs to 1.62 μs. A similar approach was demonstrated by Miyazawa *et al.* (2010) where carbon interstitials were injected through the ion implantation of carbon and a subsequent diffusion anneal. The resulting carrier lifetime was shown to be close to 20 μs, which is sufficient for 10 kV-rated PIN diodes. With these advancements in the material properties and carrier lifetimes, a forward voltage of 3.75 V at a current density of 100 A/cm^2 at room temperature was achieved for a 12.9 kV PIN diode (Sundaresan *et al.*, 2012).

Other than carrier lifetime, another critical issue for SiC PIN diodes is a concept referred to as bipolar degradation. It has been reported that SiC bipolar devices suffer from severe forward voltage drift during constant forward current stress. This has been attributed to the expansion of the basal plane dislocations in the drift layer, which causes a corresponding increase of the resistance in the drift region consequently (Lendenmann *et al.*, 2002). Several methods have been proposed and demonstrated that are designed to reduce the number of basal plane dislocations in the substrates and prevent them from propagating into a drift layer which was grown via epitaxy. An effective method was to use molten KOH to preferentially etch the basal plane dislocations, and form pits on the growth surface. Later, during epitaxial growth, these pits would close up and transform into less harmful threading edge dislocations. By implementing this method, the density of the basal plane dislocations can be reduced from 500 cm^{-2} in the substrate to less than 10 cm^{-2} in the epilayer (Zhang *et al.*, 2006). Based on these efforts, 10 kV, 50 A SiC PN diodes that have no forward voltage drift were demonstrated (Hull *et al.*, 2005). Instability or performance degradation has also been found in other SiC power devices, which can be traced back to the parasitic PN junctions in the device structure. Since the issues surrounding bipolar degradation in SiC PIN diodes have mostly been overcome, the stability of these devices has also improved. The switching of SiC PIN diodes was measured, and it was found that the reverse recovery time of an 8.6 kV SiC PIN diode was less than 200 ns, even at 275°C, which is much lower than similarly rated Si PIN diodes (Singh *et al.*, 2002). This aspect can be attributed to the diminished storage charge due to (1) the smaller carrier lifetime, (2) the fact that a thinner drift region is required for the same V_{BR}, and (3) the higher doping concentration in the drift region, which reduces the degree of high level injections.

7.5. SiC Unipolar Switches

After the commercialization of SiC rectifiers, SiC unipolar switches, such as MOSFET and JFETs, were introduced to the market between the late 2000s and early 2010s. Without the presence of storage charges in the drift region, as compared to Si IGBTs with the same voltage rating, the adoption of SiC unipolar switches is expected to further improve system efficiency. In fact, an efficiency of 99% has been demonstrated in fully SiC converters and inverters (Norling *et al.*, 2012; Mallwitz *et al.*, 2012). So far, however, SiC unipolar switches have not been as successful in the market as SiC rectifiers, where different challenges are being faced by different types of SiC unipolar switch. The characteristics of these switches will be discussed later in this section.

7.5.1. SiC MOSFETs

The Si MOSFET is the dominant component in the power device market, particularly for high-frequency devices. Therefore, it is logical to expect that SiC MOSFETs also play an important role, provided that material and cost issues are able to be solved. The most important feature of a MOSFET is the channel, which defines the gate control characteristic and the current handling capabilities. As mentioned in the previous section, it is possible to form thermally grown SiO_2 on a SiC surface, but experimental results have shown that the electron mobility in the inversion channel, calculated as the field effect mobility, is typically less than 10 cm^2/Vs, with a threshold voltage much greater than 5 V on the Si-face of 4H-SiC. Higher values were reported on other polytypes, but these, unfortunately, are less attractive for power devices based on the reasons described in the previous sections. Since the late 1990s, the improvement in electron channel mobility in 4H-SiC, particularly on the Si-face, has been one of the main focuses of research into SiC MOSFETs. With appropriate nitration treatments, such as an NO anneal after thermal oxidation, or direct oxidation using NO_2, the peak channel mobility can be improved to 37 cm^2/Vs with a threshold voltage of 5 V (Chung *et al.*, 2001). It was also found that any reduction in electron channel mobility usually came with an increase in the measured interface trap density D_{it}, particularly at the locations close to the conduction band. This can be explained by the significant coulomb scattering and the small number of free electrons existing in the channel that are able to conduct current if the number of interface traps is large (Saks and Agarwal, 2000). Figure 7.10 schematically shows the band diagram for a MOS structure corresponding to an applied gate voltage, together with the free electrons in the inversion channel, and the trapped electrons at the interface.

The total number of carriers per area n_{total} induced by a gate voltage can be estimated from Eq. (7.11), and a relationship between the true mobility and the extracted effective mobility can be expressed as Eq. (7.13).

$$qn_{\text{total}} \cong C_{\text{ox}}(V_{\text{GS}} - V_{\text{th}}), \tag{7.11}$$

$$n_{\text{total}} = n_{\text{free}} + n_{\text{trapped}}, \tag{7.12}$$

$$\mu_{\text{eff}} = \frac{Lg_{\text{d}}}{WC_{\text{ox}}(V_{\text{GS}} - V_{\text{th}})} < \mu_{\text{n}} = \frac{Lg_{\text{d}}}{Wqn_{\text{free}}}. \tag{7.13}$$

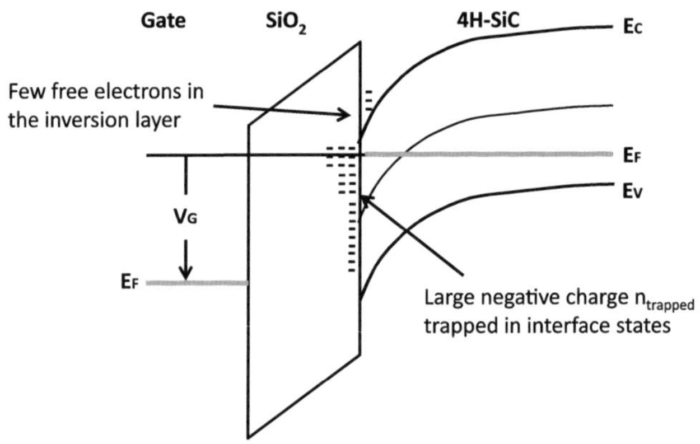

Fig. 7.10. A schematic diagram illustrating the energy band alignment and the free electrons in the channel and the trapped electrons at the interface.

Here, C_{ox} is the gate oxide capacitance per unit area, n_{free} is the number of free carriers per unit area, and $n_{trapped}$ is the number of trapped carriers per unit area. L is the gate length and W is the gate width, while g_d is the transconductance in the linear region at a small V_d, defined as I_d/V_g. When D_{it} is large, a significant proportion of the carriers will be trapped at the interface and, from Eq. (7.13), it can be concluded that the extracted mobility will be less than the true mobility. Moreover, as the temperature increases, the trapped electrons will be released, resulting in an increase in the extracted mobility and a reduction in channel resistance. It has been proposed that a more efficient approach to accurately extracting mobility is through gated hall measurement on a MOSFET.

Other approaches have been proposed that are designed to further improve the electron channel mobility, but each has its own drawbacks. Burred or implanted channels were shown to have a channel mobility as high as 140 cm^2/Vs, but the threshold voltage was reduced to nearly zero (Harada *et al.*, 2001). Phosphorus doped gate oxide using a $POCl_3$ anneal at 1000°C was shown to improve the channel mobility to 89 cm^2/Vs, but the threshold voltage was unstable and also very close to zero (Okamoto *et al.*, 2010; Sharma *et al.*, 2012). Dielectrics other than thermally grown SiO_2, such as MOCVD Al_2O_3, ALD Al_2O_3, and a $LaSiO_x/SiO_2$, dielectric stack, have also been investigated and have shown some promising results (Hino *et al.*, 2008; Lichtenwalner *et al.*, 2009; Yang *et al.*, 2015). However, the relatively smaller conduction band offsets between these dielectrics and 4H-SiC may cause a large gate leakage current in n-type MOSFETs, and pose a threat to device reliability, particularly at high temperatures (Singh, 2006). Figure 7.11 shows the relative band alignment between Si, 6H-SiC, 4H-SiC, and some of the popular dielectrics.

Assuming that it is dominated by Fowler–Nordheim tunneling, the gate leakage current can be approximated as (Schroder, 2006)

$$J_{FN} \cong A\,\hat{E}_{ox}^2 \exp\left(-\frac{B}{\hat{E}_{ox}}\right), \tag{7.14}$$

$$A = \frac{q^3(m/m_{ox})}{8\pi h\phi_B}, \tag{7.15}$$

$$B = \frac{8\pi\sqrt{2m_{ox}}}{3qh}\left(\phi_B^{3/2}\right), \tag{7.16}$$

where m is the free electron mass, m_{ox} is the effective mass of the electron in the oxide, $\hat{E}_{ox}$ is the oxide field, h is the Plank's constant, and ϕ_B is the barrier height for the electron to move from the 4H-SiC conduction band through the dielectric. Assuming an m_{ox}/m of 0.26 and assuming that ϕ_B is the conduction band offset, J_{FN} can be plotted for Al_2O_3/4H-SiC, SiO_2/4H-SiC, and SiO_2/Si,

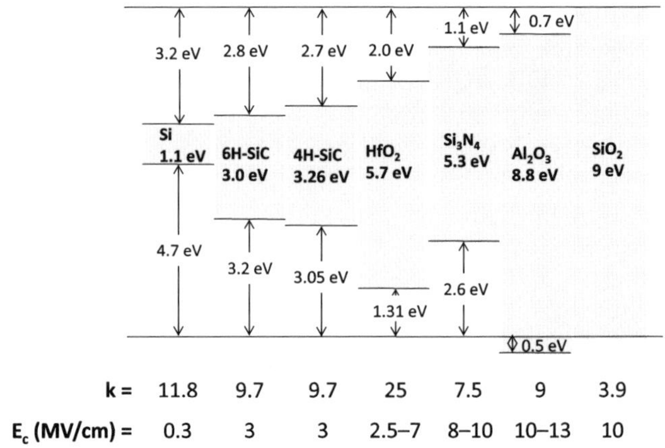

Fig. 7.11. Band alignment between Si, 6H-SiC, 4H-SiC, and some of the popular dielectrics.

as shown in Fig. 7.12. It can be seen that for Al_2O_3 on 4H-SiC, J_{FN} is increased by almost six orders of magnitude at an $\hat{E}_{ox}$ of 3 MV/cm compared to SiO_2 on 4H-SiC.

Electron channel mobilities on other crystal faces of 4H-SiC were also investigated. It was shown that electron channel mobility can be as high as 96 cm^2/Vs on the (112-0) face (Yano *et al.*, 1999), which is beneficial for trench gate UMOSFETs in 4H-SiC. An electron channel mobility of 111 cm^2/Vs was also demonstrated on the (0001-) C-face by using pyrogenic oxidation followed by a hydrogen anneal (Fukuda *et al.*, 2004).

As in Si, there are two popular gate structures for SiC power MOSFETs. The first is the planar gate structure or DMOSFET, and the second is the trench gate structure or UMOSFET. With the n-drift layer epitaxially grown on the off-axis Si-face, the electron mobility in the channel of a 4H-SiC DMOSFET is expected to be low. In addition, the p-well is typically formed through an ion implantation of aluminium, which creates many defects in the channel, and introduces further scattering centers for electrons. As a result, the channel resistance occupies a significant proportion of the total resistance in a 4H-SiC DMOSFET, contrary to its Si counterpart at the same voltage rating. Several structures have been proposed to overcome this problem.

A double epi-structure for DMOSFETs was proposed for p-base using epitaxy layers instead of implantation so as to avoid implant damage in the channel (Harada *et al.*, 2004). A self-aligned process was proposed that was designed to shrink the channel length to less than 0.5 μm, which reduced the channel

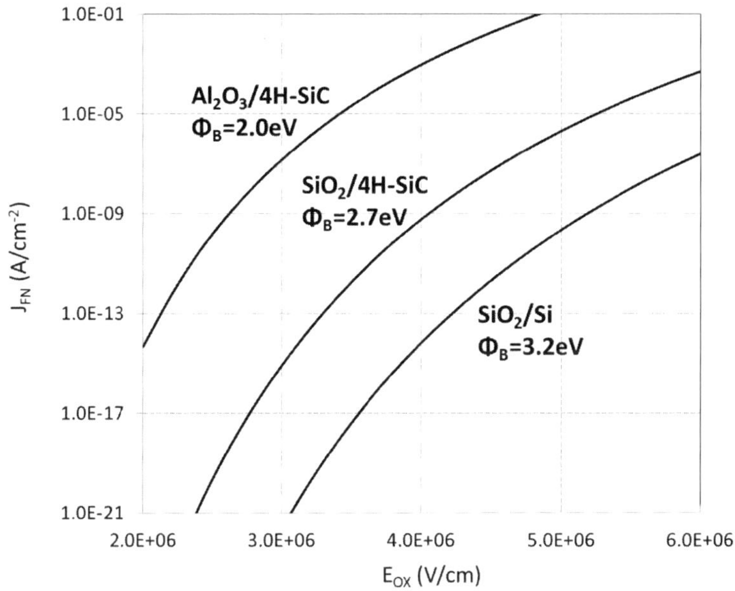

Fig. 7.12. Estimated Fowler–Nordheim tunneling current of Al_2O_3/4H-SiC, SiO_2/4H-SiC, and SiO_2/Si at different oxide fields.

resistance by a factor of between four and six (Matin *et al.*, 2004). The use of an accumulation layer close to the surface of the channel was demonstrated (Singh *et al.*, 2003). In this case, the electrons would travel with a mobility that is closer to the bulk mobility, and, in theory, has a much greater mobility than inversion mobility, but the enhanced mobility often comes at the expense of a very low, or even negative, threshold voltage, which is less favorable for power applications. A vertical 4H-SiC MOSFET was demonstrated on epitaxially grown p-well and n-drift layers on the C-face, taking advantage of the high channel mobility. As a result, an $R_{\mathrm{on,sp}}$ of 1.8 mΩcm^2 was achieved on a 10 A, 660 V device (Harada *et al.*, 2006). Today, the best-in-class 20 A, 1200 V 4H-SiC DMOSFET shows an $R_{\mathrm{on,sp}}$ of 2.7 mΩcm^2 at room temperature, and 3.9 mΩcm^2 at 150°C (Zhang *et al.*, 2015). Figure 7.13(a) shows a conventional SiC DMOSFET on the Si-face, and Fig. 7.13(b) shows an epitaxial buried channel SiC DMOSFET on the C-face. Notice that in this structure an n-type implantation was done to reduce the JFET region resistance.

A variation of a DMOSFET in 4H-SiC was recently proposed, where trenches were created in the channel region of the planar gate to form a 3D gate structure similar to that of a FinFET (Tega *et al.*, 2015). In this way, the channel density per area increases, while the channel mobilities on the sides of the

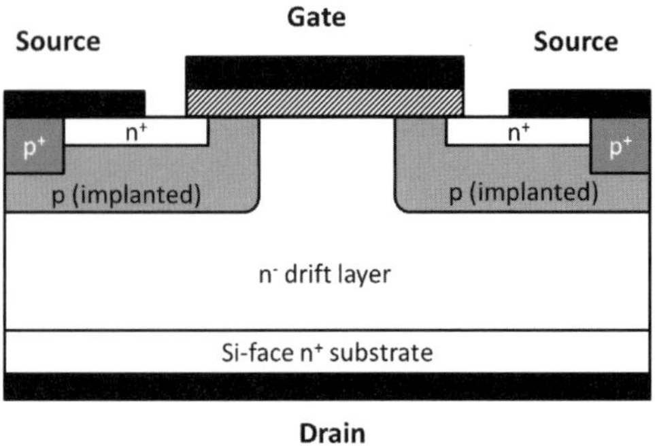

(a)

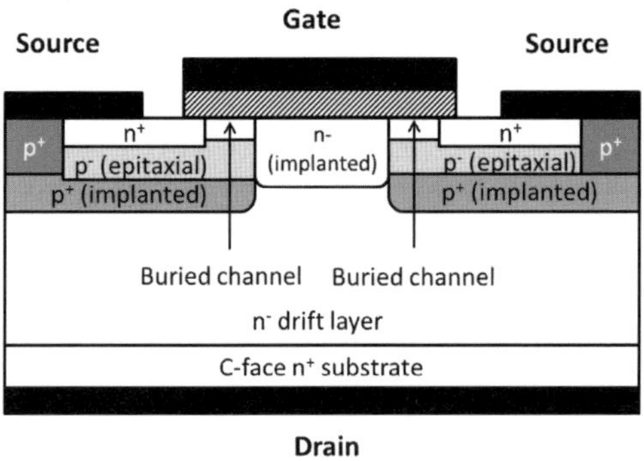

(b)

Fig. 7.13. The typical structure of (a) a conventional SiC DMOSFET on the Si-face, and (b) an epitaxial buried channel SiC DMOSFET on the C-face.

trenches are also expected to higher because they are on the (112-0) or (1-100) faces. As a result, the channel resistance reduces, and the $R_{\mathrm{on,sp}}$ improves. The internal PN body diode inside the SiC DMOSFET has the potential of being used as an antiparallel, or free-wheeling, diode, which is required in certain circuit topologies, despite the larger forward voltage drop. However, there are concerns about bipolar degradation in the same way as for 4H-SiC PIN diodes in the early years. With the improvement of material quality, as reflected by the improved reliability of PIN diodes, body diodes nowadays show stable characteristics under forward stress. An embedded JBS diode was proposed as a technique for replacing the body diode in a DMOSFET structure, further reducing the forward voltage drop and making it closer to that required for practical uses (Yen *et al.*, 2015). A 10 kV DMOSFET has been demonstrated for ultra-high voltage applications, such as smart grids (Ryu *et al.*, 2004). An $R_{\mathrm{on,sp}}$ of 123 mΩcm^2, corresponding to a $V_{\mathrm{BR}}^2/R_{\mathrm{on,sp}}$ of 813 MW/cm^2, and a

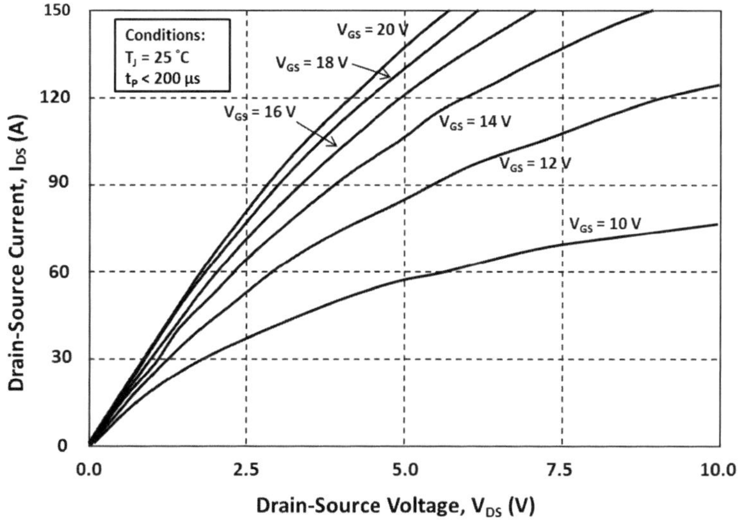

Fig. 7.14. Typical output characteristics for a 1200 V, 60 A rated SiC DMOSFET at $T_J = 25°C$.

switching time of 100 ns, were achieved. In this case, since the doping concentration of the drift region was very small, an n-type implantation was conducted in the parasitic JFET region in order to reduce the resistance.

Figure 7.14 shows the typical output characteristics for a 1200 V, 60 A rated SiC DMOSFET at junction temperature of 25°C, and Fig. 7.15 shows its output characteristics at 150°C junction temperature. In contrast to a Si power MOSFET, which always shows a clear current saturation region in its output characteristics, a state-of-the-art SiC DMOSFET usually exhibits channel length modulation when the current increases, even at low V_{DS} values. This is because the channel length in these MOSFETs is short in order to reduce the channel resistance, but, at the same time, the short channel effect inevitably emerges.

Another distinct feature of a SiC DMOSFET in comparison to a Si power MOSFET is the dependence of its on-resistances to temperature. Figure 7.16 shows a comparison of the normalized on-resistances for both Si power MOSFET and SiC DMOSFET for a temperature range from −50°C to 150°C. For the Si power MOSFET, the on-resistance increases monotonically with temperature, following the relationship that originates from the temperature dependence of the bulk mobility, which is $\sim T^{-2.3}$, since the drift region resistance in these devices dominates the on-resistance. However, this is not the case for SiC DMOSFETs. As described earlier in this section, the inversion

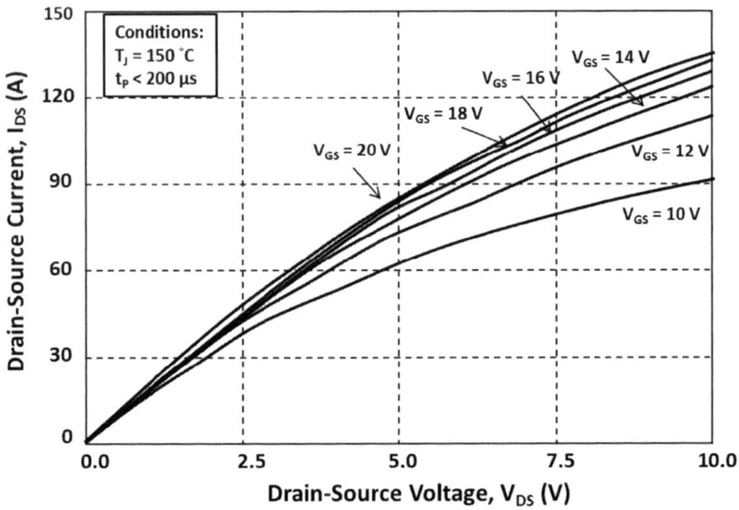

Fig. 7.15. Typical output characteristics for a 1200 V, 60 A rated SiC DMOSFET at $T_J = 150°C$.

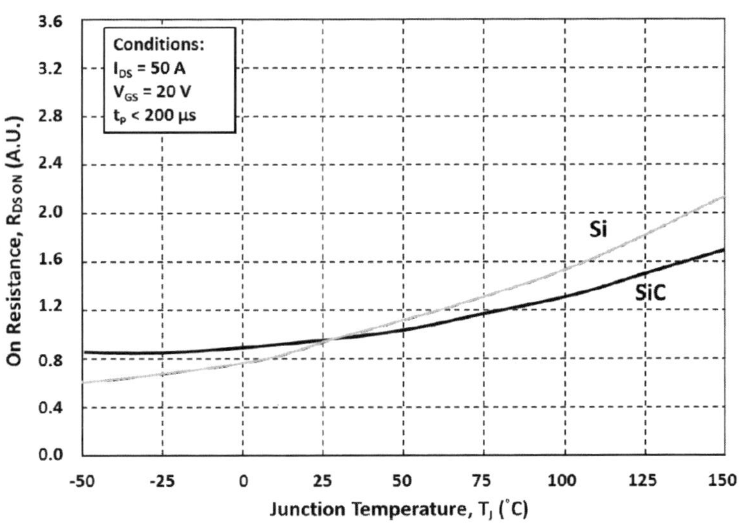

Fig. 7.16. Comparison of normalized R_{on} of a 600 V Si power MOSFET and a 1200 V SiC at different temperatures.

channel mobility in a SiC MOSFET is poor, and the electrons in the channel are trapped, and the channel resistance occupies a significant proportion of the total on-resistance. As a consequence, the on-resistance shows a more moderate dependence on temperature, which is a mixture of the channel resistance and the drift region resistance. For some reported SiC DMOSFETs, the on-resistance even decreases with temperature because the on-resistance is dominated by the channel resistance, which is reduced at high temperatures because of the release of the trapped electrons.

UMOSFETs have been developed in parallel with DMOSFETs in recent years, and early experimental results are promising, such as an $R_{\mathrm{on,sp}}$ of 311 mΩcm^2 for a 1.4 kV device (Sugawara *et al.*, 1998). However, some concerns have arisen regarding the reliability or the premature breakdown of the gate insulator in the bottom corners of the trench gate, which restricts the performance of the device. Assuming that no interface charge exists, and that the maximum allowed electric field in the gate dielectric SiO_2 is 3 MV/cm so as to satisfy reliability considerations, we can see from the simple calculation shown in Eq. (7.17) that the maximum electric field in SiC, which is located in the bottom corners of the gate trench, will be limited to a value of 1.2 MV/cm, which is less than half of the critical electric field of 4H-SiC.

$$\hat{E}_{\mathrm{SiC}} = \frac{\varepsilon_{\mathrm{SiO_2}}}{\varepsilon_{\mathrm{SiC}}} \hat{E}_{\mathrm{ox,max}} \cong 1.2(\mathrm{MV/cm}). \tag{7.17}$$

As a consequence, the maximum achievable $V_{\mathrm{BR}}^2/R_{\mathrm{on,sp}}$ will be reduced by 10 times, according to Eq. (7.2). To fully explore the potential of 4H-SiC UMOSFETs, several approaches designed to solve the problem at the gate trench corners have been proposed. In 1998, Tan *et al.* proposed a p-type protection region beneath the bottom of the gate trench that was designed to protect the corners and reduce the oxide field located there, as shown in Fig. 7.17(a). At the same time, an n-type current spreading layer was inserted between the p-base layer and the p-type protection region so as to reduce the parasitic JFET resistance. An $R_{\mathrm{on,sp}}$ of 15.7 mΩcm^2 and a V_{BR} of 1.4 kV were achieved, yielding a $V_{\mathrm{BR}}^2/R_{\mathrm{on,sp}}$ of 125 MW/cm^2, which was record high at that time. A thick bottom oxide at the bottom of the gate trench also helps in reducing the oxide field (Takaya *et al.*, 2013). Dummy deeper trenches, or a buried p+ region closer to the bottom of the trench corners will also reduce the electric field located there, as shown in Fig. 7.17(b) (Nakamura *et al.*, 2011; Wada *et al.*, 2014). A V-groove trench MOSFET that has p+ buried regions was also demonstrated, taking advantage of the superior MOS interface on the (0-33-8) face (Wada *et al.*, 2014). The state-of-the-art trench MOSFETs in 4H-SiC currently show an $R_{\mathrm{on,sp}}$ of 1.41 mΩcm^2 for 1260 V and 3.5 mΩcm^2 for 1700 V, with a $V_{\mathrm{BR}}^2/R_{\mathrm{on,sp}}$ of around 800–1100 MW/cm^2.

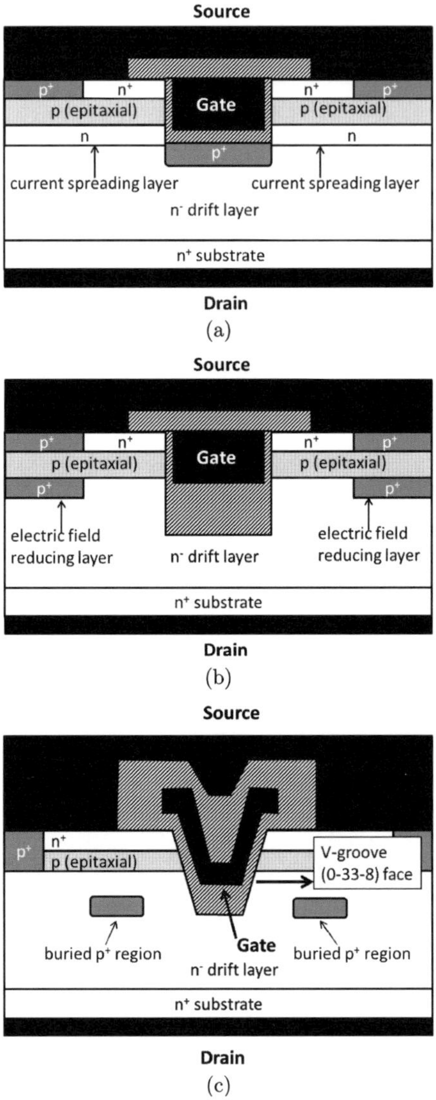

Fig. 7.17. The structure of (a) a UMOSFET with a p+ protection region beneath the bottom of the trench together with a current spreading layer, (b) a UMOSFET with a thick bottom oxide and an electric field reducing layer, and (c) a V-groove MOSFET with buried p+ regions.

4H-SiC MOSFETs at a rating of 1200 V have been commercialized by several companies. The switching behavior of these MOSFETs and their impact on the system side has been extensively investigated. For example, it was reported that with a suitable design for the gate driver, the switching loss could be reduced by 60–80%, and the efficiency in a 40 kHz, 11 kW single-phase inverter can be improved by 2% (Wang *et al.*, 2003).

7.5.2. SiC JFETs

A JEFT is another common choice for SiC unipolar devices as there are no issues surrounding the gate oxide compared to a MOSFET, and is supposedly more robust. Furthermore, the electrons travel in the channel with bulk mobility, which has a much greater mobility than the inversion channel mobility in a MOSFET. However, achieving the normally-off feature desired for the majority of power applications is more challenging. For an n-channel FET, the V_{th} should be greater than zero which means the device will be OFF when V_g is zero. The V_{th} for a JFET with a symmetric gate can be expressed as

$$V_{th} \cong V_{bi} - V_p = \frac{kT}{q} \ln\left(\frac{N_A N_D}{n_i^2}\right) - \frac{qN_D t_{ch}^2}{2\varepsilon_s}, \tag{7.18}$$

where V_{bi} is the built-in voltage of the PN junction that is formed between the p-gate and the n-channel, V_p is the pinch-off voltage of the channel, N_A is the doping concentration of the p-gate, N_D is the doping concentration of the n-channel, and t_{ch} is half the width of the n-channel. It is clear from Eq. (7.18) that the channel needs to be pinched off or depleted by V_{bi} in order to ensure V_{th} is positive. Since V_{bi} for 4H-SiC is around 3 V, much larger than the V_{bi} for Si, which is around 0.7 V, it is easier to design a normally-off 4H-SiC JFET when compared to its Si counterpart. As shown in Fig. 7.18(a), a normally-off n-channel JFET based on a trenched and implanted vertical channel structure with an $R_{on,sp}$ of 3.6 mΩcm^2 and a V_{BR} of 1700 V, yielding 830 MW/cm^2 was demonstrated by Zhao *et al.* (2003a). However, the channel width is less than 1.7 μm, posing stringent requirements on the lithography and tight controls of doping concentration. Another version of a vertical channel JFET based on the implantation of p+ into 4H-SiC to form a buried p+ grid as the controlling gate was proposed and demonstrated, as shown in Fig. 7.18(b) (Tanaka *et al.*, 2006). A 4H-SiC JFET with lateral channels defined by an epitaxial layer and an asymmetric gate was also proposed and is shown in Fig. 7.18(c) (Mitlehner *et al.*, 1999). In this case, the control of V_{th} should be easier. However, the V_p is about 30 V, which causes the device to be normally-on.

If a normally-on JFET was cascoded with an enhancement mode Si n-MOSFET, as shown in Fig. 7.19, it would mean that, from an external

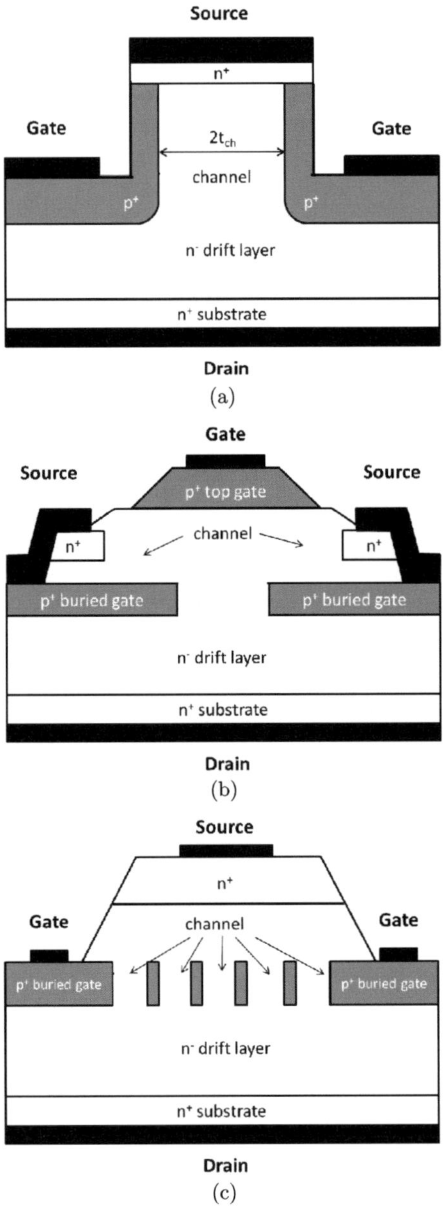

Fig. 7.18. The structure of (a) a JFET with a trenched and implanted vertical channel, (b) a JFET with a vertical channel and a p+ grid gate, and (c) a JFET with a lateral epitaxial channel.

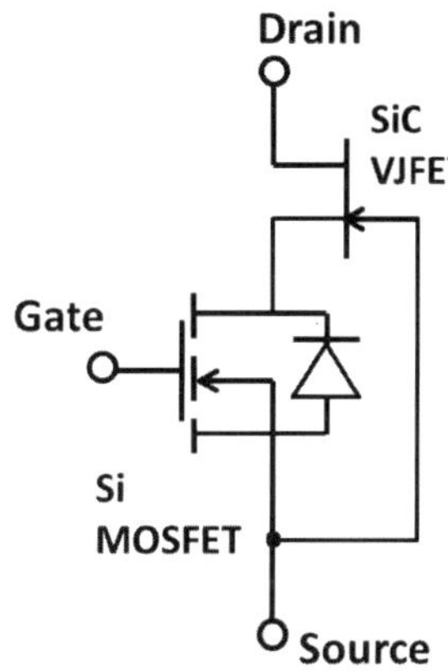

Fig. 7.19. A high-voltage normally-on SiC JFET cascoded with a low-voltage normally-off Si MOSFET.

perspective, this JFET–MOSFET combination would operate in a similar manner to a normally-off device (Baliga, 2001). Currently, SiC JFETs with rating of 1200 V have also been commercialized by a few companies as a strong competitor to SiC MOSFETs. The state-of-the-art $R_{on,sp}$ of 2.4 mΩcm^2 was reported based on an advanced version of the buried gate JFET with a V_{BR} of 1400 V and a current of 100 A (Ishikawa *et al.*, 2014). The $V_{BR}^2/R_{on,sp}$ was 816 MW/cm^2, which is comparable to that of most SiC MOSFETs. A 11 kV normally-off JFET with a trenched-and-implanted vertical channel structure was demonstrated, showing an $R_{on,sp}$ of 130 mΩcm^2 (Zhao *et al.*, 2004a). The implementation of SiC JFETs in a power electronic system shows a similar enhancement in performance to that achieved by SiC MOSFETs.

7.6. SiC Bipolar Switches

When the minority carriers are stored in the drift layer and their concentration exceeds the background doping concentration, the resistance of the drift layer will be significantly reduced in Si bipolar switches such as BJTs, IGBTs, and thyristors, a phenomenon called conductivity modulation, which has been discussed in previous chapters. However, the injection and removal of the required storage minority carriers will reduce the switching speed of the bipolar device and, consequently, increase switching losses. Therefore, Si bipolar devices are more attractive in the domain of low frequency, high power applications, but not in the high frequency domain. In the frequency spectrum, the boundary between unipolar device and bipolar devices for Si is generally considered to be around 50 kHz, while in the voltage spectrum the boundary is around 900 V.

If we follow the same assumption, it is natural to expect that SiC bipolar switches will behave in a similar manner to their Si counterparts, except that the voltage boundary would be increased to around 8000 V (Bakowski, 2000), and the frequency boundary would be relocated to a much lower value because the large forward voltage drop in the junction of a bipolar SiC device would make it less appealing. However, due to material issues, not all SiC bipolar switches behave in the same way as Si switches, as will be discussed in more detail in a subsequent section.

7.6.1. SiC BJTs and Thyristors

A schematic view of a typical cell structure for an npn power BJT in SiC is shown in Fig. 7.20. As a power switch, a BJT is usually operated in common emitter mode, where the controlling electrode is the base, with an input current I_B passing through it to produce a forward bias in the base–emitter junction. At the same time, an output current I_C will flow through the collector electrode, producing a voltage drop V_{CE} across the collector and the emitter electrodes. After the input current is reduced to zero, the output current will be terminated, making BJT a normally-off device by nature. When a BJT is off, a large voltage can be sustained between the collector and the emitter, with a depletion region developed at the base–collector junction. To reduce the control power, a large common emitter current gain β, which is defined as I_C/I_B, is preferred.

For the latest SiC BJTs, both the base and emitter layers are formed using epitaxy as the damage introduced through implantations in the active regions will reduce the carrier lifetime, which is already small in SiC when compared to Si. Typically, p+ implantation regions are created beneath the base contacts so as to improve the contact resistance. Considering an npn BJT, and using

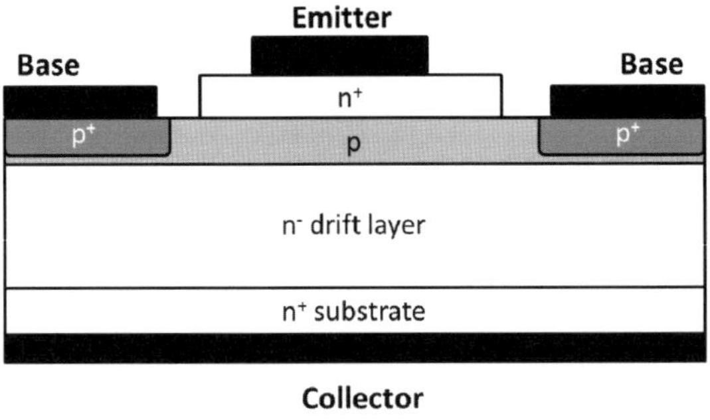

Fig. 7.20. The structure for an npn SiC BJT.

first order approximation, the efficiency of the emitter injection efficiency γ_E and the base transport factor α_T can be expressed as

$$\gamma_E \cong \frac{1}{1+\left(\frac{D_{nB}}{D_{pE}}\right)\left(\frac{L_{pE}}{W_B}\right)\left(\frac{N_{DE}}{N_{AB}}\right)}, \tag{7.19}$$

$$\alpha_T = \frac{1}{\cosh\left(W_B/L_{nB}\right)} \cong \frac{1}{1+W_B^2/2L_{nB}^2}, \tag{7.20}$$

where W_B is the base width, D, L, and N are the diffusion coefficient, the diffusion length, and the doping concentration, respectively, with the B in the subscript indicating the Base and the E indicating the Emitter. Diffusion length is defined as

$$L_{nB} = \sqrt{D_{nB}\tau_{nB}}, \tag{7.21}$$

where τ represents the carrier lifetime. The common base current gain α, which is defined as I_C/I_E, can be expressed as

$$\alpha = \gamma_E \alpha_T. \tag{7.22}$$

And its relationship with β is

$$\beta = \frac{\alpha}{1-\alpha}. \tag{7.23}$$

It is clear from the above equations that if the carrier lifetime in the base τ_{nB} is reduced, the diffusion length L_{nB} will also be reduced, yielding a smaller α_T and, hence, smaller $\alpha$and β values. Other than the bulk carrier lifetime, which is determined by the material growth technologies together with deep level reduction processes, as described in previous sections, the surface recombination lifetime is also important, as well as other mechanisms that will affect the recombination of carriers. In particular, a high density of interface traps usually appear on the surface of 4H-SiC, and their existence will facilitate surface recombination. It was found that different geometries will exhibit different surface exposures and, therefore, will have an important effect on the current (Domeij *et al.*, 2005). It was also found that by using an appropriate passivation, such as an NO anneal to reduce the interface trap density on the surface, the current gain will increase (Miyake *et al.*, 2011). It was reported by Huang *et al.* (2003) that the distance from the p+ implantation regions for the base contacts to the emitter edges will have a detrimental effect on the current because the implantations will introduce crystal damage, which acts as a recombination center close to the electron path, consequently reducing the carrier lifetime. With advancements in both device design and processing, carrier lifetime has greatly improved, and α_T is approaching unity. In this case, β can be approximated as

$$\beta \cong \left(\frac{D_{nB}}{D_{pE}}\right)\left(\frac{L_{pE}}{W_B}\right)\left(\frac{N_{DE}}{N_{AB}}\right). \tag{7.24}$$

For 4H-SiC, the acceptors are only partially ionized because of their deep levels, as discussed earlier. From Eq. (7.24), β for 4H-SiC BJTs will be increased by a factor of N_{AB}/N_{AB}^+, which is the inverse of the ionization percentage, when compared with Si BJTs. A current gain of 257 was reported for a Si-face BJT, and a current gain of 439 was reported for a C-face BJT (Miyake *et al.*, 2012a), which is about an order of magnitude higher than a typical Si power BJT.

While SiC BJTs enjoy a very large β resulting from the incomplete ionization of the acceptors in the base, one of the most important features of a bipolar device is affected. Figure 7.21 shows the typical I_C–V_{CE} curves for Si power BJTs. When a BJT is turned on as a switch, the V_{CE} is small and the device is operated in the saturation region where both the base–emitter junction and the base–collector junction are forward biased. In this case, holes are injected into the drift region, as in the PIN diodes, and conductivity modulation occurs. As a result, the I_C–V_{CE} curves clearly show a hard saturation region where the on-resistance is less than the drift region resistance.

In the previous literature, however, this has not been observed on SiC BJTs for three reasons. First, the carrier lifetime is small, and, hence, the diffusion length is short, reducing the region where conductivity is being modulated, as discussed in Sec. 3.3.1. Second, the background doping concentration for the drift layer in a SiC device is roughly two orders of magnitude higher than that in a Si device with the same V_{BR} rating, making it more difficult to reach a

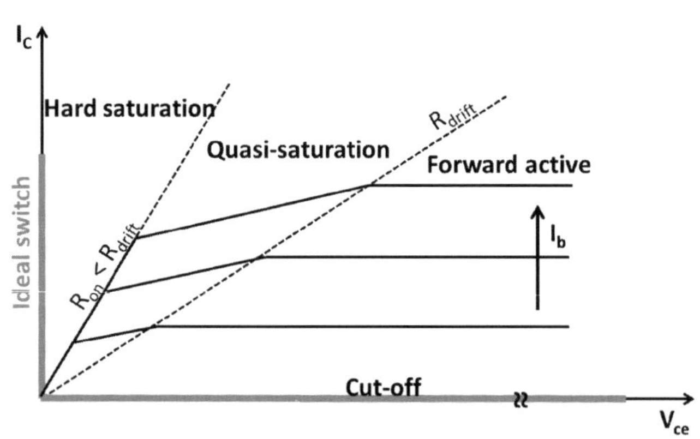

Fig. 7.21. Typical I_C–V_{CE} curves for Si power BJTs showing a region of hard saturation.

high level injection for conductivity modulation. Third, incomplete ionization in the p-layer reduces the efficiency of its ability to inject holes into the drift region. This is particularly worse in SiC BJTs since the concentration of the p-base layer is intentionally reduced in order to achieve a high current gain. Without conductivity modulation in the drift layer, SiC BJTs operate as a current controlled unipolar device. A benefit of this unipolar-like behavior is the positive temperature dependence of $V_{CE,SAT}$ or $R_{on,sp}$, which makes device paralleling easier for large current requirements. In contrast, the $V_{CE,SAT}$ for Si BJTs typically shows negative temperature dependence. Additionally, the current gain β of a SiC BJT is reduced at higher temperatures because of the increase in the ionization percentage of the acceptors in the base, while the current gain of a Si BJT increases at higher temperatures because of the increase in carrier lifetime. These distinct temperature dependences make SiC BJTs less prone to thermal runaway, which is one of the drawbacks for Si BJTs. An $R_{on,sp}$ of 2.8 mΩcm^2 and a β of 117 was reported for a SiC BJT with a $V_{BR,CEO}$ of 1570 V and a $V_{BR,CBO}$ of 1870 V at room temperature, yielding a $V_{BR}^2/R_{on,sp}$ of 1249 MW/cm^2, which is slightly better than for a SiC MOSFET (Domeij *et al.*, 2012). The current gain β was reduced to 70 and the $R_{on,sp}$ increased to 5.2 mΩcm^2 at 150°C. A 21 kV 4H-SiC BJT was reported to have a current gain β of 63 and an $R_{on,sp}$ of 321 mΩcm^2, thanks to the simplicity of the cell structure and the efficient edge termination (Miyake *et al.*, 2012b). The calculated $V_{BR}^2/R_{on,sp}$ was 1373 MW/cm^2. The switching of SiC BJTs has been tested, with a turn-on time of 32 ns and a turn-off time of 60 ns reported, which is comparable to the results shown for SiC MOSFETs, and is further evidence of negligible conductivity modulation in the drift layer (Krishnaswami *et al.*, 2005). As a consequence, SiC BJTs show a similar performance to SiC MOSETs when implemented in a power electronic system, with the benefit of having neither the concerns nor the limits imposed by the gate dielectric.

Thyristors are npnp or pnpn structures, and are typically thought of a single npn BJT and a single pnp BJT interconnected, as illustrated in the equivalent circuit shown in Fig. 7.22. When the thyristors are turned on, both BJTs will be in saturation mode and will provide base currents to each other, maintaining a self-sustaining situation called latch-up. By describing SiC BJTs as unipolar devices that do not have conductivity modulation, one might think that SiC thyristors would have the same behavior. However, this is not the case. The difference is the hole injector for the SiC thyristors, which is either the p+ substrate or a p+ epilayer, and is much more efficient than the lighter doped p-base layer for SiC BJTs. As a result, conductivity modulation of the drift layer is observed in SiC thyristors, as in SiC PIN diodes. Considering the limited availability and high resistivity of the p-type substrates, SiC thryristors are mostly built on n-type substrates using pnpn structures. A 20 kV SiC gate

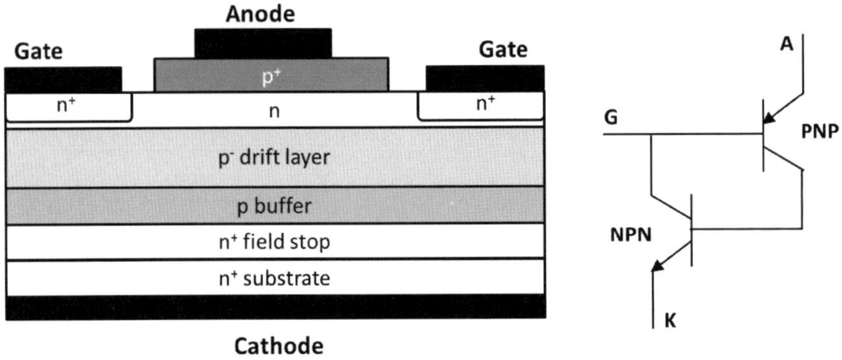

Fig. 7.22. Structure of a pnpn thyristor and its equivalent circuit model.

turn-off thyrsitor was demonstrated using a 160 μm, 2 × 10^{14}/cm^3 p-type drift layer (Cheng *et al.*, 2013). An average carrier lifetime that was greater than 4 μs was extracted, yielding a very low differential $R_{\text{on,sp}}$ of 11 mΩcm^2, which is only possible when significant conductivity occurred in the drift layer.

7.6.2. SiC IGBTs

IGBT is probably the most popular form of bipolar power device today. Similar to SiC thyristors, SiC IGBTs can benefit from conductivity modulation in the drift layer. For example, p-type hole injectors in n-channel SiC IGBTs can be heavily doped without compromising too much of the device performance. The only concern is that the epilayers of an n-channel IGBT are conventionally grown on very resistive p-type substrates. However, this has been overcome by using thin wafer technology similar to that used for advanced Si IGBTs. Rather than using a thick and resistive p-type substrate as the hole injector, several approaches have been proposed that are intended to form better hole injectors either by the implantation of p-type dopants or by growing a p-type epitaxial layer at the backside of the wafer, and then lapping it down to the desired thickness. A 12.5 kV n-channel IGBT with a 140 μm, 2 × 10^{14}/cm^3 n-type drift layer using a thin p+ epilayer as the hole injector has been reported, as shown in Fig. 7.23 (Ryu *et al.*, 2012). This device showed a differential $R_{\text{on,sp}}$ of 5.3 mΩcm^2 and a V_F of 5.5 V at 10 A. The turn-off time increased from 0.65 μs at 25°C to 2.3 μs at 175°C, which is attributed to the larger carrier lifetime and the improved p+ injector efficiency at high temperatures. It was also found that with a thicker n-type buffer layer, the turn-off time reduces and V_F increases, owing to the reduced hole injection efficiency. A 16 kV n-channel IGBT was realized with the planar gate structures on the C-face, and the wafer was lapped to a reduced total thickness of about 250 μm including an n-type

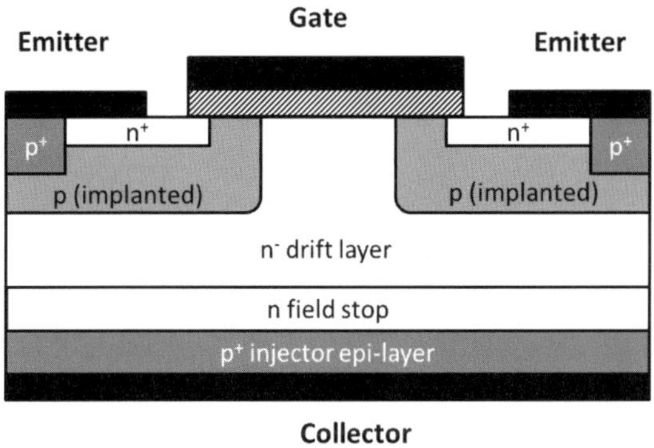

Fig. 7.23. Structure of an n-channel SiC IGBT with a p+ injector epilayer.

drift layer of 150 μm (Yonezawa *et al.*, 2013). The forward voltage was 5 V at a current density of 100 A/cm^2 with a differential $R_{on,sp}$ of 11.3 mΩcm^2. P-channel SiC IGBTs are typically less appealing than n-channel SiC IGBTs because of the very poor hole mobility in the channel, even though they have the advantage of being built on conducting n-type substrates. The low hole channel mobility results in a larger channel resistance and a smaller degree of conductivity modulation in the drift layer.

7.7. SiC Lateral Devices

In Si, lateral power devices are mainly used for power integrated circuits, ranging from a few volts to a few hundred volts, and from a few milliamps to a few amps, which is not the range where SiC devices are most appreciated. As a result, there are currently very few commercialized applications for SiC lateral power devices. However, some potential applications might emerge in the future for SiC power integrated circuits, for example, integrated gate driver circuits for ultra-high voltage SiC discrete devices. For completeness, here we summarize several high-voltage SiC lateral devices that have been reported in the past. A 460 V single RESURF 4H-SiC lateral MOSFET on the C-face was demonstrated with an $R_{on,sp}$ of 79 mΩcm^2 (Okamoto *et al.*, 2004). With a double and two-zone RESURF drift region, as shown in Fig. 7.24, the V_{BR} was improved to 1430 V and 1550 V on the Si-face and the C-face, respectively (Noborio *et al.*, 2008). The $R_{on,sp}$ was also respectively improved to 57 mΩcm^2 and 54 mΩcm^2. The $V_{BR}^2/R_{on,sp}$ was respectively 36 and 44 MW/cm^2, which is greatly beyond the silicon limit. A 4H-SiC normally-off vertical channel single RESURF lateral JFET, as shown in Fig. 7.25, was demonstrated to achieve a V_{BR} of 1000 V and an $R_{on,sp}$ of 9.1 mΩcm^2, yielding a $V_{BR}^2/R_{on,sp}$ of

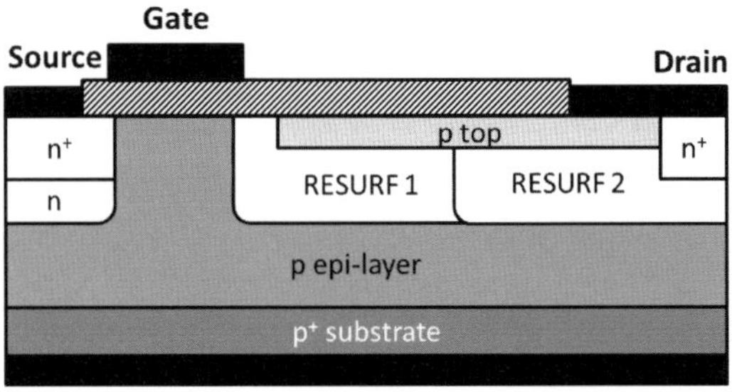

Fig. 7.24. Structure of a two-zone double RESURF SiC lateral MOSFET.

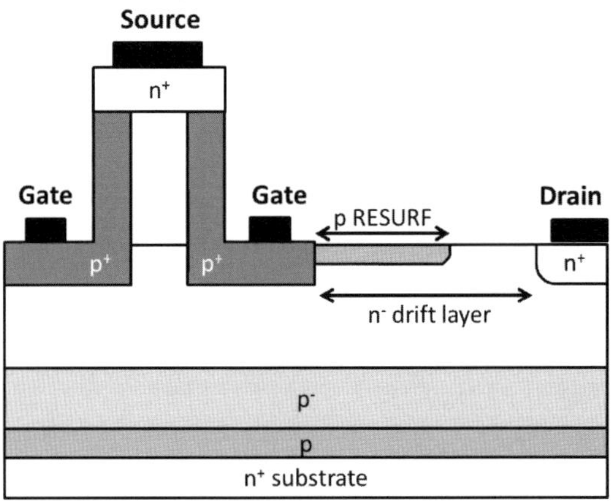

Fig. 7.25. Structure of a vertical channel single RESURF SiC lateral JFET.

116 MW/cm^2 (Zhang *et al.*, 2007). The majority of lateral SiC devices were fabricated on conducting substrates. It was proposed that the V_{BR} of SiC lateral devices can be further increased without the need for thick epilayers by using a semi-insulating substrate because, in this case, the maximum achievable V_{BR} is no longer limited in the vertical direction since the semi-insulating substrate is able to support a very high voltage. A 4200 V, 454 mΩcm^2 lateral JFET was realized based on this approach using a two-zone RESURF design (Lee *et al.*, 2012), as shown in Fig. 7.26. A lateral JFET-based power integrated circuit in 4H-SiC that provided a MHz switching capability has also been demonstrated (Zhang *et al.*, 2008). A gate driver was realized using 4H-SiC nMOS logics and processes, and is capable of switching beyond 500 kHz at 420°C, showing the potential of SiC power integrated circuits for high-temperature applications (Lamichhane *et al.*, 2014).

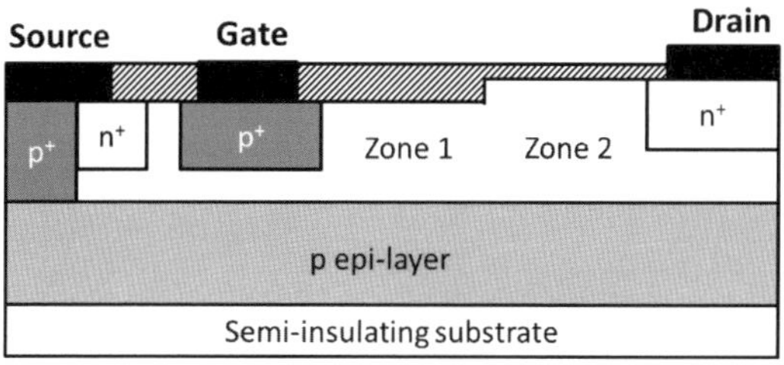

Fig. 7.26. Structure of a two-zone RESURF SiC lateral JFET on a semi-insulating substrate.

References

V. Afanasev, M. Bassler, G. Pensl and M. Schulz, Intrinsic SiC/SiO2 interface states, *Phys. Stat. Sol. (a)* **162**, 321–337 (1997).

M. Bakowski, Roadmap for SiC power devices, *JTIT* **3–4**, 19–30 (2000).

J. Baliga, Semiconductors for high-voltage, vertical channel field-effect transistors, *J. Appl. Phys.* **53**, 1759–1764 (1982).

B. Baliga, The future of power semiconductor device technology, *Proc. IEEE* **89**, 822–832 (2001).

M. Bockstedte, A. Mattausch and O. Pankratov, Solubility of nitrogen and phosphorus in 4H-SiC: A theoretical study, *Appl. Phys. Lett.* **85**, 58–60 (2004).

H. Bracht, N. A. Stolwijk, M. Laube and G. Pensl, Diffusion of boron in silicon carbide: Evidence for the kick-out mechanism, *Appl. Phys. Lett.* **77**, 3188 (2000).

P. Brosselard, N. Camara, V. Banu, X. Jorda, M. Vellvehi, P. Godignon and J. Millan, Bipolar conduction impact on electrical characteristics and reliability of 1. 2- and 3. 5-kV 4H-SiC JBS diodes, *IEEE Tran. Electron Devices* **55**, 1847–1856 (2008).

M. Capano, J. Cooper Jr., M. Melloch, A. Saxler and W. Mitchel, Ionization energies and electron mobilities in phosphorus-and nitrogen-implanted 4H-silicon carbide, *J. Appl. Phys.* **87**, 8773–8777 (2000).

M. Capano, S. Ryu, J. Cooper Jr., M. Melloch, K. Rottner, S. Karlsson, N. Nordell, A. Powell and D. Walker Jr., Surface roughening in ion implanted 4H-silicon carbide, *J. Electronic Materials* **28**, 214–218 (1999).

M. Capano, S. Ryu, M. Melloch, J. Cooper Jr. and M. Buss, Dopant activation and surface morphology of ion implanted 4H- and 6H-silicon carbide, *J. Electron. Mater.* **27**, 370–376 (1998).

X. Chen, Semiconductor power devices with alternating conductivity type high-voltage breakdown regions, US patent 5216275 A (1993).

L. Cheng, A. Agarwal, C. Capell, M. O'Loughlin, K. Lam, J. Richmond, E. Van Brunt, A. Burk, J. W. Palmour, H. O'Brien, A. Ogunniyi and C. Scozzie, 20 kV, 2 cm^2, 4H-SiC gate turn-off thyristors for advanced pulsed power applications, *IEEE PPC Digest*, 1–4 (2013).

G. Chung, C. Tin, J. Williams, K. McDonald, R. Chanana, R. Weller, S. Pantelides, L. Feldman, O. Holland, M. Das and J. Palmour, Improved inversion channel mobility for 4H-SiC MOSFETs following high temperature anneals in nitric oxide, *IEEE Electron Device Lett.* **22**, 176–178 (2001).

J. Crofton, S. Mohney, J. Williams and T. Isaacs-Smith, Finding the optimum Al–Ti alloy composition for use as an ohmic contact to p-type SiC, *Solid State Electron*, **46** 109–113 (2002).

M. Domeij, A. Konstantinov, A. Lindgren, C. Zaring, K. Gumaelius and M. Reimark, Large area 1200 V SiC BJTs with $\beta > 100$ and $\rho_{ON} < 3\,m\Omega cm^2$, *Mater. Sci. Forum* **717–720**, 1123–1126 (2012).

M. Domeij, H. Lee, E. Danielsson, C. Zetterling, M. Ostling and A. Schoner, Geometrical effects in high current gain 1100-V 4H-SiC BJTs, *IEEE Electron Device Lett.* **26**, 743–745 (2005).

G. Feng, J. Suda and T. Kimoto, Space-modulated junction termination extension for ultrahigh-voltage p-i-n diodes in 4H-SiC, *IEEE Tran. Electron Devices*, **59**, 414–418 (2012).

K. Fukuda, M. Kato, K. Kojima and J. Senzaki, Effect of gate oxidation method on electrical properties of metal-oxide-semiconductor field-effect transistors fabricated on 4H-SiC C(0001^-) face, *Appl. Phys. Lett.* **84**, 2088 (2004).

R. Ghandi, B. Buono, M. Domeij, G. Malm, C. Zetterling and M. Ostling, High-voltage 4H-SiC PiN diodes with etched junction termination extension, *IEEE Electron Device Lett.* **30**, 1170–1172 (2009).

S. Ha, P. Mieszkowski, M. Skowronski and L. Rowland, Dislocation conversion in 4H silicon carbide epitaxy, *J. Cryst. Growth* **244**, 257–266 (2002).

S. Harada, M. Kato, K. Suzuki, M. Okamoto, T. Yatsuo, K. Fukuda and K. Arai, 1.8 $m\Omega cm^2$, 10 A Power MOSFET in 4H-SiC, *IEEE Int. Electron Devices Meeting Digest*, 1–4 (2006).

S. Harada, M. Okamoto, T. Yatsuo, K. Adachi, K. Fukuda and K. Arai, 8.5 $m\Omega \cdot cm^2$ 600-V 600-V double-epitaxial MOSFETs in 4H-SiC, *IEEE Electron Device Lett.* **25**, 292–294 (2004).

S. Harada, S. Suzuki, J. Senzaki, R. Kosugi, K. Adachi, K. Fukuda and K. Arai, High channel mobility in normally-off 4H-SiC buried channel MOSFETs, *IEEE Electron Device Lett.* **22**, 272–274 (2001).

J. Hassan and J. P. Bergman, Influence of structural defects on carrier lifetime in 4H-SiC epitaxial layers: Optical lifetime mapping, *J. Appl. Phys.* **105**, 123518 (2009).

S. Hino, T. Hatayama, J. Kato, E. Tokumitsu, N. Miura and T. Oomori, High channel mobility 4H-SiC metal-oxide-semiconductor field-effect transistor with low temperature metal-organic chemical-vapor deposition grown Al2O3 gate insulator, *Appl. Phys. Lett.* **92**, 183503 (2008).

T. Hiyoshi and T. Kimoto, Reduction of deep levels and improvement of carrier lifetime in n-type 4H-SiC by thermal oxidation, *Appl. Phys. Express* **2**, 041101 (2009).

C. Huang and J. Cooper, High current gain 4H-SiC NPN bipolar junction transistors, *IEEE Electron Device Lett.* **24**, 396–398 (2003).

X. Huang, E. Van Brunt, J. Baliga and A. Huang, Orthogonal positive-level termination for chip-size SiC reverse blocking devices, *IEEE Electron Device Lett.* **33**, 1592–1594 (2012).

C. Huang, H. Hsu, K. Chu, L. Lee, M. Tsai, K. Lee and F. Zhao, Counter-doped JTE, an edge termination for HV SiC devices with increased tolerance to the surface charge, *IEEE Tran. Electron Devices* **62**, 354–358 (2015).

J. Hudgins, G. Simin, E. Santi and M. Khan, An assessment of wide bandgap semiconductors for power devices, *IEEE Transactions on Power Electronics* **18**, 907–914 (2003).

B. Hull, M. Das, J. Sumakeris, J. Richmond and S. Krishnaswami, Drift-free 10-kV, 20-A 4H-SiC PiN diodes, *J. Electron. Mater.* **34**, 341–344 (2005).

T. Ishikawa, Y. Tanaka, T. Yatsuo and K. Yano, SiC power devices for HEV/EV and a novel SiC vertical JFET, *IEEE Int. Electron Devices Meeting Digest*, 2.2.1–2.2.4 (2014).

A. Itoh and H. Matsunami, Analysis of Schottky barrier heights of metal/SiC contacts and its possible application to high-voltage rectifying devices, *Phys. Stat. Sol. (a)* **162**, 389–408 (1997).

A. Itoh, H. Akita, T. Kimoto and H. Matsunami, High-quality 4HSiC homoepitaxial layers grown by stepcontrolled epitaxy, *Appl. Phys. Lett.* **65**, 1400 (1994).

L. Jiang, R. Cheung, R. Brown and A. Mount, Inductively coupled plasma etching of SiC in SF6/O2 and etch-induced surface chemical bonding modifications, *J. Appl. Phys.* **93**, 1376–1383 (2003).

T. Kimoto, A. Itoh and H. Matsunami, Step-controlled epitaxial growth of high-quality SiC layers, *Phys. Status Solidi B* **202**, 247–262 (1997a).

T. Kimoto, A. Itoh, H. Matsunami and T. Okano, Step bunching mechanism in chemical vapor deposition of 6H– and 4H–SiC{0001}, *J. Appl. Phys.* **81**, 3494 (1997b).

T. Kimoto, N. Inoue and H. Matsunami, Nitrogen ion implantation into α-SiC epitaxial layers, *Phys. Stat. Sol. A* **162**, 263–276 (1997c).

A. Kinoshita, T. Ohyanagi, T. Yatsuo, K. Fukuda, H. Okumura and K. Arai, Fabrication of 1.2 kV, 100 A, 4H-SiC(0001) and (000-1) junction barrier Schottky diodes with almost same Schottky barrier height, *Mater. Sci. Forum* **645–648**, 893–896 (2010).

A. Konstantinov, Q. Wahab, N. Nordell and U. Lindefelt, Ionization rates and critical fields in 4H silicon carbide, *Appl. Phys. Lett.* **71**, 90–92 (1997).

R. Kosugi, Y. Sakuma, K. Kojima, S. Itoh, A. Nagata, T. Yatsuo, Y. Tanaka and H. Okumura, First experimental demonstration of SiC superjunction (SJ) structure by multi-epitaxial growth method, *Proc. Intl. Symp. Power Semiconductor Devices IC*, 346–349 (2014).

S. Krishnaswami, A. Agarwal, S. Ryu, C. Capell, J. Richmond, J. Palmour, S. Balachandran, T. Chow, S. Bayne, B. Geil, K. Jones and C. Scozzie, 1000-V, 30-A 4H-SiC BJTs with high current gain, *IEEE Electron Device Lett.* **26**, 175–177 (2005).

M. Lades, W. Kaindl, N. Kaminski, E. Niemann and G. Wachutka, Dynamics of incomplete ionized dopants and their impact on 4H/6H–SiC devices, *IEEE Tran. Electron Devices* **46**, 598–604 (1999).

R. Lamichhane, N. Ericsson, S. Frank, C. Britton, L. Marlino, A. Mantooth, M. Francis, P. Shepherd, M. Glover, S. Perez, T. McNutt, B. Whitaker and Z. Cole, A wide bandgap silicon carbide (SiC) gate driver for high-temperature and high-voltage applications, *Int. Symp. Power Semiconductor Devices ICs* 414–417 (2014).

D. Larkin, P. Neudeck, J. Powell and L. Matus, Site-competition epitaxy for superior silicon carbide electronics, *Appl. Phys. Lett.* **65**, 1659 (1994).

W. Lee, K. Chu, C. Huang, L. Lee, M. Tsai, K. Lee and F. Zhao, Design and fabrication of 4H–SiC lateral high-voltage devices on a semi-insulating substrate, *IEEE Tran. Electron Devices* **59**, 754–760 (2012).

H. Lendenmann, F. Dahlquist, P. Bergman, H. Bleichner and C. Hallin, High-power SiC diodes: characteristics, reliability and relation to material defects, *Mater. Sci. Forum* **389–393**, 1259–1264 (2002).

D. Lichtenwalner, V. Misra, S. Dhar, S. Ryu and A. Agarwal, High-mobility enhancement-mode 4H-SiC lateral field-effect transistors utilizing atomic layer deposited Al_2O_3 gate dielectric, *Appl. Phys. Lett.* **95**, 152113 (2009).

P. Losee, S. Balachandran, L. Zhu, C. Li, J. Seiler, T. Chow, I. Bhat and R. Gutmann, High-voltage 4H-SiC PiN rectifiers with single-implant, multi-zone JTE termination, *Int. Symp. Power Semiconductor Devices ICs* 301–304 (2004).

P. Losee, Y. Wang, C. Li, S. Sharma, I. Bhat, T. Chow and R. Gutmann, Comparative evaluation of anode layers on the electrical characteristics of high voltage 4H-SiC PiN diodes, *Mater. Sci. Forum* **600–603**, 1003–1006 (2009).

R. Mallwitz, C. Althof, S. Buchhold and E. Kiel, First 99% PV inverter with SiC JFETs on the market — future role of SiC, *Proc. PCIM Europe* (2012).

M. Matin, A. Saha and J. Cooper, A self-aligned process for high-voltage, short-channel vertical DMOSFETs in 4H-SiC, *IEEE Tran. Electron Devices* **51**, 1721–1725 (2004).

H. Matsunami, M. Ikeda, A. Suzuki and T. Tanaka, SiC blue LED's by liquid-phase epitaxy, *IEEE Tran. Electron Devices* **24**, 958–961 (1977).

H. Mitlehner, W. Bartsch, K. Dohnke, P. Friedrichs, R. Kaltschmidt, U. Weinert, B. Weis and D. Stephani, Dynamic characteristics of high voltage 4H-SiC vertical JFETs, *Int. Symp. Power Semiconductor Devices ICs* 339–342 (1999).

H. Miyake, T. Kimoto and J. Suda, Improvement of current gain in 4H-SiC BJTs by surface passivation with deposited oxides nitrided in N_2O or NO, *IEEE Electron Device Lett.* **32**, 285–287 (2011).

H. Miyake, T. Kimoto and J. Suda, Enhanced current gain (>250) in 4H-SiC bipolar junction transistors by a deep-level-reduction process, *Mater. Sci. Forum* **717–720**, 1117–1122 (2012a).

H. Miyake, T. Okuda, H. Niwa, T. Kimoto and J. Suda, 21 kV SiC BJTs with space-modulated junction termination extension, *IEEE Electron Device Lett.* **33**, 1598–1600 (2012b).

T. Miyazawa, M. Ito and H. Tsuchida, Evaluation of long carrier lifetimes in thick 4H silicon carbide epitaxial layers, *Appl. Phys. Lett.* **97**, 202106 (2010).

D. Morisette and J. Cooper, Theoretical comparison of SiC PiN and Schottky diodes based on power dissipation considerations, *IEEE Tran. Electron Devices* **49**, 1657–1664 (2002).

H. Nagasawa, K. Yagi, T. Kawahara and N Hatta, Reducing planar defects in 3C-SiC, *Chem. Vap. Deposition* **12**, 502–508 (2006).

T. Nakamura, T. Miyanagi, I. Kamata, T. Jikimoto and H. Tsuchida, A 4.15 kV 9.07-m$\Omega \cdot cm^2$ 4H-SiC Schottky-barrier diode using Mo contact annealed at high temperature, *IEEE Electron Device Lett.* **26**, 99–101 (2005).

T. Nakamura, Y. Nakano, M. Aketa, R. Nakamura, S. Mitani, H. Sakairi and Y. Yokotsuji, High performance SiC trench devices with ultra-low R_{on}, *IEEE Int. Electron Devices Meeting Digest*, 26. 5. 1–26. 5. 3 (2011).

Y. Nakanishi, T. Tominaga, H. Okabe, Y. Suehiro, K. Sugahara, T. Kuroiwa, Y. Toyoda, S. Yamakawa, H. Murasaki, K. Kobayashi and H. Sumitani, Properties of a SiC Schottky barrier diode fabricated with a thin substrate, *Mater. Sci. Forum* **778–780**, 820–823 (2014).

Y. Negoro, K. Katsumoto, T. Kimoto and H. Matsunami, Electronic behaviors of high-dose phosphorus-ion implanted 4H–SiC (0001), *J. Appl. Phys.* **96**, 224–228 (2004a).

Y. Negoro, T. Kimoto, H. Matsunami, F. Schmid and G. Pensl, Electrical activation of high-concentration aluminum implanted in 4H-SiC, *J. Appl. Phys.* **96**, 4916–4922 (2004b).

I. Nikitina, K. Vassilevski, N. Wright, A. Horsfall, A. O'Neill and C. Johnson, Formation and role of graphite and nickel silicide in nickel based ohmic contacts to n-type silicon carbide, *J. Appl. Phys.* **97**, 083709 (2005).

M. Noborio, J. Suda and T. Kimoto, 4H-SiC double RESURF MOSFETs with a record performance by increasing RESURF dose, *Int. Symp. Power Semiconductor Devices ICs* 263–266 (2008).

K. Norling, C. Lindholm and D. Draxelmayr, An optimized driver for SiC JFET-based switches enabling converter operation with more than 99% efficiency, *IEEE J. Solid-State Circuits* **47**, 3095–3104 (2012).

M. Okamoto, S. Suzuki, M. Kato, T. Yatsuo and K. Fukuda, Lateral RESURF MOSFET fabricated on 4H-SiC(0001∼) C-face, *IEEE Electron Device Lett.* **25**, 405–407 (2004).

D. Okamoto, H. Yano, K. Hirata, T. Hatayama and T. Fuyuki, Improved inversion channel mobility in 4H-SiC MOSFETs on Si face utilizing phosphorus-doped gate oxide, *IEEE Electron Device Lett.* **31**, 710–712 (2010).

R. Perez, D. Tournier, A. Perez-Tomas, P. Godignon, N. Mestres and J. Millan, Planar edge termination design and technology considerations for 1.7-kV 4H-SiC PiN diodes, *IEEE Tran. Electron Devices* **52**, 2309–2316 (2005).

A. Powell and L. Rowland, SiC materials — progress, status, and potential roadblocks,*IEEE Proceedings* **90**, 942–955 (2002).

S. Ryu, C. Capell, C. Jonas, L. Cheng, M. O'Loughlin, A. Burk, A. Agarwal, J. Palmour and A. Hefner, Ultra high voltage (>12 kV), high performance

4H-SiC IGBTs, *Int. Symp. Power Semiconductor Devices ICs* 257–260 (2012).

S. Ryu, S. Krishnaswami, M. O'Loughlin, J. Richmond, A. Agarwal, J. Palmour and A. Hefner, 10-kV, 123-mΩ/cm^2 4H-SiC power DMOSFETs, *IEEE Electron Device Lett.* **25**, 556–558 (2004).

N. Saks and A. Agarwal, Hall mobility and free electron density at the SiC/SiO2 interface in 4H–SiC, *Appl. Phys. Lett.* **77**, 3281 (2000).

N. Saks, A. Agarwal, S. Ryu and J. Palmour, Low-dose aluminum and boron implants in 4H and 6H silicon carbide, *J. Appl. Phys.* **90**, 2796–2805 (2001).

F. Schmid, M. Laube, G. Pensl, G. Wagner and M. Maier, Electrical activation of implanted phosphorus ions in 0001- and 11–20-oriented 4H-SiC, *J. Appl. Phys.* **91**, 9182–9186 (2002).

D. Schroder, *Semiconductor Material and Device Characterization*, 3rd Edition (Wiley-IEEE Press USA, 2006), p. 729.

S. Seshadri, G. Eldridge and A. Agarwal, Comparison of the annealing behavior of high-dose nitrogen-, aluminum-, and boron-implanted 4H–SiC, *Appl. Phys. Lett.* **72**, 2026 (1998).

Y. Sharma, A. Ahyi, T. Issacs-Smith, X. Shen, S. Pantelides, X. Zhu, L. Feldman, J. Rozen and J. Williams, Phosphorous passivation of the SiO2/4H–SiC interface, *Solid-State Electron.* **68**, 103–107 (2012).

R. Singh, D. Capell, M. Das, L. Lipkin and J. Palmour, Development of high-current 4H-SiC ACCUFET, *IEEE Tran. Electron Devices* **50**, 471–478 (2003).

R. Singh, J. Cooper, M. Melloch, T. Chow and J. Palmour, SiC power Schottky and PiN diodes, *IEEE Tran. Electron Devices* **49**, 665–672 (2002).

R. Singh, Reliability and performance limitations in SiC power devices, *Microelectron. Reliab.* **46**, 713–730 (2006).

G. Spiazzi, S. Buso, M. Citron, M. Corradin and R. Pierobon, Performance evaluation of a Schottky SiC power diode in a boost PFC application, *IEEE Trans. Power Electronics* **18**, 1249–1253 (2003).

S. Sridhara, L. Clemen, R. Devaty, W. Choyke, D. Larkin, H. Kong, T. Troffer and G. Pensl, Photoluminescence and transport studies of boron in 4H SiC, *J. Appl. Phys.* **83**, 7909–7919 (1998).

Y. Sugawara and K. Asano, 1.4 kV 4H-SiC UMOSFET with low specific on-resistance, *Proc. Intl. Symp. Power Semiconductor Devices IC*, 119–122 (1998).

S. Sundaresan, C. Sturdevant, M. Marripelly, E. Lieser and R. Singh, 12.9 kV SiC PiN diodes with low on-state drops and high carrier lifetimes, *Mater. Sci. Forum* **717–720**, 949–952 (2012).

S. Sze, Physics of semiconductor devices 2nd Ed., John Wiley & Sons, Inc. (1981).

Y. Tairov and V. Tsvetkov, Investigation of growth processes of ingots of silicon carbide single crystals, *J. Cryst. Growth* **43**, 209–212 (1978).

H. Takaya, J. Morimoto, K. Hamada, T. Yamamoto, J. Sakakibara, Y. Watanabe and N. Soejima, A 4H-SiC trench MOSFET with thick bottom oxide for improving characteristics, *Int. Symp. Power Semiconductor Devices ICs* 43–46 (2013).

J. Tan, J. Cooper and M. Melloch, High-voltage accumulation-layer UMOSFET's in 4H-SiC, *IEEE Electron Device Lett.* **19**, 487–489 (1998).

Y. Tanaka, M. Okamoto, A. Takatsuka, K. Arai, T. Yatsuo, K. Yano and M. Kasuga, *IEEE Electron Device Lett.* **27**, 908–910 (2006).

N. Tega, H. Yoshimoto, D. Hisamoto, N. Watanabe, H. Shimizu, S. Sato, Y. Mori, T. Ishigaki, M. Matsumura, K. Konishi, K. Kobayashi, T. Mine, S. Akiyama, R. Fujita, A. Shima and Y. Shimamoto, Novel trench-etched double-diffused SiC MOS (TED MOS) for overcoming tradeoff between R_{on}A and Q_{gd}, *Int. Symp. Power Semiconductor Devices ICs* 81–84 (2015).

S. Tian, Monte Carlo simulation of ion implantation in crystalline SiC with arbitrary polytypes, *IEEE Trans. Electron Devices* **55**, 1991–1996 (2008).

T. Troffer, M. Schadt, T. Frank, H. Itoh, G. Pensl, J. Heindl, H. Strunk and M. Maier, Doping of SiC by implantation of Boron and Aluminum, *Phys. Stat. Sol. (a)* **162**, 277–298 (1997).

K. Wada, T. Kimoto, K. Nishikawa and H. Matsunami, Epitaxial growth of 4H–SiC on 4° off-axis (0 0 0 1) and (0 0 0 1^-) substrates by hot-wall chemical vapor deposition, *J. Cryst. Growth* **291**, 370–374 (2006).

K. Wada, T. Masuda, Y. Saitoh, H. Tamaso, M. Furumai, K. Hiratsuka, Y. Mikamura, T. Hatayama and H. Yano, Fast switching 4H-SiC V-groove trench MOSFETs with buried P+ structure, *Int. Symp. Power Semiconductor Devices ICs* 225–228 (2014).

G. Wang, F. Wang, G. Magai, Y. Lei, A. Huang and M. Das, Performance comparison of 1200 V 100 A SiC MOSFET and 1200 V 100 A silicon IGBT, *Energy Conversion Congress and Exposition Digest*, 3230–3234 (2003).

X. Yang, B. Lee and V. Misra, High mobility 4H-SiC lateral MOSFETs using lanthanum silicate and atomic layer deposited SiO2, *IEEE Electron Device Lett.* **36**, 312–314 (2015).

H. Yano, T. Hirao, T. Kimoto, H. Matsunami, K. Asano and Y. Sugawara, High channel mobility in inversion layers of 4H-SiC MOSFETs by utilizing (112-0) face, *IEEE Electron Device Lett.* **20**, 611–613 (1999).

B. Yen, C. Hung, H. Hung, L. Lee, C. Lee, T. Yang, Y. Huang, C. Cheng and P. Chuang, 1700V/30A 4H-SiC MOSFET with low cut-in voltage

embedded diode and room temperature boron implanted termination, *Int. Symp. Power Semiconductor Devices ICs* 265–268 (2015).
Y. Yonezawa, T. Mizushima, K. Takenaka, H. Fujisawa, T. Kato, S. Harada, Y. Tanaka, M. Okamoto, M. Sometani, D. Okamoto, N. Kumagai, S. Matsunaga, T. Deguchi, M. Arai, T. Hatakeyama, Y. Makifuchi, T. Araoka, N. Oose, T. Tsutsumi, M. Yoshikawa, K. Tatera, M. Harashima, Y. Sano, E. Morisaki, M. Takei, M. Miyajima, H. Kimura, A. Otsuki, K. Fukuda, H. Okumura and T. Kimoto, Low V_f and highly reliable 16 kV ultrahigh voltage SiC flip-type n-channel implantation and epitaxial IGBT, *IEEE Int. Electron Devices Meeting Digest*, 6.6.1–6.6.4 (2013).
Q. Zhang, A. Agarwal, C. Capell, L. Cheng, M. O'Loughlin, A. Burk, J. Palmour, S. Rumyantsev, T. Saxena, M. Levinshtein, A. Ogunniyi, H. O'Brien and C. Scozzie, 12 kV, 1 cm^2 SiC GTO thyristors with negative bevel termination, *Mater. Sci. Forum* **717–720**, 1151–1154 (2012).
Z. Zhang, E. Moulton and T. Sudarshan, Mechanism of eliminating basal plane dislocations in SiC thin films by epitaxy on an etched substrate, *Appl. Phys. Lett.* **89**, 081910 (2006).
Q. Zhang, G. Wang, H. Doan, S. Ryu, B. Hull, J. Young, S. Allen and J. Palmour, Latest results on 1200 V 4H-SiC CIMOSFETs with $R_{sp,on}$ of 3.9 $m\Omega \cdot cm^2$ at 150°C, *Int. Symp. Power Semiconductor Devices ICs* 89–92 (2015).
Y. Zhang, K. Sheng, M. Su, J. Zhao, P. Alexandrov and L. Fursin, 1000-V 9.1mΩcm^2 normally off 4H-SiC lateral RESURF JFET for power integrated circuit applications, *IEEE Electron Device Lett.* **28**, 404–407 (2007).
Y. Zhang, K. Sheng, M. Su, J. Zhao, P. Alexandrov, X. Li, L. Fursin and M. Weiner, Development of 4H-SiC LJFET-based power IC, *IEEE Tran. Electron Devices* **55**, 1934–1945 (2008).
J. Zhao, K. Tone, X. Li, P. Alexandrov, L. Fursin and M. Weiner, 3.6 $m\Omega \cdot cm^2$, 1726 V 4H-SiC normally-off trenched-and-implanted vertical JFETs, *Int. Symp. Power Semiconductor Devices ICs* 50–53 (2003a).
J. Zhao, P. Alexandrov and X. Li, Demonstration of the first 10-kV 4H-SiC Schottky barrier diodes, *IEEE Electron Device Lett.* **24**, 402–404 (2003b).
J. Zhao, P. Alexandrov, J. Zhang and X. Li, Fabrication and characterization of 11-kV normally off 4H-SiC trenched-and-implanted vertical junction FET, *IEEE Electron Device Lett.* **25**, 474–476 (2004a).
J. Zhao, Recent advances of high-efficiency single crystalline silicon solar cells in processing technologies and substrate materials, *Sol. Energy Mater. Sol. Cells* **82**, 53–64 (2004b).

8

Gallium Nitride Power Devices

8.1. Introduction

Transistors for power electronic applications should behave close to an ideal switch characteristic. High-performance power transistors require high breakdown voltage, low specific on-state resistance, large maximum current, low leakage current, and fast switching time. III–V group semiconductors such as GaN and AlN are those of the most important third generation semiconductors and one of the best candidates for developing high temperature, high frequency, and high power electronic devices due to their intrinsic excellent physical properties and chemical stabilities, such as wide bandgap, high breakdown electric field, large electron saturation velocity, and good thermal conductivity. Nowadays, the progress in the development of GaN-based power devices is accelerating rapidly.

Among the GaN-based power devices, the most distinctive devices are based on the heterostructures such as AlGaN/GaN which has been studied extensively because of its importance for microwave and high power applications. AlGaN/GaN heterostructure has a large band offset and strong polarization, which endow it with large two-dimensional electron gas (2DEG) density without intentional doping. By utilizing the structure, AlGaN/GaN high electron mobility transistors (HEMTs) have much higher charge density and mobility in the conduction channel, and hence are capable of delivering breakdown voltage and on-resistance beyond the material limits of Si and SiC semiconductors for high-voltage switching applications.

In the past decade, although the characteristics of AlGaN/GaN HEMTs have improved a lot by the research done on both the device structure and the growing technique of GaN-based material, the high-performance and low-cost AlGaN/GaN HEMTs have not been readily utilized commercially. There are still many issues in the investigations of AlGaN/GaN HEMTs, such as the realization of stable normally-off MOSFET with a large threshold voltage (V_{th}), the process compatibility with CMOS technology, the large breakdown voltage versus low on-resistance, the suppression of the current collapse phenomenon, and so on. It is hence a continuing effort to further investigate and improve the AlGaN/GaN device performance in the next 5–10 years or so for the device to be commercially popular.

Normally-off operation with large V_{th} is one of the most important topics among these issues. Due to the 2DEG channel, by default the device is in a normally-on configuration. It is strongly desired in AlGaN/GaN HEMTs for a positive gate threshold voltage, e.g. greater than 3.5 V, in order to make the devices compatible with the current Si-based switching systems. Recently, fluorinated gate or gate dielectric, p-type GaN or AlGaN gate, and piezoelectric neutralization approaches were introduced to shift the threshold voltage to be positive, while these methods may affect the quality of gate structure resulting in a lower on-state current, an insufficient threshold level, and a large gate leakage current. Gate recess etching is another effective way to realize the normally-off operation, however, it usually comes at the cost of severe degradation of 2DEG channel conductance due to the interface scattering, especially when the AlGaN barrier leftover thickness is within 5 nm or less. On the other hand, floating gate is a promising approach to get the normally-off mode by charge storage without reducing the channel conductance. However, its shortcoming is on the charge retention limitation which affects the operation stability and makes them unsuitable for certain practical applications.

Recent works have experimentally confirmed that, by the process of single fluorinated gate dielectric at a low RIE power (60 W), it can adequately shift the threshold voltage. The shortcoming is that the V_{th} level is still low due to the limited fixed negative charge amount and the long distance between the fluorinated region and the 2DEG channel. Combination of partial gate recess etching to reduce 2DEG density and multiple fluorinated gate dielectric layers in metal-insulator-semiconductor (MIS) structure to keep a large amount of negative fixed charge can achieve a large V_{th} without obvious degradation in 2DEG channel conductance, and obtain a low gate leakage current density. We will take a good look at this approach in this chapter.

The process compatibility with CMOS technology is another important topic for the GaN-based device work. Nowadays, Au-containing metal schemes are extensively applied in the GaN-based HEMTs to reduce the contact resistance and prevent electrode oxidation. However, Au-containing schemes also increase the cost of the GaN-based device products. If its deposition thickness is reduced, it will result in the poor surface morphology of the contacts and sometimes degradation of the Ohmic contacts. Thus, Au-free technology is necessary in practical applications especially for the integration with current CMOS fabrication infrastructure. Recently, several approaches on Au-free contact scheme have been explored. However, the drain current reductions are significant ranging from 24% to 80% compared with the Au-treated normally-on counterparts at the same gate over-drive V_{GT} $(= V_G - V_{th})$. Thus, Au-free normally-off HEMTs are still under intense investigation since achieving both low contact resistance and high channel conductance poses challenges in the process integration.

Large breakdown voltage and high output power are both important targets in the power switches especially in the application of electric drive. Prior to the actual device fabrications, numerical simulations are usually carried out to predict the performance on design and process variations. Sentaurus TCAD software is a powerful and professional tool for electronic device simulations and is suitable for the analyses of AlGaN/GaN devices. This will be mentioned in the chapter.

8.2. AlGaN/GaN and InGaN/GaN Heterojunction Configurations

8.2.1. Theoretical Calculations of Polarization Effects

Semiconductors such as GaN, AlN, and InN are polar materials since they induce net polarization at the heterojunction interface due to the shift in the cation and anion sublattices. In unstrained zinc-blende structures, the cation and anion sublattices are arranged in such a way that there is no net polarization in the material. However, in the wurtzite crystal, *spontaneous polarization* is produced in the crystal due to the asymmetric lattice structure and ionicity of the bonds. Strain can cause a relative shift between the cation and anion sublattices and hence create net polarization at the heterointerface which is called *piezoelectric polarization*.

These high polarization and their resulting electric fields produce high interface charge density at group-III-nitride interfaces and spatial separation of the electron and hole wave functions in GaN-based quantum well structures. If the polarization-induced sheet charge density is positive, free electrons will tend to compensate the polarization-induced charge. These electrons will form a 2DEG of a certain sheet carrier concentration, assuming that the band offset in the group-III-nitride interfaces is reasonably high and that the interface roughness is low. A negative sheet charge density will at the same time cause the hole accumulation in other interfaces due to charge neutralization.

8.2.1.1. *For Spontaneous Polarization*

In Fig. 8.1, the left shows the ideal wurtzite structure where all bonds have the same length and bond angles are equal. Therefore, $P_{SP} = 0$. The right shows the

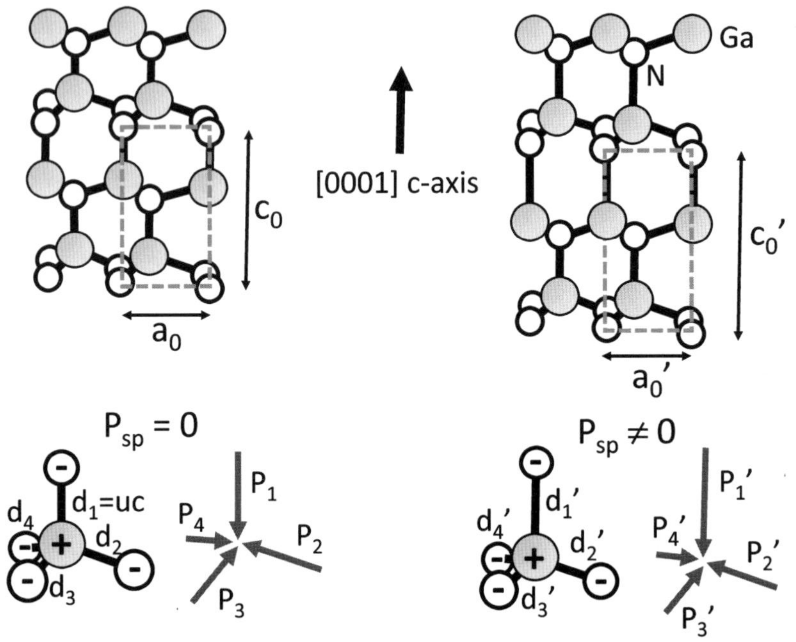

Fig. 8.1. The schematic diagram for the production of spontaneous polarization.

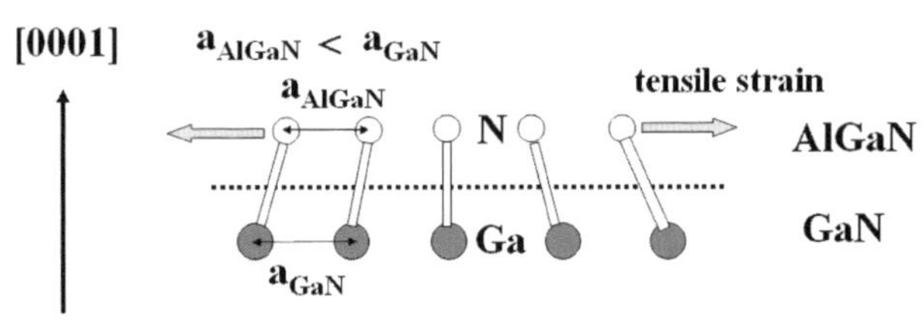

Fig. 8.2. The schematic diagram for the strain production in the AlGaN/GaN interface.

polar materials such as GaN and AlN crystals where exist asymmetric lattice structure and ionicity of the bonds. The bond along the c-axis is longer than the other bonds and the bond angles deviate from the ideal values. Therefore, the spontaneous polarization happens.

8.2.1.2. *For Piezoelectric Polarization*

Assuming the AlGaN/GaN heterostructures are grown along the [0001] axis and GaN layer is thick enough, there will be tensile strain at the interface due to the lattice constant of AlGaN being less than that of GaN, as shown in Fig. 8.2.

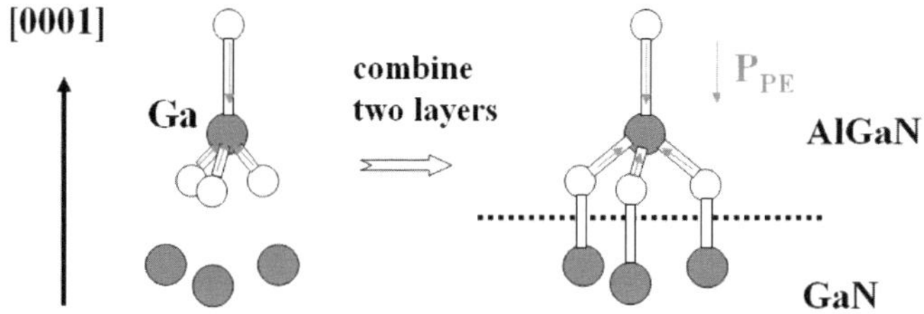

Fig. 8.3. The schematic diagram for the piezoelectric polarization production at the AlGaN/GaN interface.

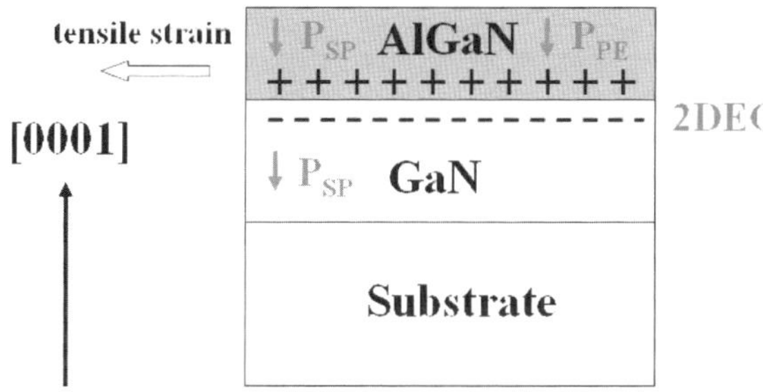

Fig. 8.4. Polarization-induced bound sheet charge, piezoelectric and spontaneous polarization of pseudomorphic AlGaN/GaN heterostructures with Ga-face polarity.

The bond lengths and angles are changed due to the strain which will result in the piezoelectric polarization, as shown in Fig. 8.3. The direction of the piezoelectric polarization will be along [0001] for the tensile strain, which is kept consistent with the spontaneous polarization direction.

8.2.1.3. *Polarization-Induced 2DEG Charge*

The polarization-induced sheet charge density and directions of spontaneous and piezoelectric polarization in Ga-faced AlGaN/GaN heterostructures are shown in Fig. 8.4. At an abrupt interface of the AlGaN/GaN heterostructure, the net polarization due to the abrupt variations of the polarization within a bilayer will cause a net polarization sheet charge density which can be defined by

$$\begin{aligned}\sigma &= P(\text{top}) - P(\text{bottom})\\ &= \{P_{\text{SP}}(\text{top}) + P_{\text{PE}}(\text{top})\} - \{P_{\text{SP}}(\text{bottom}) + P_{\text{PE}}(\text{bottom})\}.\end{aligned} \tag{8.1}$$

We can also understand the present position of 2DEG from the energy band diagram, as shown in the following Fig. 8.5. Free electrons will accumulate

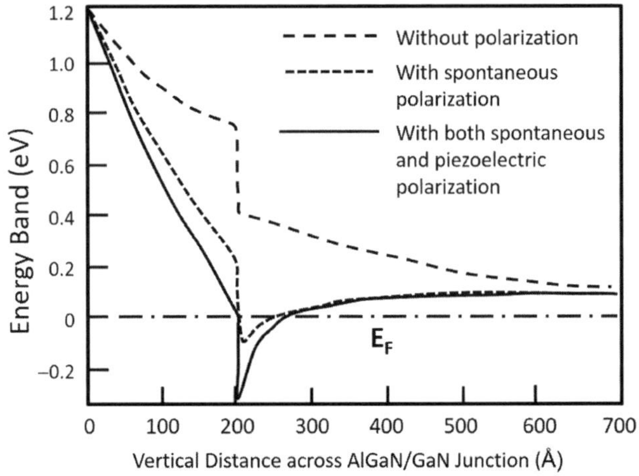

Fig. 8.5. Energy band diagram of AlGaN/GaN interface.

to balance the junction net charges, thereby forming the 2DEG layer at the AlGaN/GaN interface at the GaN side. The electrons are driven to the lowest energy position by internal electric field because of band bending. The accumulation will occur at the region where the conduction band is below the Fermi level, as shown in the cases of spontaneous polarization and of both spontaneous and piezoelectric polarization.

The bound spontaneous and piezoelectric polarization-induced sheet charge density at the $Al_xGa_{1-x}N/GaN$ heterojunction can be modeled and calculated by the following equations (Ambacher *et al.*, 1999, 2000):

$$P_{SP}^{AlGaN}(x) = (-0.052x - 0.029)\ \text{C/m}^2$$

$$P_{PE} = 2(1 - r(x))\frac{a - a_0}{a_0}\left(e_{31} - e_{33}\frac{C_{13}}{C_{33}}\right),$$

where

$$\begin{aligned} a(x) &= (-0.077x + 3.189) \times 10^{-10}\ \text{m} \\ C_{13}(x) &= (5x + 103)\ \text{GPa} \\ C_{33}(x) &= (-32x + 405)\ \text{GPa} \\ e_{31}(x) &= (-0.11x - 0.49)\ \text{C/m}^2 \\ e_{33}(x) &= (0.73x + 0.73)\ \text{C/m}^2 \\ 1\ \text{C/m}^2 &= 6.25 \times 10^{18}\ \text{e/m}^2 \end{aligned} \quad (8.2)$$

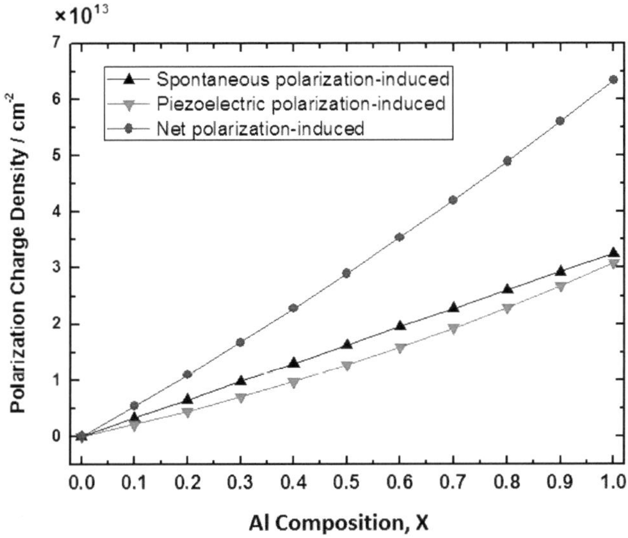

Fig. 8.6. Polarization-induced charge densities versus Al composition x, without considering the strain relaxation.

Here, x is the aluminum composition, a_o is the lattice constant of GaN, $a(x)$ is the linear interpolation of the lattice constant of relaxed $Al_xGa_{1-x}N$ with Al composition x, C_{13} and C_{33} are the elastic constant, e_{31} and e_{33} are the piezoelectric constant, and $r(x)$ is the strain relaxation factor.

For AlGaN/GaN heterostructures which are grown along [0001] direction, the charge density induced by the spontaneous and piezoelectric polarization with different Al composition without considering the strain relaxation is shown in Fig. 8.6. Both the spontaneous and piezoelectric polarization effects have the same sign for Ga polarity and tensile strain. The piezoelectric effect is non-linear with the composition x increasing, whereas the spontaneous effect is nearly linear with x increasing. Spontaneous polarization-induced charge density is a little higher than the other one which indicates both effects have similar influence on 2DEG.

Figure 8.7 shows the piezoelectric polarization-induced charge density at different Al composition at the AlGaN/GaN heterostructures when considering the relaxation factor of AlGaN layer, i.e. a thicker AlGaN layer having a higher relaxation factor. No relaxation ($r = 0$) leads to a large polarization charge density. A thicker AlGaN layer as grown has a larger relaxation factor in comparison to that of a thinner one. Therefore, how to obtain a good crystalline quality of the AlGaN thin film and at the same time to

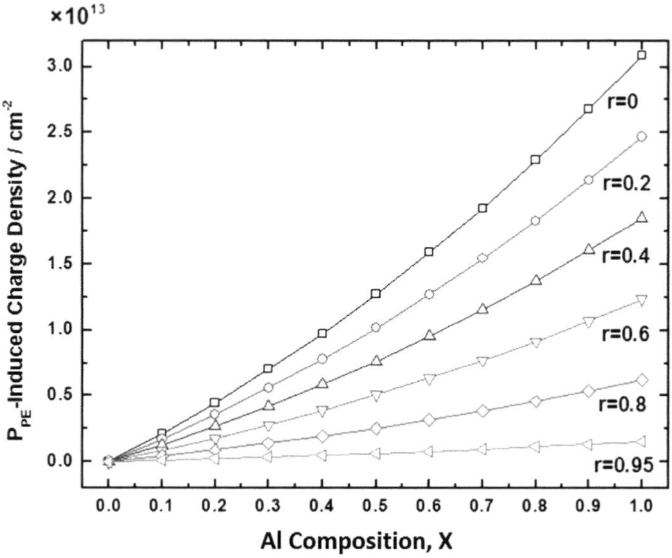

Fig. 8.7. Piezoelectric polarization-induced charge densities versus Al composition x when considering the strain relaxation factor.

retain a high piezoelectric polarization becomes a crucial issue in the film growth.

For the case of InGaN/GaN heterostructure, the set of equations is similar to that of AlGaN/GaN but with different parameters used as shown below.

$$P_{SP}^{InGaN}(x) = (-0.003x - 0.029)\ \mathrm{C/m^2}$$

$$P_{PE} = 2(1 - r(x))\frac{a - a_0}{a_0}\left(e_{31} - e_{33}\frac{C_{13}}{C_{33}}\right),$$

where

$$\begin{aligned} a(x) &= (0.344x + 3.189) \times 10^{-10}\ \mathrm{m} \\ e_{31}(x) &= (-0.08x - 0.49)\ \mathrm{C/m^2} \\ e_{33}(x) &= (0.24x + 0.73)\ \mathrm{C/m^2} \\ C_{13}(x) &= (-14x + 106)\ \mathrm{GPa} \\ C_{33}(x) &= (-174x + 398)\ \mathrm{GPa}. \end{aligned} \tag{8.3}$$

Figure 8.8 shows the polarization-induced charge density with different In composition at the $In_xGa_{1-x}N$/GaN heterostructure under fully strained condition. The spontaneous polarizations for InGaN and GaN layers are similar

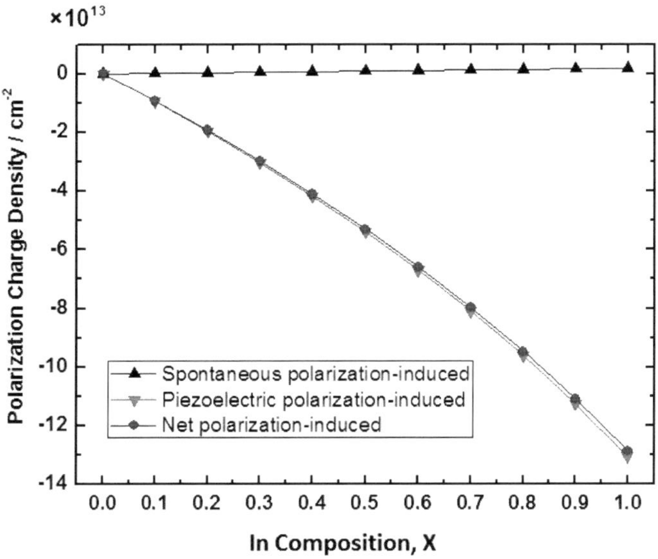

Fig. 8.8. Polarization-induced charge densities versus in composition x.

with little difference. The piezoelectric polarization-induced charges, now under compressive stress, dominate and are opposite in polarity to spontaneous polarization.

It is important to note that the net spontaneous polarization of AlGaN/GaN heterostructures is much larger than that of InGaN/GaN heterostructures, which could be attributed to the distinction in c_0/a_0 and/or u of lattice parameter.

Figure 8.9 shows the piezoelectric polarization-induced charge density with different In composition in the InGaN/GaN heterostructures when considering the relaxation factor of InGaN layer. Similarly, no relaxation ($r = 0$) or less relaxation in InGaN layer will induce a higher polarization charge density.

8.2.2. Calculation of 2DEG Sheet Carrier Density

Take the AlGaN/GaN heterostructure as an example onwards. For the determination of the sheet carrier concentration and carrier distribution profile in the 2DEG channel, one needs to solve the Schrodinger equation and Poisson equation in a self-consistent scheme. It is also necessary to specify the boundary conditions at the interfaces. However, this numerical process for a detailed solution can be complex in the calculation of the 3D concentration in the 2DEG region. Instead, we employ some theoretical equations for the

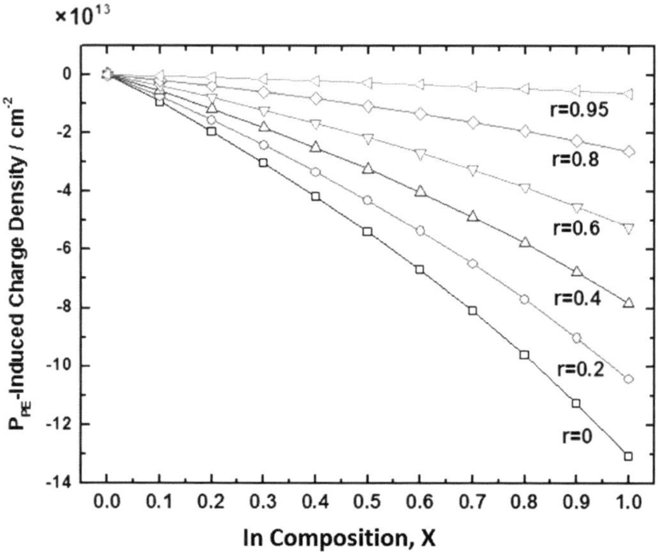

Fig. 8.9. Polarization-induced charge density versus in composition x when considering the strain relaxation factor.

calculation of 2DEG sheet density in a simpler approach for the purpose of understanding the 2DEG induced charges.

In the AlGaN/GaN heterostructure, as an example we use Ni as the Schottky contact metal. At the interface, towards the substrate, the Fermi level is set to be one half of the bandgap of GaN. The background free carrier concentration is assumed to be $10^{15}\ \mathrm{cm}^{-3}$. The 2DEG sheet electron concentration is calculated by using the total bound sheet charge $\sigma(x)$ which we have got from (8.1) and the following equations:

$$n_{\mathrm{s}}(x) = \frac{\sigma(x)}{e} - \left(\frac{\varepsilon_0 \varepsilon(x)}{de^2}\right)(e\phi_{\mathrm{b}}(x) + E_{\mathrm{F}}(x) - \Delta E_{\mathrm{c}}(x)), \tag{8.4}$$

where d is the thickness of the AlGaN barrier, $e\phi_{\mathrm{b}}$ is the Schottky barrier of gate metal contact, E_{F} is the Fermi level with respect to the GaN conduction-band-edge energy, and ΔE_{C} is the conduction band offset at AlGaN/GaN interface. To determine the sheet carrier concentration from the polarization-induced sheet charge density in the equation above, we define the following parameters:

Dielectric constant:

$$\varepsilon(x) = -0.5x + 9.5. \tag{8.5}$$

Schotty barrier:

$$e\phi_{\rm b} = (1.3x + 0.84)\, eV. \tag{8.6}$$

Fermi energy:

$$E_{\rm F}(x) = E_0(x) + \frac{\pi h^2}{m_{\rm e}^*(x)} n_{\rm s}(x), \tag{8.7}$$

where the ground subband level of the 2DEG is given by

$$E_0(x) = \left(\frac{9\pi h e^2}{8\varepsilon_0 \sqrt{8 m_{\rm e}^*(x)}} \frac{n_{\rm s}(x)}{\varepsilon(x)} \right)^{\frac{2}{3}} \tag{8.8}$$

with the effective electron mass, $m_{\rm e}^*(x) \approx 0.22 m_{\rm e}$, and the band offset

$$\Delta E_{\rm C} = 0.7(E_{\rm g}(x) - E_{\rm g}(0)), \tag{8.9}$$

where the bandgap of AlGaN is found to be

$$\begin{aligned} E_{\rm g}(x) &= x E_{\rm g}({\rm AlN}) + (1-x) E_{\rm g}({\rm GaN}) - x(1-x) \cdot 1.0\,{\rm eV} \\ &= x \cdot 6.13\,{\rm eV} + (1-x) \cdot 3.42\,{\rm eV} - x(1-x) \cdot 1.0\,{\rm eV}. \end{aligned} \tag{8.10}$$

As an example, by taking the thickness of AlGaN as 30 nm, we then calculate the diverse sheet carrier concentration with Al composition x, as shown in Fig. 8.10. From the figure, we can see that the sheet carrier concentration of 2DEG increases with a higher Al composition ratio x.

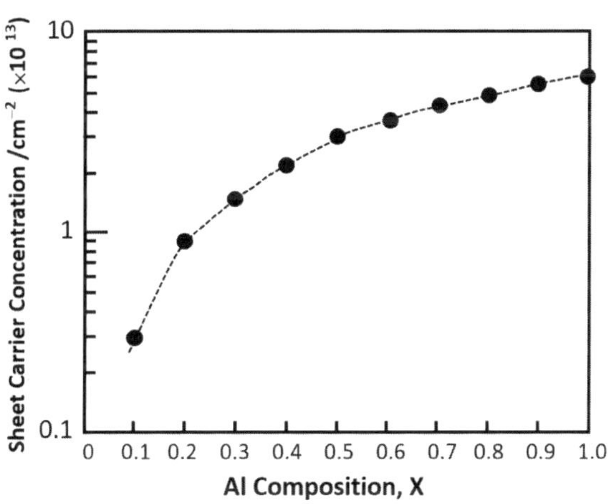

Fig. 8.10. Calculated 2DEG sheet density versus Al composition ratio x.

8.2.3. Calculation of Critical Thickness of Strained Layer

Several theories have been developed to find the critical thickness in metastable heteroepitaxial layer in relationship to the strain relaxation (Matthews and Klokholm, 1972; Fischer *et al.*, 1994; Qu *et al.*, 1998). Among them, the theories developed by Matthews and Fischer are most cited. By placing an "image dislocation" outside the crystal instead of the free-surface boundary conditions, Fischer's theory is considered as a more suitable model to predict the critical strained layer thickness. The workable equations from these two models are shown as below.

The equation for Fischer's model:

$$\left|\frac{a(x)-a_{\mathrm{sub}}}{a_{\mathrm{sub}}}\right| = \frac{b\cos\lambda}{2h_{\mathrm{c}}} \times \left(1 + \frac{1-\frac{v}{4}}{4\pi(1+v)\cos^2\lambda} \times \ln\frac{h_{\mathrm{c}}}{b}\right). \qquad (8.11)$$

The equation for Matthews' model:

$$\left|\frac{a(x)-a_{\mathrm{sub}}}{a_{\mathrm{sub}}}\right| = \frac{b}{2h_{\mathrm{c}}} \times \frac{1-\frac{v}{4}}{2\pi(1+v)} \times \ln\frac{h_{\mathrm{c}}}{b}. \qquad (8.12)$$

Here $a(x)$ is the lattice constant of upper layer, a_{sub} is the lattice constant of the bottom layer (substrate), b is the slip distance (or the magnitude of Burger's vector), λ is the angle between the Burger's vector and the dislocation line, h_{C} is the critical layer thickness, and v is the Poisson's ratio. These parameters are listed in Table 8.1.

We calculate the critical thickness of both AlGaN/GaN and InGaN/GaN structures with various Al (In) compositions using the two model equations and parameters as listed in Table 8.1. The results are shown in Fig. 8.11 for AlGaN/GaN and Fig. 8.12 for InGaN/GaN, and their numerical values are listed in Tables 8.2 and 8.3.

Table 8.1. Parameters used in the critical thickness calculation.

Parameters	AlGaN/GaN structure	InGaN/GaN structure
a_{sub} (nm)	0.3189	0.3189
a_{AlN} or a_{InN} (nm)	0.3112	0.3545
b (nm)	0.3189	0.3189
λ(°)	60	60
ν	0.36	0.36

For AlGaN/GaN structure

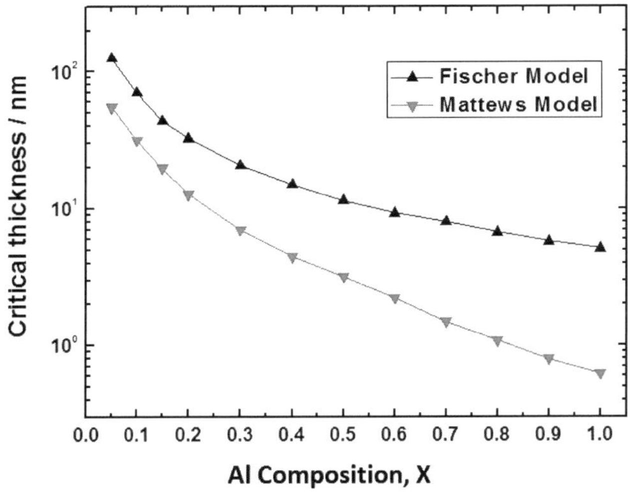

Fig. 8.11. Critical thickness of AlGaN grown on relaxed GaN substrate calculated using Fischer and Matthews models versus Al composition ratio x.

For InGaN/GaN structure

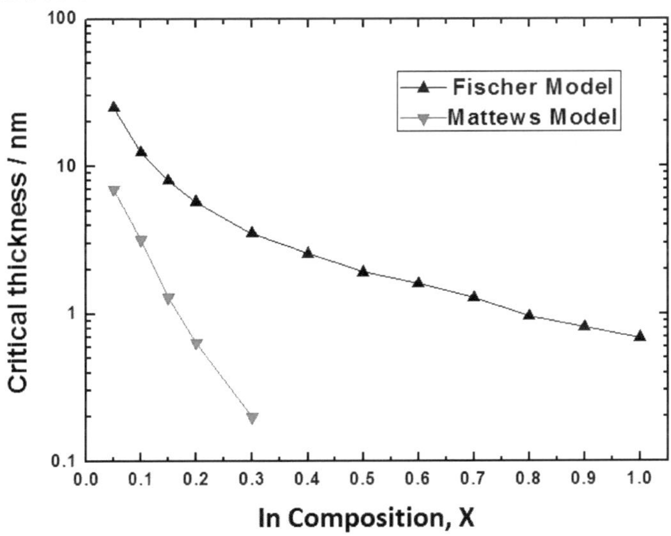

Fig. 8.12. Critical thickness of InGaN grown on relaxed GaN substrate calculated using Fischer and Matthews models versus In composition ratio x.

Table 8.2. Critical thickness of AlGaN calculated from two model equations.

x (Al)	0.05	0.1	0.15	0.2	0.3	0.4	0.5	0.6	0.7	0.8	0.9	1.0
h_C (Fischer)	125	70	44	33	21	15	12	9.3	8.0	6.7	5.7	5.1
h_C (Matthews)	55	32	20	13	7.1	4.5	3.2	2.3	1.5	1.1	0.8	0.6

Table 8.3. Critical thickness of InGaN calculated from two model equations.

x (In)	0.05	0.1	0.15	0.2	0.3	0.4	0.5	0.6	0.7	0.8	0.9	1.0
h_C (Fischer)	25	13	8.0	5.8	3.5	2.6	1.9	1.6	1.3	0.96	0.81	0.69
h_C (Matthews)	7.2	3.2	1.3	0.64	0.23							

For both cases, the critical thickness decreases with increasing Al and In concentrations. However, the calculated results from Matthews model are in general lower than those from manufacturers' datasheets or research papers. The results from Fischer model are closer to those practical data.

8.3. Simulation of GaN HEMTs

The influences of Al composition and AlGaN layer strain relaxation on the 2DEG density and distribution can be analyzed by simulations. There are two useful methods to take care of the formation of polarization charges within the device structure, namely the auto-generation approach by the internal function of the popular Sentaurus simulation tool, or by a manual charge placement method by placing suitable amount of charges near interface locations in simulation. For example, Fig. 8.13 indicates the 2DEG charges by the auto-generation method (solid-line) and the placing charge method (dashed-line) for a AlGaN/GaN HEMT device with AlGaN thickness and Al mole fraction at 30 nm and 0.25 nm, respectively. The GaN thickness is at 2 μm with the background electron concentration of 5×10^{14} cm^{-3}. Bulk acceptor traps are placed at a 1.0 eV below midband with a density of 10^{16} cm^{-3}. For the placing charge method, the 2D polarization charge at the heterointerface is manually calculated and placed as 1.29×10^{13} cm^{-2}. A negative sheet charge density of 3.2×10^{13} cm^{-2} and the corresponding sheet donor traps with the same density are preset on the AlGaN surface for the purpose of overall charge configuration. More details on the placing charges method are described in Sec. 8.3.1. Both approaches yield similar 2DEG distribution shapes and charge densities.

8.3.1. Fabrication Induced Trap Charges

In a typical unintentionally doped AlGaN/GaN HEMT device shown in Fig. 8.14, the device fabrication process normally generates interfacial traps at

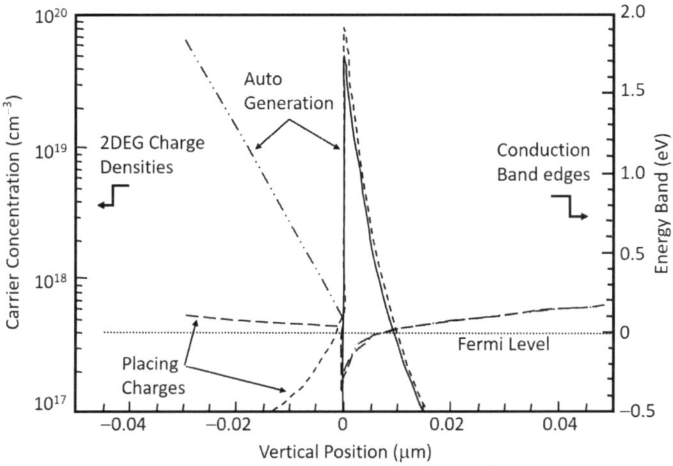

Fig. 8.13. Plots of 2DEG density distribution and associated conduction band edges with respect to the Fermi level in the AlGaN/GaN HEMT structure using two charge formation approaches. Both approaches have the similar 2DEG shapes and densities.

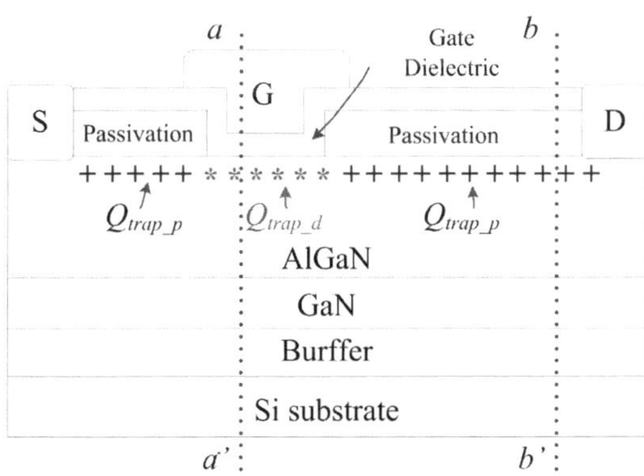

Fig. 8.14. Schematic of an AlGaN/GaN HEMT device where "+" denotes the interfacial traps $Q_{\text{trap\_p}}$ at passivation/AlGaN and the "*" denotes $Q_{\text{trap\_d}}$ at dielectric/AlGaN interface.

passivation/AlGaN ($Q_{\text{trap\_p}}$) or dielectric/AlGaN ($Q_{\text{trap\_d}}$) interfaces due to interface defects, dangling bonds or adsorbed ions. These traps will strongly affect the 2DEG density. In such a practical case, the 2DEG density could be lower than 1×10^{12} cm^{-2} in a normally-on HEMT device from its theoretical value. Experimentally, the interfacial trap density Q_{trap} values, ranging from 10^{11} to 10^{12} cm^{-2}, can be obtained from C–V or frequency–dispersion measurements. However, these interfacial traps at passivation (or dielectric)/AlGaN interface which are accounted for the lower 2DEG conductance, hysteresis in device transfer characteristic (such as I_D–V_G sweep) and current collapse have their ionization rates both time and Fermi level dependent. That makes the process to identify these fabrication-induced traps difficult.

On the other hand, these values will become different once the fabrication processes vary. Currently, for the device made with various different processes, there lacks a simple and versatile trap-charge configuration scheme to identify the definite trap charges to be placed in simulation for a quick performance evaluation. Recently, Sun *et al.* (2016) have come up with a set of trap charge ranges which correspond to the GaN processes to be suitable for device simulation using the placing charges method.

Figure 8.15 describes some common GaN device process flows and the corresponding trap charge items associated with these processes. Figure 8.16 provides the quantified values on the trap charge ranges to be placed in simulation in respect to different process conditions, where each item is described with alphabet as below:

(A) Deposition of 4 nm Al_2O_3 film on top of AlGaN. The aluminum composition is 0.24.
(B) Deposition of 8 nm Al_2O_3 film on top of AlGaN. The aluminum composition is 0.25.
(C) Deposition of 20 nm SiO_2 film on top of AlGaN. The aluminum composition is 0.25.
(D) Deposition of 20 nm AlN film on top of AlGaN. The aluminum composition is 0.3.
(E) Deposition of gate dielectric Al_2O_3 of 4 nm on AlGaN without gate recess. The aluminum composition is 0.24.
(F) Deposition of gate dielectric Al_2O_3 of 25 nm on AlGaN without gate recess. The aluminum composition is 0.25.
(G) Deposition of gate dielectric Al_2O_3 of 8 nm on AlGaN without gate recess. The aluminum composition is 0.25.
(H) Deposition of gate dielectric Al_2O_3 of 13 nm on AlGaN without gate recess. The aluminum composition is 0.27.
(I) AlGaN gate recess and gate Al_2O_3 dielectric deposition of 18 nm. The aluminum composition is 0.25.
(J) AlGaN gate recess and gate Al_2O_3 dielectric deposition of 6 nm. The aluminum composition is 0.27.
(K) AlGaN gate recess and gate Al_2O_3 dielectric deposition of 15 nm. The aluminum composition is 0.3.
(L) AlGaN gate recess and gate Al_2O_3 dielectric deposition of 25 nm. The aluminum composition is 0.3.

(M) Post dielectric deposition annealing: Al_2O_3 of 18 nm with 500°C, N_2, 1 min annealing. The aluminum composition is 0.25.
(N) Post dielectric deposition annealing: Al_2O_3 of 13 nm with 400°C, N_2, 20 min annealing. The aluminum composition is 0.27.
(O) Post dielectric deposition annealing: Al_2O_3 of 15 nm with 400°C, N_2/O_2, 2 min annealing. The aluminum composition is 0.3.
(P) Post dielectric deposition annealing: Al_2O_3 of 10 nm with 600°C, N_2, 10 min annealing. The aluminum composition is 0.26.

An example on GaN HEMT inverter configuration with its fabrication process and associated trap charge selection is illustrated in Fig. 8.17, and used in Sentaurus simulations. The normally-on GaN HEMT shown at the left-hand side is with Al_2O_3 dielectric gate, and the normally-off GaN HEMT shown at the right-hand side is with a recess gate filled with Al_2O_3 to achieve a positive threshold voltage. The simulation outcomes of using the lower limit, upper limit, and optimized Q_{trap} values are illustrated in Fig. 8.18 and with the verification of experimental measurement.

8.3.2. Normally-off HEMT Device with Field Plates

A normally-off design is favored by the power electronic industry because of its inherent safety and robustness of the device configuration. The feasible approaches to achieve normally-off operations include (a) using a deep

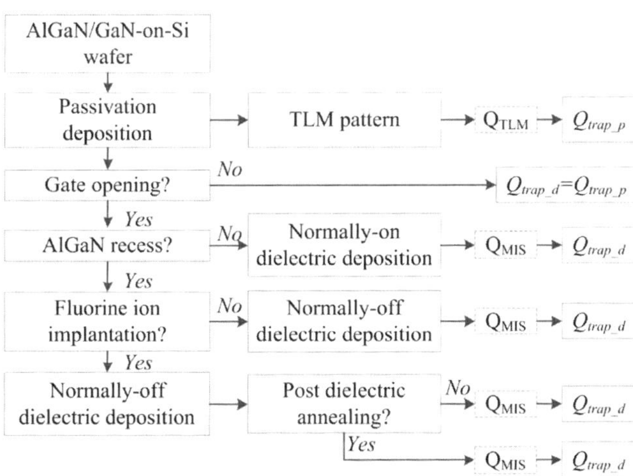

Fig. 8.15. The interfacial trap configuration scheme related to various GaN HEMT processes. The "Q_{TLM}" and "Q_{MIS}" in dashed boxes respectively denote the trap density determination process through TLM or device I_D–V_D measurement (Sun *et al.*, 2016).

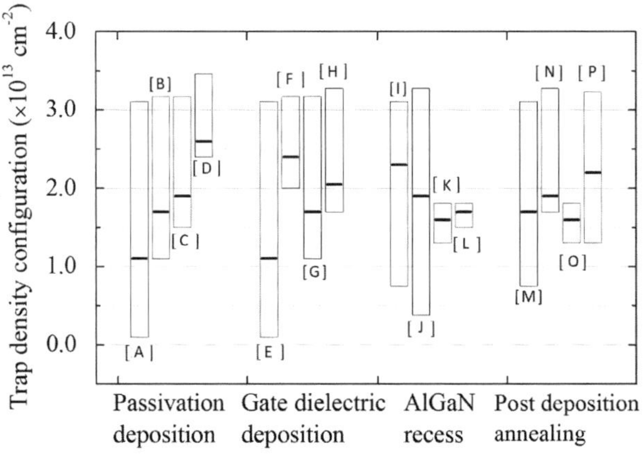

Fig. 8.16. The ranges of interfacial trap density configurations for various fabrication processes. The bars denote the respective interfacial trap density limit ranges of each reference; the line segments denote the recommended trap configurations (Sun *et al.*, 2016).

recess-gate to weaken the polarization and to reduce 2DEG charge concentration; (b) using a fluorine-treated technique to inject negative gate charges to deplete the 2DEG channel; (c) using a p-GaN neutralization structure, and so on. For this simulation example, we employ a recess-gate MIS structure with a multiple layer field plate (or multiple field plates) as shown in Fig. 8.19.

The structure has a 20 nm undoped AlGaN on the top of the 2 μm intrinsic GaN. Al composition is 0.25. The gate length is 1 μm, and the distance between gate and drain is 27 μm. The AlGaN layer in the gate region is completely recessed to realize the normally-off mode. At the same time, multiple field plate configuration is employed to improve the breakdown characteristics. The nitride field plate thicknesses are set at 50 nm, 100 nm, and 300 nm respectively. The top layer oxide thickness is 300 nm.

The simulation was first done to look at the gate threshold voltage, which is important as the threshold voltage determines whether the device is to be turned on by a gate voltage. Figure 8.20 gives the outcome of the simulation result. The threshold voltage is found to be around 2.7 V. The simulation on I_D–V_D follows with a set of gate voltages from $V_G = 2$ V to 15 V, as shown in Fig. 8.21. As expected, when V_G is below the threshold voltage, the drain current is almost zero. The on-resistance $R_{on} = 5.5\,\text{m}\Omega\cdot\text{cm}^2$ which is determined by the slope of the curve at $V_G = 15$ V.

The device breakdown voltage is set when the drain leakage current reaches 0.01 μA/mm. As shown in Fig. 8.22, the value found on the three-step field

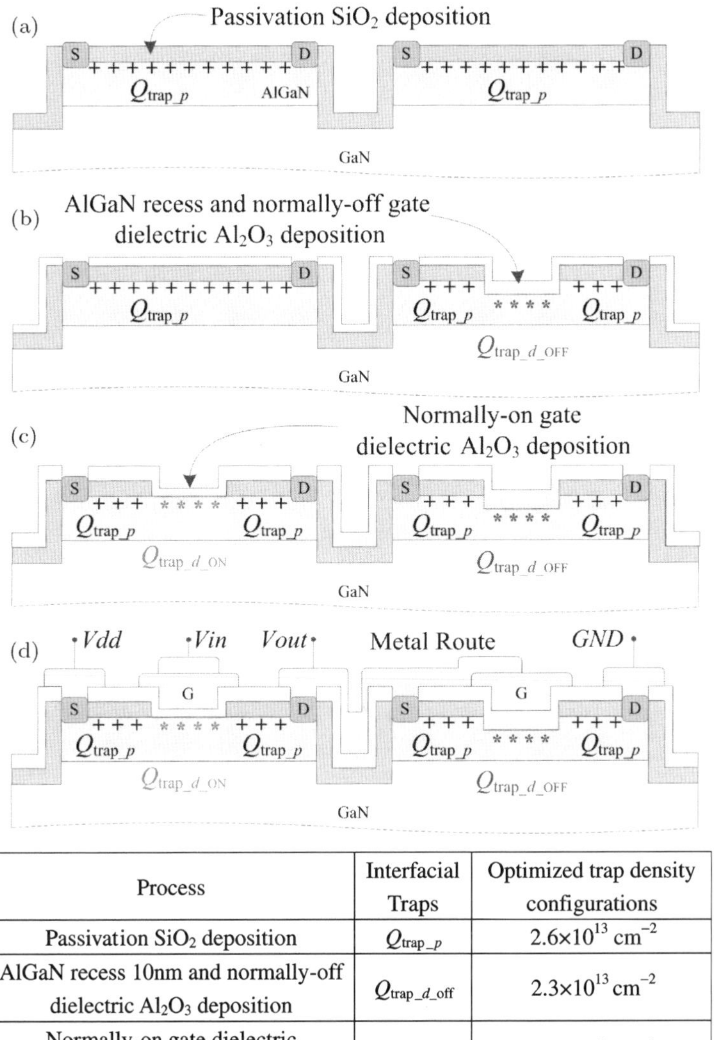

Process	Interfacial Traps	Optimized trap density configurations
Passivation SiO_2 deposition	Q_{trap_p}	2.6×10^{13} cm^{-2}
AlGaN recess 10nm and normally-off dielectric Al_2O_3 deposition	$Q_{trap\_d\_off}$	2.3×10^{13} cm^{-2}
Normally-on gate dielectric Al_2O_3 deposition	$Q_{trap\_d\_on}$	1.7×10^{13} cm^{-2}

Fig. 8.17. AlGaN/GaN inverter with key fabrication processes and the associated trap charge configuration used for the Sentaurus simulation shown in the table below. (a) Passivation SiO_2 deposition, (b) AlGaN recess and normally-off gate dielectric Al_2O_3 deposition, (c) normally-on gate dielectric Al_2O_3 deposition, (d) metal route interconnection (Sun *et al.*, 2016).

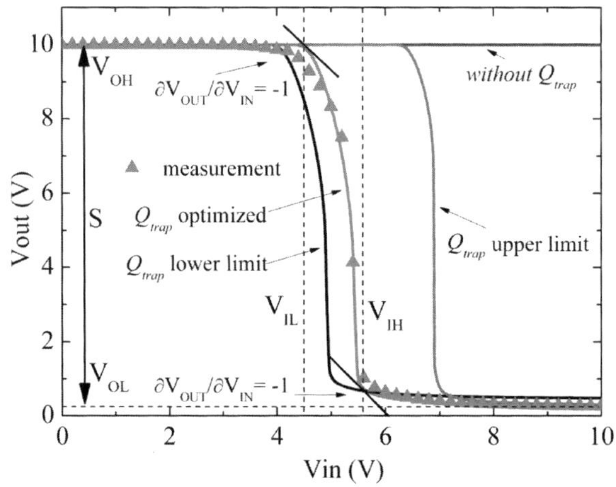

Fig. 8.18. Simulation and measurement data of the GaN inverter configuration. The output swing (S) is the change between V_{OH} and V_{OL} which is at 10 V. Without the proper trap charges placed, the Sentaurus simulation gives an incorrect outcome (Sun *et al.*, 2016).

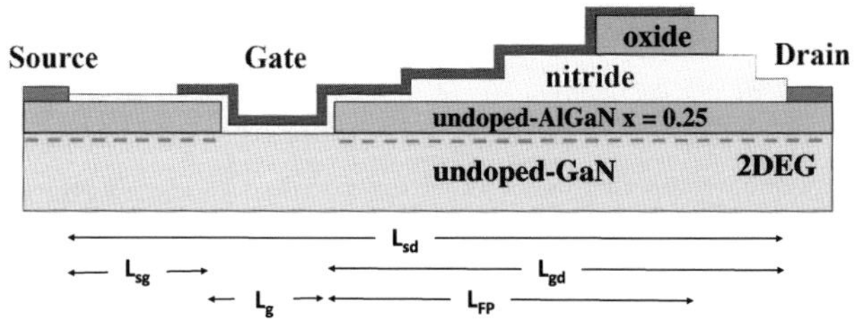

Fig. 8.19. Schematic of recess-gate HEMT structure with optimized field plate. $L_{sd} = 30\,\mu$m, $L_{gd} = 27\,\mu$m, $L_{sg} = 2\,\mu$m, $L_g = 1\,\mu$m, and $L_{FP} = 20\,\mu$m.

plate is at 1700 V and 600 V is found for the case of single field plate. The field plate serves a good purpose to divert the electric field away from the main junction at the gate right-side dielectric, and thus brings up the breakdown voltage.

8.4. Current Collapse in GaN HEMT

Current collapse in GaN HEMT is a critical concern as it limits the output power and the switching characteristics of the device when used for power electronic pulse switching. Current collapse phenomenon is generally caused by trapped charges at off-state at the AlGaN barrier surface, AlGaN/GaN

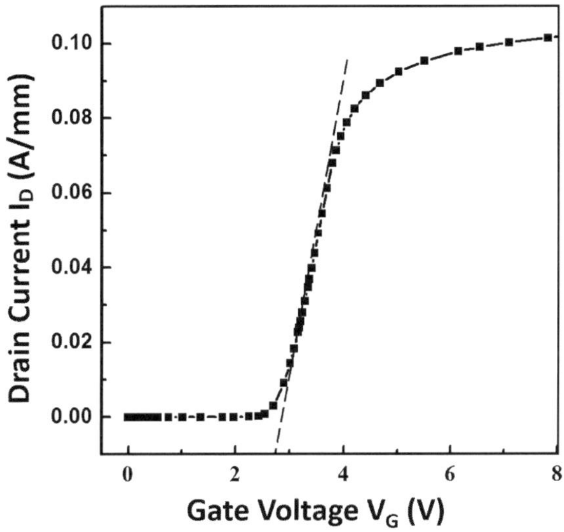

Fig. 8.20. I_D–V_G sweep at $V_D = 2\,V$. The gradient line indicates that the threshold voltage is at around 2.7 V.

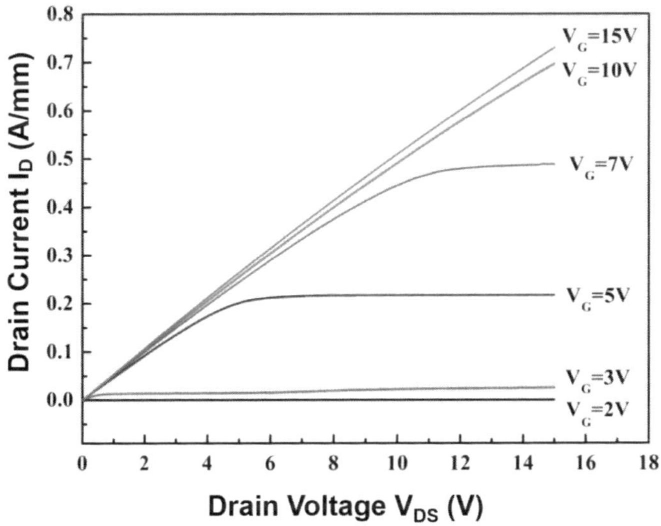

Fig. 8.21. I_D–V_D curves at different gate voltages between 2 V and 15 V.

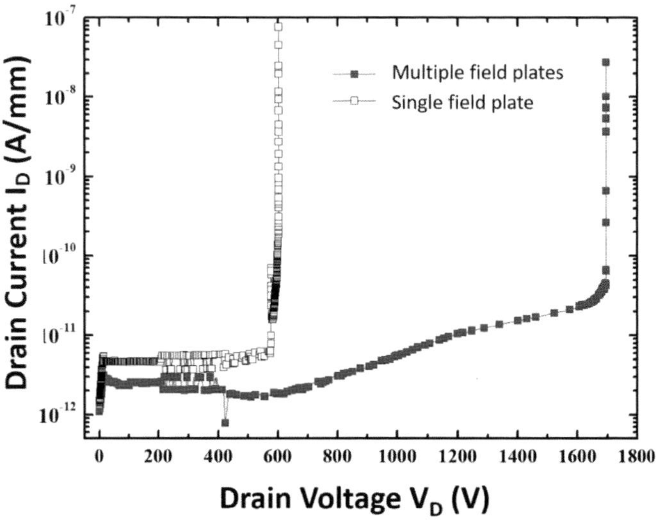

Fig. 8.22. Off-state I_D–V_D characteristics of recess-gate HEMT devices with single and optimized field plate at $V_G = 0$ V.

interface, and GaN bulk. Unlike Si- and SiC-based devices which have native passivation layers by thermal oxidation, the GaN-based devices are known to have a higher density of surface states.

The schematics of space charge locations and energy band structures in AlGaN/GaN HEMTs are shown in Fig. 8.23(a) and 8.23(b). A large number of donor-like traps, located at the top surface of AlGaN, are originated from defects such as N vacancies from surface dislocations, unsaturated bonds, plasma and thermal induced damages during the fabrication process. Even with the passivation oxide and nitride layers by PECVD, the defect density can still be high and acts as the main source of influence on 2DEG conductivity of the GaN power devices. Under the steady state without high voltage stress, they are ionized completely due to the strong polarization-induced electric field. The positive charges from the ionized donor-like traps compensate the negative surface polarization charges, resulting in a reduction of the surface electric field.

Under the off-state high stress bias, electrons are injected from the gate forming the gate leakage current will get across the metal/semiconductor interface by the tunneling and/or thermal emission process due to the high electric field at the gate corner towards the drain side. The large amount of injected electrons will move and hop along the AlGaN surface, leading to surface leakage current. Only a small number of electrons can be temporally captured by

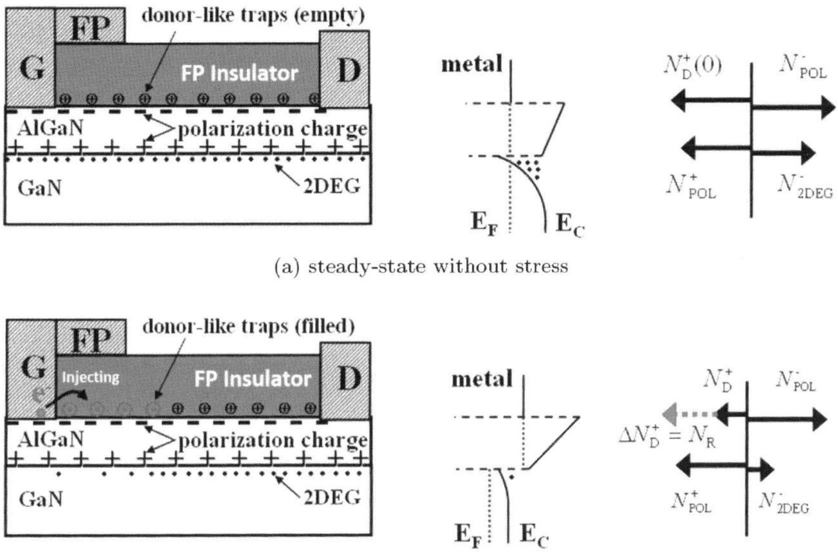

(a) steady-state without stress

(b) steady-state with voltage stress

Fig. 8.23. Schematics of space charge distribution in AlGaN/GaN HEMTs. (a) The surface donor-like trap charge, polarization charge and 2DEG charge distribution at steady state without stress, and (b) acting as the trapping centers for the injected electrons from the gate metal under high-voltage stress, the distribution has been affected. 2DEG density is now reduced.

the ionized surface donor-like traps near the gate edge region due to the low recombination coefficient under extremely high electric field, whereas most electrons are captured by the surface traps when they diffuse along the AlGaN surface at lower electric field region. The trapped electrons induces an additional electric field to repel the 2DEG electrons in channel and results in a reduction of 2DEG density. When the device is turned on, electrons start to emit from the trapped states, i.e. the de-trapping process, which is the dominant behavior for the on-state conductivity recovery.

Hence, the dynamic process of electron trapping and detrapping at AlGaN surface is considered to be crucial processes, causing current collapse not only due to the much higher trap density on the surface, but also its direct influence on the polarization of 2DEG conduction channel. The "virtual gate" theory was generally accepted to explain the current collapse phenomenon. The electrons drawn from the gate contact by the off-state electric field are trapped at the AlGaN surface defects near the gate region towards the drain side. When the device is turned on by a positive gate pulse, the trapped electrons at the AlGaN

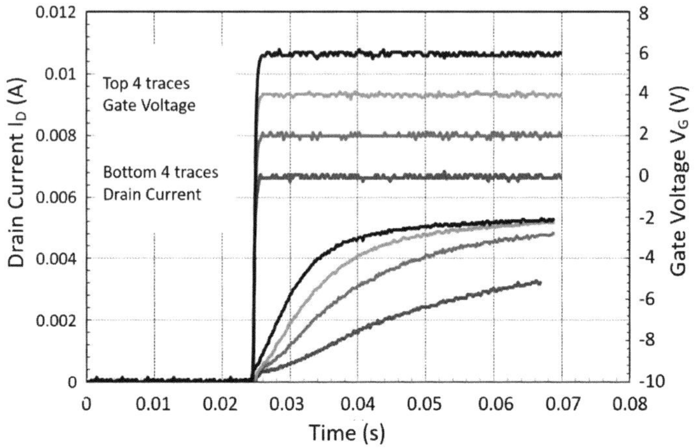

Fig. 8.24. Turn-on waveforms of an un-passivated GaN HEMT device.

surface cannot be removed in time and the induced field will remain to deplete the 2DEG channel, acting like a gate extension but with a negative potential. Thus, the increase of the 2DEG on-resistance in this process limits the on-state current and results in the current collapse phenomenon. Figure 8.24 shows a measurement on the turn-on waveforms on a GaN HEMT device at different gate voltages without surface passivation. The current collapse phenomenon is very severe.

Recent works indicated that the surface region of about 100 nm in dimension next to the gate towards the drain side, when under the influence of high electric field, is mainly responsible for the drain leakage current at off-state. Hence, this can also be considered as the main electron injection region, and the remaining low field region towards the drain side is considered as the electron diffusion and recombination region which is then responsible for causing the current collapse phenomenon. The dynamic behaviors of the field induced charge trapping can be modeled. The quantitative relationships between the electric field and surface trapped electron densities, and the spatial distributions of the trapped electrons and excess free electrons can then be determined. The spatial distribution can be used by the Sentaurus to calculate the device current–voltage characteristics at current collapse.

Under high voltage stress, electrons injected from the gate will get through the barrier of the metal/semiconductor contact by tunneling and/or thermal emission process. The large amount of electrons will hop along the AlGaN surface, leading to surface leakage current. Near the gate edge, only a small number of electrons will be captured by the ionized surface donor-like traps due

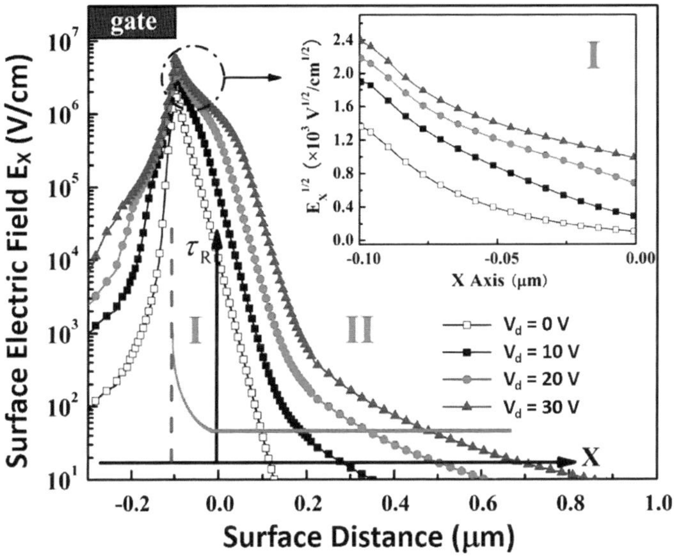

Fig. 8.25. Distribution plots of horizontal electric field at AlGaN surface (at 1 nm from top) near the gate towards the drain side under different drain voltage. "I" and "II" represent high field and low field regions where the exponential and constant distribution for electron recombination time are employed respectively. The inset shows the square-root of electric field in region "I" (Huang *et al.*, 2014a).

to low recombination coefficient under extremely high electric field (referring to region "I" in Fig. 8.25), whereas most electrons will be captured by the surface traps when they diffuse along AlGaN surface at lower electric field region (see region "II" in Fig. 8.25). The electrons trapped at the AlGaN surface, serving like a gate extension with negative bias, induce an electric field to deplete the 2DEG electrons in channel and results in a reduction of 2DEG density.

The dynamics of electron recombination and emission at AlGaN surface is governed by the following equation:

$$\frac{\partial N_{\mathrm{R}}}{\partial t} = C_{\mathrm{n,D}} N_{\mathrm{D}}^{+} n - \frac{N_{\mathrm{D}}^{+}(0) - N_{\mathrm{D}}^{+}}{\tau_{\mathrm{E}}}, \tag{8.13}$$

where N_{R} is the density of trapped electrons in recombination with the ionized donor-like traps and N_{D}^{+} is the remaining ionized trap density, both of which are variables related to the time t. $N_{\mathrm{D}}^{+}(0) = N_{\mathrm{R}} + N_{\mathrm{D}}^{+}$ is the initial ionized trap density, namely, the steady-state density without surface field. $C_{\mathrm{n,D}} \cong 1/[\tau_{\mathrm{R}} N_{\mathrm{D}}^{+}(0)]$ is the electron recombination coefficient which is mainly determined by the electron recombination lifetime $\tau_{\mathrm{R}} \cdot n$ is the excess free electron density at AlGaN surface injected from gate contact which is much higher than the background electron density n_0. τ_E is the lifetime of electron emission.

In conjunction with the boundary condition: $N_D^+ = N_D^+(0)$ at $t = 0$, Eq. (8.13) can be solved and the following expression can be derived.

$$N_R = -\frac{C_{n,D}N_D^+(0)n\tau_E}{1 + C_{n,D}n\tau_E} \exp\left[-\left(C_{n,D}n + \frac{1}{\tau_E}\right)t\right] + \frac{C_{n,D}N_D^+(0)n\tau_E}{1 + C_{n,D}n\tau_E}. \quad (8.14)$$

Under steady stress state (i.e. $t \gg \tau_E$ and τ_R), we obtain

$$N_R = \frac{\frac{\tau_E}{\tau_R}n}{1 + \frac{\tau_E}{N_D^+(0)\tau_R}n}. \quad (8.15)$$

The average horizontal electric field along the AlGaN surface at region II is low, as shown in Fig. 8.25. Taking into account the low electron mobility due to the huge surface scattering, the surface electron drift velocity is rather low in moving towards the drain side. The usage of the diffusion-recombination model gives a possible solution to describe the distribution of excess free electron density $n(x)$ along AlGaN surface. The steady diffusion equation is shown as

$$D_n \frac{d^2n(x)}{dx^2} = \frac{n(x)}{\tau_R}. \quad (8.16)$$

Solving (8.16) in conjunction with the boundary conditions: $n(x) = n(0)$ at $x = 0$ and $n(x) = n_0$ at $x \gg L_n$, we can obtain the excess electron density distribution as

$$n(x) = (n(0) - n_0)\exp\left(-\frac{x}{L_n}\right) + n_0, \quad (8.17)$$

where $L_n = (D_n\tau_R)^{1/2}$ is the electron diffusion length, and D_n is the electron diffusion coefficient. Thus, it is possible to get the surface trapped electron density distribution with the position x under steady-state high-voltage stress by substituting (8.17) in (8.15), as shown below.

$$N_R(x) = \frac{\frac{\tau_E}{\tau_R}\left\{(n(0) - n_0)\exp\left(-\frac{x}{L_n}\right) + n_0\right\}}{1 + \frac{\tau_E}{N_D^+(0)\tau_R}\left\{(n(0) - n_0)\exp\left(-\frac{x}{L_n}\right) + n_0\right\}}. \quad (8.18)$$

When the device is turned on, electron injection from the gate is reduced greatly because of the dramatic fall in the electric field strength at the gate corner. The sum of surface free electrons and the trapped electrons remains unchanged at the beginning as the electron emission and recombination processes still persist.

However, after the initial stage, the electron recombination process becomes much weaker due to the removal of the injected electrons from the gate. The electron emission from the trapped state, i.e. the de-trapping process, becomes the dominant behavior in the on-state recovery process. The process can be described by

$$\frac{\partial N_R(x,t)}{\partial t} = \frac{-N_R(x,t)}{\tau_{E,on}}, \tag{8.19}$$

where $\tau_{E,on}$ is the lifetime of electron emission in the on-state recovery process. Solving Eq. (8.19) in conjunction with the boundary condition: $N_R(x, t = 0) = N_R(x)$ and $N_R(x, t = \infty) = 0$ at $t \gg \tau_{E,on}$, we can obtain the quantity of the trapped electrons as a function of time t as

$$N_R(x,t) = N_R(x)\exp\left(-\frac{t}{\tau_{E,on}}\right). \tag{8.20}$$

AlGaN surface donor-like traps with an energy level of 0.37 eV below conduction band are employed here to obtain physical quantities. During the off-state stressed condition, the life time $\tau_R = (1/(\sigma_{n,R} v_n N_D^+))\exp((E_C - E_D)/(kT))$ and $\tau_E = (1/(\sigma_{n,E} v_n N_C))\exp((E_C - E_D)/(kT))$, where the electron capture cross-section $\sigma_{n,R} = 1.0 \times 10^{-13}\,\text{cm}^2$, emission cross-section $\sigma_{n,E} = 1.2 \times 10^{-16}\,\text{cm}^2$, electron thermal velocity $v_n = 6 \times 10^6\,\text{cm} \cdot \text{s}^{-1}$, conduction band state density $N_C = 2.2 \times 10^{18}\,\text{cm}^{-3}$, and surface electron mobility $\mu_n = 100\,\text{cm}^2/\text{Vs}$. $N_D^+(0) = 2.5 \times 10^{20}\,\text{cm}^{-3}$ and $n_0 = 1.0 \times 10^{10}\,\text{cm}^{-3}$ at AlGaN surface are used in the calculation. $\tau_R = 1.0 \times 10^{-8}\,\text{s}$ and $\tau_E = 9.6 \times 10^{-4}\,\text{s}$ are obtained for region "II". In the high field region "I", as shown in the inset of Fig. 8.24, the square-root of field is found to be approximately linear with the distance x. Therefore, an exponential relationship can be used for the electron recombination lifetime in this region, considering Poole–Frenkel theory in which trap barrier reduction is proportional to the square-root of field. The electron diffusion length of 1.61 μm is obtained from the expression $L_n = ((kT/q)\mu_n \tau_R)^{1/2}$ at room temperature. The injection electron density $n(0)$ acts as the key variable in the calculation. In the off-state, the measured gate leakage current was found at the order of 10^{-7} A/mm when the drain bias is below 100 V. The excess free electron density $n(0)$ at the order of $10^{14}\,\text{cm}^{-3}$ can be found by calculation using the current expression $I_g = qn(0)v_s d$, where v_s is the electron saturation velocity at AlGaN surface. Thin AlGaN surface area of thickness $d = 1.0\,\text{nm}$ is considered here. Then the trapped and excess electron density distributions can be determined by (8.17) and (8.18).

Figure 8.26(a) shows the distribution of calculated densities of trapped electrons and excess free electrons at steady-state under 10 V, 20 V, and 30 V stress. The trapped electron spatial distribution is determined from the amount

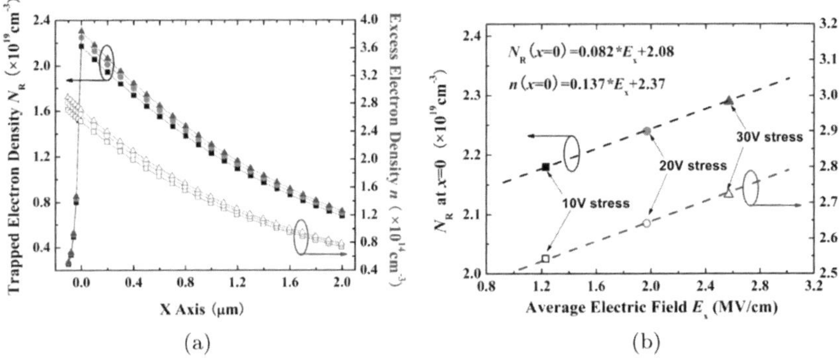

Fig. 8.26. (a) The variation of trapped electron density N_R and excess free electron density n with the position x under stress voltage 10 V (square points), 20 V (round points), and 30 V (triangle points), and (b) the variation of N_R and n at $x = 0$ with average horizontal electric field of high field region "I". The dashed lines show a good linear fitting for the trapped electron and excess electron density with average field.

of excess surface free electrons by leakage current. A simple linear relationship can well describe and fit the peak charge density changing with the average surface horizontal field at the gate corner, as shown in Fig. 8.26(b). The injected excess and trapped electrons increase with the increasing field strength in the high field region and their quantitative relationship provides a reliable basis to analyze the current collapse phenomena in numerical detail.

A HEMT device without field plate and without passivation was measured to verify the current collapse model. The schematic structure and optical microscopy image of the fabricated device are shown in Fig. 8.27(a) and 8.27(b) with device dimensions of $L_g = 2\,\mu$m, $L_{gs} = 5\,\mu$m, $L_{gd} = 5\,\mu$m, and $W_g = 100\,\mu$m. The thickness of Al_2O_3 gate insulator is 15 nm. No passivation or surface cap layer protection is employed in all the devices to be helpful to observe and investigate the current collapse phenomena. With the trapped electron charge distribution modeled, the Sentaurus TCAD tool is used to calculate the I–V characteristics and the transient drain current recovery process of the HEMT devices. Material parameters used in the simulation, such as 2DEG density, electron mobility, and electron saturation velocity, as listed in Table 8.4, are first calibrated by benchmarking with I_d–V_d and I_d–V_g characteristics of HEMT devices under zero stress by laboratory measurement. In the current collapse simulation, settings of various trapped electron density along AlGaN surface are determined by the modeled distribution shape obtained from Fig. 8.26(a). The timings of voltage pulses in performing current collapse measurement are schematically shown in Fig. 8.27(c). During

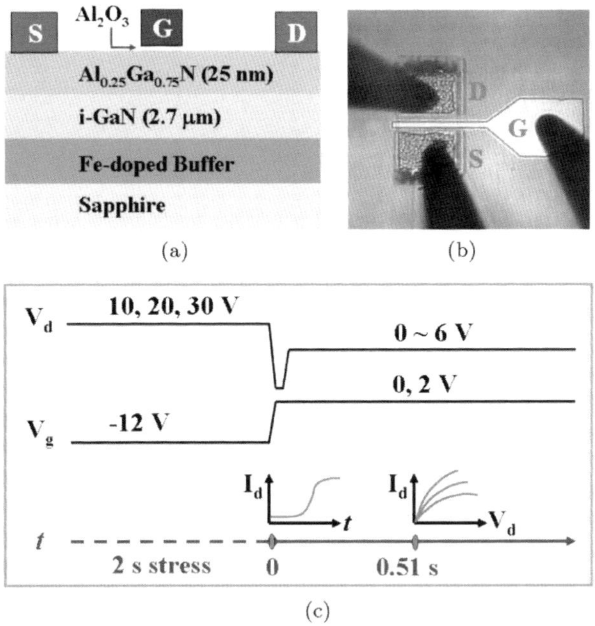

Fig. 8.27. (a) Schematic of AlGaN/GaN HEMT structure, (b) optical microscopy image of the devices in testing, and (c) timings and parameters setting in current collapse measurement. Transient drain current recovery processes are measured immediately after removing 2 s off-state stress bias and I_d–V_d curves are measured at 0.51 s after removing the stress bias.

Table 8.4. Physical parameters used in simulations.

Physical parameters	**Values used in simulations**
Gate metal work function	4.8 eV
2DEG sheet density	$4.6 \times 10^{12}\,cm^{-2}$
Electron mobility in GaN	$800\,cm^2/Vs$
Electron saturation velocity in GaN	1.25×10^7 cm/s
Ionized surface donor-like trap density	$2.5 \times 10^{20}\,cm^{-3}$

the off-state, a gate bias of -12 V is applied with a drain stress bias ranging from 10 V to 30 V for 2 s. After removing the stress drain bias, the on-state drain current is measured at 0.51 s after removing the stress bias. In order to avoid the self-heating effect, a low drain bias of less than 6 V and short-time voltage steps are employed at on-state. In the drain current transient recovery,

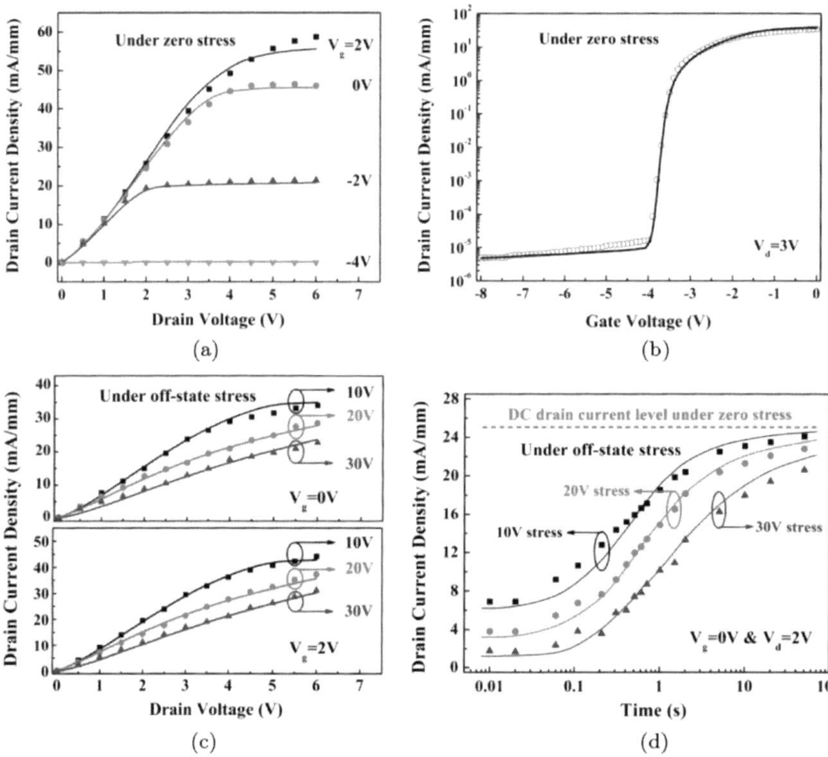

Fig. 8.28. Comparison of simulated (a) I_d–V_d, and (b) I_d–V_g characteristics under zero stress with the experimental data points. Comparison of simulated, (c) I_d–V_d characteristics at $V_g = 0$ or 2 V, and (d) transient drain current recovery process as a function of time after different off-state stress bias with the experimental data points. The flat dashed line in, (d) shows DC drain current level without current collapse. Solid points and solid lines represent the experimental data points and simulation data respectively. Huang *et al.* (2014a).

the drain current at $V_g = 0$ V and $V_d = 2$ V is measured after removing the stress bias.

Figure 8.28 shows the experimental data and analytical data by simulations. Figure 8.28(d) shows the entire drain current recovery process as a function of time after different off-state drain voltage stress. The transient drain current recovery time is long and can be up to several minutes due to slow electron emission from the traps, especially for the case of 30 V voltage stress. Therefore, good surface passivation treatment in device formation is necessary to reduce the current collapse and improve the device switching characteristics.

8.5. Reduction of Current Collapse with Gate Field Plate

The current collapse can be suppressed by approaches such as proper passivation of AlGaN surface and moderation of surface electric field. The good effect of using field plate (FP) design in strip structured HEMTs to reduce current collapse has been experimentally confirmed. As the gate leakage current is controlled strongly by the electric field near the gate corner towards the drain side in the off-state, the electric field reduction can potentially help mitigate this issue. The FP design has been utilized to effectively alleviate the electric field crowding at the gate corner of the drain side. The gate leakage current is hence reduced which is helpful in restraining the current collapse by decreasing the amount of the trapped electrons at the AlGaN surface (as seen in Fig. 8.29). Furthermore, the additional horizontal electric field near the end of the FP will enhance the electron de-trapping process and accelerate the electrons hopping along AlGaN surface which further reduces the trapped electron density and hence results in the notable suppression of current collapse.

The AlGaN/GaN HEMTs without and with FP designs were fabricated with gate width $W_g = 120\,\mu$m and gate length $L_g = 3\,\mu$m. The 20 nm Si_3N_4 and 300 nm SiO_2 bilayers were deposited as FP insulators. Both strip and circular structures were employed to demonstrate the effects of FPs on the dynamic on-resistance and recovery time of current collapse. The schematic of the device structure with specific parameters is shown in Fig. 8.30(a) and the sample optical microscopy images are shown in Fig. 8.30(b) and 8.30(c). Stressed I_d–V_d curves were measured at time = 1 s after 2 s off-state voltage stresses at $V_d = 10$ V and $V_d = 50$ V. The transient on-resistance is calculated by measuring the drain current as a function of time at $V_d = 2$ V and $V_g = 0$ V after stress. The timings in performing current collapse measurements are schematically shown in Fig. 8.30(d).

Figure 8.31 shows the measured on-resistance increase ratios after 2 s off-state gate and drain voltage stress on different FP designs in both strip and circular structures for various L_{gd} and L_{gd} lengths. It is seen that the on-resistance increase ratios are suppressed in all devices with employment of gate FP which is mainly attributed to the reduction of peak surface electric field to get electric field distribution more evenly over a longer distance along the device surface. For a fixed FP insulator type and thickness, the optimized FP length is closely related with L_{gd}. Short FP design is more suitable for the devices with a short L_{gd} due to its more uniform electric field distribution which is beneficial to reduced the surface electron trapping. It is found that employment of 2 μm FP will result in a lower on-resistance increase ratio for the devices with a shorter L_{gd} length such as 7 um, while employment of 4 μm FP can better suppress the increase of the on-resistance in other devices with

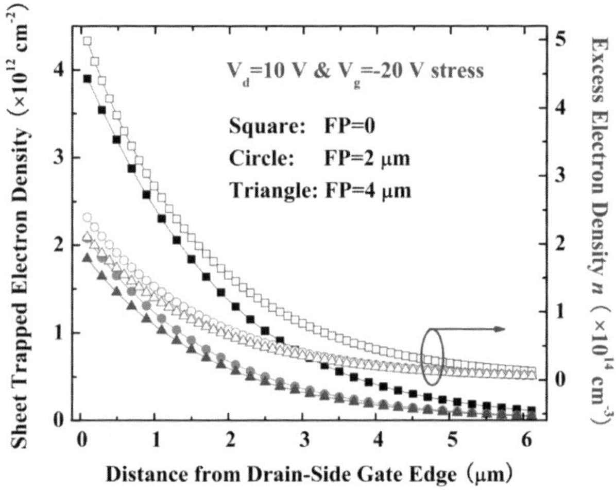

(a)

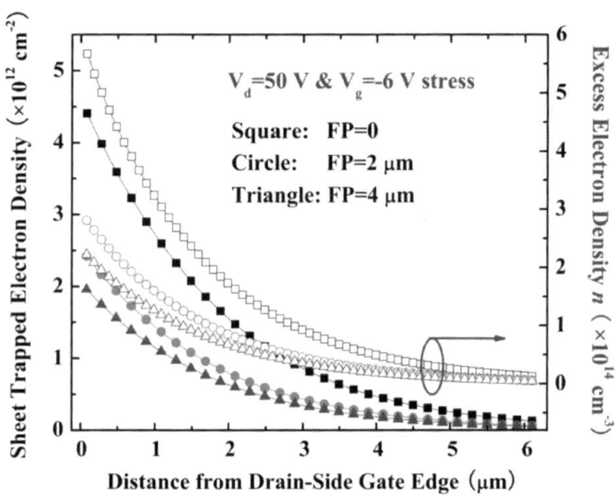

(b)

Fig. 8.29. The variations of sheet trapped electron density and excess free electron density n along the AlGaN surface near the gate towards the drain side under off-state stress for devices with and without field plates. The drain-side gate edge is located at $x = 0$. Huang *et al.* (2014a).

a longer L_{gd}. Without FP design, circular structures will be a better choice in achieving a less increase of the on-resistance. The critical channel area in the circular structures is surrounded by the source and is far away from the mesa edge which leads to a reduced leakage current at the mesa edges and hence a reduced electron trapping amount along the mesa edge.

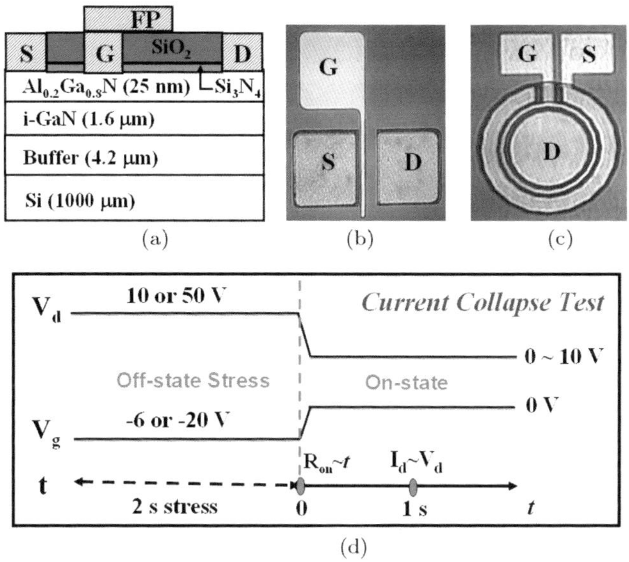

Fig. 8.30. (a) Schematic of AlGaN/GaN HEMTs with gate FP, (b) and (c) optical microscopy images of the devices with strip and circular structures, (d) timings and parameters setting in current collapse measurements. Transient on-resistance recovery was measured immediately after removing 2 s off-state stress bias and device stressed I_d–V_d was measured at 1 s after removing the stress bias.

Figure 8.32 shows the distribution plots of simulated horizontal electric field at AlGaN surface near the gate towards the drain side under drain voltage stress. The number of the injected electrons decreases as the peak electric field is reduced with FP design by alleviating the field crowding at the gate corner which reduces the electron injection and results in a distinct reduction in the increase of the on-resistance. At the same time, the surface field extended by the field plate moves the surface electrons further away from the gate edge. It will also enhance the electrons hopping along AlGaN surface and hence reduce the surface trapped electron density, eventually resulting in a better suppression of current collapse.

Figure 8.33 shows the quasi steady-state I_d–V_d characteristics by simulation by placing proper trapped electron density distribution according to the calculated data by the model. Experimental data points are also shown. Figure 8.34 shows the transient on-resistance recovery process as a function of time after different off-state voltage stresses. The transient drain current recovery time for the device without FP design is long and it can be up to several minutes due to the slow electron de-trapping from the traps. Therefore, good FP design in device formation is necessary to improve the device switching characteristics.

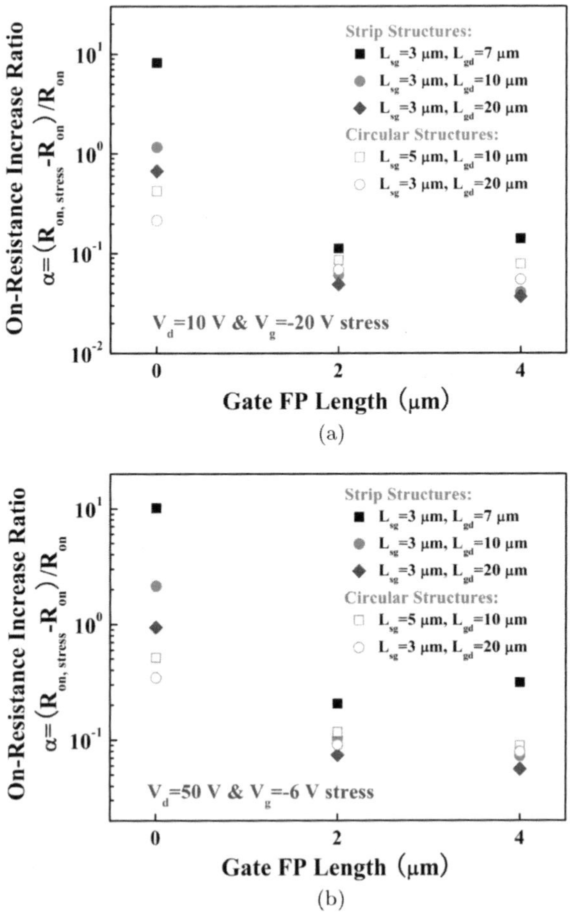

Fig. 8.31. Measured on-resistance increase ratios after 2 s off-state gate and drain voltage stresses on FP length for strip and circular structure devices with various L_{sg} and L_{gd}. R_{on} represents the on-resistance at zero stress and $R_{on,stress}$ represents the on-resistance after off-state bias stress at 10 V and 50 V. Huang *et al.* (2014a).

8.6. Temperature Effects

Stressed I_d–V_d measurements on unpassivated AlGaN/GaN HEMT were carried out at chuck with a heating control unit. Figure 8.35 shows the I_d–V_d characteristics under stressed and unstressed conditions at three different temperatures: 300 K, 400 K, and 500 K. As seen in the figure, the current collapse phenomenon has a strong dependence on temperature. Current collapse

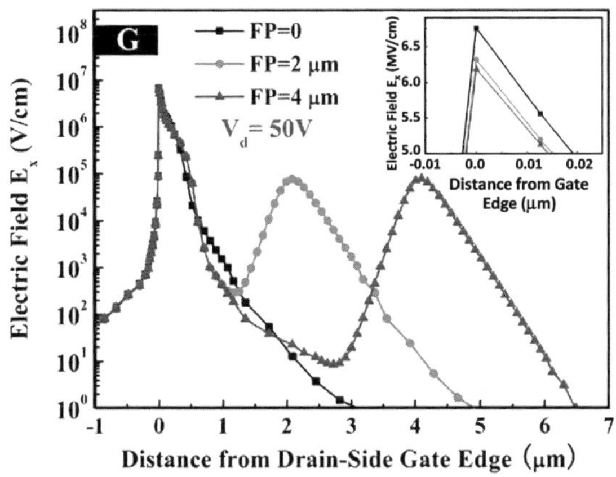

Fig. 8.32. Distribution plots of simulated horizontal electric field at AlGaN surface (within 1 nm depth) near the gate towards the drain side at drain stress of 50 V. The insets show the comparison of the zoom-in peak electric fields for the devices without or with field plates. The drain-side gate edge is located at $x = 0$. Huang *et al.* (2014a).

was more significant at low temperature. When temperature reaches 500 K, nearly no obvious current collapse was observed.

Table 8.5 summarizes the performance degradation of AlGaN/GaN HEMT devices, where ΔR_{on} denotes the on-state resistance increment of the stressed device compared with that prior to stress; $R_{\text{on,DC}}$ denotes the on-state resistance of the device prior to stress; ΔI_{sat} denotes the saturation current reduction of the stressed device compared with that prior to stress; $I_{\text{sat,DC}}$ is the saturation drain current value of the device prior to stress. The variations of on-state resistance and saturation current compared to the steady state value prior to stress are denoted by percentage which illustrates the trend of device performance degradation with temperature variation. As temperature rises from 300 K to 500 K, the increment of the on-state resistance reduces from 12.83% to 0.57%, and the reduction of saturation drain current decreases from 26.32% to 0.01%. It can be concluded that, at a high temperature, the surface trapped charge easily de-trap and the remaining quantity will be much less than that at low temperature.

8.7. Normally-off Operations in AlGaN/GaN HEMTs

Normally-off operation with a high V_{th} is strongly desired in AlGaN/GaN HEMTs for high power switching applications in order to make the devices compatible with the current power electronic systems. Various approaches have

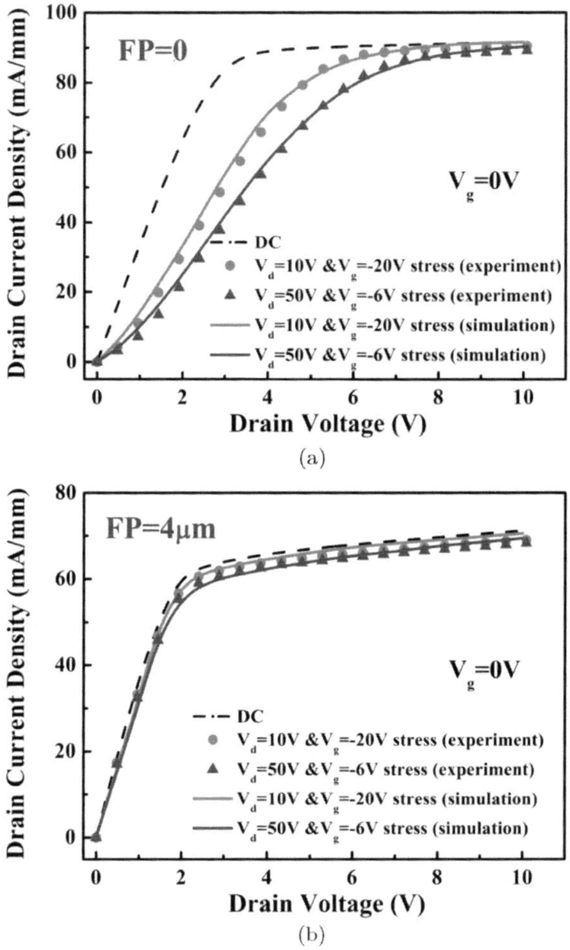

Fig. 8.33. I_d–V_d characteristics of the HEMT devices (a) without and (b) with the 4 μm FP measured after off-state drain voltage stress of 10 V and 50 V. Huang *et al.* (2014a).

been developed to realize the normally-off operation. For example, fluorine implantation in combination with high work function metal gate is introduced to shift the threshold voltage V_{th} towards positive. P-type GaN on AlGaN is also employed to increase V_{th}, but it is typically less than 1.5 V. Gate recess etching is another effective approach to realize the normally-off operation, however, it usually comes at the cost of severe degradation of channel conductance due to the interface scattering, especially when AlGaN barrier leftover thickness is less than 5 nm. Recent works have experimentally confirmed that

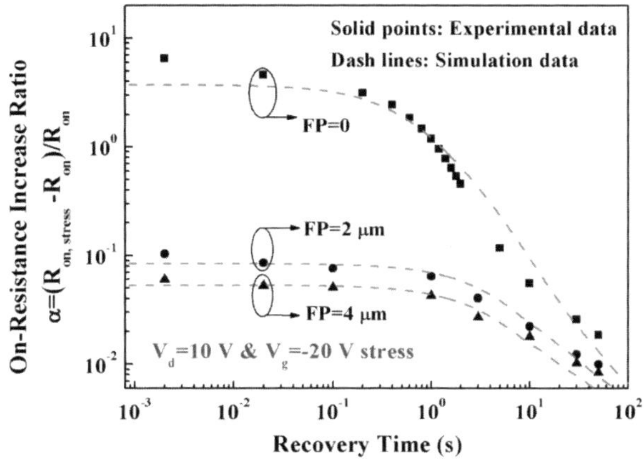

(a)

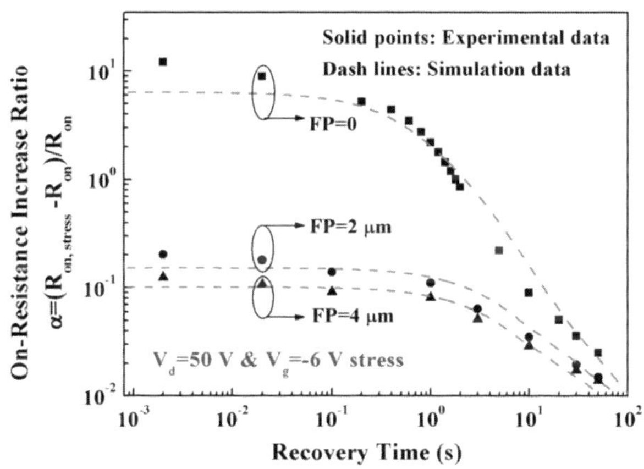

(b)

Fig. 8.34. Transient recovery process of on-resistance after the off-state drain voltage stresses of (a) 10 V and (b) 50 V which were applied for 2 s. On-resistance is calculated at $V_d = 2$ V and $V_g = 0$ V. Huang *et al.* (2014a).

the floating gates or charged traps in the gate dielectric is a promising approach to maintain normally-off mode without reducing the channel conductance. The shortcoming now is mainly on the charge retention limitation at high temperature which affects the operation stability at high power applications.

Employment of partially gate recess etching can reduce 2DEG density without severe degrading in the channel conductance, while gate insulator by plasma ionized treatment can further increase V_{th} to become more positive.

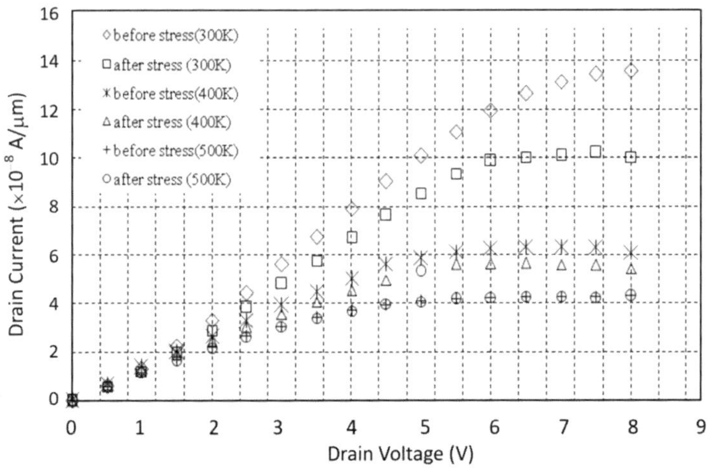

Fig. 8.35. DC (before stress) and pulsed (after stress) characterization of unpassivated HEMT device at 300 K, 400 K, and 500 K.

Table 8.5. Current collapse at different device temperatures.

Temperature (K)	300	350	400	450	500
$\Delta R_{on} / R_{on,DC}$	12.83%	11.01%	7.9%	3.11%	0.57%
$\Delta I_{sat} / I_{sat,DC}$	26.32%	10.80%	9.63%	5.86%	0.01%

Combination of these two approaches will be promising to realize normally-off operation with a good V_{th}. Herewith, the fabrication of Au-free normally-off AlGaN/GaN MIS-HEMTs with partial gate recess and bilayer fluorine treatments on gate dielectric is explained as an example to showcase the mentioned technology. Partial gate recess trench is made to lower the 2DEG density and positively move V_{th} without severe degradation in 2DEG channel mobility. Following that, bilayer fluorine treatments on both the gate dielectric surface and AlGaN barrier surface under very low bias power are employed to further increase V_{th}.

The starting AlGaN/GaN epi-wafer has a p-type (111) Si substrate, a 2-μm buffer layer, a 2-μm undoped GaN, and a 20-nm undoped $Al_{0.25}Ga_{0.75}N$ barrier layer, as shown in Fig. 8.36(a). The sheet 2DEG density of 9.2×10^{12} cm^{-2} with a carrier mobility of 1430 cm^2/Vs was found by room temperature Hall measurements. Mesa isolation was first formed using Cl_2/BCl_3 plasma RIE etching, followed by the source/drain ohmic contact formation by e-beam evaporated Ti/Al/W (40/100/60 nm) annealed at 875°C for 30 s in N_2. A 200-nm

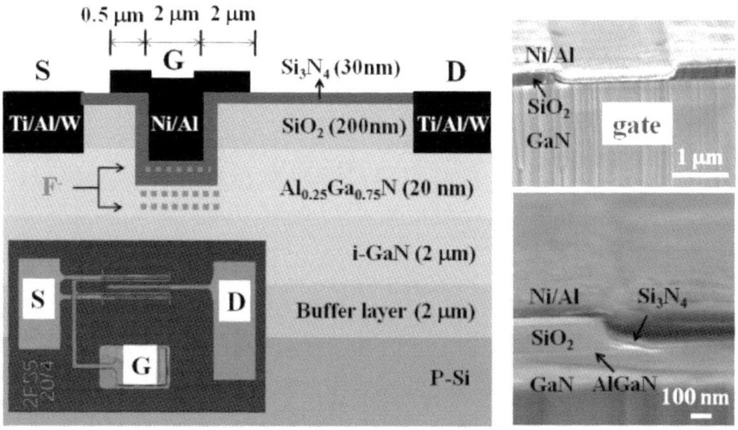

(a) (b)

Fig. 8.36. (a) Cross-sectional schematic of the normally-off AlGaN/GaN MIS-HEMT with the recessed gate and F^- trap charges in AlGaN barrier and gate dielectric layer. The inset shows the optical microscopy image. (b) Cross-sectional SEM images of the device gate and around the gate edge (Huang *et al.*, 2014b).

Table 8.6. Various process conditions (B–E) for the normally-off GaN gate plasma treatments.

Samples	Gate plasma treatment conditions
A	Normally-on (control)
B	(120 s Cl^- + 300 s F^-) at AlGaN
C	(120 s Cl^- + 480 s F^-) at AlGaN
D	(120 s Cl^- + 720 s F^-) at AlGaN
E	(120 s Cl^- + 480 s F^-) at AlGaN and 180 s F^- at Si_3N_4

PECVD-SiO_2 was deposited for surface passivation and a recessed gate window with 2-μm length and 300-μm width was defined. SiO_2 was removed by wet etching followed by Cl^- (Cl_2/Ar at 100 W) and F^- (CHF_3 at 30 W) plasma treatments of different stages and time durations to realize the targeted normally-off operation. Various process conditions A–E are listed in Table 8.6 and the samples are named after the process conditions.

A 30-nm thick Si_3N_4 was deposited by PECVD as the gate dielectric. The idea for this approach is that F^- induced negative-charge is trapped in two stages at the thin AlGaN barrier and the Si_3N_4 layer, respectively, without encroaching into the 2DEG region in order to substantially preserve the channel mobility. To protect the dielectric, a very low power of 30 W was used during the fluorine treatment and the reduction of the gate dielectric thickness was kept within 5 nm. For a fair comparison, 30 s Cl^- treatment for sample A is applied to obtain the similar surface state of AlGaN barrier in the normally-on control device. Ni/Al (50/150 nm) metals were evaporated for the gate contact. Finally, a 600-nm SiO_2 was deposited for device passivation followed by pad window opening and routing using sputtered Al.

Figure 8.36(b) shows the cross-sectional SEM images of the device gate and around the gate edge in which sloped gate sidewall and smooth gate dielectric layer are made. All the samples were made on the same wafer and underwent the same process treatments except for different plasma treatments at the gate region as indicated in Table 8.6.

The Atomic Force Microscopy (AFM) image, as shown in Fig. 8.37(b), was taken right after the AlGaN RIE recess (120 s Cl^- + 480 s F^-). It is used to study the impact of the low-power FPT on the top barrier by comparison with 8.37(a). The AFM analysis indicates a small change in rms value of surface roughness from 0.41 nm to 0.56 nm at the top barrier surface after plasma treatment. This demonstrates 100 W chlorine and 30 W fluorine treatments under the prescribed process conditions are suitable for the device gate recess processing.

Figure 8.38 shows the I_D–V_G transfer characteristics of the fabricated MIS-HEMTs under different Cl^- and F^- plasma treatment conditions. V_{th} values are determined as the gate bias intercept by extrapolating the linear portion of the plot. Typical normally-on device (sample A) with $V_{th} = -5.8$ V is obtained. Samples B–E were under 120 s RIE Cl^- plasma treatment to partially trench

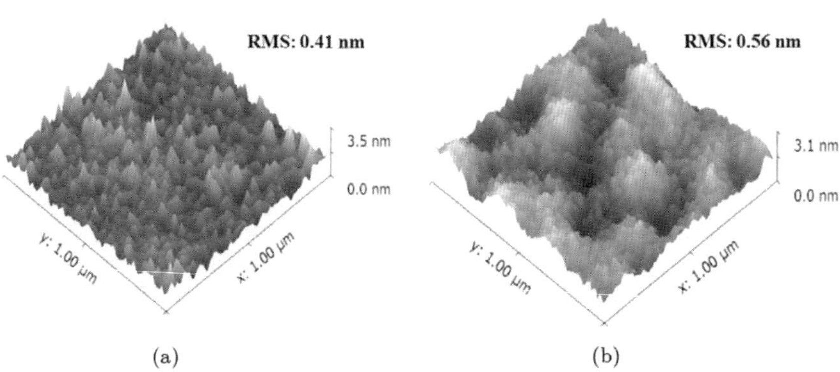

Fig. 8.37. AFM images at the top AlGaN barrier surface (a) without and (b) with low-power chlorine (120 s) and fluorine (480 s) plasma treatments.

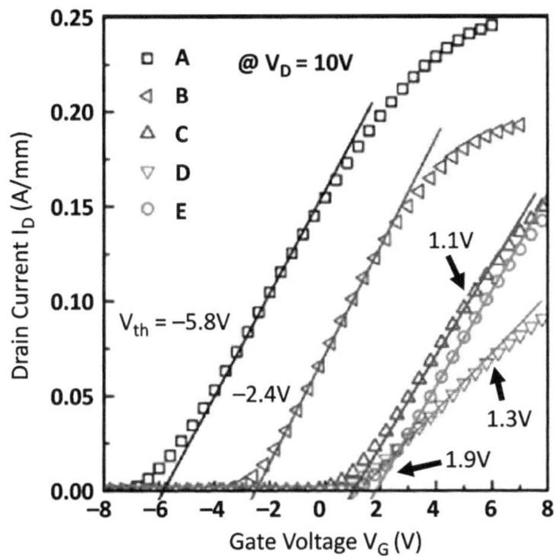

Fig. 8.38. Transfer I_D–V_G curves of the fabricated MIS-HEMTs measured at $V_D = 10$ V.

the AlGaN barrier for about 10.6 nm. Various RIE F^- plasma treatments at low bias power were used to further enhance the positive V_{th} for samples B–E. Positive V_{th} of 1.1 V and 1.3 V were achieved by combining partial gate recess and F^- treatments in AlGaN barrier for 480 s and 720 s, respectively. It is worth to mention that sample E under 480 s F^- plasma treatment at the AlGaN layer plus 180 s F^- plasma treatment at a low power on Si_3N_4 layer gives the best V_{th} (=1.9 V) and no obvious degradation of drain current owing to F^- trap charge accumulation only on the Si_3N_4 surface layer and no penetration into 2DEG region under a very low power condition.

Figure 8.39 gives the performance comparison with other reported Au-free HEMTs. The performance of the fabricated normally-off MIS-HEMTs is attractive by having maintained the drain current similar to the normally-on sample while achieving a high V_{th} of 1.9 V at the same time. It is clearly feasible to fabricate Au-free normally-off HEMTs by employing the partial gate recess with two-stage trapped-charge gate structure.

8.8. High Threshold Voltage Normally-off MIS-HEMTs

The gate threshold voltage of around 2 V may be insufficient for power electronic applications. The switching of inductive circuit induces voltage transients which can interfere in the normal operations of the switching device if its

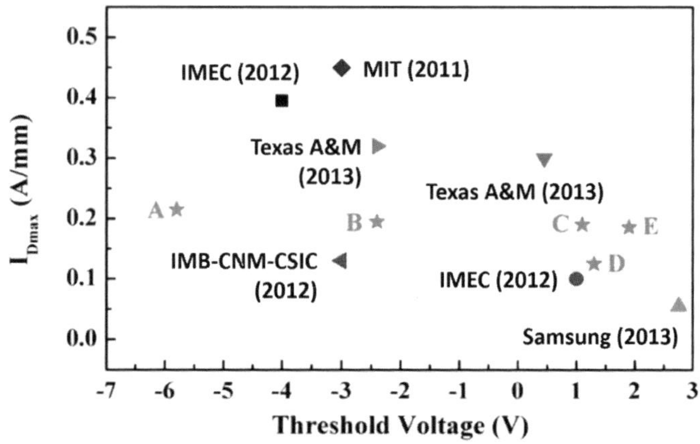

Fig. 8.39. Performance comparison among Au-free GaN-based HEMTs. The stars (A–E) represent the measured current ratings of devices made of processes A–E (Table 8.6) and their corresponding threshold voltages.

threshold voltage is insufficient. For a practical consideration, it is desirous and necessary to have a high threshold voltage for normally-off power MIS-HEMT devices.

Partial AlGaN barrier trench (~50%) using low power chlorine-based ICP or RIE etching can effectively reduce 2DEG density and shift the V_{th} towards higher positive without obvious degradation in 2DEG channel mobility. Multiple FPT on the barrier surface or within the gate dielectric under low RIE power will keep the fluorine ions near the material surface and at the same time it increases the V_{th} to be more positive without introducing too much fluorine damage in the 2DEG channel. The technique can achieve both a high threshold voltage of above 5 V and a low gate leakage current at pA/mm level.

The cross-sectional schematic, fabrication flow diagram, and optical microscopy image of the fabricated MIS-HEMTs are shown in Fig. 8.40. The fabrication process begins with the mesa isolation by selectively etching down 300 nm of the epitaxial layers. Source and drain ohmic contacts are achieved by depositing Ti/Al/Ni/Au (25/125/45/55 nm) metals using E-beam system and annealing at 850°C for 30 s in N_2. A 150 nm SiO_2 is deposited by PECVD for surface passivation. Then a recessed gate window of 3 μm length and 200 μm width is defined by photolithography. Gate trench of 10 nm on AlGaN barrier (~50% thickness) is realized by Cl^- plasma treatment in ICP-RIE system, followed by RIE F^- plasma treatment for 180 s under a low power of 30 W.

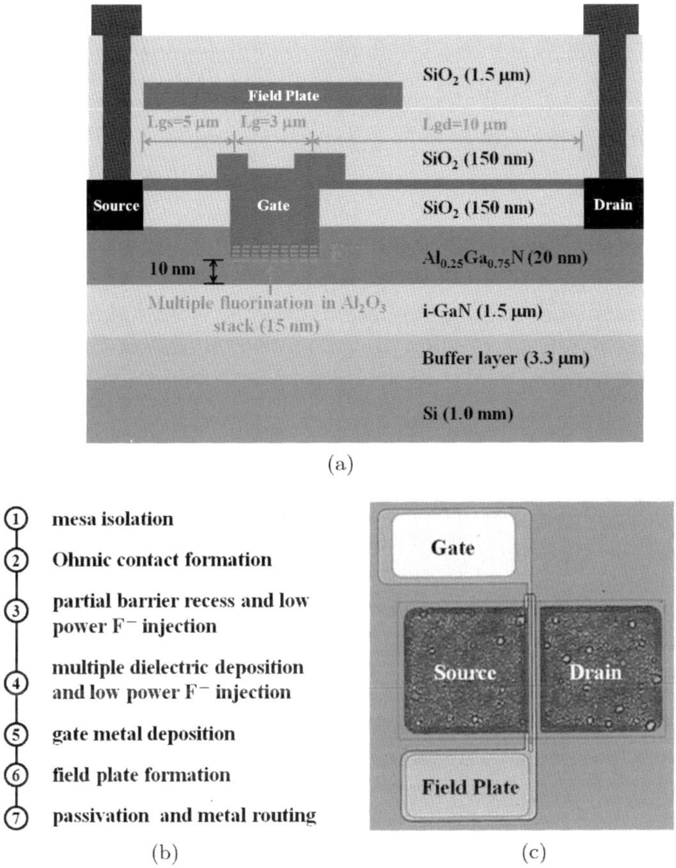

Fig. 8.40. (a) Device schematic with multiple layers of fluorinated dielectrics for gate formation, (b) fabrication process flow, and (c) optical microscopy images of the proposed normally-off AlGaN/GaN MIS-HEMTs with the triple fluorinated dielectrics.

The innovative idea in this technique is that low power (30 W) RIE fluorination plasma treatment with minimum ionic scattering damage was implemented after the deposition of each ALD-Al_2O_3 gate dielectric layer. The ALD growth and subsequent RIE processes are carried out for three rounds, followed by a good ALD-Al_2O_3 cap on the top without any plasma treatment. The total Al_2O_3 thickness is kept around 15 nm. Ni/Au/Ti metals are deposited by an E-beam evaporator system as the gate electrode, and then 350 nm PECVD-SiO_2 layer is deposited on the top to serve as the field plate dielectric. Thick Ti/Au/Ti metal is deposited as the field plate and 1.5 μm PECVD-SiO_2 is employed for the passivation dielectric. Finally, pad windows are opened and

Ti/Au metal routing is carried out to complete the fabrication. The optical microscopy image of the fabricated normally-off AlGaN/GaN MIS-HEMTs is shown in Fig. 8.40(c). The various field plate combinations such as gate extension of 1.5 μm and 2.5 μm toward drain and field plate length of 1.5 μm, 2.5 μm, and 4.5 μm are optimized in the design process (Wang *et al.*, 2015a).

Among the five die samples, Die A is the normally-on device; Die B is trench-only structure where half of the Al_2O_3 layer is removed to reduce the induced 2DEG concentration; Dies C, D, and E are treated with RIE fluorination process on Al_2O_3 dielectric layer with different doses as indicated in Table 8.7. The dielectric layer thickness is kept the same at around 18 nm. The total equivalent amount of fluorine ions acting at the Al_2O_3/AlGaN interface for the electrostatic effect can be modeled by the following equation:

$$Q_{\mathrm{EQ}} = Q_1 + Q_2\frac{d_2+d_3+d_4}{d_1+d_2+d_3+d_4} + Q_3\frac{d_3+d_4}{d_1+d_2+d_3+d_4} + Q_4\frac{d_4}{d_1+d_2+d_3+d_4}, \quad (8.21)$$

where Q_1 is the amount of charge located right at the Al_2O_3/AlGaN interface, Q_2 is the amount of charge located at a distance d_1 away from the interface, so on and so forth for Q_3 and Q_4.

The threshold voltage can then be modeled after knowing the total equivalent fluorination charge Q_{EQ}, as below:

$$V_{\mathrm{th}} = \frac{\Phi_{\mathrm{b}} - \Delta E_1 - \Delta E_2 - \Delta E_3}{q} - \frac{qQ_{\mathrm{2DEG}}(t_{\mathrm{AlGaN}} - t_{\mathrm{recess}})}{\varepsilon_{\mathrm{AlGaN}}} - \frac{q(Q_{\mathrm{2DEG}} + Q_{\mathrm{EQ}} + Q_{\mathrm{T}})}{\varepsilon_{\mathrm{Al_2O_3}}}\sum_{\mathrm{n}=1}^{4} t_{\mathrm{n}}, \quad (8.22)$$

Table 8.7. Fluorination treatments on Dies C, D and E.

Sample Dies	**C**	**D**	**E**
1st F^- treatment conc. [Q_1] (cm^{-2})	-1.40×10^{12}	0	0
2nd F^- treatment conc. [Q_2] (cm^{-2})	-8.60×10^{12}	-8.60×10^{12}	-6.51×10^{12}
3rd F^- treatment conc. [Q_3] (cm^{-2})	-8.60×10^{12}	-8.60×10^{12}	-1.21×10^{13}
4th F^- treatment conc. [Q_4] (cm^{-2})	-4.19×10^{12}	-1.12×10^{13}	-1.30×10^{13}
Equivalent F^- conc. [Q_{EQ}] (cm^{-2})	-1.52×10^{13}	-1.69×10^{13}	-1.82×10^{13}
ALD-Al_2O_3 d_1, d_2, d_3, d_4 (nm)	7.4, 6, 4.3, 8	7.4, 6, 7.6, 8	6.5, 7.3, 7.6, 8
Total dielectric thickness after RIE fluorination treatments (nm)	18.00	18.15	18.10

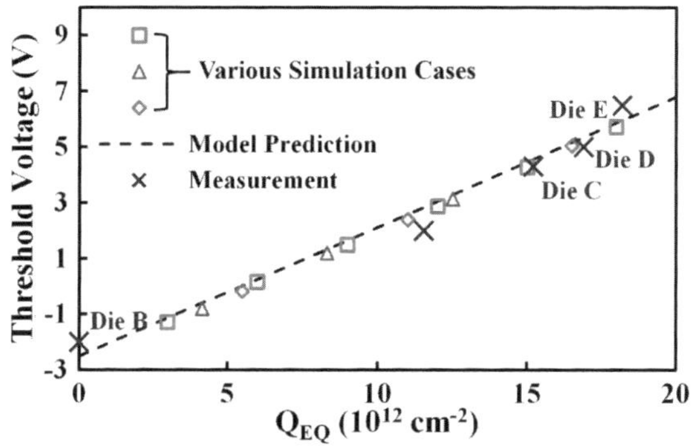

Fig. 8.41. The relationship between threshold voltage (V_{th}) and the total equivalent fluorine charge (Q_{EQ}). The data by model prediction (dash line), Sentaurus simulations (square, triangle, and diamond symbols), and laboratory measurements (cross symbol).

where the barrier between Ni and Al_2O_3 is $\Phi_b = 3.5\,eV$, ΔE_1, and ΔE_3 are the conduction band offset between Al_2O_3/AlGaN and AlGaN/GaN respectively, which add up to be 2.1 eV. ΔE_2 is the conduction band offset from the Fermi level before fluorine treatments, which is $-0.16\,eV$. Permittivities are $\varepsilon_{AlGaN} = 9.2\varepsilon_0$ and $\varepsilon_{Al_2O_3} = 7\varepsilon_0$. Q_{2DEG} is around $6 \times 10^{12}\,cm^{-2}$ for the 2DEG density at AlGaN/GaN interface when AlGaN thickness is at 10 nm. Figure 8.41 gives the threshold voltage values obtained by theoretical calculation by Eq. (8.22), by the Sentaurus simulation and by the measurements on fabricated devices. In general, they match quite well.

Figure 8.42 shows the measured I_D–V_G, and g_m of various samples fabricated, namely Dies A–E at $V_D = 1\,V$. The threshold voltage is obtained by linearly extrapolating the tangent at g_m maximum. The recess-only Die B shifts the V_{th} from -4.8 V to -2.0 V, showing effectiveness of partial recess to 2DEG density reduction. The fluorination plasma treated Dies C–E shift the V_{th} to various positive values, where Die E has the highest V_{th} of $+6.5$ V that agrees with the design. Consistent V_{th} measurements in saturation region are also obtained for all dies. Meanwhile, the maximum g_m for Die E (28 mS/mm) is similar to the untreated normally-on Die A (31.9 mS/mm), indicating very trivial degradation on the sub-threshold swing.

Figure 8.43 shows the I_D–V_G and I_G–V_G characteristics of Dies A–E in logarithmic scale. Very low gate leakage of 10^{-6} mA/mm at $V_G = 12\,V$ is observed for Die E, indicating that the quality of the multilayer fluorinated gate dielectric was preserved. However, F^- ions treatment on the AlGaN barrier (Die C) provides a path for trap-assisted leakage and thus worsens the gate quality. Good stability on V_{th} and I_G has been verified experimentally

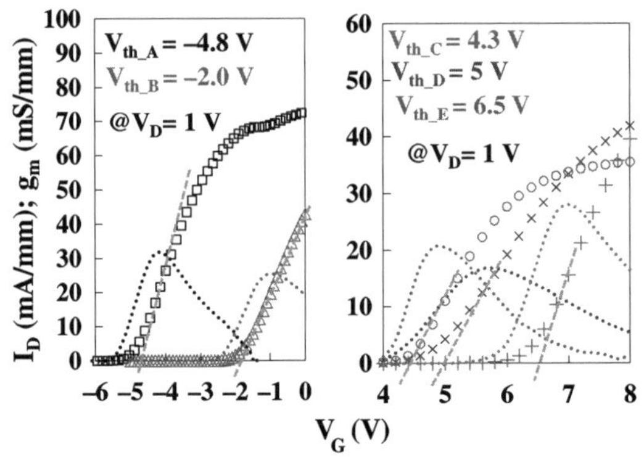

Fig. 8.42. Comparison of measured I_D–V_G (data points) on devices Dies A–E at $V_D = 1$ V. V_{th} is obtained by linear extrapolating the I_D–V_G curves at the maximum transconductance (dotted lines). Wang *et al.* (2015a).

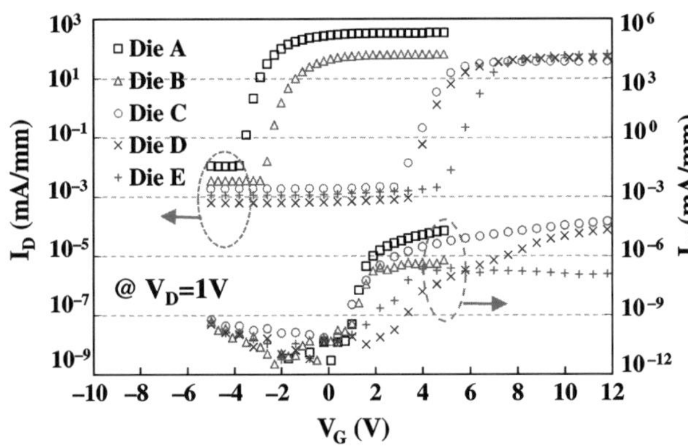

Fig. 8.43. I_D–V_G characteristics and I_G–V_G of Dies A–E at $V_D = 1$ V in logarithmic scale. Low off-state gate leakage current of 10^{-6} mA/mm is observed for Die E. Wang *et al.* (2015a).

through repetitive measurements. A moderate 60% increase in the dynamic on-resistance (R_{on}) was measured after off-biasing V_D at 200 V for 10 s.

Figure 8.44(a) shows the on-state I_D–V_D curves of Dies A–E at gate overdrive voltage ($V_{GT} = V_G - V_{th}$) of 10 V. A moderate 29% degradation in I_{DMAX} (from 490 mA/mm to 346 mA/mm) is observed for Dies D and E compared to normally-on Die A. However, about 60% degradation is observed on Die C, indicating that the F^- treatment directly on the AlGaN surface may have deteriorated the AlGaN/GaN interface and degraded the 2DEG carrier

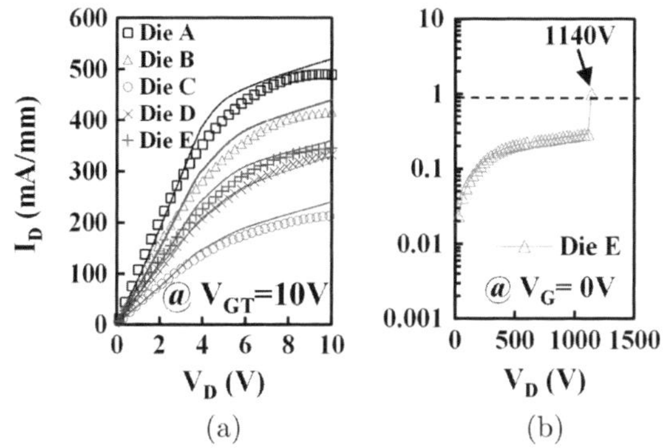

Fig. 8.44. (a) On-state I_D–V_D measured data points and simulation (solid lines) of Dies A–E at $V_{GT} = 10$ V. (b) Breakdown characteristics for Dies E at $V_G = 0$ V. Wang *et al.* (2015a).

Table 8.8. The 2DEG electron mobility for Dies A–E from simulations.

Dies	**A**	**B**	**C**	**D**	**E**
Mobility (cm^2/Vs)	1450	1200	600	1000	1050

mobility. The 2DEG carrier mobilities at the gate region shown in Table 8.8 are obtained from the fitting by Sentaurus simulations for Dies A–E in Fig. 8.44(a) matching the on-state I–V characteristics. The low 2DEG mobility in Die C is observed. In Fig. 8.44(b), the off-state breakdown characteristics of Dies E with $L_{GD} = 15\,\mu$m and $L_{FP} = 2.5\,\mu$m at $V_G = 0$ V are shown. A maximum breakdown voltage of 1140 V is obtained for Die E. It agreed with the designed rating confirming the suitable gate formation without any premature breakdown.

8.9. Argon Pre-processed Fluorination Plasma Treatment

Fluorination Plasma Treatment (FPT) is a promising gate processing method applied to achieve a high threshold voltage for normally-off operations. As described in Sec. 8.8, by applying multiple RIE based FPTs on Al_2O_3 gate dielectric along with partial gate recess, it can introduce a large amount of trapped negative charge in the gate dielectric which can then achieve a large V_{th}, e.g. as high as 6.5 V with a satisfactory I_{DMAX}. However, the multiple low-power FPT process is unable to provide sufficient deep-level trap states that can keep electrons staying at the trapped states at high temperature. This causes a major disadvantage as GaN power device will need to handle high voltage and high current switching at high temperature. In this section, an improved thermal stability scheme is proposed for gate engineering. A short-period of Argon (Ar) plasma treatment (APT) prior to the FPT is applied to effectively induce trap states at deeper energy levels within the Al_2O_3 dielectrics. A single ICP-RIE based FPT follows after APT to complete the process. These fluorination-induced negative charges at the deeper trap states can provide better thermal stability.

The mechanism of V_{th} shift can be explained by the conduction band energy diagram along the gate stack region as shown in Fig. 8.45. In this figure, the barrier Φ_b between Ni and Al_2O_3 is 3.5 eV. E_1 and E_3 are the conduction band offset between Al_2O_3/AlGaN and AlGaN/GaN respectively, totally 2.1 eV. E_2 is the conduction band offset from the Fermi level without fluorine treatment, which is −0.16 eV. When there is no FPT applied to the Al_2O_3 gate dielectrics, the conduction band at the AlGaN/GaN interface is below the Fermi level (E_F) and the 2DEG is induced. Therefore, V_{th} is negative due to the intrinsic existence of the 2DEG. When the Al_2O_3 gate dielectric is fluorinated, significant amount of negative charge will reside in the Al_2O_3 at the trap sites and shift the conduction band at the AlGaN/GaN interface above the Fermi level. Hence, the 2DEG will be depleted by the negative charge at the gate

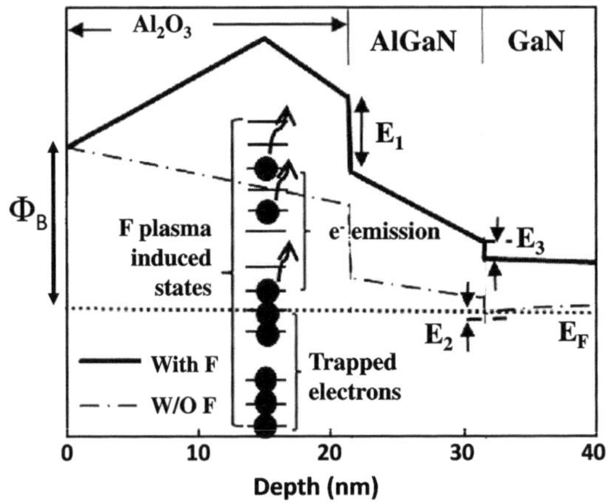

Fig. 8.45. The conduction band energy (E_C) schematic diagram along the gate stack region for devices with (solid line) and without (dashed line) FPT at $V_G = 0$ V.

region and the device will have its positive threshold voltage. It is notable that if these negative charges are resided above the E_F, it will be emitted within a period of time. The emission time constant is directly related to the trap energy level that the charge is resided, which will be further discussed.

The RF plasma power controls the state of fluorine radicals bombarding the sample surface. It increases the total amount of trap charges within Al_2O_3 and also extends the trap sites' energy levels to be slightly deeper. However, high RF power in the RIE system will have the negative impact to damage the 2DEG channel quality. On the other hand, the ICP-RIE based FPT is able to control the amount of dissociated fluorine radicals and the ion bombardment energy independently by the coil and cathode power settings, respectively. To further push the traps towards deeper energy levels during FPT, a short-period of APT is carried out on the gate dielectric surface prior to the FPT. Such a process induces shallow surface damage with broken Al–O bonds which can increase the amount of Al–F bonds and make the associated trap energy levels deeper than those without APT.

The fabrication process flow and the cross-sectional schematic of the device are shown in Figs. 8.46 and 8.47. The prototype device has L_G, L_{GS}, L_{FP}, and L_{GD} of 3 μm, 5 μm, 1.5 μm, and 5 μm, respectively. The AlGaN/GaN-on-Si wafer has the 2DEG carrier density and mobility of $8.5 \times 10^{12}\,cm^{-2}$ and $1450\,cm^2/Vs$ respectively. The process begins with mesa isolation by BCl_3-based ICP-RIE and SiO_2 dielectric deposition by PECVD. Ti/Al/Ni/Au (25/125/45/55 nm) source/drain ohmic contacts are made by RTA at 850°C for 30 s. About 10 nm (50%) of AlGaN is removed to define the gate region by low power BCl_3-based ICP-RIE to reduce the 2DEG concentration without damaging the AlGaN/GaN interface, and to preserve carrier mobility in the 2DEG channel. This is followed by 6.5 nm of ALD-Al_2O_3 dielectric layer deposition made at 250°C. The device goes through a 20 s ICP-Ar plasma treatment with a coil power of 50–100 W. Then, ICP-CHF_3 plasma treatment is carried out with cathode/coil power settings of 10/200 W for 4 min. The process parameters used for the gate formation and the resulting surface roughness rms obtained from the AFM are listed in Table 8.9 and Fig. 8.48. The APT recipe implemented is based on the existing Al_2O_3 etching recipe using $BCl_3/Cl_2/Ar$ ICP-RIE. It uses the same coil power and Ar gas flow as the Al_2O_3 etching recipe to ensure the amount of dissociated Ar^+ is sufficient. Minor adjustment of the cathode power from the etching recipe can be made to seek for optimized results. It is found that by controlling the cathode power of the Argon treatment below 75 W, the device is able to preserve the surface roughness rms within 1 nm. Finally, Ni/Au (15/150 nm) gate-metal deposition follows with annealing at 400°C for 5 min.

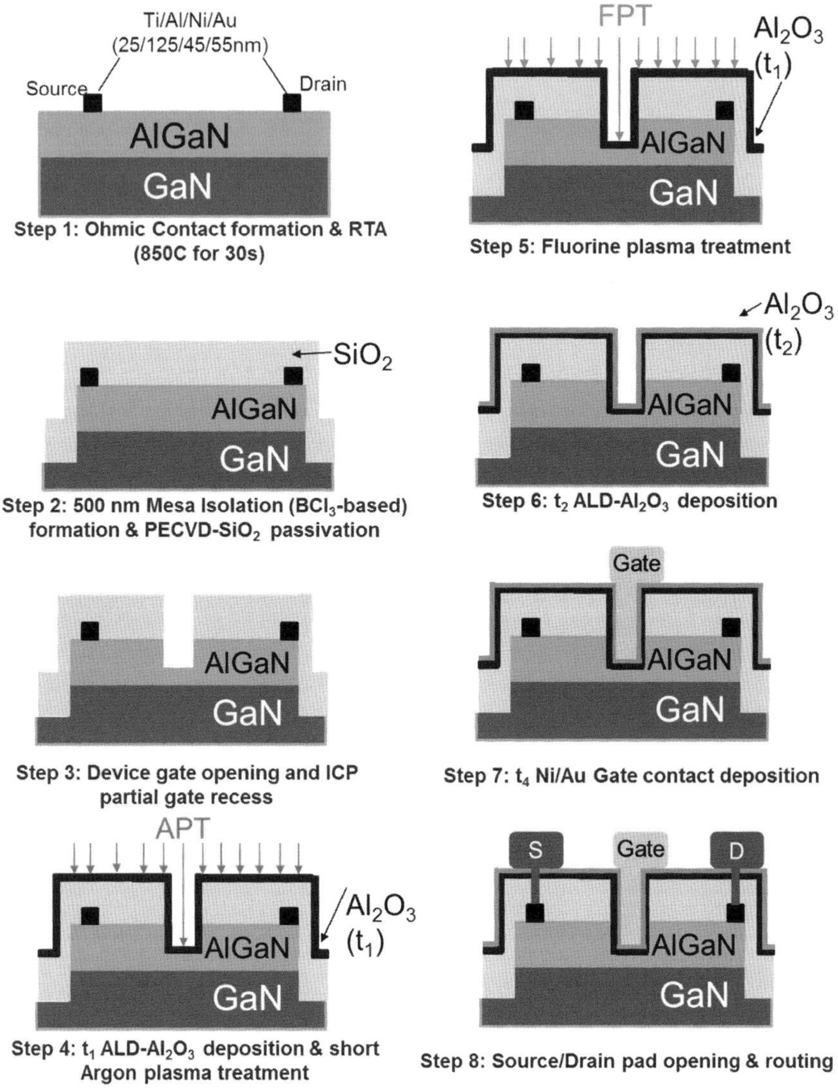

Fig. 8.46. The fabrication process flow of a normally-off Al_2O_3/AlGaN/GaN MIS-HEMT with the Argon pre-fluorination treatment (APT then FPT) gate process.

Figure 8.49(a) compared the measured I_D–V_G characteristics of the prototype Dies A and B without APT, and Fig. 8.49(b) for Dies C, D, and E with APT. A V_{th} of 4.4 V is obtained for Die D with the Argon coil power at 75 W. By increasing the Argon coil power to 100 W (Die E), it is able to further increase

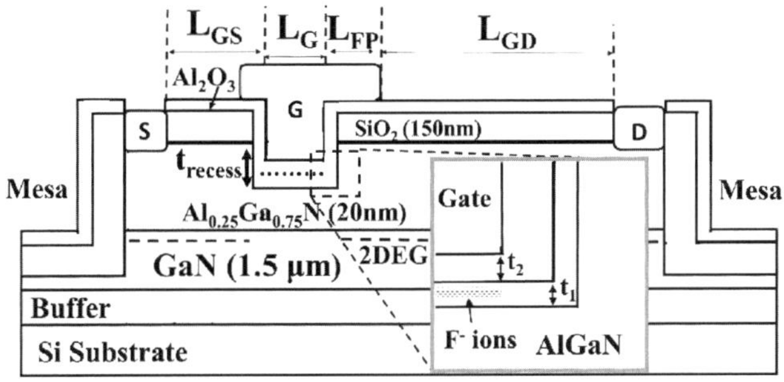

Fig. 8.47. The device cross-sectional schematics of the normally-off AlGaN/GaN MIS-HEMT. The inset magnifies the gate region, where the fluorine-induced charges (F^-) and the thickness of gate dielectric stack (t_1 and t_2) are shown.

Table 8.9. Processing parameters for the gate dielectric of prototypes: Dies A–E.

	Die A	Die B	Die C	Die D	Die E
ALD-Al_2O_3 Thickness (t_1) (nm)			6.5		
Argon Coil Power (W)	N/A	N/A	50	75	100
Fluorination Coil Power (W)	N/A		200		
ALD-Al_2O_3 Thickness (t_2) (nm)			15		
AFM surface RMS Roughness (nm)	0.38	0.53	0.61	0.78	2.29

the V_{th} to 5 V. The preservation of gate leakage can be seen in Fig. 8.49(c) with moderate Argon coil power applied. In Fig. 8.49(d), a comparable I_{DMAX} of about 320 mA/mm is measured for Dies B–D, showing the preservation of 2DEG channel conductivity after Argon bombardment and FPT. As for the Die E, its Argon coil power is too high which roughens the treated surface, damages the channel, and worsens the I_{DMAX} and the gate leakage current.

According to Fig. 8.50(a), proper APT can also reduce the ratio between the dynamic on-state resistance and static on-state resistance, $R_{dynamic}/R_{on}$ where $R_{dynamic}$ is the current collapse dynamic resistance measured after drain stress applied for 10 s at off state. It can improve the breakdown voltage as well as shown in Fig. 8.50(b) measured at $V_G = -5$ V for untreated Die A and $V_G = 0$ V for plasma-treated Dies B–E. These improvements imply the effective reduction of the shallow trap states by the APT process. For instance, Die D is able to have about 7.3% smaller $R_{dynamic}/R_{on}$ (1.27 versus 1.37) and 40% higher breakdown voltage (940 V versus 670 V) than Die B for FPT-only.

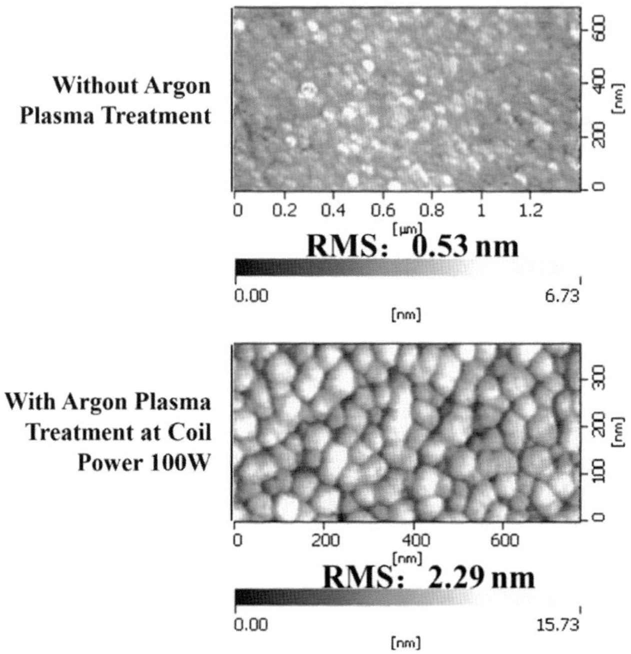

Fig. 8.48. AFM image of the Al_2O_3 surface with and without Argon plasma treatment (top: without APT; bottom: with APT at coilpower of 100 W).

However, if the APT coil power is too high (Die E), the off-state leakage will significantly increase by about one order of magnitude and the breakdown voltage will drop to 800 V. Therefore, a good control of the coil power for the APT process is necessary to ensure a good performance for APT-FPT GaN normally-off devices.

8.10. High Temperature Threshold Voltage Stability

The sweep of I_D–V_G hysteresis of Die B (FPT only) and Die C (APT+FPT) at 25°C and 200°C are shown in Figs. 8.51(a) and (b). At 200°C, much smaller hysteresis of about 0.8 V for Device C is obtained than 2.2 V for Device B. This is about 63.6% reduction in hysteresis. It indicates that less population of shallow traps within Al_2O_3, that emitted trapped charge at 200°C, is observed when the gate dielectric is treated with APT and FPT.

To further elaborate the V_{th} thermal stability and its relationship with trap distribution, the high temperature V_{th} stability of Dies B and D is reported

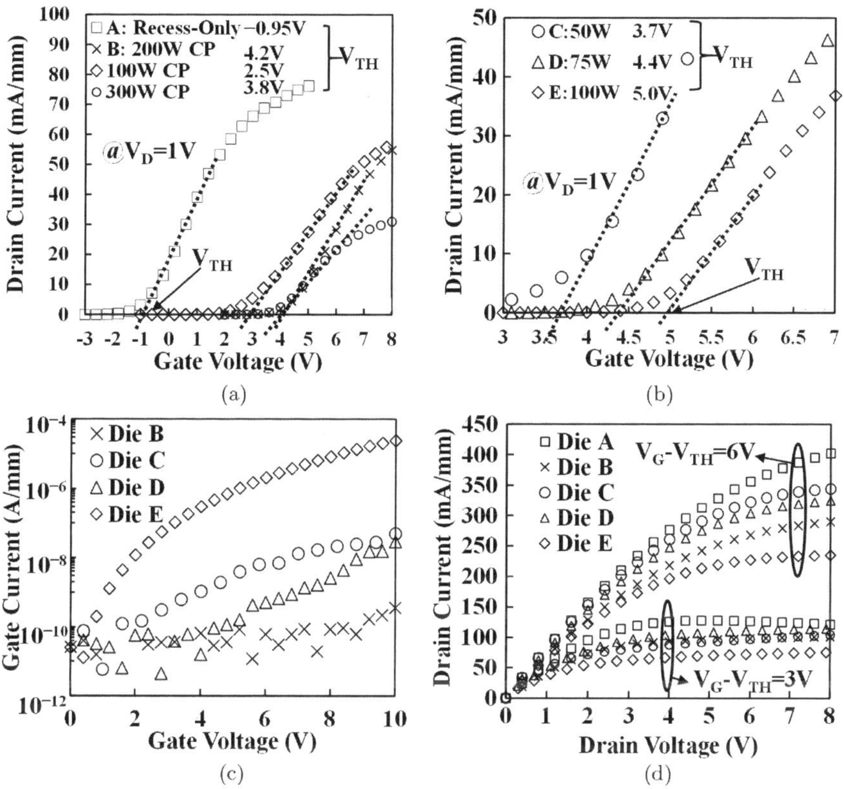

Fig. 8.49. Measurements of (a) I_D–V_G of Dies A and B with different ICP-fluorination coil power of 100–300 W and, (b) I_D–V_G of Dies C–E with Argon plasma pre-treatment, (c) I_G–V_G at $V_D = 1$ V, (d) I_D–V_D of Dies A–E with $V_G - V_{th} = 3$ V and 6 V.

in Fig. 8.52(a) and 8.52(b). The pre-FPT APT is able to improve the V_{th} at 200°C significantly to be at 2.5 V, implying the negative charges induced from FPT at Die D are mostly residing at deeper level within the Al_2O_3 energy band.

Specifically, the electron emission levels at fluorinated traps corresponding to different temperature can be modeled by Eq. (8.23), where E_{T_MAX} is the deepest energy level of the trapped charge to be emitted from, k_b is the Boltzmann's constant, T is the device temperature. $\gamma_n = (\nu_{th}/T^{0.5})(N_c/T^{1.5}) = 3.25 \times 10^{21}(m_n/m_0)$ where m_n/m_0 (=0.16) is the relative electron effective mass within Al_2O_3, ν_{th} is the thermal velocity of Al_2O_3 obtainable from $\nu_{th} = (3k_bT/(m_n m_0))^{0.5}$, N_C is the effective density of states in Al_2O_3 obtainable from $N_C = 2(2\pi m_n m_0 k_b T/h^2)^{1.5}$, h is the Planck's constant. σ_{Al2O3} is

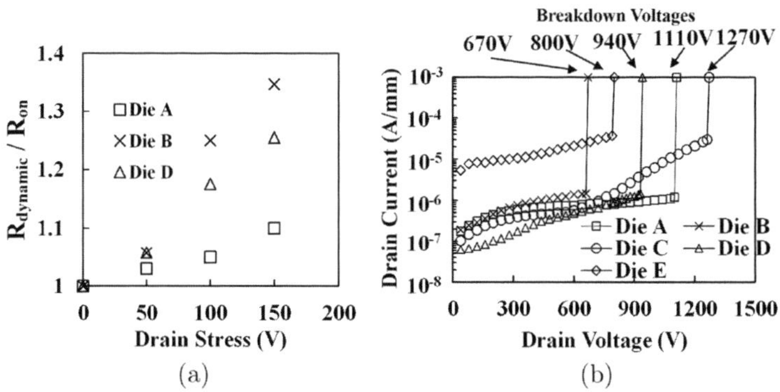

Fig. 8.50. (a) The measured ratio between the dynamic on-resistance ($R_{dynamic}$) and the static on-resistance (R_{on}) of Dies A, B, and D after drain stress applied for 10 s at off-state, (b) off-state I_D–V_D breakdown measurement of Dies B–E at $V_G = 0$ V and Die A at $V_G = -5$ V.

the capture cross-section of the electrons in Al_2O_3, which is about 10^{-16} cm^2, and t = 50 s is the device heating time to allow for device surface temperature stabilization to reach quasi-steady state. It is reasonable to assume that all of the charges trapped in levels shallower than E_{T_MAX} will be emitted by thermal energy, and those at levels deeper than E_{T_MAX} will remain trapped. Once the E_{T_MAX} for a given temperature is found as indicated in Fig. 8.52(d), the amount of sheet negative charge (Q_N) remaining within the Al_2O_3 can be calculated from the change in V_{th} referring to the V_{th} at 25°C and the V_{th} model depicted in Eq. (8.24). In Eq. (8.24), the barrier between Ni and Al_2O_3 is $\Phi_b = 3.5$ eV, and ΔE_1 and ΔE_3 are the conduction band offset between AlGaN/GaN and Al_2O_3/AlGaN respectively, totally 2.1 eV. ΔE_2 is the offset of the conduction band from the Fermi level before FPTs, which is -0.16 eV. Permittivities of AlGaN and Al_2O_3 are $\varepsilon_{AlGaN} = 9.2\varepsilon_0$ and $\varepsilon_{Al2O3} = 7\varepsilon_0$, respectively. The second and the third terms in Eq. (8.24) illustrate the electrostatic potentials across the AlGaN and Al_2O_3 layers, respectively. $Q_{2DEG} = 6 \times 10^{12}$ cm^{-2} is the 2DEG density at AlGaN/GaN interface when the thickness of AlGaN is at 10 nm. $Q_T = -1.5 \times 10^{12}$ cm^{-2} is the fixed charge density at Al_2O_3/AlGaN interface Wang *et al.* (2016).

To estimate densities (D_N) of trap states formed by plasma treatments in Al_2O_3 bandgap, the pulsed I_D–V_G measurement by stressing the gate at different levels of negative gate base voltage (V_B) was carried out (Fig. 8.52(c)). Based on the Poole–Frenkel trap emission theory shown in Eq. (8.25), V_B is able to provide an external electric field (ξ) that enhances the emission of

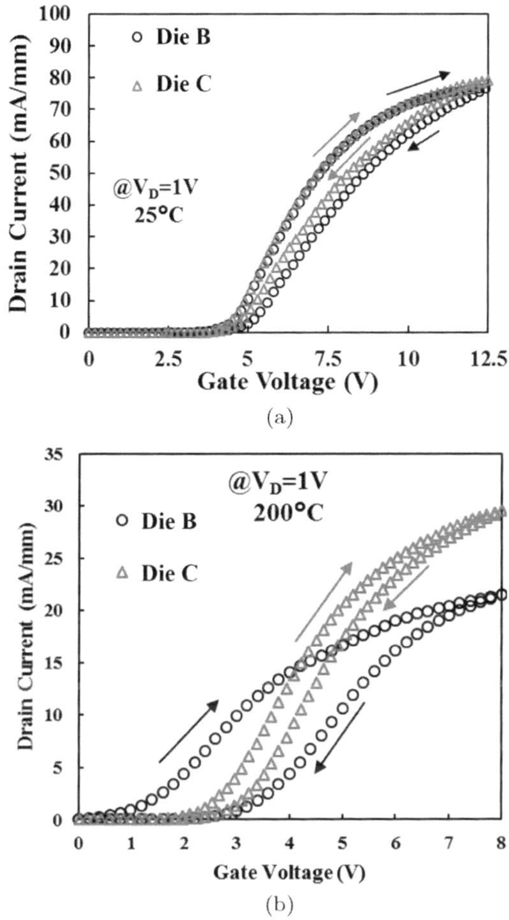

Fig. 8.51. The measured I_D–V_G hysteresis of Device B (FPT only) and Device C (APT+FPT) at $V_D = 1$ V when the ambient temperature is (a) 25°C and (b) 200°C.

negative charges which are trapped at above certain energy levels. Thus, a larger reduction of V_{th} from the sweep is observed with increasing $|V_B|$ value. Specifically, pulsed I_D–V_G measurement with V_B from 0 V to −40 V held for 50 s at room temperature is performed as shown in Fig. 8.52(c). In Eq. (8.25), $Q_{N,0}$ and $Q_{N,VB}$, which are the trapped charges within Al_2O_3 before and after applying the gate stress, can be derived from the change in V_{th} with Eq. (8.24). Hence, the corresponding E_{T_MAX} relationships of both the high temperature emission and gate stressing approaches are shown in Fig. 8.52(d).

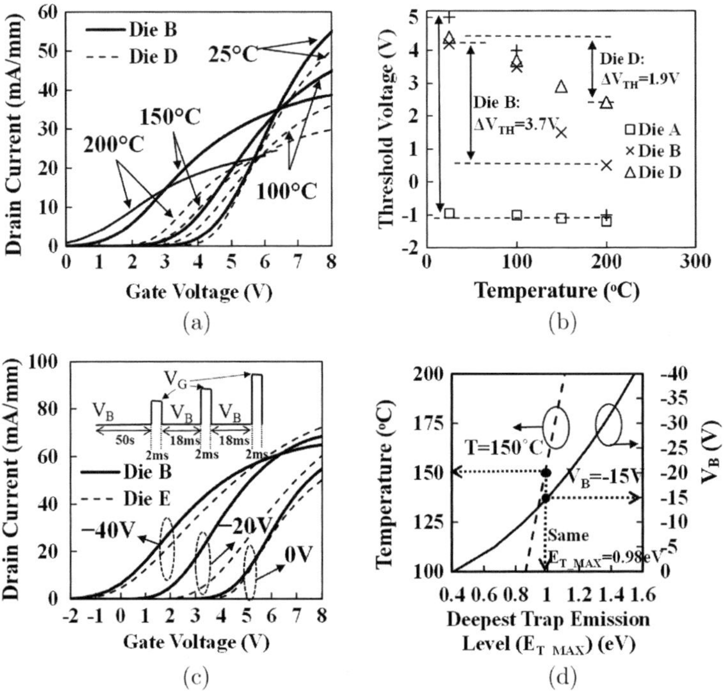

Fig. 8.52. Measured I_D–V_G sweeps of Die B and D for $V_D = 1\,\text{V}$ at (a) ambient temperatures of 25°C, 100°C, 150°C, and 200°C; (b) The ΔV_{th} of Dies A, B, D at high temperatures; (c) measured I_D–V_G sweeps with stressed gate voltage (V_B) from 0 V to −40 V; (d) The deepest trap emission level (E_{T_MAX}) in relationships with the stressed gate bias and the ambient temperature. Wang *et al.* (2016).

This relationship is useful to measure the distribution of trap charges at various trap levels within the gate dielectric bandgap.

$$E_{T\_MAX} = k_b T \ln(\gamma_n \sigma_{Al2O3} T^2 t), \tag{8.23}$$

$$V_{th} = \frac{\Phi_b - E_1 - E_2 - E_3}{q} - \frac{qQ_{2DEG}(t_{AlGaN} - t_{recess})}{\varepsilon_{AlGaN}\varepsilon_0} - \frac{q((Q_{2DEG} + Q_T)t_{Al2O3} - Q_N t_2)}{\varepsilon_{Al_2O_3}\varepsilon_0}, \tag{8.24}$$

$$E_{T\_MAX} = \sqrt{q\xi/\pi\varepsilon_{Al2O3}\varepsilon_0} - (kT/q)\ln(1 - Q_{N,0}/Q_{N,VB}), \tag{8.25}$$

where

$$\xi = (V_G/t_{Al2O3} - \phi_b/qt_{Al2O3}).$$

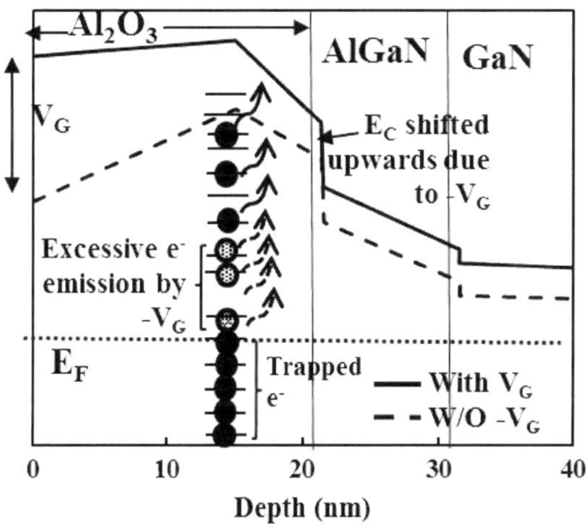

Fig. 8.53. The energy band diagram along the gate stack region for fluorinated gate dielectrics with (solid line) and without (dashed line) negative gate stress.

The trapping and emission of electrons at the fluorination plasma induced states within the Al_2O_3 energy band is illustrated in Fig. 8.53. When a negative V_G is applied, the external field will pull the E_C upwards, resulting in more emission of trapped negative charge above the E_F within a short period of time. In other words, the trap emission level can be controlled by the applied V_G. Such an effect is similar to that of increasing the ambient temperature for the emission of trapped electrons.

The amount of intrinsic gate charges Q_N at room temperature can be accurately calculated from the measured V_{th} and by Eq. (8.24) as shown in Fig. 8.54(a). In Fig. 8.54(b), the D_N mapping at corresponded energy level with and without APT (Die D and B, respectively) are found by the gate stressing technique. The change in Q_N due to gate stressing test (ΔQ_{N2}) of different V_B can be extracted by integrating a sixth-order polynomial function that fitted with the D_N data-points with the rms error less than 2%. ΔQ_{N2} can be directly compared with the change in Q_N caused by high-temperature (ΔQ_{N1}) with the same corresponded E_{T_MAX}, as shown in the Table 8.10. A good fit between ΔQ_{N1} and ΔQ_{N2} indicated the effectiveness of the gate stressing technique on D_N extraction.

8.11. GaN-Based Inverter Configuration

The enhancement-mode AlGaN/GaN MIS-HEMT with high V_{th} and good conductivity is good to form a monolithic inverter by integrating with

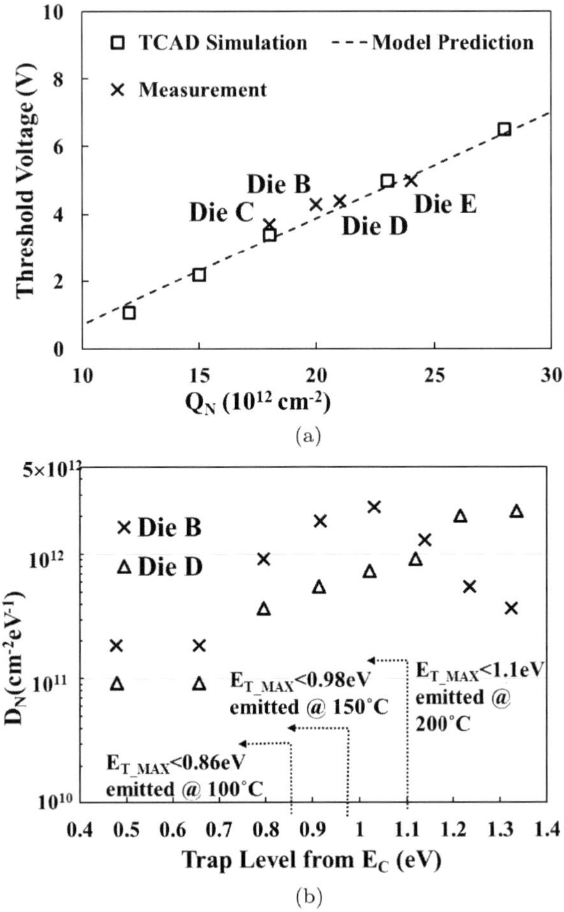

Fig. 8.54. (a) The trapped charges Q_N of Dies B–E obtained by Eq. (8.24) with measured V_{th}; (b) the distribution of fluorine-induced trap states in the A_2O_3 energy level away from the E_C. The corresponding thermal emission levels are also indicated. Wang *et al.* (2016).

a depletion-mode (V_{th} < 0 V, D-Mode) normally-on MIS-HEMT. The cross-sectional schematic of the devices with metal routing and the equivalent circuit diagram of the inverter are shown in Fig. 8.55. In this configuration, the normally-off HEMT is acted as the switch, while the normally-on HEMT is acted as the resistive load. Due to the absence of well-performing GaN-based PMOS transistor, such a depletion-load NMOS configuration is a possible substitution of complementary configuration to form the GaN-based logic

Table 8.10. V_{th} reduction (ΔV_{th}) and Q_N reduction (ΔQ_{N1}) at high temperature obtained from the measurement for fabricated Dies B and D.

Temperature	100°C		150°C		200°C	
E_{T_MAX}	0.86 eV		0.98 eV		1.1 eV	
Dies	B	D	B	D	B	D
$\lvert\Delta V_{th}(V)\rvert$	0.9	0.6	2.7	1.4	3.7	1.9
ΔQ_{N1} (10^{12} cm^{-2})	2.3	1.1	5.4	2.6	7.1	3.5

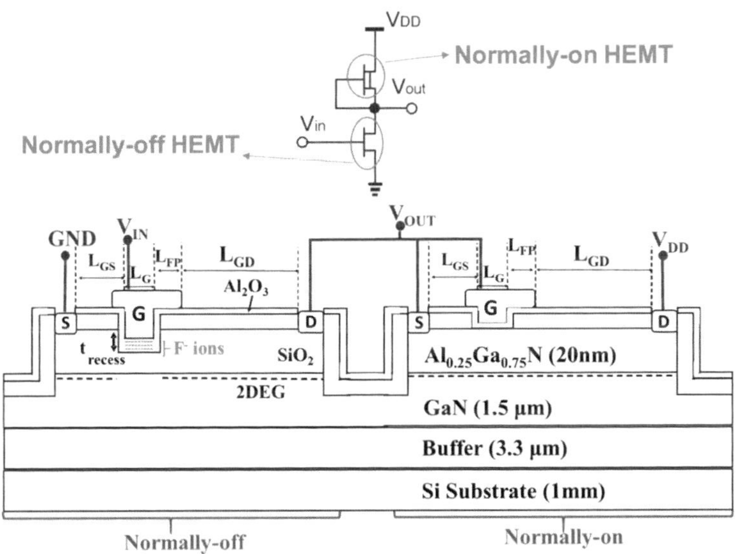

Fig. 8.55. Cross-sectional schematics of an inverter with normally-on and normally-off devices in serial configuration.

inverter. The specific steps and processing parameters used for gate fabrication are summarized in Table 8.11.

The static input–output voltage transfer characteristics (VTC) of the inverter for different device width ratio between normally-off and normally-on HEMTs (W_{ratio}) are shown in Fig. 8.56. In the VTC characteristics, the definition of input-low (V_{IL}, lower voltage when slope $= -1$), output-low (V_{OL}, lowest output voltage), input-high (V_{IH}, higher voltage when slope $= -1$), output-high (V_{OH}, highest output voltage) are also shown. It is observed that

Table 8.11. The fabrication process related to the inverter circuit.

Step	Process	Parameters
1	AlGaN partial recess at normally-on region	Cl_2 ICP @ 30 W for 30 s
2	ALD-Al_2O_3	6.5 nm
3	F^- treatment	CHF_3 RIE @ 30 W for 140 s F^- Conc.~= $-6.51 \times 10^{12}\,cm^{-2}$
4	ALD-Al_2O_3	7.3 nm
5	F^- treatment	CHF_3 RIE @ 30 W for 260 s F^- Conc.~= $-1.21 \times 10^{13}\,cm^{-2}$
6	ALD-Al_2O_3	7.6 nm
7	F^- treatment	CHF_3 RIE @ 30 W for 280 s F^- Conc.~= $-1.30 \times 10^{13}\,cm^{-2}$
8	Fluorinated Al_2O_3 removal at the normally-on region	BCl_3 ICP @ 200 W for 180 s
9	ALD-Al_2O_3	8 nm
10	Gate metal deposition	Ni/Au (15/150 nm)

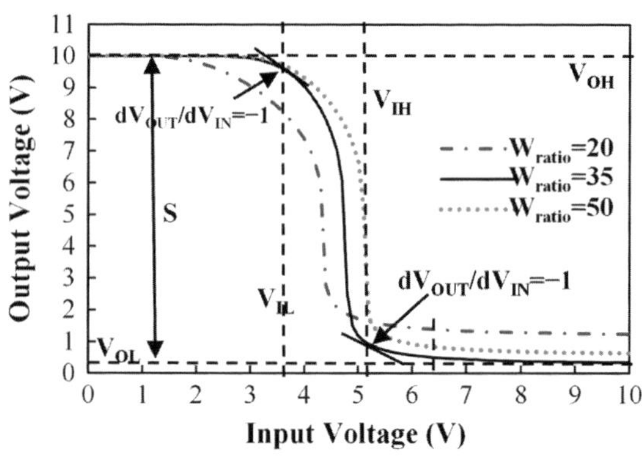

Fig. 8.56. VTC of the GaN inverter with W_{ratio} of 20, 35, and 50, where V_{OL}, V_{IL}, V_{OH}, and V_{IH} are indicated (Wang *et al.*, 2015b).

$W_{ratio} = 35$ resulted in the highest output swing S, noise margins N_{ML}, and N_{MH}. A high S/V_{INMAX} of 96.6% is achieved with W_{ratio} of 35.

The dynamic performance of the inverter with $W_{ratio} = 35$ at f = 100 kHz has been demonstrated in Fig. 8.57. Successful and fast inverter switching from 0 V to 10 V is observed on input and output signals in Fig. 8.57(a). In Fig. 8.57(b), the zoom-in portion on the rise of output waveform is shown. It is found that there is a trade-off between the response times when switching

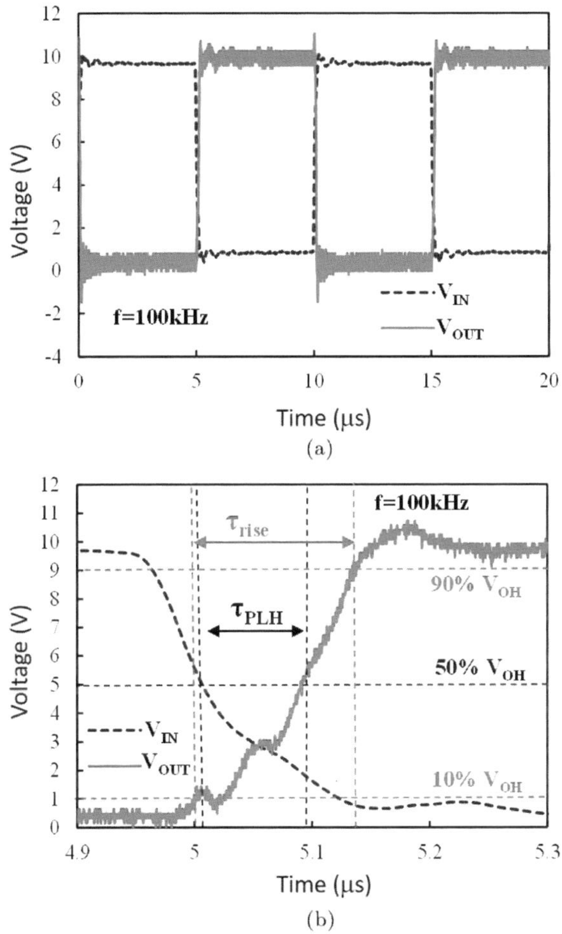

Fig. 8.57. Dynamic performance of the GaN inverter with $W_{ratio} = 35$ at $f = 100\,kHz$ with timespan of (a) 0–20 μs which shows complete switching periods and, (b) 4.9–5.3 μs that presents the switch-on transient for V_{OUT}.

on and off at different W_{ratio}. The rise time is around 120 ns for the fabricated inverter which is suitable for use in the high power and high frequency DC/DC converter.

8.12. Summary

In this chapter, the concept on heterojunction polarization and the formation of GaN HEMT devices with the 2DEG channel were reviewed. This included the calculations of polarization charges induced by AlGaN/GaN and InGaN/GaN heterojunctions, the Sentaurus device simulations with trap charges, the field plates and current collapse phenomena, the normally-on and normally-off MIS-HEMT structures and formation principles. The normally-off gate structure and its formation were elaborated with more details, including the thermal stability of the fluorinated MIS-HEMT devices and their physical mechanism on trap states.

References

O. Ambacher, J. Smart, J. R. Shealy, N. G. Weimann, K. Chu, M. Murphy, W. J. Schaff, L. F. Eastman, R. Dimitrov, L. Wittmer, M. Stutzmann, W. Rieger and J. Hilsenbeck, Two-dimensional electron gases induced by spontaneous and piezoelectric polarization charges in N- and Ga-face AlGaN/GaN heterostructures, *J. Appl. Phys.*, **85**(6), 3222–3233 (1999).

O. Ambacher, B. Foutz, J. Smart, J. R. Shealy, N. G. Weimann, K. Chu, M. Murphy, A. J. Sierakowski, W. J. Schaff, L. F. Eastman, R. Dimitrov, A. Mitchell and M. Stutzmann, Two-dimensional electron gases induced by spontaneous and piezoelectric polarization in undoped and doped AlGaN/GaN heterostructures, *J. Appl. Phys.*, **87**(1), 334–344 (2000).

A. Fischer, H. Kuhne and H. Richter, "New Approach in Equilibrium Theory for Strained Layer Relaxation," *Phys. Rev. Lett.*, **73**(20), 2712–2715 (1994).

H. Huang, Y. C. Liang, G. S. Samudra, T.-F. Chang and C.-F. Huang, Effects of gate field plates on the surface state related current collapse in AlGaN/GaN HEMTs, *IEEE Trans. Power Electron.*, **29**(5), 2164–2173 (2014a).

H. Huang, Y. C. Liang, G. S. Samudra and C. L. N. Low, Au-Free normally-off AlGaN/GaN-on-Si MIS-HEMTs using combined partially recessed and fluorinated trap-charge gate structures, *IEEE Electr. Dev. Lett.*, **35**(5), 569–571 (2014b).

J. W. Matthews and E. Klokholm, Fracture of brittle epitaxial films under the influence of misfit stress, *Mater. Res. Bull.*, **7**(3), 213–221 (1972).

J. Qu, J. Li and G. Zhang, AlGaN/GaN heterostructure grown by metalorganic vapor phase epitaxy, *Solid State Commun.*, **107**(9), 467–470 (1998).

R. Sun, Y. C. Liang, Y.-C. Yeo, Y.-H. Wang and C. Zhao, Realistic trap configuration scheme with fabrication processes in consideration for the simulations of AlGaN/GaN MIS-HEMT devices, *IEEE Journal of Emerging and Selected Topics in Power Electronics*, **4**(3), 720–729 (2016) (United States). doi: 10.1109/JESTPE.2016.2549959.

Y.-H. Wang, Y. C. Liang, G. S. Samudra, H. Huang, B.-J. Huang, S.-H. Huang, T.-F. Chang, C.-F. Huang, W.-H. Kuo and G.-Q. Lo, 6.5 V high threshold voltage AlGaN/GaN power metal-insulator-semiconductor-HEMT using multi-layer fluorinated gate stack, *IEEE Elect. Dev. Lett.*, **36**(4), 381–383 (2015a).

Y.-H. Wang, Y. C. Liang, G. S. Samudra, B.-J. Huang, Y.-C. Liao, C.-F. Huang, W.-H. Kuo and G.-Q. Lo, High output swing monolithic inverter with E-D mode MIS-HEMTs for GaN power integrated circuits, *11th IEEE International Conference on Power Electronics and Drive Systems*, Sydney, Australia, pp. 585–588 (2015b).

Y.-H. Wang, Y. C. Liang, G. S. Samudra, C.-F. Huang, W.-H. Kuo and G.-Q. Lo, The physical mechanism on the threshold voltage temperature stability improvement for GaN HEMTs with pre-fluorination argon treatment, *Appl. Phys. Lett.*, **108**(23), 233507(1–5), (2016) (United States). doi: 10.1063/1.4953573.

9

Fabrication and Modeling of Power Devices

In the previous chapters, the reader has been exposed to a variety of power semiconductor devices, both well-known ones and recently developed ones. These devices are actually realized in silicon using several basic process steps. These process steps are described here in brief. They include surface topography steps such as lithography, etching, deposition, and epitaxy. Also included are doping and thermal steps such as ion implantation, diffusion, and oxidation. More details can be found in the texts dedicated to the microelectronics processing area (Sze, 1988; Chang and Sze, 1996; Wolf and Tauber, 1990; Cambell, 1996; Plummer *et al.*, 2000). In order to increase probability of success after fabrication, it is advisable to use modeling and simulation software tools to verify device performance that emulates actual fabrication using realistic processing steps. These modeling methods and underlying basic principles are subsequently discussed. The chapter concludes with discussion on gate oxide improvements and use of selective epitaxy. Letters n, p, P and N will be interchangeably used to represent doping type throughout Chaps. 9, 10 and 11.

9.1. Unit Process Steps

9.1.1. Lithography

The patterns for each layer on the layout have to be transferred to Silicon or other materials like SiC or AGaN or other semiconductors. This is achieved by the lithography step in processing. The basic principle is to use a photosensitive material called resist for exposing the layout geometries. The resist is deposited on silicon wafer, which possibly has one or more materials grown or deposited on it, as shown in Fig. 9.1.

The resist is deposited on Silicon wafer by spinning the wafer at high speed with droplets of the dissolved resist. Wafer is put on a vacuum chuck with resist

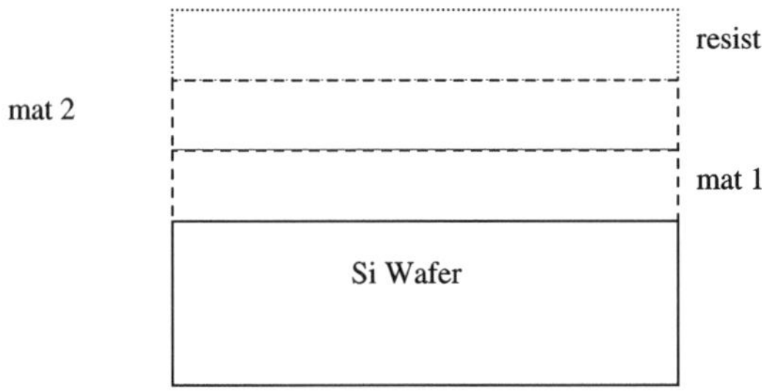

Fig. 9.1. Silicon cross section after deposition.

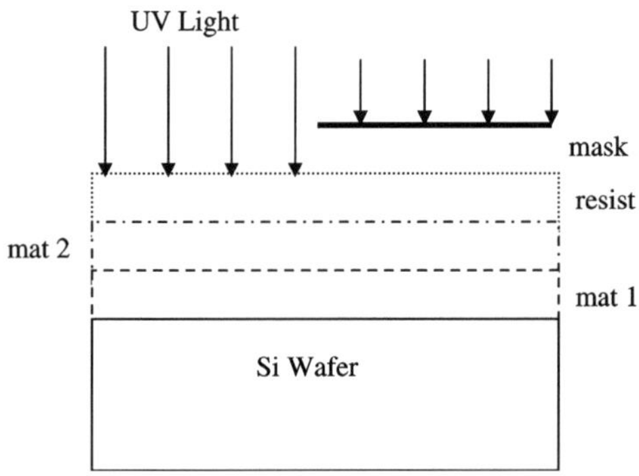

Fig. 9.2. The process of exposure.

droplets. Wafer is then spun with speeds in the range of 1000–5000 rpm. The resist thickness $z = \frac{kp^2}{\sqrt{w}}$ where p is percentage of solid resist in the solvent, k is a constant and w is revolution speed. The thickness is independent of spinning time which is typically 30–60s. The solvent is then evaporated by baking the wafer at about 50°–80°C. This step is known as soft bake. The resist is typically 0.1–1 μm thick.

After resist coating, ultraviolet light is shone through the mask plate as shown in Fig. 9.2. This light normally comes from a step and repeat or step and scan system. In the systems, a large part of wafer is exposed in steps and the mask is usually larger by a factor of 5 so that reducing optics is used. These systems ensure uniform intensity over the large extent of wafer and proper focusing on the resist. The wavelength of light used in exposure, λ, ranges from g-line at 0.436 μm, i-line at 0.365 μm, k_rF excimer laser at 0.248 μm or A_rF at 0.193 μm depending on the system. However, g-line or i-line steppers are sufficient for most power device applications. Power device fabrication masks rarely require resolution enhancement methods such as phase shift mask or annular illumination or double patterning. However, non-aggressive optical proximity corrections based on experimental studies in the fabrication facility to be used are likely to be helpful. In optical proximity correction, masks are distorted so that actual fabricated structure will be as desired. Both the actual feature and whether the feature is isolated or in a densely packed region affect this correction. For example, if it is known that an isolated etched trench is wider by about 0.2 μm, the trench mask may be made 0.8 μm wide if a trench of actual width 1 μm micron is desired.

The positive resist exposed to UV light softens and can be dissolved in a developer to expose mat 2 (Fig. 9.3). Now, mats 2 and 1 in the exposed area can be selectively etched to do processing in the masked silicon.

However, due to diffraction of light around mask edges and slight over or under exposures, it is difficult to get an exact replica of an open feature transferred to the resist. Also, the resist edge after the development may not be vertical. Many times, surface on which resist is deposited is nonplanar. In this case, the exposed and developed patterns are quite complex. The reason is that the light incident on the resist and the light reflected from the Si substrate or polysilicon form standing waves in the resist. The standing wave pattern is identical everywhere in the case of a flat substrate. The pattern is complicated if substrate surface is uneven. Hence different parts of the resist get different exposures. This standing wave effect can be substantially reduced using post-exposure bake as the photoactive compound in the resist usually diffuses at a reasonable temperature. Also, use of bottom antireflective coatings (BARC) may be beneficial. However, uneven surfaces, on which resist is patterned, may lead to systematic feature variations. In the deep UV resists,

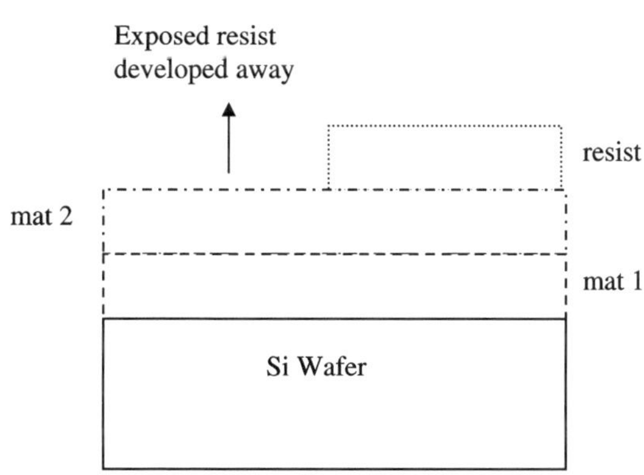

Fig. 9.3. Cross section after development.

the actual exposure takes place during post exposure bake through a reaction. Hence post exposure bake in the deep UV lithography is critically important. Such factors inherently limit photolithography features to few tenths of a micron unless resolution enhancement techniques are used.

9.1.2. Etching

The normal step, which follows lithography, is etching. Etching is a selective removal of material from the wafer. The basic principle is that the materials to be etched will only be removed from exposed areas and will not be removed from the areas protected by unexposed resist. There is a variety of chemistries available for etching commonly used materials in silicon or other technology such as nitride, oxide, metal, and polysilicon. Normally, etch rate determines the processing time. The processing conditions, etching method, and etching chemistry primarily determine the etch rate.

Ideally, any material other than the material to be etched should not be affected in this process. The silicon wafer is in the state shown after mat 2 has been etched subsequent to the development previously shown (Fig. 9.4). Note that the edge of the etched material is not vertical and there is significant undercut in mat 2 under resist. This is normally the case when wet etch is done by spraying wafers with or dipping wafers in the etching chemical.

This undercut in etching is a problem which increases some features and decreases the others. Etch rate for layers below mat 2 could be significant depending on etch selectivity which is defined as the ratio of the rate of material to be etched (mat 2) to that of the material blow it (mat 1). In such cases, precision time control or use of etch stop layers below material to be etched becomes essential.

With advances in plasma etching technology, it is possible to maintain vertical edges or well-controlled angle at the edges for etching features. Also,

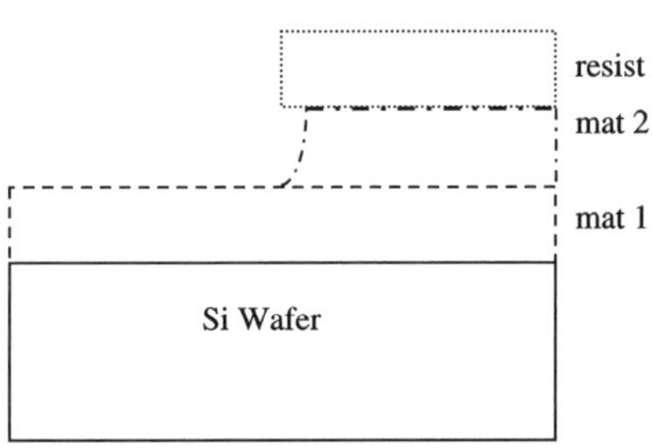

Fig. 9.4. Profile after etching mat 2.

very deep and narrow features can be etched. In some cases, etching process may leave polymer residues which are hard to remove. Hence, a recipe needs to be optimized depending on the chemistry of the reaction. Good etch selectivity between oxide and polysilicon/silicon is possible. However, it may not be that good between polysilicon and silicon, as well as nitride and oxide pairs. In a harsh etching environment, the resist is again baked after development and this step is known as hard bake so that it does not heavily erode or Peel off during the etch.

There is a variety of dry etching methods but Reactive Ion Etching that uses specific chemistries is the most popular technology. Typical reactive gases are halogen based compounds or molecules such CF_4, SF_6, Cl_2, HBr [Oehrlein and Rembetski, 1992]. They are mixed with O_2, He, Ar, H_2 to enhance etch rate or facilitate removal of solid by-products. These chemistries are useful in many materials involved in semiconductor processing. The etching is done in a process chamber under vacuum typically 200 mTorr. A plasma gas discharge is produced by applying high RF voltage up to kV level between the substrate and the chamber wall. The plasma contains reactive radicals which react with Si or SiO_2 and form volatile product which are pumped out.

9.1.3. Deposition

There is a variety of chemistries available for depositing commonly used materials in silicon/SiC/GaN technology such as nitride, oxide, metal, and polysilicon. Normally, deposition rate determines the processing time. The processing conditions such as total chamber pressure, temperature, reactant concentration, deposition method, and deposition chemistry primarily determine the rate.

One would like to have uniform cover of deposited material on the surface of the wafer. It is quite easy to get uniformity if the surface on which deposition is done is planar. However, a typical deposition pattern with a CVD isotropic deposition system on a nonplanar substrate will be as shown in Fig. 9.5.

One can easily notice nonuniformity of the deposited layer. In fact, the layer is very thin on the vertical walls of the initial surface. This may pose some problems for metal layer reliability as metal layers normally carry a fixed current and a thinner layer will have a much higher current density. If the width of the hole is reduced, an isotropically deposited material can actually fill the trench as the middle tall feature and sidewalls merge as shown in Fig. 9.6. It is also possible to achieve a thin layer conformal deposition of the materials used in technology as shown in Fig. 9.7. This is achieved when reactants are able to migrate on the surface rapidly. Atomic Layer Deposition (ALD) also

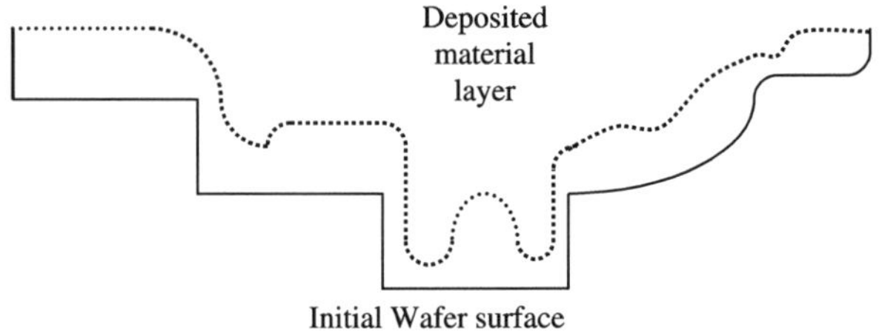

Fig. 9.5. Profile after deposition on nonplanar surface.

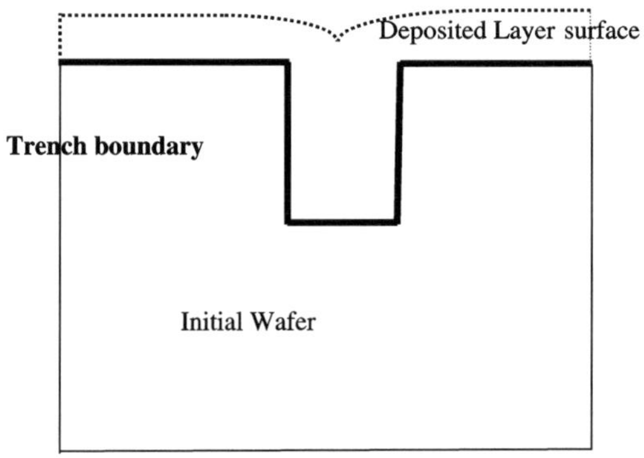

Fig. 9.6. Filling of trench using deposition.

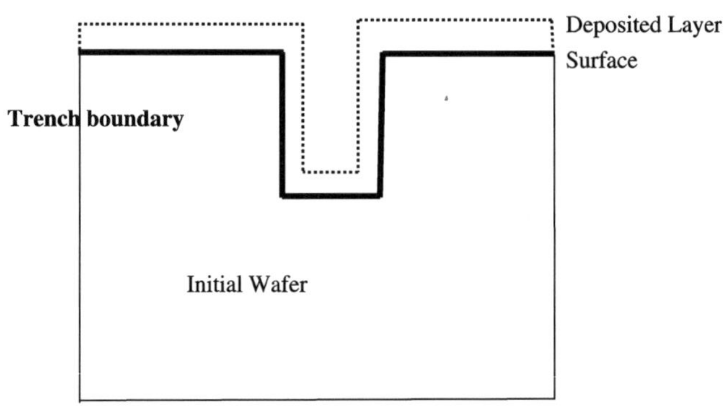

Fig. 9.7. Conformal deposition.

yields conformal deposition of many dielectric films of extremely high quality and reliability but the rate of deposition is low.

9.1.4. Oxidation

This is one of the most important steps in silicon processing as oxide can be used as a block for impurity diffusion, as a dielectric for capacitance or interlayer insulation or electric field redistribution, or as a protective layer to guard silicon against contamination. Oxide can also be selectively grown using nitride as a mask, as the nitride oxidation is much slower than silicon oxidation, or some other methods. Thus the areas of silicon covered by nitride will not oxidize.

The oxide is normally grown in dry oxygen, steam, or high-pressure oxygen in 800–1200°C range. The main reactions are

$$\mathrm{Si} + \mathrm{O_2} \rightarrow \mathrm{SiO_2}. \tag{9.1}$$

$$\mathrm{Si} + 2\mathrm{H_2O} \rightarrow \mathrm{SiO_2} + 2\mathrm{H_2}. \tag{9.2}$$

The wet oxidation is much faster than dry oxidation at a given temperature and pressure. Oxidation occurs, via a flow of oxidants through existing oxide layer to the oxide–Si interface. There are certain general guidelines on which methods to use.

The oxidation is a function of both the ambient and temperature as well as crystallographic orientation of silicon. For accurate growth rate and shape calculation, simulation tools are required. However, simple analytic calculations are also feasible. The analytic growth rate calculation for thermal oxidation is based on Deal–Grove model (Deal and Grove, 1965) that uses the concept of oxidant diffusion through oxide. Under Deal–Grove model, the oxide thickness d_{ox} is given by

$$d_{\mathrm{ox}}^2 + A d_{\mathrm{ox}} = B(t + \tau), \tag{9.3}$$

where B/A and B are the linear and parabolic rate constants and τ is the time required to grow the initial oxide under the present ambient. The rate constants B/A and B are reasonably described with Arrhenius relationships of the form

$$B = C_{\mathrm{P}} \exp\left[-\frac{\hat{E}_{\mathrm{P}}}{kT}\right], \tag{9.4}$$

$$\frac{B}{A} = C_{\mathrm{L}} \exp\left[-\frac{\hat{E}_{\mathrm{L}}}{kT}\right]. \tag{9.5}$$

For a pressure of one atmosphere, Table 9.1 (Plummer *et al.*, 2000) lists typical values of the relevant constants.

Table 9.1. Oxidation parameters.

Ambient	B	$B/A(111)$ silicon	$B/A(100)$ silicon
Dry oxygen	$C_P = 772\,\mu m^2\,h^{-1}$	$C_L = 6.23 \times 10^6\,\mu m\,h^{-1}$	$C_L = 3.71 \times 10^6\,\mu m\,h^{-1}$
	$\hat{E}_P = 1.23\,eV$	$\hat{E}_L = 2.0\,eV$	$\hat{E}_L = 2.0\,eV$
Steam	$C_P = 386\,\mu m^2\,h^{-1}$	$C_L = 1.63 \times 10^8\,\mu m\,h^{-1}$	$C_L = 0.97 \times 10^8\,\mu m\,h^{-1}$
	$\hat{E}_P = 0.78\,eV$	$\hat{E}_L = 2.05\,eV$	$\hat{E}_L = 2.05\,eV$

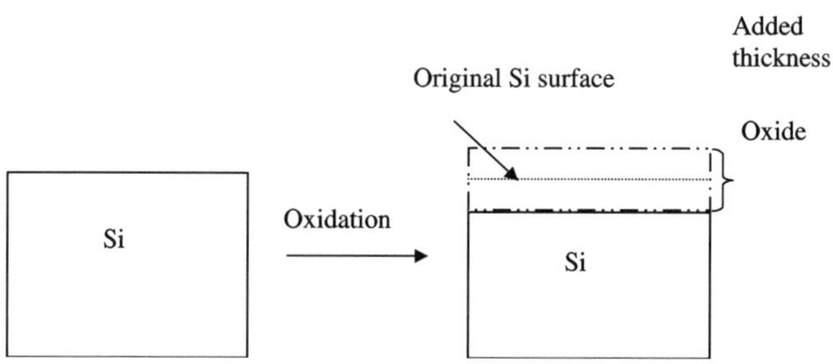

Fig. 9.8. Increased elevation due to oxidation.

When very high quality, thin oxide is needed such as the gate oxide in MOS processing, only dry oxidation is used. Nowadays, in short channel technologies, nitrated oxides are used. In wet oxidation, there is a possibility of creating OH groups which makes oxide porous so that it becomes susceptible to premature breakdown.

When good thick oxide is needed for purposes of isolation, e.g. the field oxide in MOS processing, dry oxidation followed by wet oxidation is used. This reduces the time for oxidation by a factor of about 10. This is the only method for growth of good quality thick oxides of thickness around 1 μm in practice. The high-pressure oxidation also can be used to enhance the growth rate.

Deposited oxide layers are normally used for intermetal isolation, as metals such as aluminum cannot sustain high oxidation temperatures and there is no Si easily available for oxidation.

There are many consequences of oxidation which alter processing somewhat. These will now be discussed briefly.

9.1.4.1. *Nonplanarity*

While oxidizing, about 50% of Si layer is consumed. Hence the oxide layer will have higher elevation than original Si and will eventually lead to nonplanar starting surfaces (Fig. 9.8).

If the alignment of the original Si and new oxide surface is important, recessed oxide growth is often used. In this method, silicon layer, of about half the thickness of the oxide to be grown, is etched and then oxide grown so that the surface is at about the same level as original Si.

9.1.4.2. *Bird's Beak Formation*

Nitride is an effective mask for oxidation. If it were a perfect mask, the oxidation after patterning nitride will look as shown in Fig. 9.9.

In reality, due to high stresses near oxide–nitride interface, oxide encroaches under nitride and lifts it off to produce bird's beak structure (Fig. 9.10).

9.1.4.3. *Impurity Segregation*

Due to high temperature during oxidation, the impurities tend to diffuse deeper. The redistribution, however, is affected by the presence of oxide–Si interface. The concentration of boron near the interface is reduced in Si compared to the normal concentration (Fig. 9.11). This leads to boron penetration from p+ polysilicon to the channel of p-channel devices in a typical shot channel MOS technology. To suppress this effect, nitrated oxides are used. The

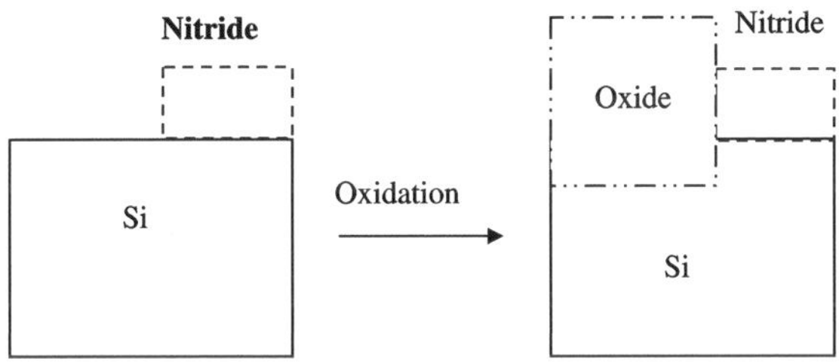

Fig. 9.9. Selective oxidation with perfect nitride.

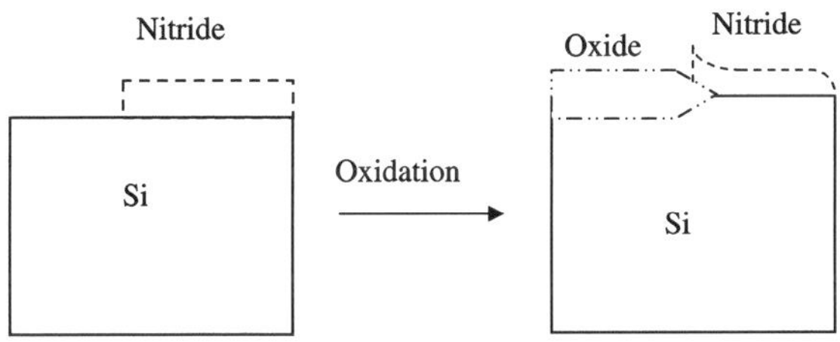

Fig. 9.10. Bird's beak formation.

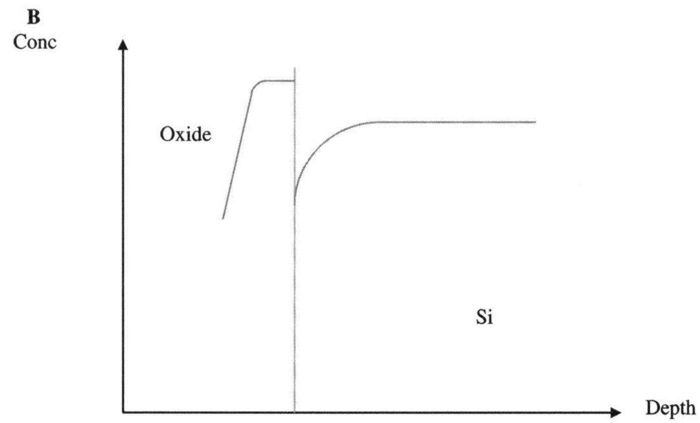

Fig. 9.11. Boron redistribution due to segregation.

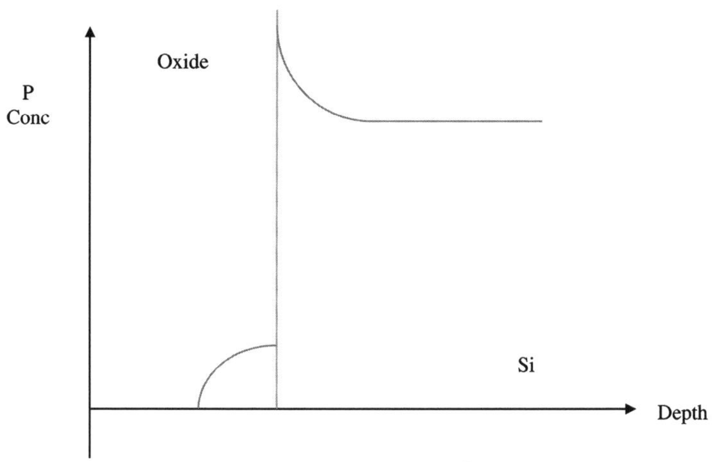

Fig. 9.12. Phosphorus redistribution due to segregation.

concentration of P and As is increased as shown in Fig. 9.12. High-pressure oxidation alters the level of segregation.

9.1.5. Ion Implantation

This is a process of introducing impurity or other ions into Si by giving them enough energy to penetrate into Si. Higher energy ions will penetrate deeper into silicon. Hence the ion energy and the number of ions to be introduced measured by dose determine the impurity profile. If the scattering of impurity projectiles is random, it is possible to find the average depth of penetration (range R_P), the standard deviation in the depth (ΔR_P) (also known as straggle), and the standard deviation perpendicular to beam direction (ΔR_L). The typical values of range are in 0.01–2 μm interval.

Accurate implant profiles that take into account detailed physics can only be estimated using process simulators. In 1D, the simplest model treats the depth profile as a Gaussian with peak located at a distance given by the range R_P and straggle (standard deviation) $\sigma_P = \Delta R_P$. The data on R_P and σ_P is available and is explained by Lindhard, Scharff, and Schiott (LSS) theory (Lindhard *et al.*, 1964). Under this profile model,

$$n(x) = \frac{\Phi}{\sqrt{2\pi}\sigma_P} \exp\left[-\frac{(x - R_P)^2}{2\sigma_P^2}\right], \tag{9.6}$$

where $n(x)$ is the dopant concentration at depth x and Φ is the implant dose. R_P and σ_P are the function of the implant energy and the implanted material and are given in Figs. 9.13 and 9.14 (Trapp *et al.*, 1993). The peak concentration is

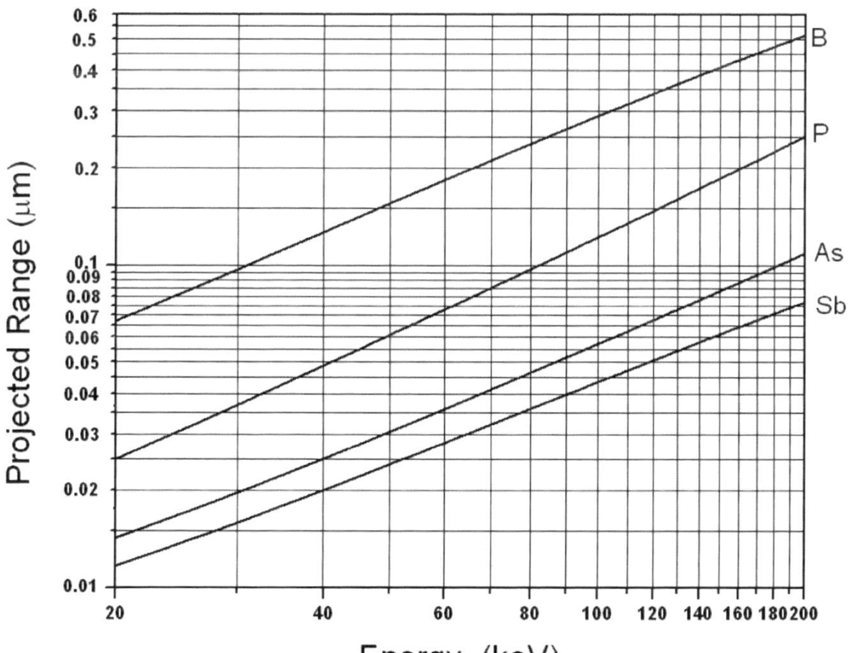

Fig. 9.13. Projected range (R_P) as a function of implantation energy.

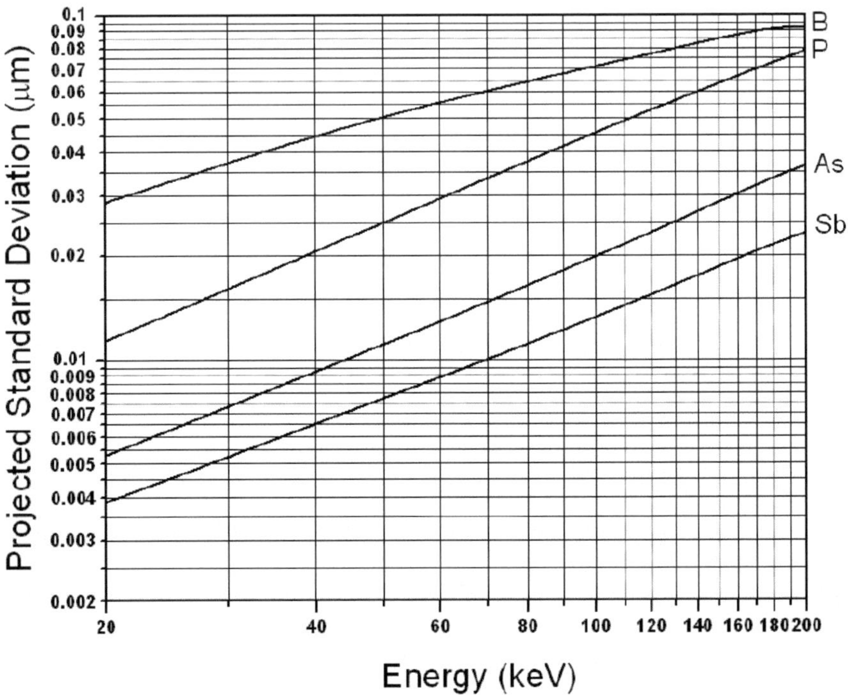

Fig. 9.14. Projected standard deviation (ΔR_p) as a function of implantation energy.

given by

$$n_0 = \frac{\Phi}{\sqrt{2\pi}\sigma_p} \approx \frac{0.4\Phi}{\sigma_p}. \tag{9.7}$$

For P, As, and Sb, the distribution of ions is a Gaussian. For boron, it is found that the distribution has a long tail which is modeled by a Pearson IV or V distribution function. High-energy oxygen implants used to be popular for forming oxide deep inside silicon for silicon on insulator (SOI) devices.

Precise control of doping profile and integrated doping concentration is possible by ion implants. However, it is not totally free of problems, which will be discussed in brief.

9.1.5.1. *Channeling*

The incident ion beam could sometimes be perpendicular to one of the crystallographic planes in Si. In this case, due to less scattering along the path, the ion may penetrate much deeper than the normal depth predicted by random

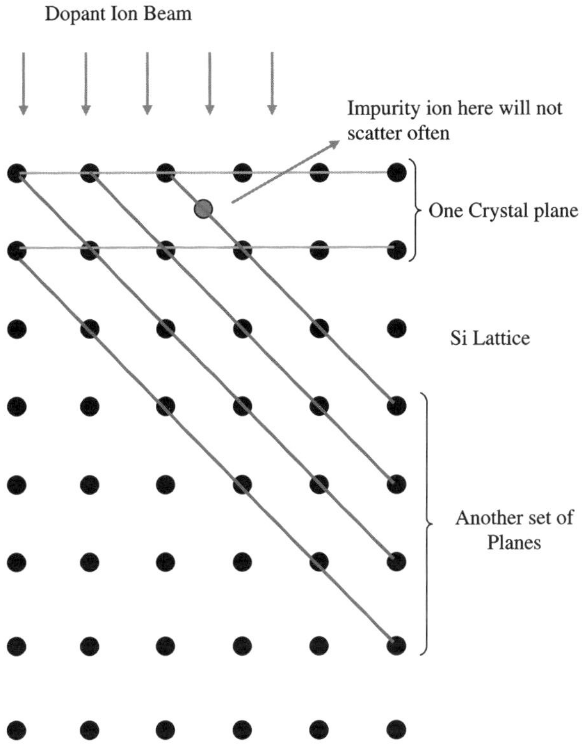

Fig. 9.15. Channeling in single crystal Si.

scattering (Fig. 9.15). To avoid channeling, the beam may be sent at 3–7° to the normal.

9.1.5.2. *Damage*

Crystalline structure of Si is damaged by implants as silicon atoms are scattered randomly and displaced from their normal positions. This damage is to the extent that silicon becomes amorphous in a short span. Hence a post-implant anneal becomes essential to give enough thermal energy to Si to restore original crystal structure. Normally, an anneal of about 30 minutes at 850–900°C is adequate to remove the crystalline damage and activate dopants by moving them into substitutional sites. Nowadays, rapid thermal annealing (RTA), spike anneal or flash anneal is often used as thermal budget is quite limited for short channel MOSFETs. RTA for 30 seconds at around 980°C is typically sufficient. This anneal, however, will redistribute the impurities by diffusion and some of the implant advantages are adversely affected.

The removal of the damage alone is not the only reason for the thermal step. The dopant activation is also carried out in this step. The dopant activation is normally limited by the solid solubility of the dopant in Si and increases with increasing temperature. The highest achievable activation is normally on the order of 10^{20} cm^{-3}.

9.1.6. Epitaxy

Epitaxy is the ordered crystalline growth which bears a definite relation to underlying monocrystalline substrate. It is normally used to grow silicon layers in which devices will be fabricated. The main advantage of epitaxy is that the doping of the epitaxial layer is not much affected by the substrate doping. Hence it is straightforward to get abrupt changes in the doping with the epitaxial growth. There are many methods to grow these layers such as gas phase epitaxy, liquid phase epitaxy, and molecular beam epitaxy. Fairly high growth rates at reasonably low temperature are possible. Also, these layers can be grown selectively on crystalline substrate where the layer does not grow on the areas covered with amorphous materials like oxide or nitride.

9.1.7. Diffusion

This is a natural phenomenon due to the tendency for impurities to move from a region of high concentration to a region of low concentration. Hence if some impurities are introduced near silicon surface, they will diffuse deeper into silicon at higher temperature. Diffusion was a very popular way of selectively introducing impurities in silicon before implants came along. Due to slow impurity diffusion in SiO_2, oxide was used as a mask for selected introduction of impurities (Fig. 9.16).

Even if this technique is not so much used for introducing dopants in the present implant age, the diffusion is still a part of processing life. The

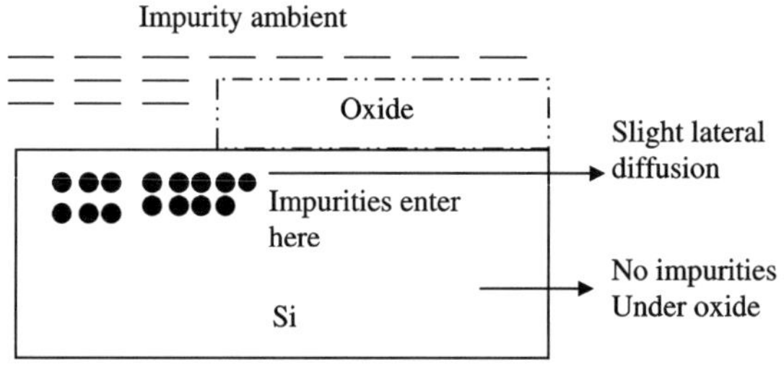

Fig. 9.16. Oxide as diffusion mask.

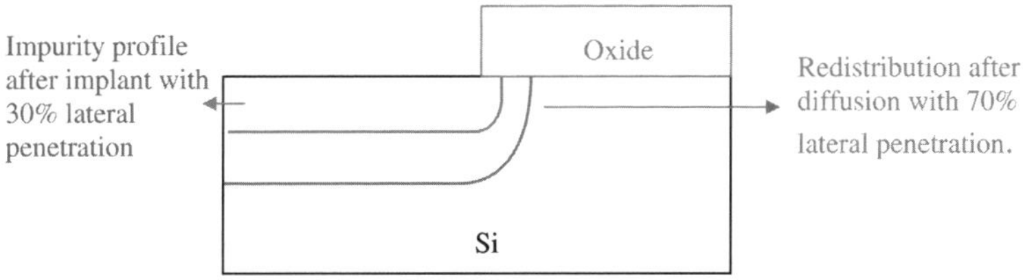

Fig. 9.17. Redistribution of implanted impurities after anneal.

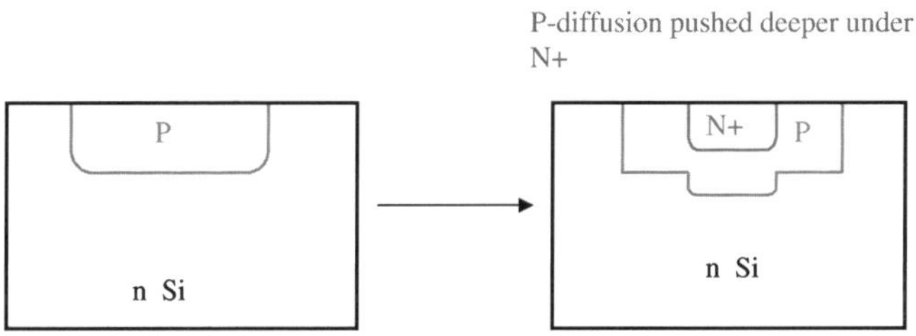

Fig. 9.18. Effect of impurity on diffusion.

post-implant anneal makes the implanted impurities redistribute by diffusion as shown in Fig. 9.17.

The diffusion takes place at any time when silicon is heated. The presence of lateral diffusion at about 70% of depth decides some of the critical device features and inherently limits some of the device features and properties.

For very deep diffusions, the profile is a near Gaussian. For very shallow ones, it is a near error function. The characteristic depths are determined by the diffusion coefficient D which is an exponential function of the temperature and the time t of diffusion. $\sqrt{Dt}$ is typically the characteristic depth for the distribution. The diffusion coefficient D is a very strong function of concentration of vacancies and interstitial within the crystal. It also depends on the presence of other impurities and the other materials. One such effect we saw in segregation before. The other one is commonly known as emitter push effect because this structure occurs in the fabrication of bipolar junction transistor. Here, phosphorus is diffused into a region where boron was previously doped as shown in Fig. 9.18. Boron is pushed deeper into the region where phosphorus is present. The diffusion also affects the steepness of epitaxial layer — substrate junction.

Simple estimation of 1D depth profile after diffusion with a very thin initial layer of dopants is possible. D in this simple model depends on impurity type

Table 9.2. Diffusion parameters for common dopants.

Element	D_0 (cm^2/s)	$\hat{E}_a$ (eV)
B	10.5	3.69
Ga	3.60	3.51
In	16.5	3.90
P	10.5	3.69
As	0.32	3.56
Sb	5.60	3.95

and temperature.

$$D = D_0 \exp\left[-\frac{\hat{E}_a}{kT}\right], \tag{9.8}$$

where $\hat{E}_a$ is the activation energy, k is the Boltzmann constant (here, $k = 8.62 \times 10^{-5}$ eV/k). D_0 and $\hat{E}_a$ are tabulated in Table 9.2 (Sze, 1988). The parameters do not depend on spatial coordinates.

In specific case when a wafer is exposed to a gas source of a dopant so that diffusion with a constant surface concentration C_s occurs, the dopant concentration $C(x)$ at a depth x is given by

$$C(x) = C_s \operatorname{erfc}\left(\frac{x}{2\sqrt{Dt}}\right) = C_s\left\{1 - \frac{2}{\sqrt{\pi}}\int_0^{\frac{x}{2\sqrt{Dt}}} \exp(-z^2)dz\right\}, \tag{9.9}$$

where erfc is the complementary error function.

In case the dopants are first introduced in a thin layer with integrated dopant concentration Φ,

$$C(x) = \frac{\Phi}{\sqrt{\pi Dt}} \exp\left[-\frac{x^2}{4Dt}\right]. \tag{9.10}$$

If this is done by a very shallow ion implantation, Φ is basically the dose.

If the implant depth is significant, the combined implant and diffusion profile in an inert ambient annealing as a function of depth is given by

$$n(x) = \frac{n_0}{\sqrt{\left(1 + \frac{2Dt}{\sigma_p^2}\right)}} \exp\left(-\frac{(x - R_p)^2}{2(\sigma_p^2 + 2Dt)}\right). \tag{9.11}$$

This simple profile does not take into account point defect-diffusion linkage, electric field effects, different contributions to D from a variety of D_0 and

$\hat{E}_a$ and effect of the presence of other impurities (Fig. 9.18). Here, boron is pushed deeper under heavy phosphorus doping and is called as dopant push effect. Accurate profiles taking into account all the effects can only be obtained with process simulators (Synopsys; Silvaco).

9.2. Basic Models for the Simulation of Unit Process Steps

9.2.1. Thermal Oxidation Models

9.2.1.1. *1 D Model*

As process simulation programs solve the problem numerically, the oxidation is treated as an incremental step and differential form of Eq. (9.3) is useful.

$$2d_{\mathrm{ox}}\Delta d_{\mathrm{ox}} + A\Delta d_{\mathrm{ox}} = B\Delta t, \tag{9.12}$$

where an additional oxide of thickness Δd_{ox} grows in time Δt. Hence the present oxide thickness is

$$d_{\mathrm{ox}} = d_{\mathrm{ox\_old}} + \left[\frac{B}{2d_{\mathrm{ox\_old}} + A}\right]\Delta t. \tag{9.13}$$

This will give the same oxide thickness as the normal Deal–Grove model (Deal and Grove, 1965) if no corrections are made and the time step is small. However, this equation is normally modified to include a term that contributes only in the thin oxide regime.

i.e.

$$d_{\mathrm{ox}} = d_{\mathrm{ox\_old}} + \left[\frac{B}{2d_{\mathrm{ox\_old}} + A} + r_{\mathrm{thinox}}\right]\Delta t, \tag{9.14}$$

where

$$r_{\mathrm{thinox}} = \alpha \exp\left(-\frac{\hat{E}}{kT}\right)\left(\frac{Pi}{1\ \mathrm{atm}}\right)^{P_{\mathrm{TH}}} \exp\left(\frac{-d_{\mathrm{ox}}}{L}\right). \tag{9.15}$$

Here, T is the oxidation temperature and Pi is the oxygen partial pressure. α, $\hat{E}$, P, L all depend on the ambient and crystallographic orientation. This correction becomes negligible once d_{ox} exceeds 3L due to exponentially decreasing term.

The rate constants B/A (linear) and B (parabolic) are continuous functions of T, oxygen partial pressure, average carrier concentration at Si surface (100 nm average) and chlorine concentration and/or surface impurity concentration. The functional dependence is quite elaborate and is not included here. It can be found in program manuals (Synopsys; Silvaco).

Similarly, the polysilicon oxidation is modeled in a similar fashion with changed parameters if needed.

Silicon nitride oxidation, however, is modeled in a very different way, i.e.

$$d_{\text{oxnitride}} = r_{\text{nitride}} \left(t_0 + \Delta t\right)^F, \tag{9.16}$$

where r_{nitride} is the oxidation rate, F is a constant and

$$t_0 = \left(\frac{d_{\text{ox\_old}}}{r_{\text{nitride}}}\right)^{\frac{1}{F}}. \tag{9.17}$$

Obviously, nitride oxidation is very slow as it is an effective mask for selective oxidation.

9.2.1.2. *2D Models*

1D model cannot give the shape of oxide around a nitride edge. Also, it cannot correctly describe the oxidation of a nonplanar structure.

In 2D, away from nitride or a mask edge (presumably at ∞) the oxide growth equation is the same as 1D, i.e.

$$\frac{d(d_{\text{ox}\infty})}{dt} = \frac{\text{B}}{2d_{\text{ox}\infty} + A} + r_{\text{thinox}}, \tag{9.18}$$

where $d_{\text{ox}\infty}$ is the oxide thickness far away from the nitride edge. However, in the simplest 2D oxidation analytic model, the 2D oxide growth rate is modeled by

$$\frac{d(d_{\text{ox}}(x))}{dt} = \frac{1}{2}\text{erfc}\left(\frac{\sqrt{2}}{s}\frac{x_0 - x}{d_{\text{ox}\infty} - d_{\text{oxinitial}}}\right)\frac{d(d_{\text{ox}\infty})}{\text{dt}}, \tag{9.19}$$

where x_0 is the half oxide thickness point, normally coinciding with the nitride edge in a standard MOS device, the parameter s controls the length of the Bird's beak in a standard MOS technology, $d_{\text{oxinitial}}$ is the stress relief oxide thickness, and erfc is the complementary error function.

To get accurate oxide shape, one has to invoke numerical models. The important models are the viscous flow model and the elastic model. The viscous flow model dominates at higher temperatures when oxide can flow easily. At lower temperatures, the newly grown oxide forces the previously grown/deposited oxide and nitride to orient (deform) according to the laws of elasticity. There are models which combine viscous and elastic models. These models are numerical and use the finite element method. They generally account for oxide shape and crystal orientation changes at the interface.

Calculation also can give approximate or accurate values of stress buildup. All simulation packages typically have models that account for several effects at a reasonable simulation speed and over all temperature ranges. Such models are important for the detailed study of silicon trench oxidation or oxide reflow in planarization in power devices. Oxide can actually be doped with boron and phosphorus to form borophospho silicate glass (BPSG) which can reflow at much lower temperature around 850°C. This is quite useful in planarizing surfaces if needed.

9.2.2. Diffusion Models

Simple analytic models for diffusion profile calculation are described here. Usually, 1D or multidimensional diffusion profiles are found by solving

$$\frac{\partial C}{\partial t} = D\nabla^2 C, \tag{9.20}$$

where D is the diffusion coefficient. The parameters do not depend on spatial coordinates. C is the impurity concentration.

In 1D, (9.20) becomes

$$\frac{\partial C}{\partial t} = D\frac{\partial^2 C}{\partial x^2}. \tag{9.21}$$

This simple profile does not take into account many effects. The diffusion coefficient D is a very strong function of the concentration of point defects in silicon, namely, vacancies and interstitials. These are quite often generated because of implant damage and other physical phenomena during thermal steps. To include the effects of point defect–diffusion linkage, electric field effects, different contributions to D from a variety of D_0 and $\hat{E}_a$, Eq. (9.20) is modified to

$$\frac{\partial C}{\partial t} = \vec{\nabla} \bullet (D\vec{\nabla}C). \tag{9.22}$$

Since D depends on the electric field and the point defect concentrations, it is now specially changing. Also, point defect diffusion equations need to be solved as their concentration changes with time, making D time dependent. A point to note that the point defects diffuse much faster and much deeper than the dopants and hence much deeper grid to the tune of hundreds of microns is required for their simulation although grid spacing can be large. However, advanced model can only be numerically solved in simulators (Synopsys; Silvaco).

For arsenic this calculation is different due to clustering especially in high concentration regime.

The segregation effect between material i and j modifies the dopant flux from i to j by

$$J = h\left(\frac{C_i}{m} - C_j\right), \tag{9.23}$$

where C_i and C_j are chemical interface concentrations in materials i and j respectively, h is the interface transport coefficient and m is the equilibrium interface segregation coefficient. The segregation affects impurity concentrations significantly as shown previously in Figs. 9.11 and 9.12.

9.2.3. Ion Implantation Models

9.2.3.1. *Analytic Models*

In 1D the simplest model treats the profile as a Gaussian with peak located at a distance given by range R_{p} and straggle (standard deviation) σ_{p} as discussed before. Analytic models in simulators have options of using Pearson IV or Dual Pearson IV distributions to accurately describe profiles to account for channeling tail and other physical effects. Two additional parameters which describe the tilt of the profile (skewness γ) and the flatness of the peak (kurtosis β) are required to characterize Pearson IV distribution. These are also accompanied with analytic model for damage to correctly deal with transient enhanced diffusion. For more accurate profiles with progressive damage evolution and detailed damage and impurity distribution, Monte Carlo simulation is needed. In simulation of most process steps in power devices, this model is not needed. But it is useful when a shorter channel MOS structure is used.

9.2.3.2. *Monte Carlo Ion Implant Model*

In this model, the trajectories of motion of individual particles are tracked to find the implant profile. Random numbers determine current state of lattice, scattering by nuclei, and direction of flight of implanted particle p.

During a collision-free distance without nuclear scattering, the projectile ion looses energy due to inelastic electronic scattering but this does not alter the direction of motion much due to light electrons. This loss dominates when projectile energy is high.

The treatment of amorphous materials such as resist, oxide, nitride, and damaged silicon is simpler. However, Si needs to be treated differently before severe damage. There are damage models normally included in process simulators with this objective.

Once the projectile ion looses sufficient energy, it comes to a halt. By summing the number of particles at the locations, it is easy to find doping density at a given location. The calculation of scattering related parameters is modified to incorporate crystalline structure of silicon. Thus, the channeling is automatically modeled by this technique.

The number of projectile ions whose motion is tracked must be above 10,000 and profile accuracy is improved by increasing the number at the expense of computer time.

9.2.4. Optical Lithography

The simulation of optical lithography involves three steps, namely, Optical computations — calculation of intensity distribution in the resist and elsewhere; Exposure — modeling of changing composition of resist with exposure, and Development — rate of dissolution of resist depending on exposure state and solvent properties.

9.2.4.1. *Optical Computations*

The basic approach to calculate the optical intensity in a 2D system assumes separable expression, i.e.

$$I(x, y; t) = I_{\text{incident}}(x) I_{\text{sw}}(y; t), \tag{9.24}$$

where $I_{\text{incident}}(x)$ is dependent on the imaging system, i.e. wavelength, source, mask and quality of lenses, etc. x-direction is along the surface, y is along the depth, and t is the time.

For $I_{\text{incident}}(x)$ calculation, there are several parameters of interest. One of the important parameter is the illumination wavelength λ. The other important parameter is the numerical aperture of the objective lenses (NA_0). NA is the measure of the ability of a lens to capture diffracted rays. Of course, the mask pattern is also playing a major role in determining the intensity distribution along with defocus denoted by δ. The final important parameter is the degree of coherence σ. An incoherent source is a source of infinite dimension ($\sigma = \infty$) and coherent source ($\sigma = 0$) is a point source. A large value of sigma gives a smooth intensity distribution around an edge but has a poor intensity low in the opaque area. Low σ typically has a large overshoot in open areas that dampens slowly and may give proximity effect problems. However, open area to adjacent opaque area intensity transition is quite sharp in low σ case. Normal acceptable range for σ is between 0.4 and 0.9.

9.2.4.2. *Exposure*

The exposure modeling depends on the treatment of the changes in the optical properties of resist with depth and time based on $I_{\text{sw}}(y; t)$ and photochemical changes. The positive photoresist normally contain a base resin, a solvent, a photoactive compound (PAC), and a reaction product. PAC reduces the development rate of the resist and is hence called as inhibitor. Three parameters: (A) exposure dependent absorption, (B) exposure independent absorption (by resist) and (C) representing rate of exposure or PAC destruction define exposure.

9.2.4.3. *Development*

The final relative inhibitor (PAC) concentration $m(x, y)$ after an exposure gives the exposure state. The development process is modeled as a surface controlled etching reaction, that is determined by the local value of m. If $R(x, y)$ is the development rate at a point $P(x, y)$, the position of P at a time $t+\Delta t$ is given by

$$P(x, y; t + \Delta t) = P(x, y; t) + \int_0^{\Delta t} R(x, y)\,\hat{n}\,dt, \tag{9.25}$$

where $\hat{n}$ is the surface normal unit vector. $R(x, y)$ is the development rate in μm/minute calculated from one of the known models.

9.2.5. Etching

Due to complex process involved during etching and deposition, the problem is treated as a geometrical problem in which material segments move in response to the processes involved. Any material surface is represented by a set of points and each segment by line joining the points. Depending on the shape, the number of points will change as the surface evolves. Several different models are available for etching in simulators.

In isotropic etching, every point on the surface of the structure exposed to the etchant moves at the same rate normal to the surface tangent at the point. The rate maybe different for different materials, i.e.

$$\text{Isotropic etch rate} = \vec{R}_{\text{i}}(x, y) = -\hat{n}\,R_{\text{i}}, \tag{9.26}$$

where $\hat{n}$ is the unit normal pointing outwards. This type of etching is normally present in wet etch.

In directional etching, only those points on the structure, which are illuminated by a columnar beam of incoming energetic ions, are eroded. If the beam makes an angle of ϕ with the y-axis, the

$$\text{Directional etch rate} = \vec{R}_{\text{d}}(x, y) = R_{\text{d}}(-\vec{x}\sin\phi + \vec{y}\cos\phi). \tag{9.27}$$

This model is applicable in some plasma assisted processes and is the other extreme of the isotropic case.

In angle dependent etching, each point on the surface moves in accordance with its orientation and angle dependent sputtering yield. This yield depends on the angle at which ions strike the surface.

9.2.6. Deposition

Several models for the deposition are available and used in the simulation.

In the isotropic deposition model, the material is deposited uniformly over the surface. This situation occurs in chemical vapor deposition (CVD) process where reactants migrates rapidly along surface resulting in uniform reactant concentration regardless of the surface topography. The deposition rate is opposite to the isotropic etch and is given by

$$\vec{R_{\mathrm{i}}}(x, y) = \hat{n}\, R_y. \tag{9.28}$$

The CVD deposition of SiO_2, Si_3N_4, polysilicon follow this pattern.

There are other deposition models such as cosine law deposition and planetary source deposition available.

9.3. Advances in the Processes for Power Devices

9.3.1. Modifications to Improve Gate Oxide Reliability and Breakdown Performance

In power device fabrication, trench gate structure or planar structure are used to define MOS device regions. Such a planar structure is shown in Fig. 9.19. While performing polysilicon etch to define gate stack, the corners of the

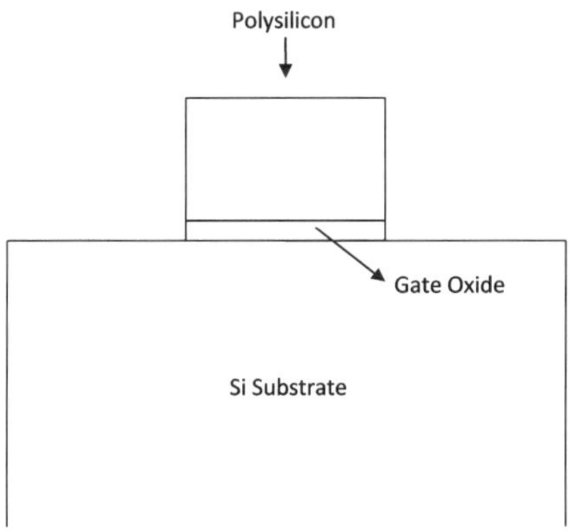

Fig. 9.19. Planar device gate stack.

gate oxide are damaged due to harsh etching environment. If the gate stack was left as it is, there will be a premature breakdown of the gate oxide at these edges and the device will fail to meet specifications. If one decides to increase the gate oxide thickness to surmount this problem, the device current is reduced and the device effective resistance increases which is not acceptable. Polysilicon oxidation is an important additional process step, which surmounts this problem with improved gate oxide reliability without much penalty on the device resistance.

In polysilicon reoxidation step, an oxidation is performed after the gate stack etch. Due to tendency to form bird's beak type structure (Sec. 9.1.4.2), the oxide will be thicker at the polysilicon edges as shown in Fig. 9.20. Oxygen diffuses through the oxide below polysilicon to a distance approximately equal to the additional oxide thickness grown on polysilicon sidewall and thickens the oxide in this region. Typical oxide thicknesses used in power devices in this step are around 10 nm. Increased gate oxide thickness at the edges also improves the hot carrier reliability with reduced gate current. Hot carriers are normally generated near the gate edge at the drain end due to high electric field when large voltage is applied to the drain. Thicker oxide reduces the amount of carriers reaching the gate as the carriers now have to surmount much thicker oxide barrier.

Similar steps also need to be taken when trench gate structure is used as shown in Fig. 9.21. After a trench etch is done, a slight wet etch is performed to round the edges at the bottom of the trench. Otherwise, the electric field at the corners will be too large and premature breakdown can happen at the corner. The oxide at the corner is thicker after oxidation and improves breakdown voltage. Further steps in forming the gate involve isotropic polysilicon deposition to fill the trench and chemical mechanical polishing to planarize

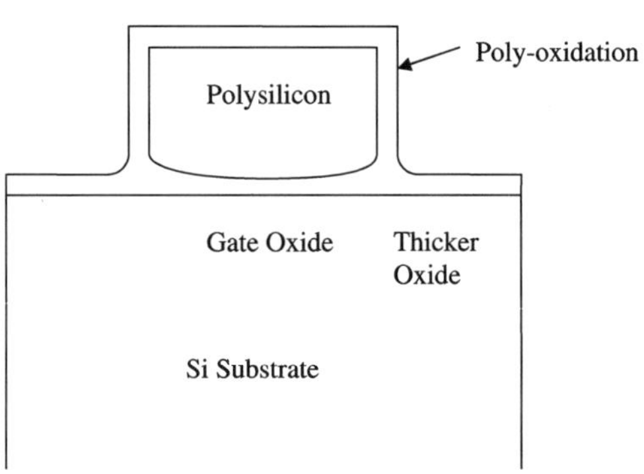

Fig. 9.20. Oxide thickening with poly-reoxidation.

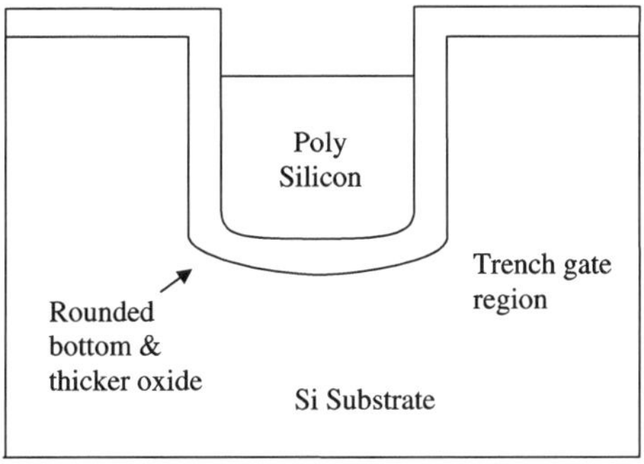

Fig. 9.21. Trench gate structure with polysilicon overetch and oxide rounding at the bottom.

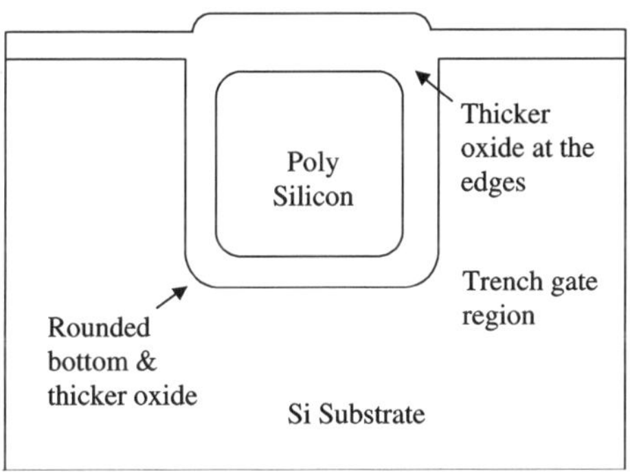

Fig. 9.22. Oxide thickening at edges with poly-reoxidation.

the surface. Again, in order to improve the oxide thickness at the top edges, polysilicon is over etched by 50 nm or so (Fig. 9.21) and oxidized to thicker oxide at the top edges as seen in Fig. 9.22.

9.3.2. Use of Selective Epitaxial Growth for Performance Enhancement

The superjunction devices described in Chap. 6 require uniformly doped p- and n-columns. The best quality columns of this type can be formed using selective epitaxy. An ideal version of PFVDMOS structure shown in Fig. 6.21 can be implemented using the selective epitaxy process taking advantage of the presence of oxide at the top and sidewall of the n-type drift region. During selective epitaxy, silicon can be *in situ* doped with the required p-type doping concentration and it will not grow on the top and sidewall of the oxide. Hence p-type uniformly doped vertical column of the superjunction structure can be formed. The process step is described well in (Chang and Sze, 1996).

Surface preparation before epitaxial growth is of utmost importance (Celik and Ozturk, 1999; Carroll *et al.*, 2000). An *ex situ* thorough clean just before introducing wafer into ultra-high vacuum epitaxial reactor chamber is essential so that the silicon surface has no native oxide and it is atomically flat. Usually, a final clean with 1:1000 Hf(49%):deionized water is sufficient. Once the wafer is loaded into the chamber, an *in situ* bake at 800°C for 2 minutes in hydrogen ambient or 10 seconds in high vacuum removes any native oxide or carbon contamination. If hydrogen is present, surface dangling bonds are passivated improving surface quality. Once the clean is finished, epitaxy is done using dichlorosilane and chlorine mix or alternating cycles. In the latter alternating cycle technique, a thin layer of single crystal silicon is grown on the substrate and amorphous silicon deposits on the oxide or nitride during growth cycle. During chlorine cycle, a part of the single crystal film is etched. Since amorphous silicon etches much faster, it is completely removed achieving selective growth. If this clean is not complete, there may be nucleation around amorphous centers randomly occurring in the silicon area and serious pitting will occur which is localized near these centers similar to one shown in Fig. 9.23.

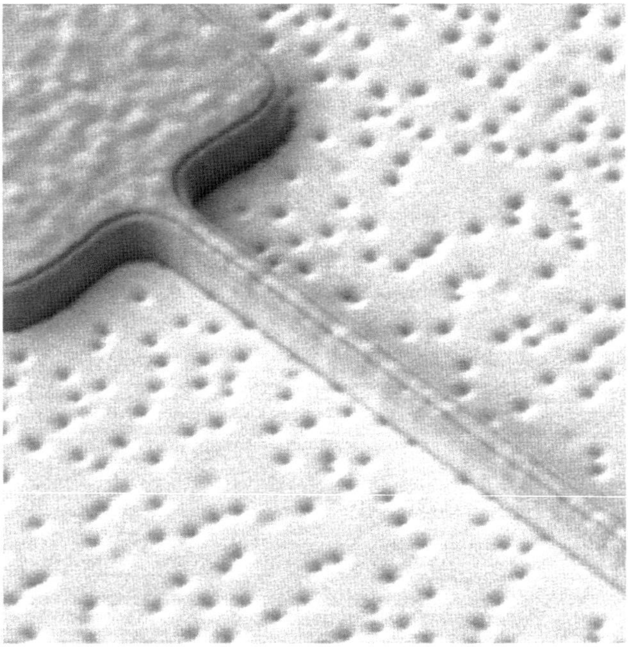

Fig. 9.23. Pitting due to improper cleaning of the surface.

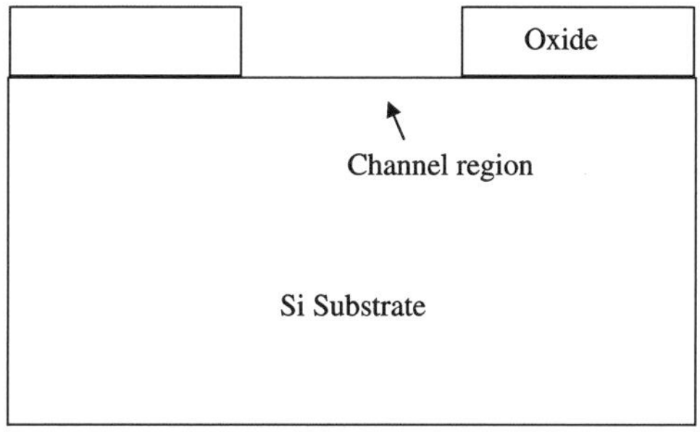

Fig. 9.24. Initial structure to etch silicon.

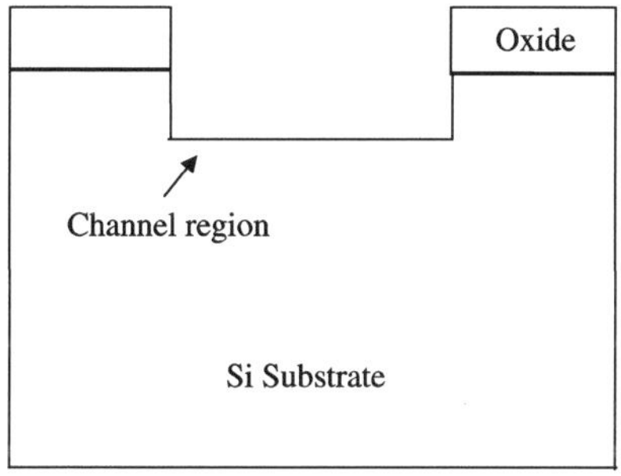

Fig. 9.25. silicon in the channel region is etched.

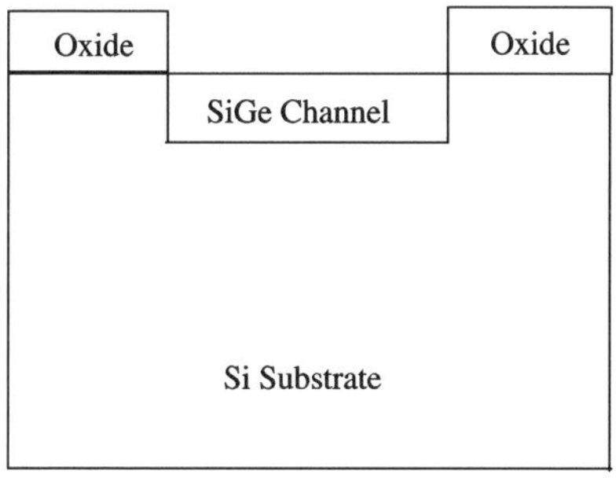

Fig. 9.26. SiGe channel region is grown with selective epitaxy.

One important use of selective epitaxy is in forming a thin 50 nm SiGe layer only in the channel region by using a process sequence depicted in Figs. 9.24–9.26. A special care needs to be taken for uniform surface as a slight facet with dipping of SiGe near sidewall oxide interface may form as silicon may be etched faster in that region. If SiGe material is used everywhere, the breakdown voltage is significantly reduced as the critical field for breakdown in SiGe is much smaller than that in silicon. On the other hand, use of SiGe in the channel region reduces device resistance as the electron mobility in SiGe is much higher. Hence by selectively growing SiGe in the channel region, lower device resistance can be achieved without disturbing the drift region and break down voltage.

References

S. A. Cambell, *The Science and Engineering of Microelectronic Fabrication* (Oxford University Press, 1996).

M. S. Carroll, J. C. Sturm and M. Yang, Low temperature preparation of oxygen and carbon free surfaces for Si and SiGe epitaxial growth by rapid thermal chemical vapor deposition, *J. Electrochem. Soc.* **147**(12), 4652–4659 (2000).

C. Y. Chang and S. M. Sze, *ULSI Technology* (McGraw Hill, 1996).

S. M. Celik and M. C. Ozturk, Low thermal budget surface preparation for selective epitaxy, *J. Electrochem. Soc.* **146**(4), 1557–1564 (1999).

B. E. Deal and A. S. Grove, General relationship for thermal oxidation of silicon, *J. Appl. Phys.* **36**, 3770 (1965).

J. Lindhard, M. Scharff and H. E. Schiott, *Mat. Fys. Medd.* **33**(14) (1964).

G. S. Oehrlein and J. F. Rembetski, Plasma-based dry etching techniques in the silicon integrated circuit, *IBM J. Res. Development* **36**, 140–157 (1992).

J. D. Plummer, M. D. Deal and P. B. Griffin, *Silicon VLSI Technology* (Prentice Hall, 2000).

Silvaco process and device simulation tools — ATHENA and ATLAS.

Synopsys process and device simulation tools — TSUPREM4 and MEDICI.

S. M. Sze, *VLSI Technology*, 2nd edn. (McGraw Hill, 1988).

B. Trapp and K. Lopp (eds.), *Semiconductor Technology Handbook* (Technology Associate Publisher, 1993), p. 72

S. Wolf and R. N. Tauber, *Silicon Processing for the VLSI Era*, Vol. 1, 2nd edn., 2002 and Vol. 2, 1990 (Lattice Press).

10

Practical Case Studies in Silicon Power Devices

After introduction to fabrication steps involved in implementation of power semiconductor devices, we move on to two detailed case studies. Using Poly-Flanked Vertical Double-diffused Metal Oxide Semiconductor (PFVDMOS) presented in Sec. 6.5 of Chap. 6 and Tunable Oxide Bypass MOS (TOBMOS) presented in Sec. 6.8 of Chap. 6 power devices as vehicles of demonstration, use of the simulation tools and unit process steps described in Chap. 9 will be demonstrated through detailed description of process integration, process and device simulation, design, and measurements. Throughout this chapter, specific issues regarding power device fabrication with built-in robustness, performance enhancement, and self-alignment are discussed. In order to increase probability of success after fabrication, it is advisable to use modeling and simulation software tools to verify device performance fabricated using realistic processing steps. Simulation results on the examples are also presented. This is then followed by layout strategies and design of splits while carrying out development work using polysilicon flanked vertical double-diffused metal oxide semiconductor (PFVDMOS) transistor as well as tunable oxide bypass MOS power device.

10.1. Case Study I: Process Integration and Design of PFVDMOS

The superjunction concept behind minimizing R_{on} for a given breakdown voltage in LMOS and VMOS devices has been explained in Chap. 6. The first device successfully fabricated using the novel idea was a vertical MOS device (Deboy *et al.*, 1998). However, the 600 V COOLMOS™ device suffers several disadvantages. The multi-epitaxy process adopted in COOLMOS™

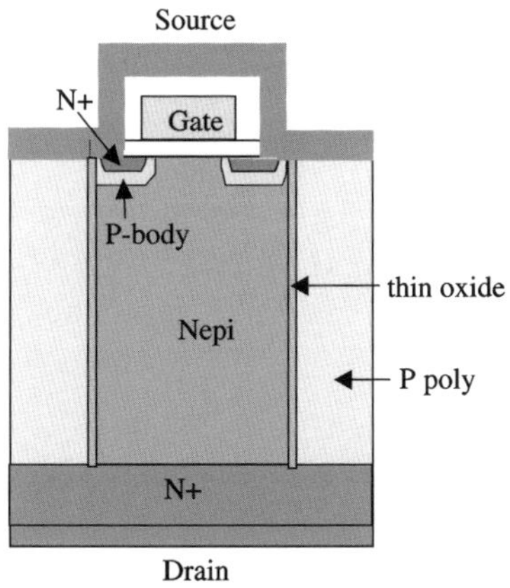

Fig. 10.1. Basic PFVDMOS device structure.

design is expensive and cannot achieve a uniform distribution of dopants as interdiffusion between n- and p-columns will occur due to the high temperature cycles in multi-epitaxy process and device realization. Thus, an optimal doping concentration as specified by the superjunction theory cannot be achieved.

With this problem in mind, a novel technology of PFVDMOS was developed. In this device, each n-drift region is flanked by two p-poly regions, hence the name **P**oly-**F**lanked **VDMOS (PFVDMOS)**, as shown in Fig. 10.1.

PFVDMOS structure has the following advantages:

(i) The use of trench etching and Polysilicon refill enables simpler process steps and increase process robustness.
(ii) The use of thin oxide as diffusion barrier yields more uniform doping profile and thus enables shrinking of p- and n-column size as there is no interdiffusion and makes feasible much increase of n/p doping.
(iii) As established by simulation results, addition of thin sidewall oxide does not affect the device operation as long as p-poly is contacted to p-body/source and n-epi is contacted to N+/drain as shown in Fig. 10.1.

10.1.1. Process Integration to Implement PFVDMOS Device

A process flow for PFVDMOS rated at 250 V has been drawn up prior to device fabrication. One important distinction between the process flow of PFVDMOS and conventional VDMOS is the insertion of additional process steps to realize the p-poly in PFVDMOS (Fig. 10.1).

The additional p-poly column can be realized by the process steps given below at the start of device fabrication:

(i) Starting from N+ substrate, n-epi with suitable doping and thickness is grown.
(ii) A blanket thick oxide is deposited.
(iii) A trench of the required dimension is etched in the oxide and silicon.
(iv) A thin layer of oxide is grown/deposited on the silicon sidewall to act as diffusion barrier.
(v) Oxide is etched away at the bottom of the trench just above the n+ region with reactive ion etching to facilitate formation of a junction between p-poly and n+.
(vi) A polysilicon is deposited to refill the trench.
(vii) Ion implantation is used to supply the necessary doping.
(viii) Drive-in with suitable time and temperature is used to obtain the required doping profile.
(ix) Etch back of oxide and polysilicon is done to planarize the surface.
(x) From this point onward, conventional process (i.e. gate, p-body and source formation) is used to complete the device fabrication. The first step is gate oxidation followed by the *in situ* heavily doped poly-gate formation using gate mask.
(xi) P-body deep implant of boron is carried out with patterned resist still present to block implant in unwanted regions, and then the resist is stripped. This way p-body implant is robustly aligned to the gate edge.
(xii) P-body drive-in that also anneals the implant damage follows.
(xiii) Source arsenic implant and post-implant anneal: this arsenic implant is blocked by the gate polysilicon and hence is self-aligned to the gate edge.
(xiv) BPSG deposition and contact pattern to open contact areas. Borophospho silicate glass (BPSG) is boron and phosphorus doped oxide that helps in planarization as it can reflow at lower temperature around 900°C.
(xv) Metallization to complete device formation.

Process and device simulators are normally used to determine process parameters and assess device performance.

Since MOSFET in off-state is essentially a diode, short loops involving diode fabrication have been tried where MOSFET elements are excluded. A Merged-PiN (MPS) diode like device has also been attempted by depositing Schottky contact directly after the p-poly formation without going through Gate poly/P-body/Source N+/P+ process steps. To provide control device for comparison with PFVDMOS, both conventional VDMOS and PiN diode have also been incorporated into the mask with two types of gate length at 2.5 and 3 μm, respectively. Completed floor plan is shown in Fig. 10.2.

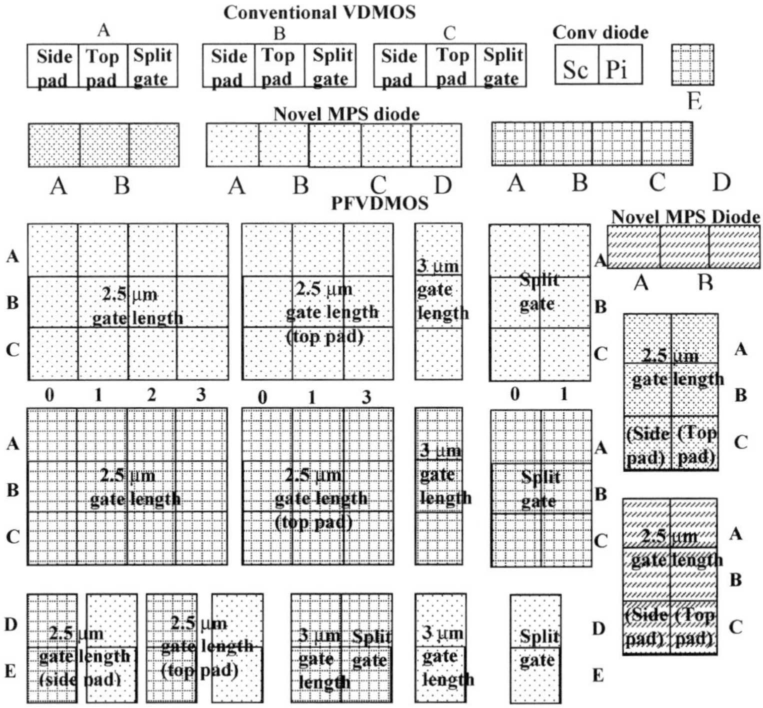

Fig. 10.2. PFVDMOS floor plan. The boxes are poly trench with 0, 0.1, 0.2, and 0.3 μm undercut, respectively. A, B, C, D, and E are variations for oxide isolation trench; 0, 1, 2, 3 are variations to ensure proper N+/source contact and p-body spacing.

Mask layouts have to conform to design rules and implement desired device structures.

After assessing the feasibility of PFVDMOS with fabrication of PiN diode structure, actual device fabrication was carried out. Both PFVDMOS and conventional devices were fabricated on 20 μm thick, 7×10^{15} cm^{-3}, n/n+-epi wafer. The starting lot consists of a total of eight wafers with several variations in design. Listed below are the details of these eight wafers. Unless otherwise stated, there is no oxide below p-poly.

- Wafer # 1: Schottky diode (bottom p-poly oxide still remains)
- Wafer # 2: MPS diode
- Wafer # 3: Schottky diode (bottom p-poly oxide still remains)
- Wafer # 4: MPS diode

- Wafer # 5: PFVDMOS
- Wafer # 6: PFVDMOS
- Wafer # 7: PFVDMOS with special pad design
- Wafer # 8: PFVDMOS with special pad and lower dose N LDD implant

All diode wafers had a simpler 4 mask process with Oxide Isolation, P-Poly trench, Contact, and Metallization. PFVDMOS with standard pad wafers was a 10 mask process while PFVDMOS with special pad was a 12 mask process. The two additional masks in special pad design were to put additional via and second metallization so that the gate and source pads could be seated on top of the device area in separate metallization layers. This reduces parasitic resistance contribution to on-resistance and saves device area at the expense of increased complexity by the two additional masks.

The MOS part follows a standard DMOS fabrication process. The whole process sequence along with important Scanning Electron Microscope (SEM) cross-sections as well as considerations is described in detail.

10.1.2. Simulation and Process Parameter Determination of PFVDMOS Device

Although process flow gives individual process steps, practical process parameters to achieve the desired doping profiles and oxide thicknesses can only be obtained through simulations. Process simulation using TSUPREM4 (Synopsys) was first carried out using the above process flow. Then, device simulator MEDICI (Synopsys) was used to obtain both on- and off-state characteristics of PFVDMOS. The input file with explanations is first described to indicate model choice and input deck setup followed by results of these studies.

10.1.2.1. *Process Parameters and Doping Profiles*

The novel device as proposed in Fig. 10.1 was simulated. The device was designed to support a breakdown voltage of 250 V. The smallest possible n- and p-layer width, that met the requirement that all the source contacts, channels must be seated inside the n-epi layer of 5 μm, respectively was chosen. Using the superjunction theory equations given in Chap. 6, the doping concentration in p- and n-layers were calculated to be at 7×10^{15} cm^{-3}. Using these values, both process and device simulation (Synopsys) were used to study and compare the DC characteristics of PFVDMOS and conventional DMOS.

This is the input file for process simulator with detailed explanations. Important process steps were optimized by changing the process parameters to obtain the best results. The simulation commands are in capital letters and descriptions are in mixed case.

$ TSUPREM4 – PFVDMOS Process Simulation file. Time is in minutes and temperatures in degree centigrade.

PFVDMOS specification (P Poly: Dose = 8E12, tilt = 10; P Poly drive-in: time = 500, temp = 1150; Source implant: Energy = 200)

$ Specify x mesh. In the ohmic n^+ region, specify large spacing of 1 μm. In the device region, specify spacing of 0.25 μm for a reasonable grid. Since program has adaptive triangular grid, it will introduce grid points where there is a lot of change. Y-direction is all ohmic n+ drain region, so the spacing is large.

```
LINE  X  LOCATION = −5      SPACING = 1.0
LINE  X  LOCATION = −3.75   SPACING = 0.25
LINE  X  LOCATION = 0       SPACING = 1.0
LINE  X  LOCATION = 3.75    SPACING = 0.25
LINE  X  LOCATION = 5.0     SPACING = 1.0

LINE  Y  LOCATION = −3.5    SPACING = 1.0
LINE  Y  LOCATION = −3      SPACING = 1.0
```

$Initialize the structure (n+ sub) with low resistivity of 0.02 $\Omega \cdot$ cm.
INITIALIZE PHOSP = 0.02 RESIST

$ n-epi deposition for n-type superjunction drift region. The region is phosphorus doped with desired concentration and has thickness of 18.5 μm and 40 grid spaces within. The thickness is finalized at the minimum value at which desired breakdown voltage can be attained from device simulator MEDICI.

DEPOSIT SIL THICK = 18.5 SPAC = 40 PHOS = 7E15 DY = 0.1

$ Masking Oxide
DEPOSIT OXIDE THICK = 0.5 SPAC = 2

$ p-type Polysilicon Trench definition. First oxide and then epi-silicon is etched to create this trench. Structure is ideal as TSUPREM4 has only simple lithography, etching, and deposition models.

```
ETCH   OXIDE START   X=-2.5,   Y=-30
ETCH   CONTINUE      X=-2.5,   Y=-7.0
ETCH   CONTINUE      X=2.5,    Y=-7.0
ETCH   DONE          X=2.5,    Y=-30

ETCH   SIL START     X=-2.5,   Y=-30
ETCH   CONTINUE      X=-2.5,   Y=-7.0
ETCH   CONTINUE      X=2.5     Y=-7.0
ETCH   DONE          X=2.5     Y=-30
```

$ Plot grid and profiles for complete structure
PLOT.2D SCALE GRID

$ Sidewall oxidation to grow oxide of sufficient thickness that acts as an inter-diffusion barrier. Since edges are straight, compress model is adequate as there is no nitride. If stress calculation is needed, VISCOELA (Viscoelastic) model can be used.
METHOD COMPRESS
DIFFUSE TIME=100 TEMP=1000 DRYO2

$ Etch bottom oxide. This removes only the oxide at the bottom of the trench completely. Original top oxide being thicker, the sidewall oxide is not affected as the etch is only in the vertical direction.
ETCH OXIDE THICK=0.1

$ Poly deposition of 1 μm thickness which does not fill the trench. This is done so that poly can be doped in the depth of the trench using tilted implant later. If trench is filled at this stage, the p-type dopants will be only in the top region and may not be able to reach the bottom poly even after prolonged diffusion.
DEPOSIT POLY THICK=1 SPAC=5

$ Plot grid and profiles for complete structure
PLOT.2D SCALE GRID

$ P-poly Boron tilted implant to dope it with the right dose to give correct final boron concentration. Dose may need to be optimized. To avoid shadow region issues with 11°, implant, use accurate Monte Carlo model that individually tracks each ion profile.

```
IMPLANT   BORON   DOSE=8E12   ENERGY=90   TILT=11
   MONTE N.ION=10000
IMPLANT   BORON   DOSE=8E12   ENERGY=90   TILT=-11
   MONTE N.ION=10000
```

$Save intermediate output file so that simulation can be resumed from this point. This is needed to find adequate diffusion time that makes boron concentration uniform in p-type poly-trench in the final structure.
SAVEFILE OUT.FILE = PPOLY

$ Now do isotropic poly deposition to fill up the trench. Dopants in previous 1 μm poly are distributed everywhere uniformly in the next thermal cycle.
DEPOSIT POLY THICK = 3 SPAC = 10

$ Cap oxide deposition to prevent out-diffusion of dopants toward surface.
DEPOSIT OXIDE THICK = 0.5 SPAC = 2

$ P-poly drive-in to achieve dopant uniformity. Basically, diffusion is long enough to remove any concentration gradients. Since there are no concentration gradients, it becomes uniform. Minimum time to achieve this must be selected as n+ drain region will diffuse into n-epi reducing its effective thickness if the time is too long. The simplest model with simple treatment of point defects is used (PD.FERMI). This is adequate as diffusions are long enough that point defects reach their equilibrium concentration quite early. For short diffusions, PD.TRANS should be used, and for VLSI devices with ultra-shallow junctions, PD.FULL/ACT.FULL should be used for high accuracy.
DIFFUSE TIME = 500 TEMP = 1150

$ Now the top cap layer of oxide is etched so that p-body and source regions can be formed later.
ETCH OXIDE THICK = 0.7

$ Etch top poly-layer so that poly is only left in the trench.
ETCH POLY THICK = 4.5

$ Etch original masking oxide that was used to form trench.
ETCH OXIDE THICK = 0.7

$ This completes the formation of p- and n-alternating columns.

$ Plot grid and profiles for complete structure. Different materials are given different colors. 2D doping contours are plotted between 10^{14} and 10^{18} with five lines in each decade of concentration.

```
SELECT    Z = DOPING    TITLE = "MOSFET"
PLOT.2D   SCALE
COLOR     SILICON       COLOR = 7
COLOR     OXIDE         COLOR = 5
```

```
COLOR      POLY                        COLOR=3
COLOR      ALUMINUM                    COLOR=2
FOREACH    X (14 TO 18 STEP 0.2)
  CONTOUR   VALUE=(10^ X)        COLOR=4   LINE.TYP=2
  CONTOUR   VALUE=(-(10^ X))     COLOR=2   LINE.TYP=5
END
```

$ Vertical profile at X = 1 μm is plotted to check uniformity near the center of poly. To avoid interpolation, it is better to use grid line at x = 1 instead of using any x.

```
SELECT  Z=LOG10(doping)  TITLE="Vertical Profiles at x=1"
LABEL="Log(conc)"
PLOT.1D  X.V=0 COLOR=2
```

$ Again save profile so that simulation can be resumed without doing 5 hours diffusion simulation.

```
SAVEFILE  OUT.FILE=PDIFF
```

$ Simulate only half of the structure. This is because there are two DMOS devices symmetrically placed around the center at X = 0. No need to simulate both. This saves half simulation time.

```
STRUCT  TRUNCATE  LEFT X=0

METHOD  COMPRESS
```

$ Grow gate oxide. Time is decided by the desired thickness and ambient of dry oxygen is dictated by high gate oxide breakdown voltage needed in power devices and good quality of the oxide.

```
DIFFUSE  TIME=76 TEMP=1000  DRYO2
```

$ 0.6 μm Gate polysilicon deposition and pattern using gate mask. Poly remains to the right of source/p-body region. Poly should be thick enough to block source implant so that only lateral source profile defines the channel length. Also, this assures that channel length of the device is the same despite mis-alignment of gate poly mask. Source and p-body implants are automatically self-aligned with the poly gate edge as resist defines poly feature.

```
DEPOSIT  POLY THICK=0.6  SPACES=1
ETCH  POLY  P1.X=3.5  LEFT
```

$ Plot grid and profiles for complete structure

```
PLOT.2D  SCALE GRID
```

$ Since p-body implant is deeper, it is better to leave resist on top of the poly. Hence this step is added to emulate resist as there is no detailed lithography simulation in TSUPREM4. Gate mask is patterned.

```
DEPOSIT   PHOTORESIST THICK = 1.2   SPACES = 2
ETCH   PHOTORESIST LEFT   P1.X = 3.5
```

$ P-body implant with boron at high energy and reasonable dose is done. Dose is determined by the level of p-type doping needed to get desired MOS threshold voltage.

```
IMPLANT   BORON DOSE = 4E13 ENERGY = 90
```

$ Remove resist before anneal to drive-in boron as the resist cannot stand high temperature.

```
ETCH   PHOTORESIST ALL
```

$ P-body boron drive-in is now done to anneal implant damage and get enough lateral diffusion of boron in the n-epi to get reasonable channel length of the device.

```
DIFFUSE   TIME = 20 TEMP = 1100

SAVEFILE   OUT.FILE = PBODY
```

$ Source resist deposit and pattern is done so that the implant does not enter other regions not in the device. The mask opens poly in some area so that alignment of source to the gate edge is assured.

```
DEPOSIT   PHOTORESIST THICK = 1.2   SPACES = 2

ETCH   PHOTORESIST START   X = 2.75   Y = −30
ETCH   CONTINUE            X = 2.75   Y = −3.0
ETCH   CONTINUE            X = 3.5    Y = −3.0

ETCH   DONE           X = 3.5    Y = −30
ETCH   OXIDE START    X = 2.75   Y = −30
ETCH   CONTINUE       X = 2.75   Y = −3.0
ETCH   CONTINUE       X = 3.5    Y = −3.0
ETCH   DONE           X = 3.5    Y = −30
```

$ Source implant is now done with high energy and high dose to give low resistivity source region.

```
IMPLANT   AS   DOSE = 5E15   ENERGY = 200
```

$ Remove resist before anneal.

```
ETCH PHOTORESIST ALL
```

$ Source anneal to cure implant damage.
DIFFUSE TIME = 30 TEMP = 900

SAVEFILE OUT.FILE = VDMOS0

$ 1 μm thick BoroPhospoSilicateGlass is deposited and contact holes are now etched to connect all source, p-poly and p-body regions using aluminum. BPSG reflow step can be added here if needed to planarize the structure.

```
DEPOSIT   OXIDE THICK = 1
ETCH      OXIDE START     X = 0     Y = −30
ETCH      CONTINUE        X = 0     Y = −21.75
ETCH      CONTINUE        X = 2.9   Y = −21.75
ETCH      DONE            X = 2.9   Y = −30
```

$ Aluminum metallization to connect regions.
DEPOSIT ALUMINUM THICK = 1.7

ETCH ALUMINUM RIGHT P1.X = 3.5

$ Passivation oxide is now deposited. Since only device is simulated here, bonding pad formation steps can actually be skipped here as pads never overlap device areas.
DEPOSIT OXIDE THICK = 0.1 SPAC = 1

SAVEFILE OUT.FILE = VDMOS1

$ Reflect to obtain complete structure so that full structure with both DMOS devices is generated.
STRUCT REFLECT RIGHT

$ Define electrodes. A single point electrode is adequate as region connected to the point will become an electrode. Since metal connects to two poly sidewalls at X = 4.25 μm and X = 14 μm at oxide–poly interface, put an electrode there. Also, put source electrode at the same level at the left end where Al connects source. All of Al automatically becomes source. Also connect p-poly column to source and p-body at X = 8 and X = 18. Add drain contact at the bottom n+ region.

```
ELECTROD   NAME = Gate   X = 4.25   Y = −22.5
ELECTROD   NAME = Gate   X = 14     Y = −22.5
```

```
ELECTROD   NAME=Source       X=8
ELECTROD   NAME=Source       X=0    Y=-22.5
ELECTROD   NAME=Source       X=18
ELECTROD   NAME=Drain BOT

$ Final device details are saved for further device simulation in MEDICI.
SAVEFILE OUT.FILE=VDMOS.spu4 MEDICI ELEC.BOT

$ Plot grid and profiles for complete structure

SELECT     Z=DOPING                  TITLE="SOI-MOSFET"
PLOT.2D    SCALE
COLOR      SILICON                   COLOR=7
COLOR      OXIDE                     COLOR=5
COLOR      POLY                      COLOR=3
COLOR      ALUMINUM                  COLOR=2
FOREACH    X (14 TO 21 STEP 0.2)

  CONTOUR   VALUE=(10^X)        COLOR=4   LINE.TYP=2
  CONTOUR   VALUE=(-(10^X))     COLOR=2   LINE.TYP=5
END
PLOT.2D         GRID SCALE      C.GRID=2
```

The simulated final doping concentrations of both conventional VDMOS and PFVDMOS drift region of 18 μm thickness are shown in Fig. 10.3. From the figure, it is clear that the doping concentration of p- and n-layers in PFVDMOS are matched around the value of 7×10^{15} cm^{-3}. The p-poly layer concentration is uniform in the region of interest due to long thermal cycle used to remove any concentration gradients. The "bulge" in the doping profile near the edge of the p-poly actually arose due to the difference in incident angles of the tilted implants on the sidewall and the bottom wall. From the figure, it is also apparent that the doping concentration of the n-drift region in PFVDMOS was nearly an order of magnitude higher than the n-drift region in conventional VDMOS (7×10^{15} cm^{-3} vs 8×10^{14} cm^{-3}). This would obviously translate into much reduced R_{on} in PFVDMOS structure. Clearly, process uncertainties can give rise to variations in these uniform concentrations. However, these should be less than 10% as both the epi-growth and implant dose are fairly well controlled. After achieving the desired doping profiles using process simulators such as TSUPREM4 or ATHENA, the next step is to evaluate off-state and on-state performance of the device using device simulators such as MEDICI or ATLAS (Synopsys) (Silvaco).

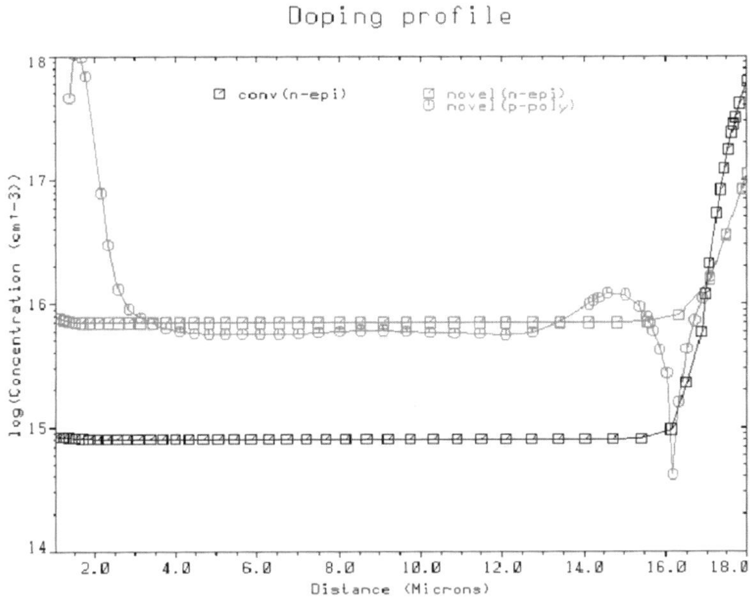

Fig. 10.3. Doping profiles in conventional DMOS and novel PFVDMOS device.

10.1.2.2. *Simulation of Off-State Performance*

Off-state simulation was performed with MEDICI using the following input file where the details are included in comments as before in Sec. 10.1.2.1.

$ MEDICI-off-state simulation.

$ Read mesh from TSUPREM4 save file. This way a triangular grid from process simulation is imported. Also read doping profiles.
MESH IN.FILE = VDMOS.spu4 TSUPREM4 PROFILE ^POLY.ELE

$ Plot the device again to ensure everything looks right.

PLOT.2D BOUND JUNC DEPL FILL SCALE

PLOT.2D SCALE GRID BOUND TITLE = "Initial Grid"

PLOT.2D SCALE BOUND FILL TITLE = "Impurity Contour"
CONTOUR DOPING LOG MIN = 14 MAX = 22 DEL = 0.1 COLOR = 2
CONTOUR DOPING LOG MIN = −14 MAX = −22 DEL = 0.1 COLOR = 1

$ Check doping profile for verification purposes.

```
PLOT.1D  DOPING  X.START = 10  X.END = 10  Y.START = −22
Y.END = −5
+  Y.LOG  POINTBOT = 1E14  TOP = 1E20
+  COLOR = 2  TITLE = "doping profile"
```

$ Model statement. Models have to be carefully selected. IMPACT.I model includes impact ionization due to high electric field that is responsible for breakdown in the devices. CONMOB models electron and hole mobility changes with local doping concentration. FLDMOB models mobility reduction due to high parallel electric field to account for the velocity saturation effect. SRFMOB2 models effects of oxide interface roughness and perpendicular gate field on mobility. Other three parameters model three physical effects of concentration-dependent Shockley–Read–Hall recombination, Auger recombination, and bandgap narrowing.

```
MODELS IMPACT. I CONMOB FLDMOB CONSRH AUGER BGN
SRFMOB2
```

$ Symbolic and method statement. By default, Gummel method is chosen as carriers are 0 and only Poisson's equation is numerically solved in 2D. This is always a starting step. The voltages then can be progressively increased.

```
SYMB      CARR = 0
METHOD    ICCG DAMPED
```

$ Initial solution with all voltages zero is found.

```
SOLVE  V(Source) = 0.0  V(Drain) = 0.0
```

$ Obtain solution using previous one with 2-carriers and change method to Newton as reasonable initial guess is available.

```
SYMB  CARR = 2 NEWTON
```

$ Bvdss simulation sequence. Save voltage and current data in log file BRDATA. Initially take small drain voltage steps. Then increase steps. Otherwise solution cannot be obtained. Once two good low voltage solutions are available, good initial guess is generated using linear projection of the solutions.

```
LOG OUT.FILE = BRDATA
SOLVE ELECTROD = Drain V(Drain) = 0 VSTEP = 0.1 NSTEP = 2
SOLVE ELECTROD = Drain V(Drain) = 0.5 VSTEP = 0.5 NSTEP = 2
SOLVE ELECTROD = Drain V(Drain) = 1 VSTEP = 5 NSTEP = 2
```

$ Continue solving using continuation method till breakdown is reached. Continuation method automatically selects the drain voltage steps.

```
SOLVE ELEC = Drain CONTINU C.VSTEP = 0.001 C.VMAX = 800
C.IMAX = 1E-9 C.TOL = 0.1
```

$ Plot breakdown $I - V$ curve with a scale of 0–300 V.

```
PLOT.1D X.AX = V(Drain) Y.AX = I(Drain) POINTS ^ORDER TOP =
8E-10 LEFT = 0 RIGHT = 300 TITLE = "Vbr"
```

$ Full flowlines, V and impact ionization for last solution to see where breakdown occurs. For superjuncion concept, it should not occur at p-body and drain junction. Plot potential contours to see if field is uniformly distributed.

```
PLOT.2D BOUND JUNC DEPL FILL SCALE   TITLE = "Struct at
   Breakdown"
CONTOUR POTENTIA DEL.V = 10 COLOR = 2
CONTOUR FLOWLINE COLOR = 1
CONTOUR II.GEN   COLOR = 4
```

$ Lot of other physical quantities such as *E*-field vector, hole distribution, electron distribution, etc. can be plotted.

The results of simulation are shown in Fig. 10.4. It is clear from the figure that both devices with concentrations shown in Fig. 10.3 were capable of supporting a breakdown voltage of 280 V. Notice that the peak ionization contour was concentrated at the p–n junction in the conventional case, whereas ionization contour was evenly spread out in the n-layer in PFVDMOS indicating effectiveness of superjunction structure. Remembering that the n-doping concentration in PFVDMOS was about an order of magnitude higher than conventional VDMOS, this result is very significant as it means that a higher n-drift concentration can still be used in PFVDMOS without compromising the breakdown voltage. Thus, PFVDMOS has provided a way of overcoming the ideal silicon limit (Chap. 6) for conventional devices (Gan *et al.*, 2001, 2002).

10.1.2.3. *On-State Simulation Results*

In order to verify the fact that PFVDMOS has indeed a better on-state performance than conventional VDMOS, on-state simulations have to be carried out on both conventional and PFVDMOS devices. The input file is similar to the one in Sec. 10.1.2.2. In this case, the gate voltage is first stepped to a value, followed by the drain voltage. IMPCT.I is not required in the model as

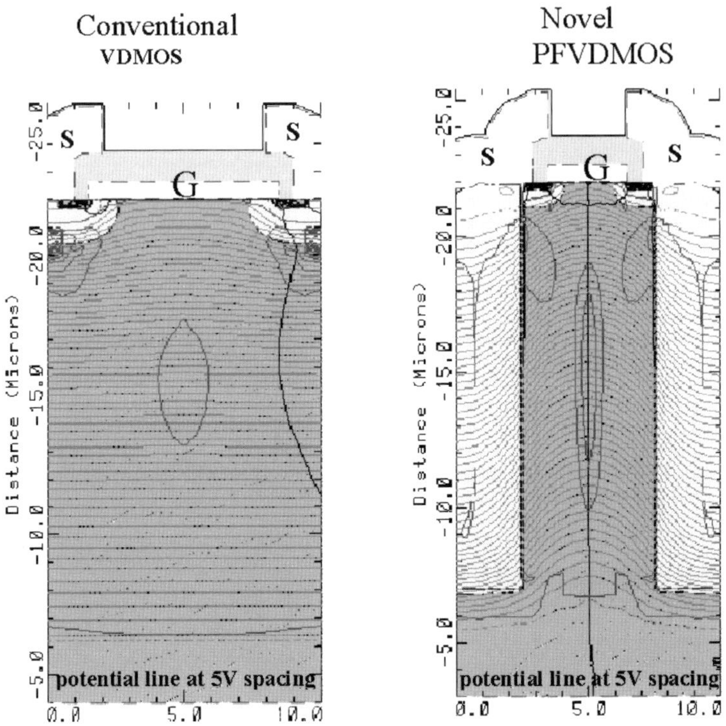

Fig. 10.4. Simulated conventional and PFVDMOS devices at breakdown (280 V).

voltages are low. The results for the two devices are compared in Fig. 10.5. From the figure, it is clear that PFVDMOS has better on-state characteristics compared to conventional VDMOS. A summary of DC characteristics is given in Table 10.1 below. From the table, it is clear that $R_{\text{on,sp}}$ of PFVDMOS is only one-third of the corresponding value in conventional VDMOS. Since $R_{\text{on,sp}}$ of the PFVDMOS device is even better than the ideal limit, it can be safely concluded that PFVDMOS has indeed broken the ideal silicon limit (Chap. 6).

10.1.2.4. *Process Sensitivity and Edge Termination*

There are always some variations in doping profiles and oxide thicknesses due to inherent statistical process variations. The performance of superjunction devices is highly sensitive to any charge imbalance present in the device. A detailed analysis on these based on a nominal breakdown voltage of 600 V has been reported by Shenoy *et al.* (1999). A similar sensitivity analysis of

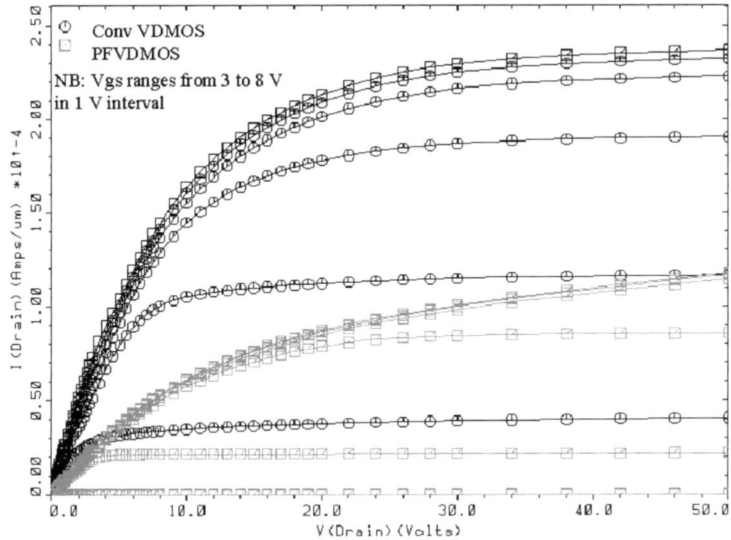

Fig. 10.5. On-state simulation results of conventional and PFVDMOS.

Table 10.1. Comparisons of simulated DC characteristics between conventional and PFVDMOS.

	PFVDMOS	Conventional VDMOS
BV_{dss}/V	280	280
$R_{on,sp}/m\Omega \cdot cm^2$	3.7	11.7

charge imbalance on PFVDMOS device off-state performance has also been performed using simulators, and this is shown in Fig. 10.6.

From the figure, it is clear that to obtain a breakdown voltage of above 250 V, a charge mismatch tolerance of about ±7% is required, whereas for a breakdown voltage of above 180 V, a charge mismatch of about ±15% can be tolerated. Thus, 180 V PFVDMOS should be achievable with the current technology, and with a perfect process control, a 250 V device can even be achieved.

One of the greatest advantages of simulations, apart from optimization of device structures, is evaluation of effects of possible scenarios and potential problems. A possible problem inherent in PFVDMOS process flow is the formation of a narrow air gap in the middle of a poly trench due to noncon-

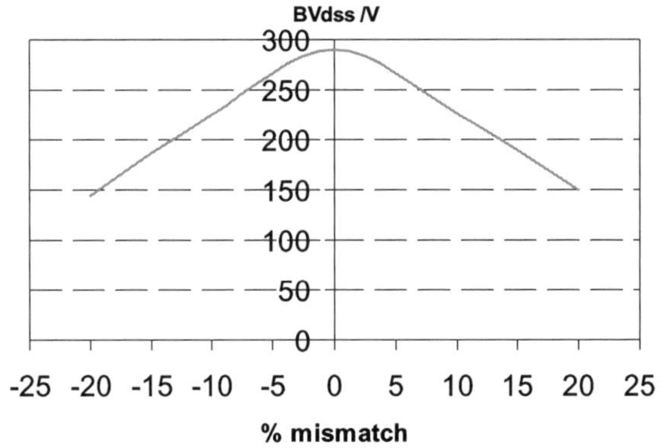

Fig. 10.6. Effects of charge mismatch on PFVDMOS device breakdown voltage.

formal deposition of poly. The effect of this was investigated and the result has shown that a narrow 0.1 μm air gap at the middle of the poly trench does affect the device off-state performance much as it drops the breakdown voltage to 230 V from 280 V. The degradation is mainly due to the loss of charge in the air gap resulting in a slight mismatch in p- and n-column charge. Since the current technology cannot achieve a perfect charge balance, this degradation can generally be treated as an additional charge mismatch.

In contrast to the conventional case, the background doping of PFVDMOS is much higher (in the present case, it is 7×10^{15} cm^{-3} instead of 8×10^{14} cm^{-3}). Thus, even though the active area of the PFVDMOS device is capable of supporting a high breakdown voltage, the rest of the silicon area fails to achieve this. If no proper edge termination is found, the device will breakdown prematurely at the edge of the device. In order to prevent this problem, a proper edge termination scheme must be adopted. An edge termination scheme with oxide trench filled with p+ poly surrounding the edge of the active device, as shown in Fig. 10.7, was designed. By using the termination scheme, breakdown voltage of novel VDMOS as simulated earlier is attainable as confirmed through simulation results.

10.1.3. Experimental Results

10.1.3.1. *Process Flow and Run Sheets*

As explained in the previous section, after assessing the feasibility of PFVDMOS and determining process parameters with simulators, actual device fabrication was carried out. Both PFVDMOS and conventional devices were fabricated on 20 μm thick, 7×10^{15} cm^{-3}, n/n+ epi-wafer.

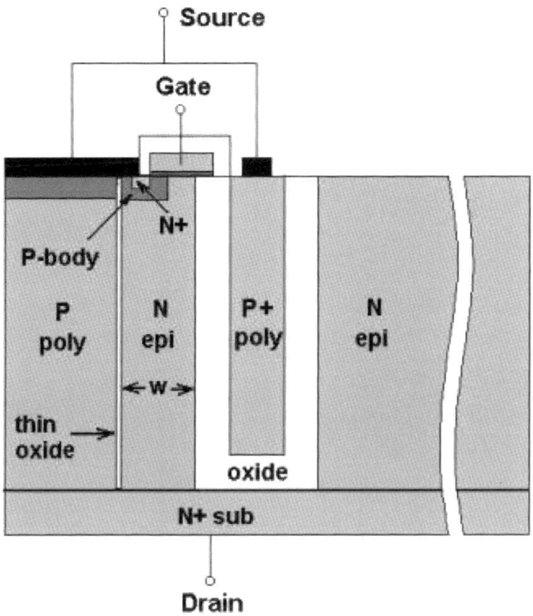

Fig. 10.7. Edge termination scheme.

There are 11 masks in total used for the fabrication of PFVDMOS. These are listed below.

(1) Mask 1: Oxide isolation
(2) Mask 2: Trench termination
(3) Mask 3: P-poly trench
(4) Mask 4: Poly gate
(5) Mask 5: P-body
(6) Mask 6: N+ source
(7) Mask 7: P+ P-body contact
(8) Mask 8: Contact
(9) Mask 9: Metal 1
(10) Mask 10: Via
(11) Mask 11: Metal 2.

Variations were introduced to Masks 1 and 2 to take into account possible undercuts during trench-etching. Other variations were made to Masks 3, 5, 6, and 7 mainly to ensure proper source/N+ contact and also dielectric isolation between source metal and gate poly while ensuring maximum p-body spacing to reduce resistance contribution from the narrow neck region formed between

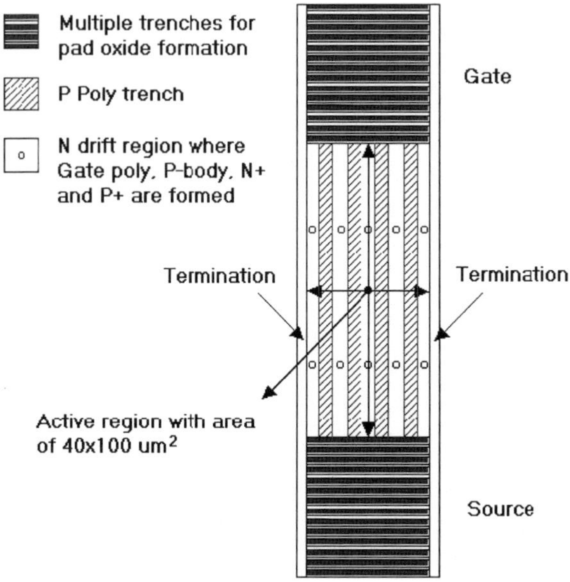

Fig. 10.8. Partial mask layout showing the top view of PF-VDMOS structure.

two p-body regions (Fig. 10.1). Sufficient splits always have to be designed so that despite process variations, some devices will work. Here, the most tricky step is of p-poly doping. Hence, some splits have to be designed with implant angle and dose, so that best charge balance can be found.

The layout of a PFVDMOS device is shown in Fig. 10.8.

The run sheet of the process looks very different from the regular process flow as it includes all cleaning/inspection steps. As a sample, a few entries in the run sheet for PFVDMOS are included in Table 10.2.

10.1.3.2. *On- and Off-State Measurement on First Run of PFVDMOS*

Figures 10.9 and 10.10 show the typical measured off-state and on-state results obtained from various wafers. Generally, the results have shown that the fabricated devices were functional even though their off-state results were less than satisfactory (about 100 V for the best results as shown in Fig. 10.9 compared with the expected value of 280 V). Conventional devices have lower breakdown voltage, about 50 V for PiN structure. Also, the leakage current of PFVDMOS is about 1–2 orders of magnitude higher than PiN diode, and is undesirable as high leakage current would lead to higher static power loss. This increase in leakage current could be attributed to the numerous p-poly

Table 10.2. Typical process run sheet.

Step	Process description	# wfs	Date and time/signoff				Remarks
			TRACK IN		TRACK OUT		
			Date and time	Init.	Date and time	Init.	
1.	Lot start • 8 wafers: n−/n+ epi, (100), thick 18–22 μm $\rho = 0.6$–$0.8\,\Omega \cdot$ cm • 10 wafer: n(100) $\rho = 10$–$80\,\Omega \cdot$ cm, for test wafer	18					
2.	Wafer marking and pretreatment Mark backsides of wafers. (Lot and Wafer No.), #1–8 for the epi. wafers; #9–18 for the test wafer						
3.	Mask 0: Fiducial mask RT: 1.7 μm, PR/CT/SB Prg#: 1/2/1, ST Prg.#:, Mask ID Exp: 180 mJ/cm^2 PEB/DP/HB Prg. #: 1/2/1 Inspection:						Stepping distance: 9800 μm × 8890 μm Fiducial: (2500 μm, 200 μm)
4.	Fiducial etch Si _6000 Å Sequence: MARK Time: 160 s Inspection:						

(*Continued*)

Table 10.2. (*Continued*)

Step	Process description	# wfs	Date and time/signoff				Remarks
			TRACK IN		TRACK OUT		
			Date and time	Init.	Date and time	Init.	
5.	Resist strip Sequence: PRS Piranha II 10+10 m — DR(6X) — SRD#6 Meas. Alphastep (Check fiducial depth) Etch depth (C)7534 (E)7450 A (spec.: 5500–6500 A)						
6.	SRO-1 Growth_600 A (ASM-B2) SC-1 10 min — DR(6X) Piranha I 100°C 5 min — DR(12X) 15:1HF#1, 10 s — DR(6X) — SRD#6 Recipe: SRO-600 Tube #B2 Meas. Rudolph pgm#FOXIDE<500T Tox(tw) = 582 A, (Spec: 580–620 A) Nitride-1 deposition_2500 A (ASM-C2) Recipe: NIT_1500, Tube #C2						

	Meas. Rudolph pgm#FNIT<6000TN Tnit(tw) = 2450.8 A, (Spec: 2300–2700 A) Darkfield Inspection:
7.	LPCVD oxide 0.5 μm (ASM_C4) Recipe: , Tube: C4 For #1 and #2 wafers only Meas. Rudolph pgm#FTEOS Tox(tw) = _______ A, (Spec: 4500–5500A)
8.	LPCVD polysilicon deposition _0.8 μm (low stress poly) Recipe: Tube: C3 Tox (tw) = 7930 A (Spec:μm) Inspection:
9.	POCL3 doping Recipe: POCL-930 15:1HF#2, 90 s etching until dewet — DR(10X) — SRD#6

(Continued)

Table 10.2. (*Continued*)

Step	Process description	# wfs	Date and time/signoff				Remarks
			TRACK IN		TRACK OUT		
			Date and time	Init.	Date and time	Init.	
10.	Boron tilted implant Recipe: B/90K/8E12 Species: _______, Dose: _______ Energy: _______, Tilt angle: 11° with two twists implants (0° and 180°)						#03, #04 Tilt: 9° #05, #06 Tilt: 10°
33.	Poly-Si deposition — 2 μm (ASM-C1) Piranha I 100°C 10 min — DR(12X) — SRD#6						
33.5	Recipe: , Tube: C1 15/1 HF 1 min, 10 W stress Meas. Rudolph pgm#FPOLYSI T = 19407 A Tpoly(tw) 9825 A, Spec: (1.8–2.2 μm)						
34.	Boron diffusion Recipe: Note: 1. Thermal budget:1150°C, 500 min 2. 1000 A oxide grows before diffusion Meas. Rudolph pgm#FOXIDE>500T Tox (tw) = 1293 A, (Spec: A)						

57. BPSG reflow (ASM A3)
    Paranha I 100°C, 5 min — DR(6X) — SRD#6
    Recipe: , Tube: A3
    900°C, N2, 30 min

58. Si-AL PVD deposition 2.5 μm
    SC1 80°C 10 min — DR(6X) — 15:
    1HF#2, 20 s — DR (6X) — SRD#6
    Recipe:
    Rsheet =____________

59. Mask 9 Metal 1 (B Field)
    RT: 2.0 μm Msk ID:
    PR/CT/SB Prg #: 1/8/1
    Step Prg. #: , Exp: 250 mJ/cm^2
    PEB/DP/HB Prg #: 1/2/1
    OL Prg #: manual
    Spec: $\pm$0.20 μm, W#:

(*Continued*)

Table 10.2. (*Continued*)

Step	Process description	# wfs	Date and time/signoff				Remarks
			TRACK IN		TRACK OUT		
			Date and time	Init.	Date and time	Init.	
	T L C R B X ____ ____ ____ ____ ____ Y ____ ____ ____ ____ ____ Inspection:						
60.	Dry etch Sequence: 2.5μ-AL (e/p) Time:____620____s Inspection:						
61.	Resist strip Sequence: PRS ACT 690, 65°C, 20min — DR(6X) — SRD#6 Inspection:						Testing Diode first, then test MOSFET later
62.	Pad dielectric deposition Sequence: Pad Thk: (scb) (c)________(e)__________						

63.	Mask 10 passivation (DK-FIELD) RT: 2.0 μm, Msk ID: PR/CT/SB Prg #: 1/8/1 Step Prg. #: , Exp: 250 mJ/cm^2 PEB/DP/HB Prg #: 1/2/1 Inspection:
64.	Pad etch and resist strip Sequence: Pad e/p (alg#41) time:________s Meas. Rudolph pgm# Patterned-thin SiO_2 Thk: (scb) (c)________(e)__________ (spec <50 A) ACT690, 65°C, 20 min — DR(6X) — SRD#6 Inspection:
65(a)	Alloy (Tube, A4) Sequence: Temp = 420°C, $t = 30$ min, $N_2 + H_2$
65(b)	Backstrips

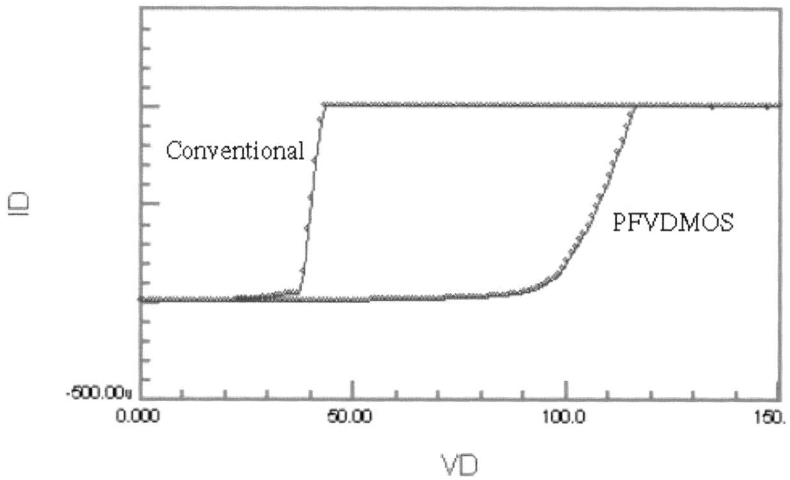

Fig. 10.9. Measured off-state characteristics of PFVDMOS.

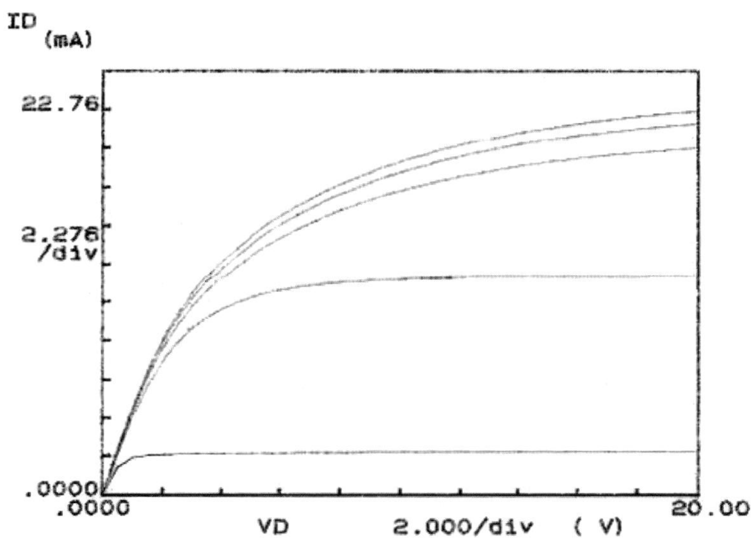

Fig. 10.10. Measured PFVDMOS on-state characteristics with V_{gs} from 0 to 7 V in 1 V step.

columns in PFVDMOS that have large defect densities. Since both conventional and PFVDMOS are fabricated on the same substrate, only on-state result of PFVDMOS was characterized and is shown in Fig. 10.10. With the device area of $(40 \times 400)\ \mu m^2$, the measured $R_{on,sp}$ was calculated to be about $35\ m\Omega \cdot cm^2$. This is much larger than the simulated values as the drain contact at the top of the wafer was probed resulting in a 2X increase in the on-state resistance. A more accurate R_{on} can be obtained by introducing a deep n+ which makes a low-resistance contact from the pad to the n+ substrate.

10.1.3.3. *Investigation of Causes Behind Degraded Performance*

As seen from Fig. 10.9, even at the best measured breakdown voltage of about 100 V, PFVDMOS still did not reach the expected value of 280 V. It is important to analyze the reasons behind the measured on- and off-state performances and then do necessary changes to achieve improvements. Both hot spot and SEM images (Fig. 10.11) were examined to investigate the cause of much worse performance than the prediction of simulators.

Hot spot images revealed that breakdown originated from n-epi regions, especially from those at the edge of the device. An examination of SEM images has suggested that this might be due to the over-etching of p-poly resulting in very uneven structures (Fig. 10.11(a)) and source metal encroaching into p-poly Fig. 10.11(b). This would then cause field crowding to occur at the corner of p-poly as shown in Fig. 10.12.

Other possible site of problem might be at the n-epi next to the oxide column. Remembering that phosphorus would segregate out of oxide into silicon during wet oxidation, it is expected that n-epi next to the oxide column would have higher n-doping. If doping of p-poly is smaller than n-epi doping, these n-epi regions with higher doping would breakdown first. In fact simulation with p-poly doping near sidewall oxide $(3 \times 10^{15}\ cm^{-3})$ < n-epi doping $(7 \times 10^{15}\ cm^{-3})$ has shown that breakdown voltage has significantly degraded to 110 V with the failure point located at the middle of n-epi. Such verification with simulations is critical in ascertaining the causes of the loss in performance after fabrication.

Even though previous simulation results have shown that the poly can be doped to the required profile, due to charge-trapping effect at the poly grain boundary, the actual free carrier profile might deviate largely from the doping introduced. This is especially so when doping in poly is low as in the present case. In fact, there exists a critical dopant concentration N* in poly below which virtually every dopant is trapped in the poly grain boundary (Kamins, 1998). At a poly grain size of $1\ \mu m$, N* is about $1 \times 10^{16}\ cm^{-3}$. Thus, it is possible that majority of the dopants introduced into p-poly are trapped in the

(a)

(b)

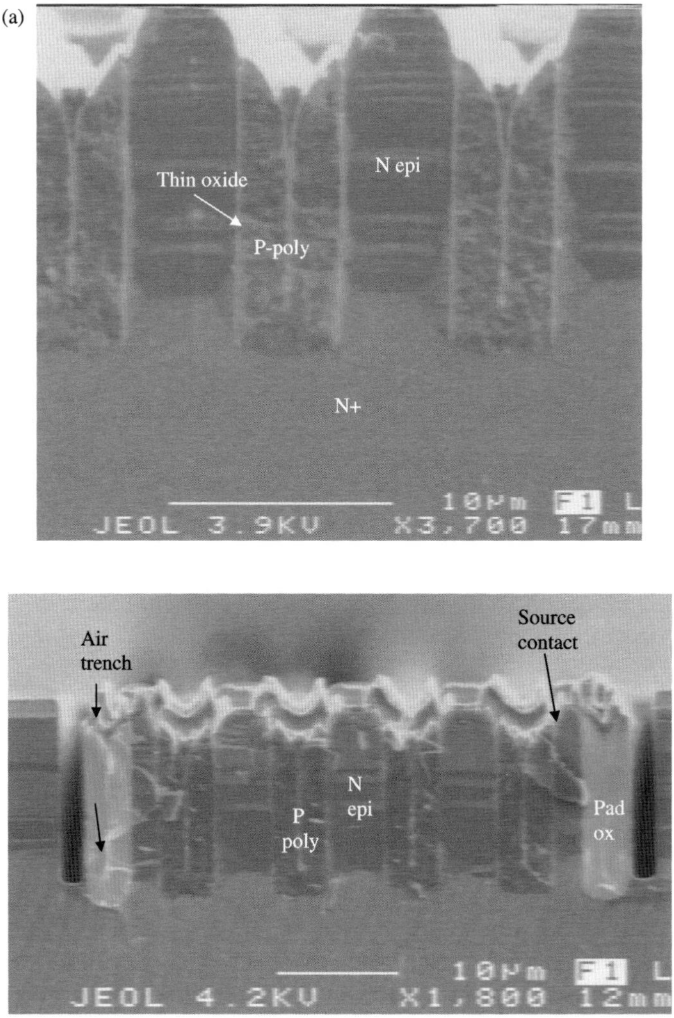

Fig. 10.11. (a) SEM images of PFVDMOS columns, (b) SEM images of PFVDMOS.

grain boundary. Thus, even though these trapped charges might be released in local high field region during off-state, the amount of free charge available in poly need not be as expected. This will lead to a charge imbalance and reduced breakdown voltage (Fig. 10.6).

From the above discussions on the likely causes of device failure, the process problem of source metal encroachment into p-poly and the dopant segregation problem can be overcome by using a better recipe in planarization and other

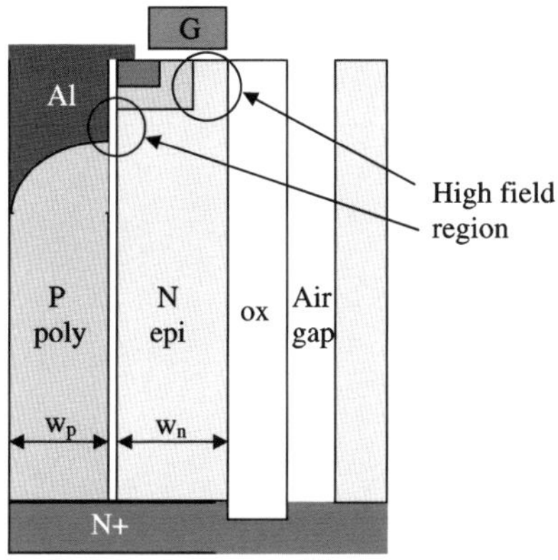

Fig. 10.12. Field crowding effect.

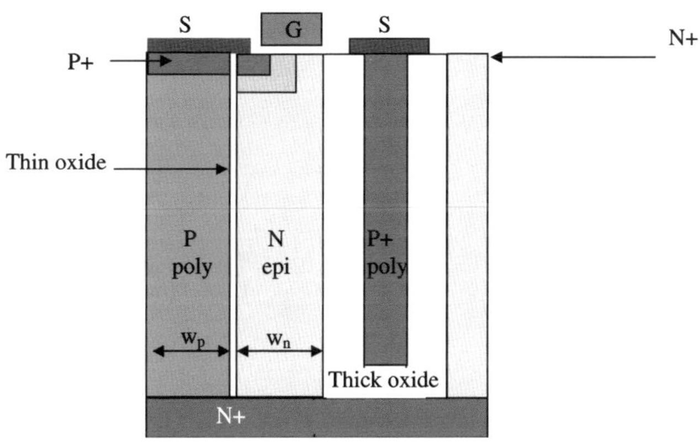

Fig. 10.13. PFVDMOS with Metal-Thick-Oxide (MTO) as edge termination.

active edge termination that reduces electric field at the n-epi edge. Both these remedies were implemented and the results will follow. The third likely cause of dopant-trapping in the poly is more fundamental and difficult to solve. The modified scheme uses better planarization and modified isolation, as shown in Fig. 10.13. Much better planarization is clearly seen from the SEM image of Fig. 10.14.

10.1.3.4. *Measurement on Modified PFVDMOS Devices*

Figure 10.15 shows measured off-state results of PFVDMOS with MTO edge edge termination of Fig. 10.13 that could achieve a breakdown voltage at 184 V

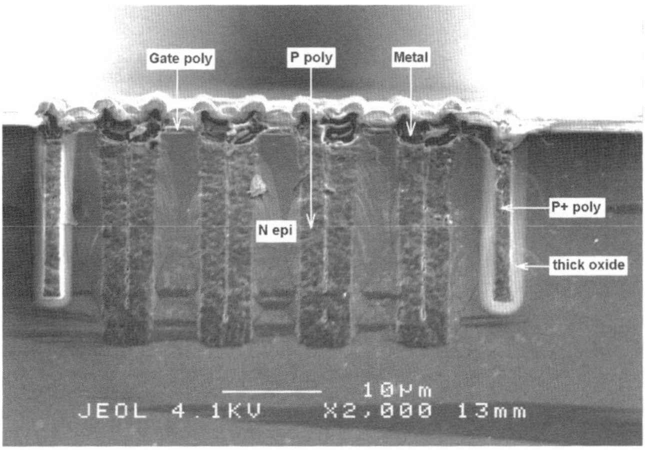

Fig. 10.14. The SEM picture showing the PFVDMOS structure with thick oxide-bypassed termination and better planarization compared to Fig. 10.11.

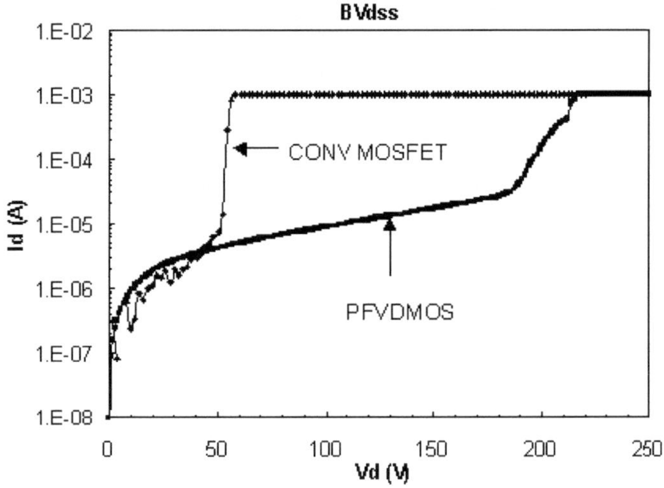

Fig. 10.15. Measurement results on device breakdown for conventional VDMOS and PFVDMOS devices.

at a reverse current level of 30 μA. The measured $R_{\text{on,sp}}$ was calculated to be about 3.68 m$\Omega \cdot$ cm^2. As can be seen in Fig. 10.15, the device suffers from high leakage current ($\sim$10 μA) compared to similar-sized "normal" device ($\sim$1 μA).

At the breakdown voltage of 184 V, the PFVDMOS device still fell short of simulated breakdown voltage of 280 V. This was found to be due to the failure of edge termination as the 1.5 μm oxide thickness used in the MTO could only sustain a breakdown voltage of about 200 V. By increasing the oxide thickness further, this problem might be solved but a more fundamental and difficult problem with PFVDMOS is its relatively high leakage current that could arise due to charge-trapping in p-poly. Thus, methods that could minimize this charge-trapping must be employed for PFVDMOS to be viable. These include low-temperature poly deposition that results in large grain formation and fewer grain boundaries, hydrogen passivation that terminates dangling bonds in poly-grain boundary, and selective epitaxy growth (SEG) that allows silicon-epitaxy to occur selectively only on exposed silicon area without the simultaneous growth on areas covered with oxide. The latter approach is particularly attractive as it could solve the poly-grain boundary problem while at the same time prevents interdiffusion between n- and p-columns by having a thin oxide layer between n- and p-columns. However, the process cost will be much higher and throughput will be adversely affected as SEG is a single wafer process.

Even though all the above methods might be used to minimize charge-trapping, the methods are under active research. Thus the use of the above methods to the PFVDMOS device might entail a lot of uncertainties. While not excluding their use in future, a more effective approach would be to find an alternative that is more robust and simpler to implement.

10.1.3.5. *Alternatives to PFVDMOS*

Remembering that the current research focus of superjunction device is geared toward achieving precise charge balance that might be difficult to achieve with current technology, other ways that might do away with this stringent requirement must be explored. In this regard, a new device structure that will shift the emphasis from controlling the doping to controlling the oxide thickness that yields much better control has been developed and called as Oxide Bypass MOSFET and introduced in Chap. 6. The structure is further modified to Tunable Oxide Bypass MOSFET to control off-state characteristics even better and improve on-state performance (Liang *et al.*, 2001; Yang *et al.*, 2003) as described in Sec. 6.8. We explore more details in the case of 100 V TOBUMOS device in the remainder of this chapter.

10.2. Case Study II: Tunable Oxide Bypass MOSFETS

The principle and theory behind OB structures have been presented in Chap. 6, Sec. 6.6. Both OB and TOB structures are shown in Fig. 6.32. In the standard Oxide Bypass Vertical Doublediffused Metal Oxide Semiconductor (OBVDMOS) structure, poly-contact of the Oxide-Bypass, p-body and n+ source are all shorted using a metal line, as shown in Fig. 6.32 (left). However, if the poly-contact is separated as shown in Figs. 6.32 (right) and 10.16, it potentially provides an additional control voltage that can affect the device performance. In principle, using simulations, an optimum device can be designed to achieve the lowest on-state resistance at a given breakdown voltage. However, process variations are inevitable and will alter desired device parameters. As described in Sec. 6.8, TOBMOS provides an avenue to compensate process variations to achieve optimum performance.

Due to the consideration of application in automotive industry, 100 V V_{br} TOBUMOS device shown in Fig. 10.16 was selected for fabrication. In UMOS structure, the MOSFET is implemented using a trench gate and the inversion layer forms on the vertical wall of the n-drift region.

10.2.1. 100 V TOBUMOS Fabrication

Due to the requirements for current power MOSFETs applications, 100 V Power MOSFETs are especially effective in both conventional and high-density isolated DC/DC power supplies for communications, automotive electrical system and industrial markets. As most of the SJ devices cannot break the ideal silicon limit to show better performance in this voltage range, TOBUMOS

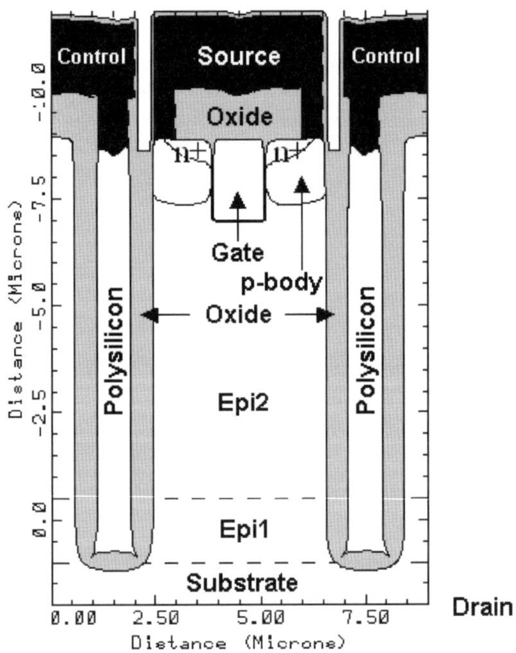

Fig. 10.16. TOBUMOS structure.

becomes more attractive for medium power applications. This part covers the development of 100 V TOBUMOS, which includes the numerical simulation to arrive at optimum device structure with enough process margin, proposed fabrication process flow, and mask layout design.

10.2.2. Simulation on 100 V TOBUMOS

A 100 V TOBUMOS structure (shown in Fig. 10.16) was generated and simulated using synopsys process simulator and device simulator (Synopsys) (Silvaco). The simulation was done on the device with dual epi-layers. The width (W) of n-drift region is 4 μm. The oxide thickness in the OB region is 0.525 μm. The first phosphorous epi-layer (Epi1) above 0.003 $\Omega \cdot$ cm antimony-doped substrate acts as the impurity out-diffusion source and has resistivity of 0.02 $\Omega \cdot$ cm. As this layer has higher resistivity and lower doping, out-diffusion of dopants into the second Epi2 layer is suppressed and more uniform doping profile can be maintained. Epi2 layer mostly sustains high reverse-blocking voltage.

In the consideration of possible variability of parameters between simulation and fabrication, detailed simulation of TOBUMOS at certain ranges of epi-thickness and resistivity was performed. The outcome of V_{br} and $R_{on,sp}$ of TOBUMOS at 0 V control bias and maximum allowed positive bias is given in Table 10.3.

The bias on the control electrode of TOBUMOS is used to tune the $R_{on,sp}$ vs V_{br} point to the farthest better performance location from ideal silicon limit line. Because most of the cases listed in Table 7.6 can break the silicon limit line, it is safe to choose the second epi-thickness of 8.5 μm with resistivity of 0.55 and 0.7 $\Omega \cdot$ cm, respectively. The latter case with resistivity of 0.7 $\Omega \cdot$ cm is selected to test the performance of TOBUMOS, while the former case is selected for optimal OBUMOS comparison.

According to Table 10.3, the relationship between V_{br} and resistivity for TOBUMOS on 8.5 μm epi-layer is shown in Fig. 10.17. It is clear that, with the optimal epi-resistivity of 0.55 $\Omega \cdot$ cm, there is no advantage of tunable voltage applied on the control electrode. With the nonoptimal epi-resistivity of 0.7 $\Omega \cdot cm^2$, positive control bias is able to reach up to 17 V. Hence, the ability to tune the device performance using the control voltage helps in expanding the process and yield window, but it is not so useful if nominal device has reached optimum performance.

The minimum $R_{on,sp}$ obtained for the device with such epi-resistivity is $6.04 \times 10^{-4}\ \Omega \cdot cm^2$. Most importantly, $R_{on,sp}$ vs V_{br} relationship at this point is better than that on the optimal OBUMOS with epi-resistivity of 0.55 $\Omega \cdot$ cm.

Table 10.3. Relationship of epi-resistivity (ρ_{epi}), epi-thickness (T_{epi}), control bias, V_{br} and $R_{on,sp}$ at $V_{GS} = 15$ V and $V_{DS} = 0.1$ V listed for 100 V TOBUMOS at $W = 4\,\mu$m and $t_{ox} = 0.525\,\mu$m.

ρ_{epi} ($\Omega \cdot$ cm)	$T_{epi} = 8.0\,\mu$m			$T_{epi} = 8.5\,\mu$m			$T_{epi} = 9.0\,\mu$m		
	Control (V)	V_{br} (V)	$R_{on,sp}$ ($\Omega \cdot cm^2$)	Control (V)	V_{br} (V)	$R_{on,sp}$ ($\Omega \cdot cm^2$)	Control (V)	V_{br} (V)	$R_{on,sp}$ ($\Omega \cdot cm^2$)
0.5	0	76.9	4.81×10^{-4}	0	80.3	4.28×10^{-4}	0	74.3	5.58×10^{-4}
	—	—	—	—	—	—	—	—	—
0.55	0	92.9	4.31×10^{-4}	0	100.6	5.57×10^{-4}	0	103.2	5.97×10^{-4}
	—	—	—	—	—	—	—	—	—
0.6	0	99.2	5.55×10^{-4}	0	100.5	5.99×10^{-4}	0	101.3	6.41×10^{-4}
	2	101.0	5.45×10^{-4}	7	106.3	5.69×10^{-4}	8	108.5	6.03×10^{-4}
0.65	0	97.3	5.86×10^{-4}	0	98.8	6.39×10^{-4}	0	99.6	6.86×10^{-4}
	7	103.8	5.57×10^{-4}	13	110.9	5.85×10^{-4}	12	111.0	6.27×10^{-4}
0.7	0	96.3	6.22×10^{-4}	0	97.3	6.77×10^{-4}	0	97.9	7.23×10^{-4}
	9	104.6	5.83×10^{-4}	17	113.2	6.04×10^{-4}	24	120.7	6.17×10^{-4}
0.75	0	94.3	6.60×10^{-4}	0	96.3	7.14×10^{-4}	0	96.7	7.65×10^{-4}
	17	110.8	5.93×10^{-4}	21	116.0	6.21×10^{-4}	23	118.6	6.51×10^{-4}

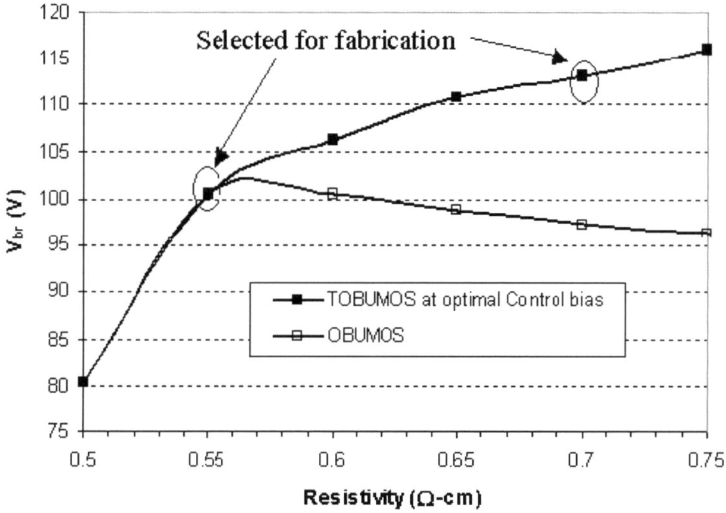

Fig. 10.17. V_{br} vs resistivity plots for OBUMOS and TOBUMOS at the second epi-depth of 8.5 μm.

10.2.3. Process Flow and Cross-Sections

The simplified process flow of 100 V TOBUMOS is best described in Fig. 10.18 using evolution of a set of TOBMOS cross-sections as the process proceeds.

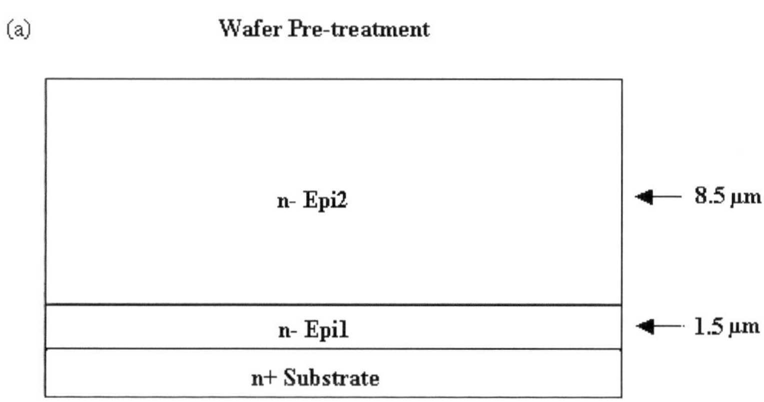

Fig. 10.18(a). Starting wafers with required epi-layers and wafer cleaning before fabrication.

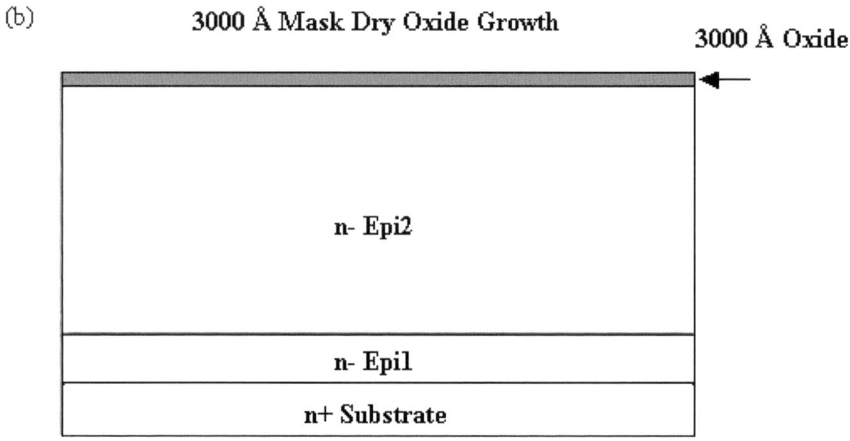

Fig. 10.18(b). Grow oxide to protect active device area where there will not be any trench.

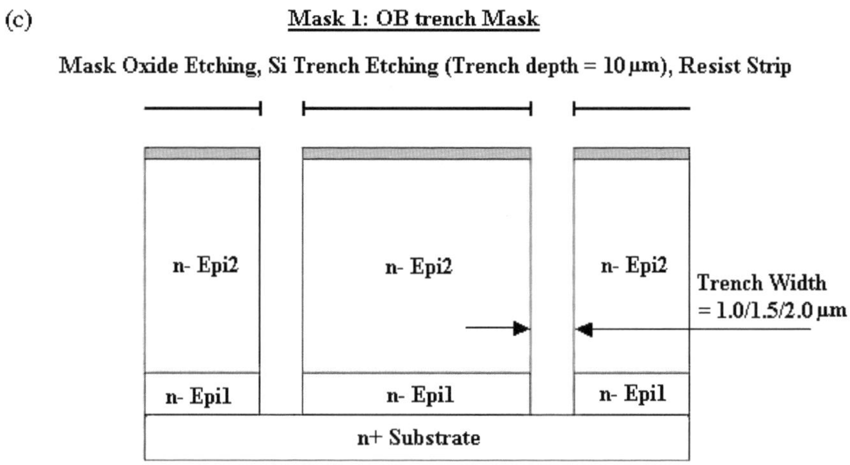

Fig. 10.18(c). Etch trench to form OB oxide (Must be wide to accommodate 500 nm oxide each side and poly).

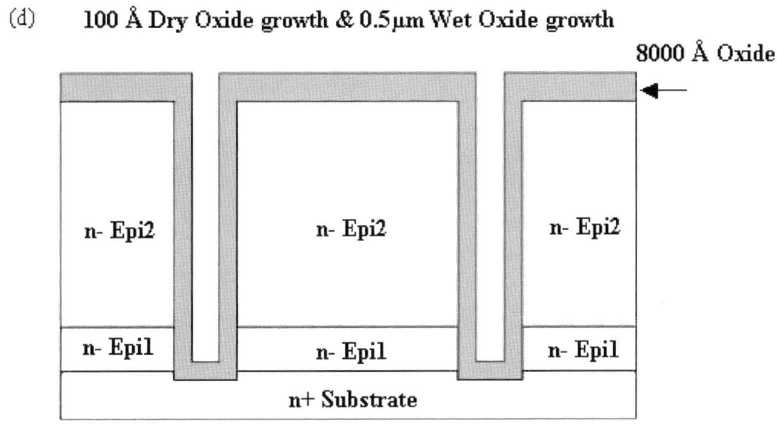

Fig. 10.18(d). Grow 10 nm high quality dry oxide and then remaining OB oxide leaving space for poly-fill.

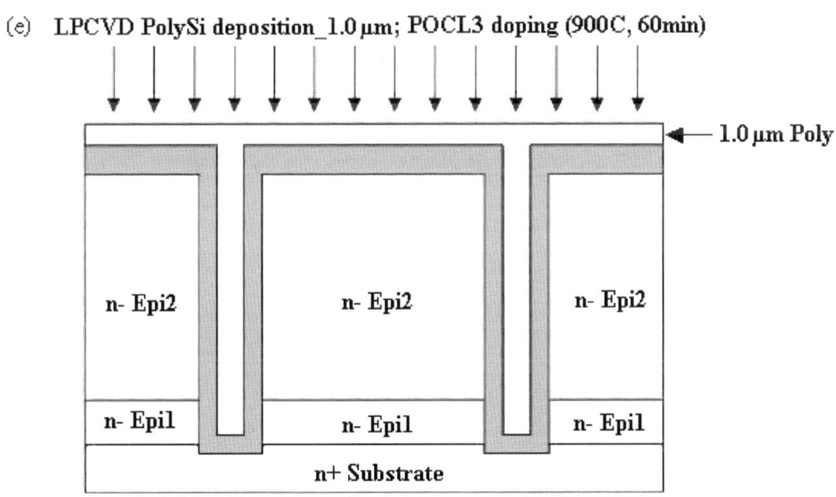

Fig. 10.18(e). 1000 nm poly is deposited and doped with phosphorus to make low-resistance connection.

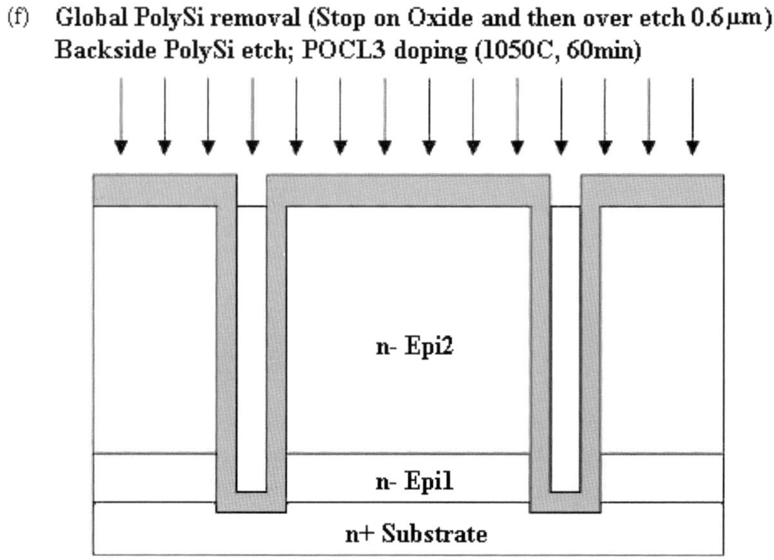

Fig. 10.18(f). Etch poly from the surface.

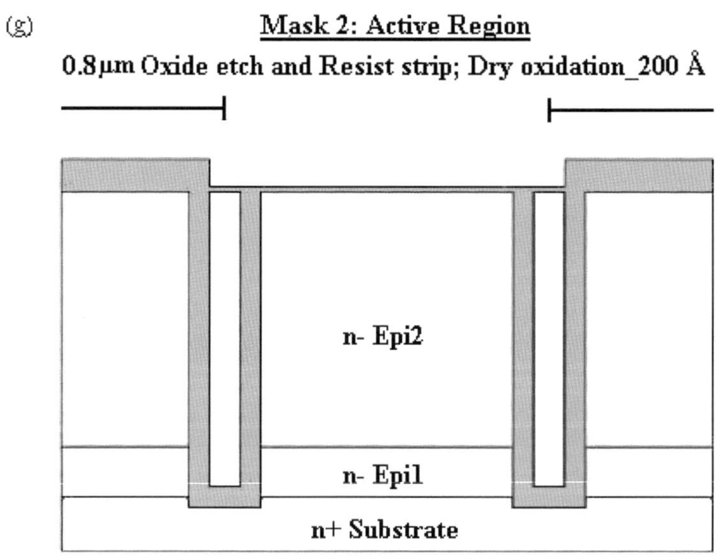

Fig. 10.18(g). Etch oxide from device area and oxidize again to begin process on active device areas.

(h) **Boron implant (2×10^{13} cm^{-2}, 90KeV, tilt = 7), Resist Strip Drive-in (Time = 100min, Temp = 1125), Dry Oxidation_200 Å**

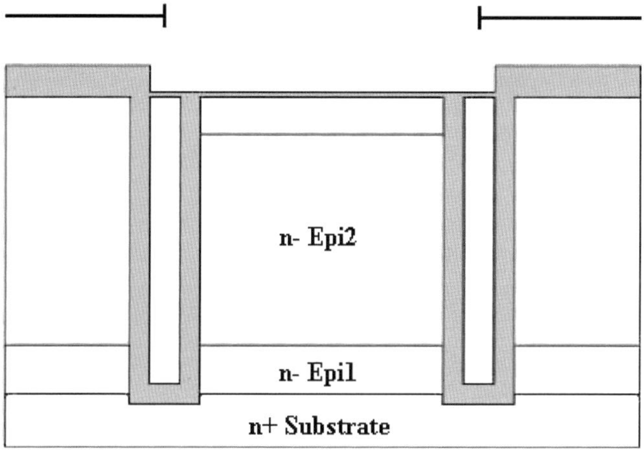

Fig. 10.18(h). Epi-layers form the drain region. Define p-body region of UMOS device with boron implant. This step is very robust as it is a blanket implant in the device area.

(i) Mask3: n+ Mask: As implant (5×10^{15} cm^{-2}, 120KeV, Tilt = 7); Resist Strip
Mask4: p+ Mask: BF2 implant (3×10^{15} cm^{-2}, 80KeV, Tilt = 7); Resist Strip; SRO (200 A) + LPCVD TEOS (2500 A)

(In one 2D cross section, only one of n+ or p+ Mask can be seen)

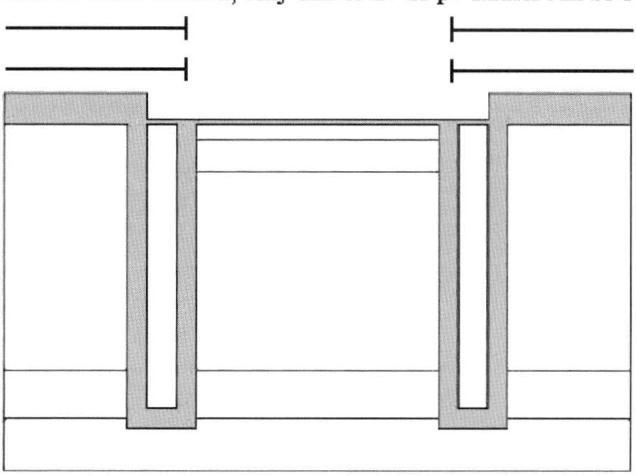

Fig. 10.18(i). Do As implant that defines source. Also do boron difluoride implant to define shallow heavily doped p-body contact regions. This p-type implant is hidden in this 2D view.

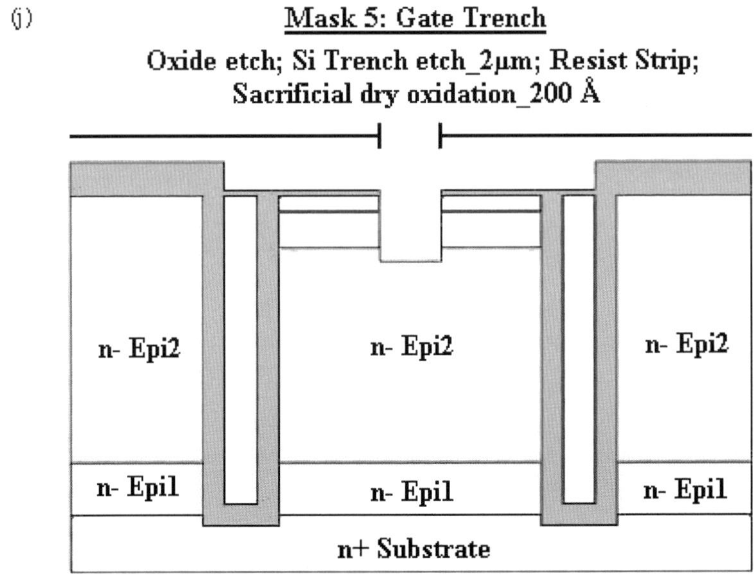

Fig. 10.18(j). Expose gate area and etch oxide and shallow silicon trench to define MOS device. Oxidize trench and wet etch to remove plasma etch damage before high-quality gate oxide is grown.

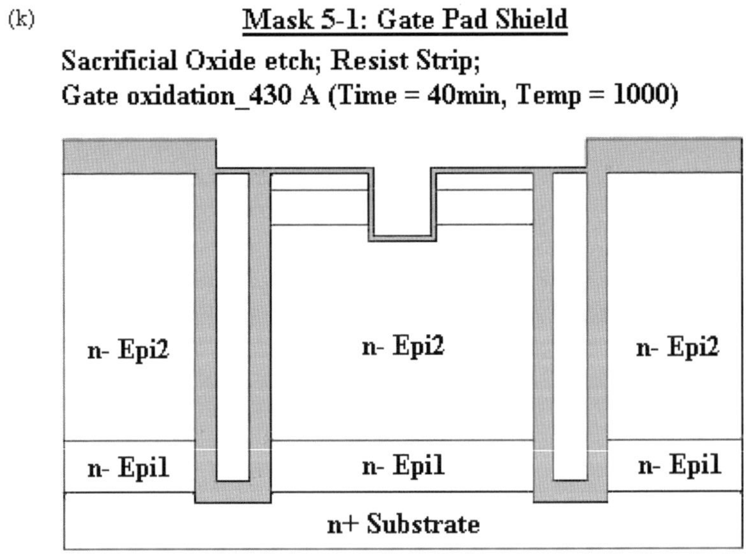

Fig. 10.18(k). Grow high-quality oxide that covers trench walls. Oxides normally thinner on sidewall and right processing conditions based on simulation should be selected.

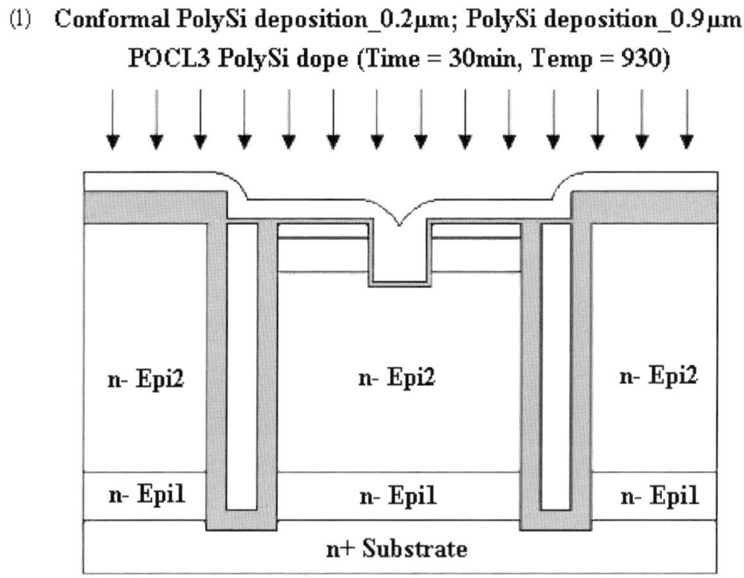

Fig. 10.18(l). Deposit polysilicon gate material and do a heavy phosphorus doping for low-resistance gate.

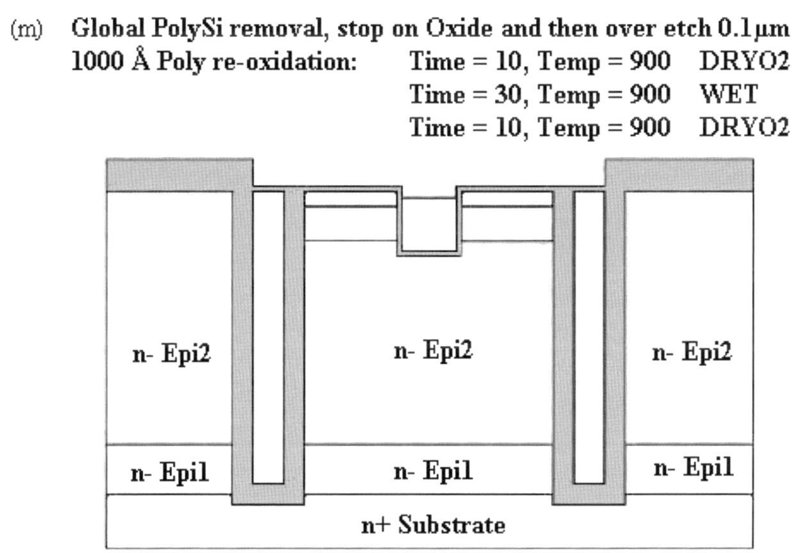

Fig. 10.18(m). Etch poly to stop on oxide. Etch further below trench level to avoid accidental shorts, still maintaining enough source overlap of the gate. Oxidize again to reach the surface level.

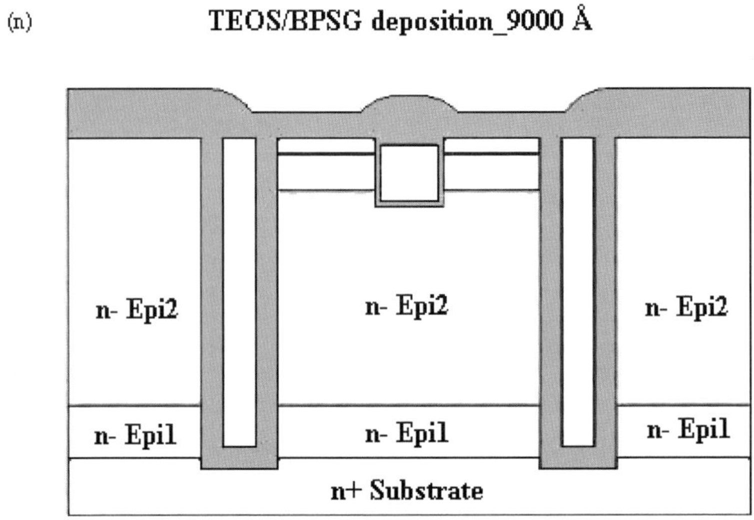

Fig. 10.18(n). Deposit thick borophosphosilicate glass (BPSG) as low-capacitance dielectric below metal. BPSG reflows at a lower temperature that helps in planarization.

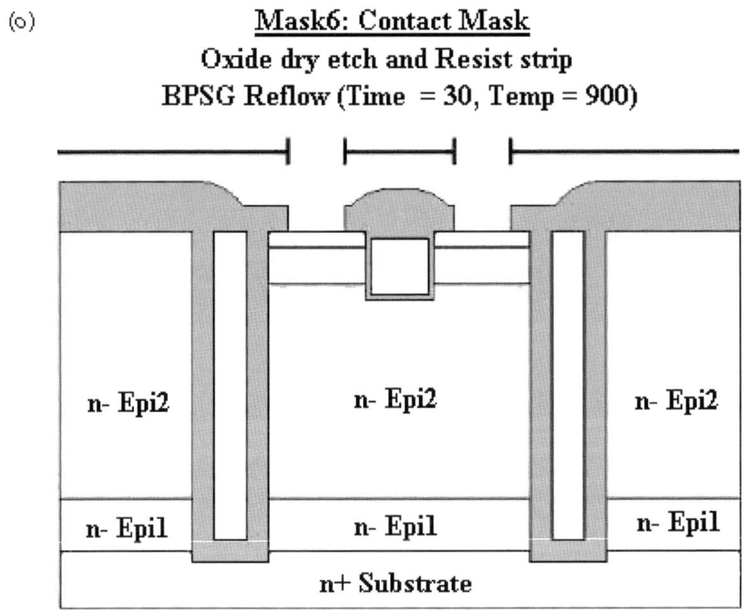

Fig. 10.18(o). Reflow BPSG for planarization. Expose contact mask to open metal connection areas. Only source contact that connects two sides of the trench device is seen here.

(p) PVD Barrier Metal Deposition; RTP Silicidation; PVD AL_1.7 μm

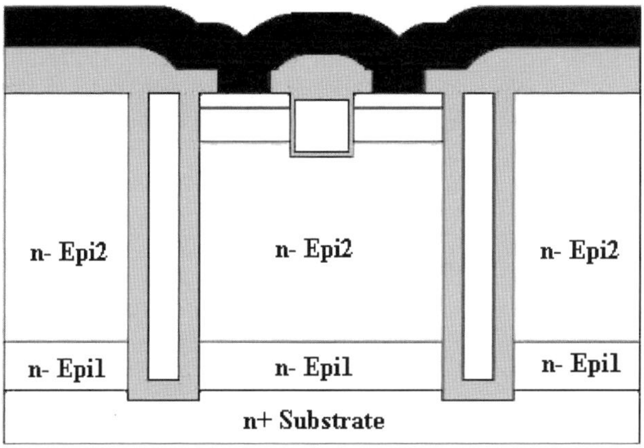

Fig. 10.18(p). Deposit and pattern metal (Aluminium) to define actual connections.

(q) Mask7: Al Etch Mask

Al dry etch and Resist strip; Passivation Oxide deposition

No etch regions are seen in this 2D crooss section as here Al covers whole device

Mask8: Passivation Mask

Pad opening and Resist strip; Backside strip; Alloy

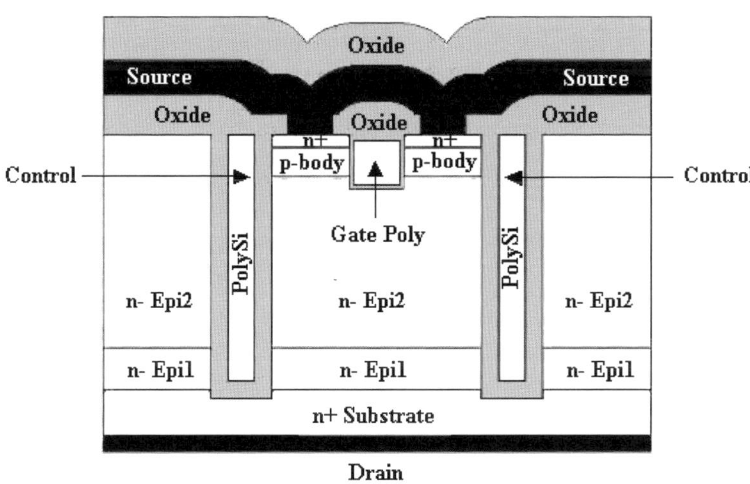

Fig. 10.18(q). Deposit thick passivation oxide to protect all device areas. Expose passivation mask and etch passivation oxide to expose only the metal pads for probing.

Fig. 10.18. Schematic cross-sections of TUBMOS device.

10.2.4. Key Precautions in TOBUMOS Fabrication

10.2.4.1. *Oxide Profile in OB Region*

The breakdown voltage is affected by the oxide thickness and profile at the trench bottoms. Because the breakdown always happens near the corner of the bottom trench oxide, a thicker oxide at the bottom than that at the sidewall will increase the breakdown voltage. On the other hand, if the trench corner is rounded by isotropic etching before oxidation, the oxide grows uniformly around the trench during the oxidation. Thus device off-state performance can be improved as well.

Figure 10.19(a) shows the desired oxide profile after fabrication where the trench corner is rounded by short-time isotropic etching. Compared to Fig. 10.19(a), 10.19(b) does not have such a rounding treatment before oxidation. The oxide in Fig. 10.19(b) becomes thin at the trench corner after the oxidation. In this case, electric field generated by the reverse bias is likely to be crowded at the corner region, which leads to a premature breakdown.

10.2.4.2. *Optimum Doping Profile in Drift Region*

As given in Fig. 10.17 and Chap. 6, owing to OB structure and tunable control voltage, 100 V TOBUMOS simulation has successfully shown the superior performance over conventional MOSFETs and SJ devices. This simulation was done on $0.02\,\Omega \cdot \text{cm}^2$ phosphorous-doped substrate. Phosphorous in the substrate is likely to diffuse into the epi-layer during thermal cycles, which is known as out-diffusion. Therefore after the thermal fabrication processes, the doping profile in the epi-layer of the device becomes nonuniform. The doping concentration is gradually reduced from the bottom to the top.

It has been proposed that a lower on-resistance can be obtained by using a nonuniform epitaxial doping profile to obtain the same breakdown voltage (Baliga, 1998). Some experiments trying different impurities and doping

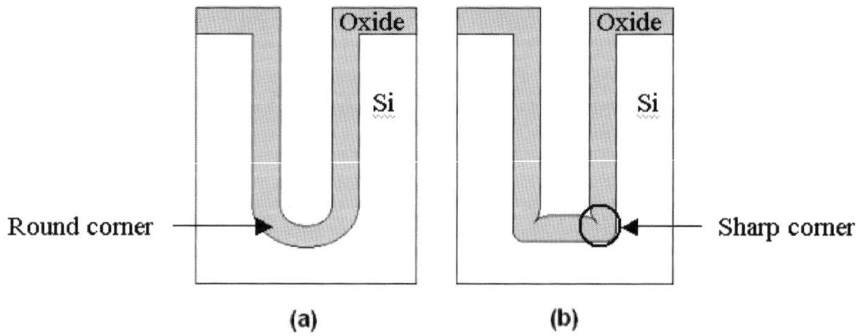

Fig. 10.19. Comparison of two different trench profiles of after oxidation.

concentrations have been done to find the optimum epi-doping profile for 100 V TOBUMOS. It was established through simulations that, by using 0.02 $\Omega \cdot$ cm phosphorous-doped substrate, the best performance of 100 V TOBUMOS can be obtained. However, the wafer with phosphorous-doped substrate is commercially unavailable. In this case, wafers with multi-epi-layers are selected. Same performance can be achieved by using this wafer instead of the previously designed one. As shown in Figs. 10.16 and 10.20, the new wafer contains two epi-layers. The first epi-layer right above the substrate is doped with phosphorous, which has the resistivity of $\rho_{\text{epi1}} = 0.02\,\Omega \cdot \text{cm}$. The resistivity for the second epi-layer above the first epi-layer is in the range between 0.55 and 0.75 $\Omega \cdot$ cm. The substrate material and doping are allowed to vary in a large range to explore if this helps in optimization. Here we choose antimony-doped substrate with $\rho_{\text{sub}} = 0.003\,\Omega \cdot \text{cm}$. Figure 10.20 shows the post-fabrication doping profile of 100V TOBUMOS with epi1 thickness = 1.5 μm, epi2 thickness = 8.5 μm, $t_{\text{ox}} = 0.525\,\mu\text{m}$, half-drift region width = 2 μm, and half OB structure width = 1μm (Fig. 10.25). The simulation result for this structure is $V_{\text{br}} = 100.6$ V and breakdown occurs at (2.43 μm, −2.4 μm), with $R_{\text{on,sp}} = 5.57 \times 10^{-4}\,\Omega \cdot \text{cm}^2$.

10.2.4.3. *Oxide Over Etch at Termination*

As in the case of PFVDMOS, the first OBUMOS fabrication yielded low breakdown voltage compared to prediction. Also, some of the gates were found to be open circuits in the sense that the gate pad did not connect to the trench gate and

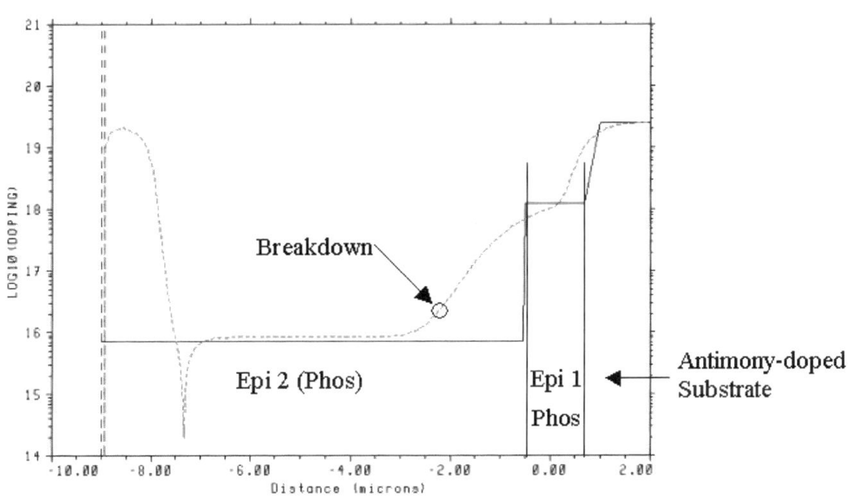

Fig. 10.20. Vertical doping profile along the drift region and substrate pre-fabrication (solid line) and post-fabrication (dashed line).

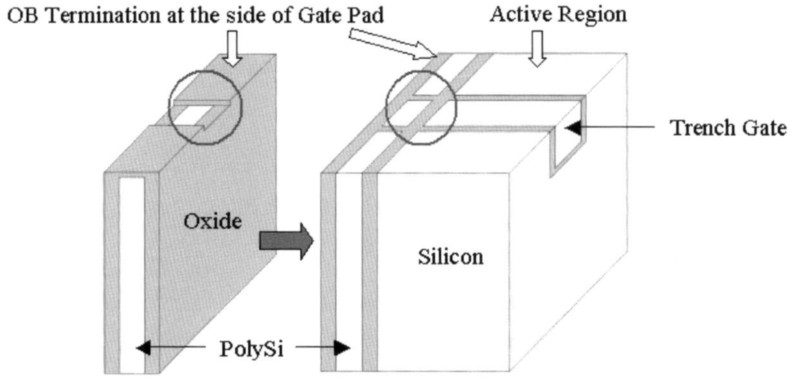

Fig. 10.21. Failed structures with slight oxide etch at gate region before and after gate formation.

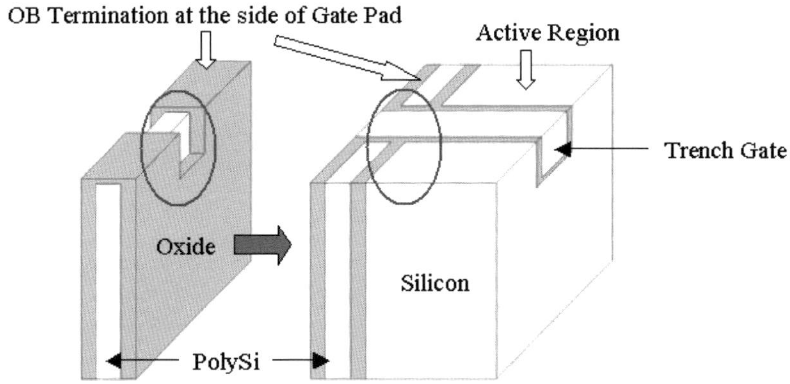

Fig. 10.22. Functional structures with oxide over etch of $1 \sim 2\,\mu$m at gate region before and after gate formation.

the device never turned on. This was most likely due to the discontinuity of Gate poly at OB termination, as shown in Fig. 10.21; the Gate is electrically open since the Gate contact is only on the OB termination. Therefore, oxide etch in OB region is required to be deep enough to allow Gate poly to go across. Normally, 1–2 μm oxide over etch (see Fig. 10.22) is desired for 2 μm Gate trench.

To check the oxide over etch depth at the termination mentioned above, a test structure shown in Fig. 10.23, was designed for SEM. Thus, by simply cutting the wafer along any horizontal line, the cross-sectional view of Gate trench at OB termination can be seen clearly using SEM. Design of such test structure is of utmost importance so that proper completion of a designed

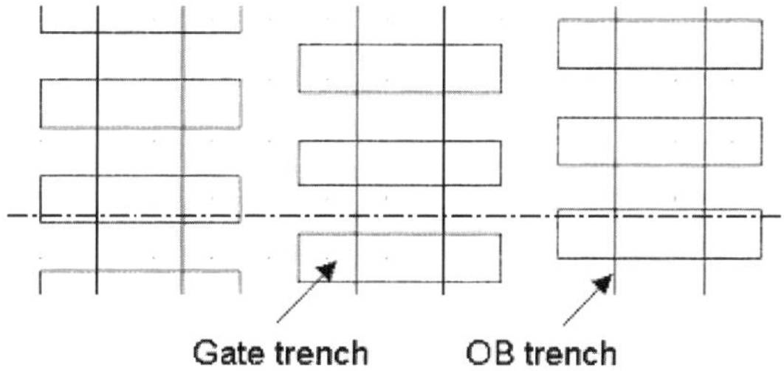

Fig. 10.23. Gate trench test structure.

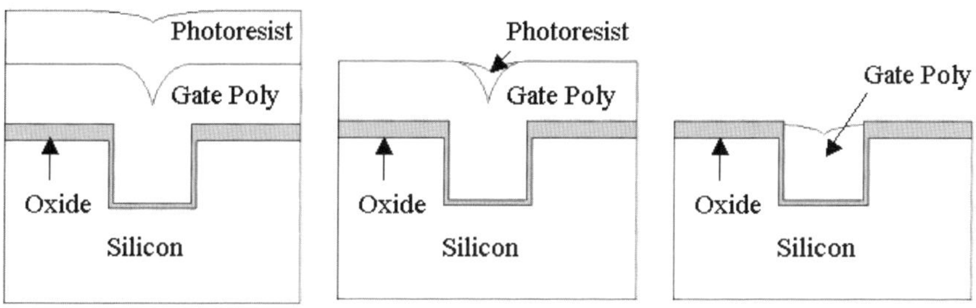

Fig. 10.24. Gate poly removal by using resist-assisted etchback technology.

process step can be verified by running a test wafer initially, and corrected if required.

10.2.4.4. *Resist-Assisted Etchback Applied on PolySi Removal*

Chemical Mechanical Polishing (CMP) is nowadays used extensively for a blanket etch that planarizes structures. Although in principle, polysilicon CMP is a straightforward technology applied on Gate poly etching for UMOS devices, it is actually quite complex in practice. Proper controls must be guaranteed to yield a uniformly polished surface. Instead, resist-assisted etchback is a simple way to remove Gate poly, and the outcome is a relatively planar surface. As shown in Fig. 10.24, photoresist is deposited onto polysilicon. By using photoresist to fill up the notch of polysilicon, a smooth surface profile can be patterned easily by adjusting relative etch rates of polysilicon and the resist.

10.2.4.5. *Additional Precautions*

To avoid problems encountered in the first fabrication, some additional precautions were taken.

(1) Some variations should be made in the consideration of mask alignment and process offset issues.

(2) The trench Gate and Gate oxide formation should be performed with extreme care, because the quality of the Gate oxide surface and the shape of Gate trench at the termination area will affect the on-state performance significantly.
(3) The distance between Gate trench mask and Source contact mask should be more than 0.8 μm, to avoid the short of Gate and Source.
(4) There should be enough space between the Source and Control contact.

With these concerns in mind, the final process and masks for 100 V TOBUMOS were designed.

10.2.5. Device Structure and Mask Layout Design

The proposed device cross-section view and fabrication masks are shown in Figs. 10.25 and 10.26, respectively. The abbreviations used in the figures are elaborated just after the figures. Compared to conventional UMOS fabrication steps, only one additional mask (POLY1 Mask) is required. Figure 10.26 shows the top views of mask layout with different POLY1 masks. They were designed to check the effect of two different schemes of termination on OB functionality.

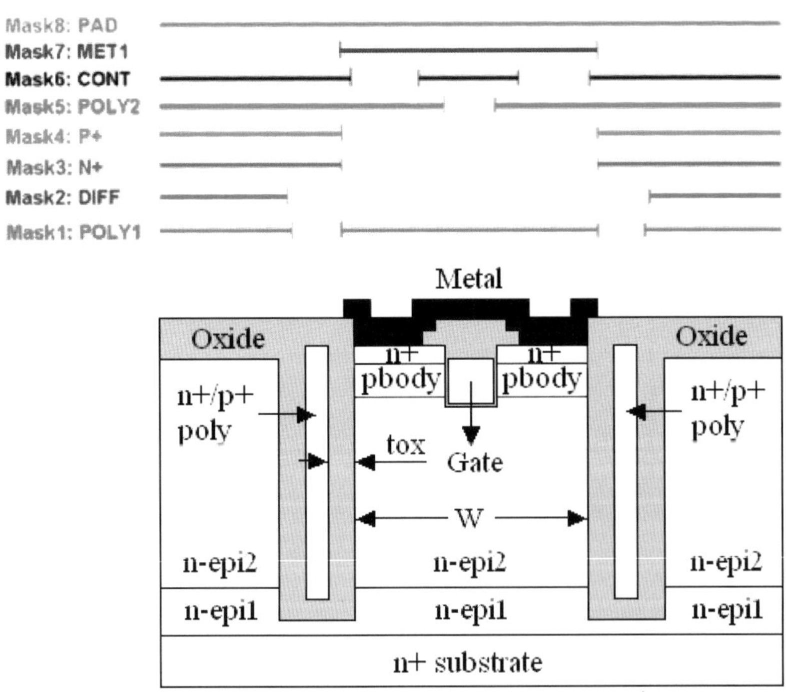

Fig. 10.25. Cross-sectional view of 100 V TOBUMOS together with masks.

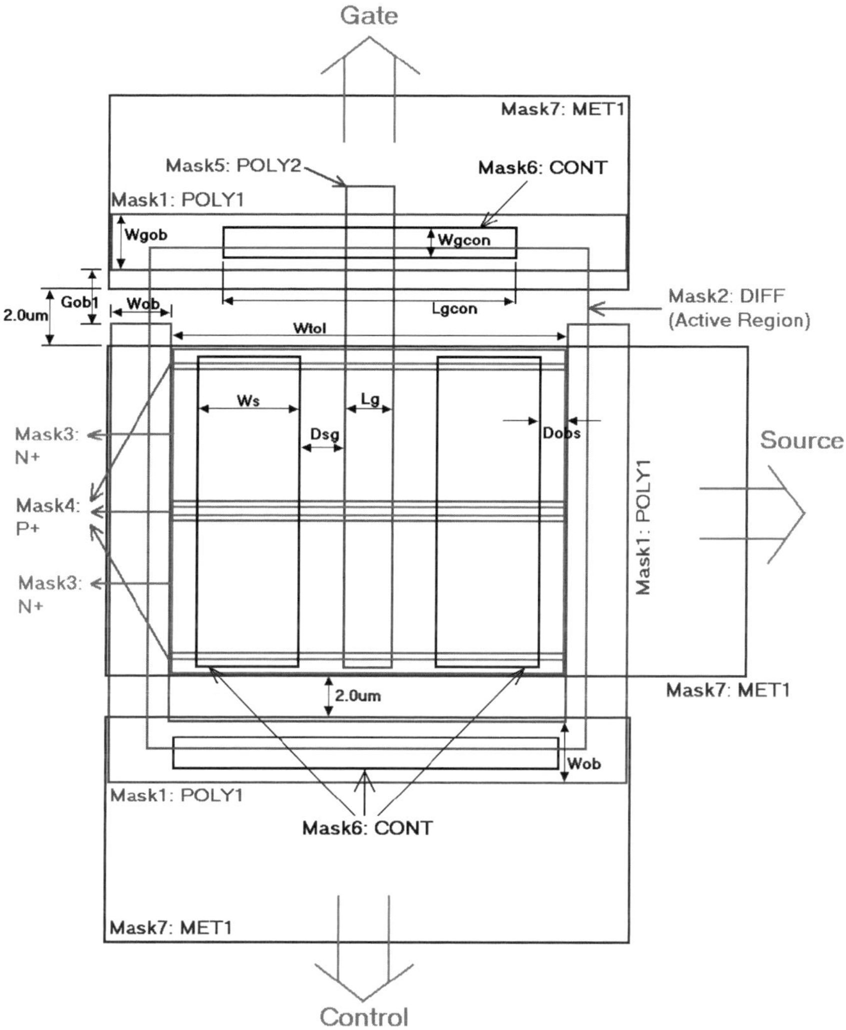

(a) Termination type 1

Fig. 10.26. Top view (mask) of 100 V TOBUMOS with two types of OB trenches at top termination.

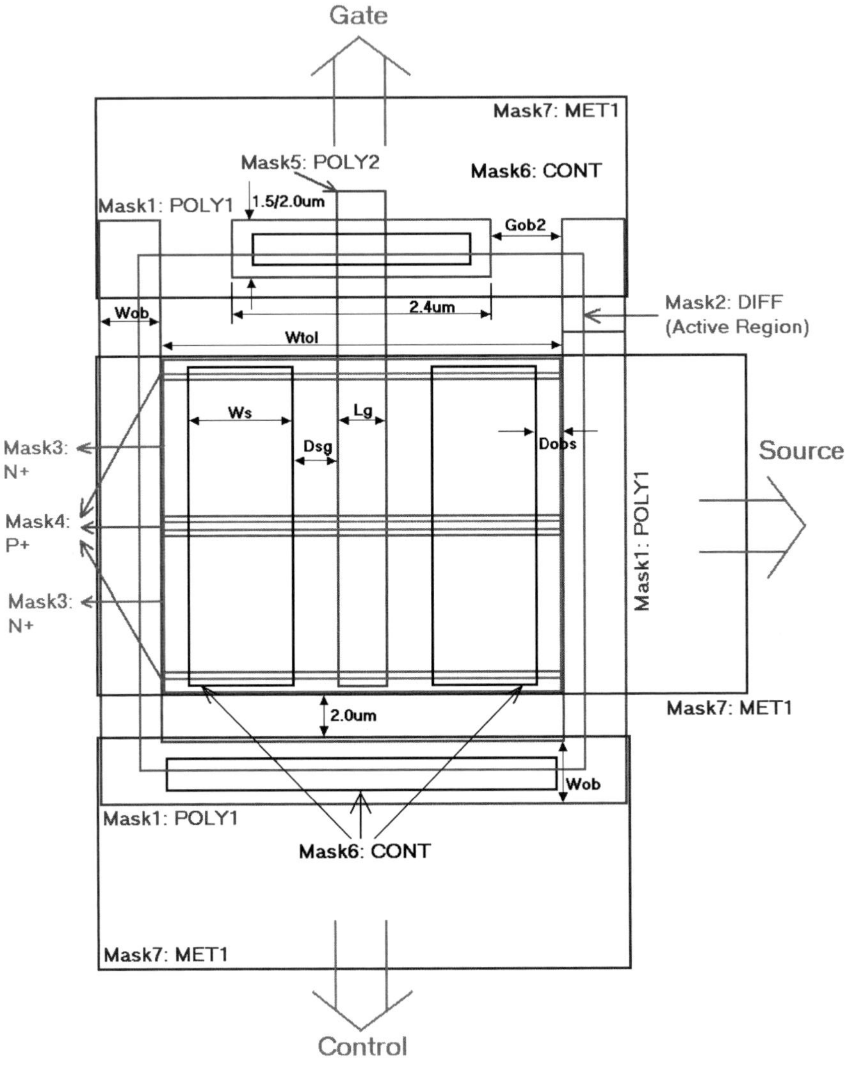

(b) Termination type 2

Fig. 10.26. (*Continued*)

Table 10.4. Parameter variations of TOBUMOS mask layout.

Parameter	Wgcon	Lgcon	Wob	Wgob	Gob1	Gob2
Range (μm)	0.8/0.9/1.0	2.8–7.0	1.0/1.5/2.0	1.5/2.0	1.25/1.75/2.0	0.8–1.8
Parameter	Wtol	Lg	Ws	Dsg	Dobs	
Range (μm)	3.0–6.0	0.8/1.0	0.6/0.8/1.0	0.3–1.05	0.1–0.7	

It is to be noted that Control voltage contact of TOBUMOS is only located below the OB trench in the layout due to the space limitation in lateral dimension. This will not affect the device performance, because only voltage bias is applied on Control contact. There is no current conduction in the highly doped polysilicon region of OB structure. There are totally nine masks in the process, and no additional mask is needed compared to the fabrication of OBUMOS.

There are totally four electrodes for TOBUMOS. Except for Drain electrode, which is deposited at the bottom of the wafer, all the other three electrodes are on the top of the wafer. The schematic of mask layout and parameter variation are shown as in Fig. 10.26 and Table 10.4, respectively. Separated contact windows are made on Source contact and Control contact as needed by TOBUMOS device structure.

The explanations for the parameters and abbreviations shown in Fig. 10.26 are as follows:

Wgcon: Gate contact width
Lgcon: Gate contact length
Wob: OB trench width
Wgob: Width of OB trench on the Gate side

Gob1: Gap of OB trench for termination type 1
Gob2: Gap of OB trench for termination type 2
Wtol: Total width between neighboring OB trenches
Lg: Gate length
Ws: Width of Source contact
Dsg: Distance between Source contact and Gate trench
Dobs: Distance between OB trench and Source contact.

10.2.6. Mask Floorplan and Splits for 100 V TOBUMOS Fabrication

The mask floor plan and wafer splits were designed so that in spite of process variations, there will be good device yield. Layout floor plan achieves three

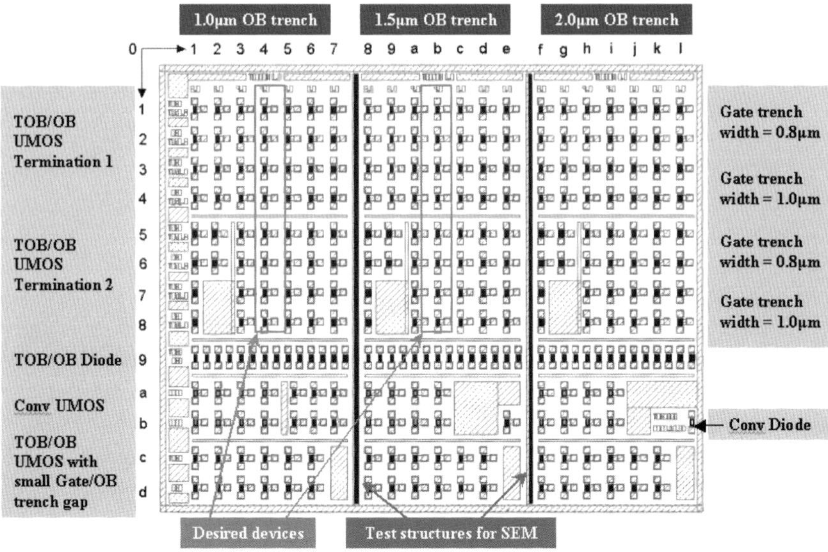

Fig. 10.27. Mask floorplan design for 100 V TOBUMOS fabrication.

different OB trench widths and has TOB diode structures without any n^+ source or trench gate so that V_{br} tenability without MOS structure interference can be assessed. These results will be presented in the measurements.

In real fabrication, mask misalignment (0.2–0.3 μm) and some parameter variations must be considered into mask layout design, as shown in Fig. 10.27. The desired devices with exact parameters as in simulation locate in columns 4 and b only and are enclosed in rectangles. Variations such as n-drift region width (Wtol) from 3.0 μm to 6.0 μm are made within each region (column 1–7, column 8–e, or column f–l), at 0.5 μm interval. Conventional UMOS and Diode are also fabricated on the same wafer for comparisons.

The following are the wafer splits used that give varying epi-layer resistivities to evaluate tuning and different OB oxide thicknesses and p-body dose in the MOSFET channel region in case there is process variation that changes desired OB thickness or threshold voltage of the trench gate MOSFET.

#01–#10: Dual epi wafers, Antimony-doped substrate
(Epi layer1: 0.02 $\Omega \cdot$ cm Phos, 1.5 μm; Epi layer2: 0.55 $\Omega \cdot$ cm Phos, 8.5 μm)
#11–#20: Dual epi wafers, Antimony-doped substrate
(Epi layer1: 0.02 $\Omega \cdot$ cm Phos, 1.5 μm; Epi layer2: 0.70 $\Omega \cdot$ cm Phos, 8.5 μm)
#21–#29: Test wafers.

Table 10.5. Oxide thickness and p-body implant splits for TOBUMOS fabrication.

Wafer no.	Oxide thickness — OB	Boron implant dose — p-body
01 & 11	0.4 μm	2×10^{13} cm^{-2}
02 & 12	0.4 μm	2×10^{13} cm^{-2}
03 & 13	0.4 μm	4×10^{13} cm^{-2}
04 & 14	0.5 μm	2×10^{13} cm^{-2}
05 & 15	0.5 μm	2×10^{13} cm^{-2}
06 & 16	0.5 μm	2×10^{13} cm^{-2}
07 & 17	0.5 μm	4×10^{13} cm^{-2}
08 & 18	0.6 μm	2×10^{13} cm^{-2}
09 & 19	0.6 μm	2×10^{13} cm^{-2}
10 & 20	0.6 μm	4×10^{13} cm^{-2}

Considering the parameter offsets in fabrication, some wafers are split for oxide thickness in OB region and p-body implantation dosage variations, as in Table 10.5.

10.2.7. 100V TOBUMOS Measurement Results and Discussions

The fabrication of TOBUMOS was performed on two types of dual epi silicon wafers. As described previously, simulated TOBUMOS devices show good performance on both on- and off-state in comparison to conventional power MOSFETs. It is verified through simulations on tunable OB-Diode that the tuning effect functions well on improving the breakdown voltage. In the same way, Drain current at on-state can be improved as well under certain control bias. This effect allows the non-optimized OB devices to have a superior performance after being tuned. This subsection deals with the details of fabrication and analyses on measurement results.

10.2.7.1. *Physical Parameter Measurements on Fabricated TOBUMOS*

Figure 10.28 is the SEM picture showing the cross-sectional view of TOBUMOS after OB trench etching. The 10 μm deep trench sidewall is straight and it has a rounded trench corner at bottom as expected due to the additional short isotropic etching step.

The SEM picture of the test structure for oxide over etch at termination is shown in Fig. 10.29. As described in Fig. 10.22, at termination region, 1–2 μm oxide over etching is required to make the connection of PolySi in both Gate trench and OB trench regions. It is measured that, the oxide over etching depth

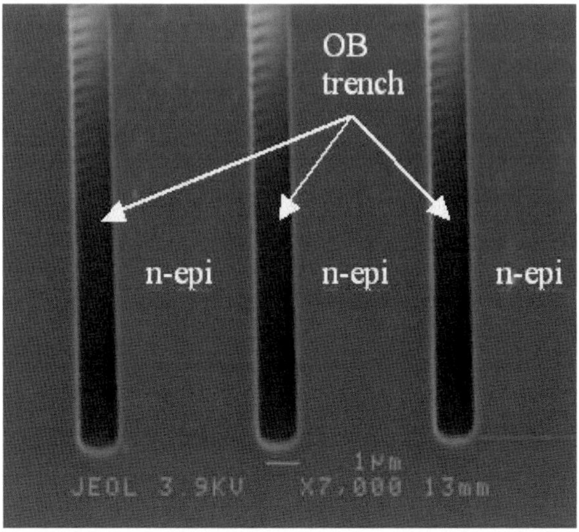

Fig. 10.28. SEM picture for TOBUMOS cross-section after OB trench etching.

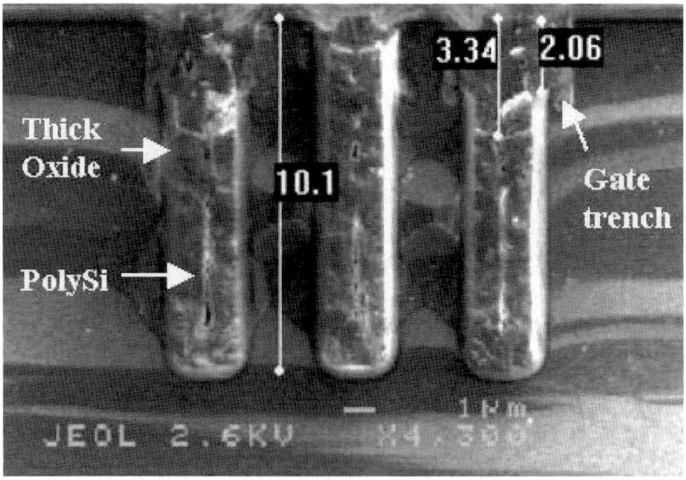

Fig. 10.29. SEM picture for Gate trench test structure.

is about 2.06 μm, while the measured Si trench depth in the Gate region is about 2.5 μm. When performing the Si etching in the Gate region, the etch rate for PolySi is higher than that of Si. This results in a deeper PolySi trench depth of 3.34 μm, as shown in Fig. 10.29.

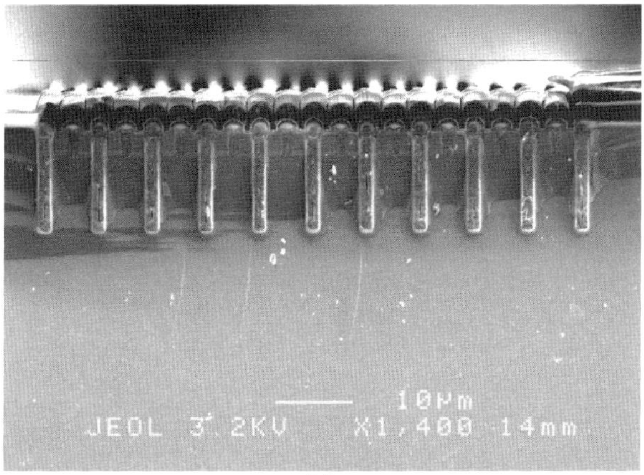

(a)

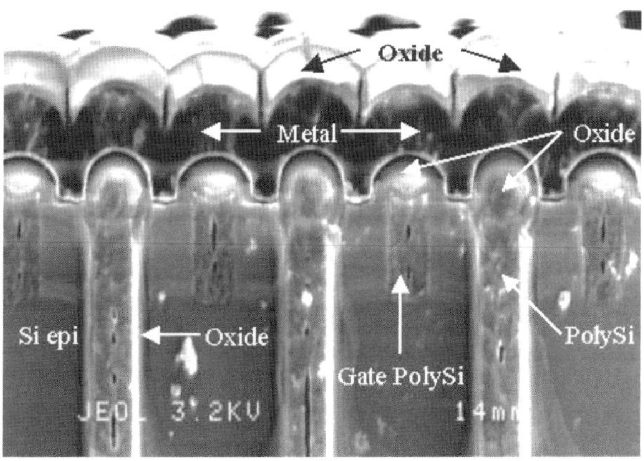

(b)

Fig. 10.30. SEM pictures for fabricated TOBUMOS.

SEM pictures were taken for the final TOBUMOS structure, as shown in Fig. 10.30. This cross-sectional view is obtained by cutting the device along the line perpendicular to the Gate trench in the active region. The entire structure and enlarged picture for the Gate region are shown in Fig. 10.30(a) and 10.30(b), respectively. It is clear that the Source metal strides over the active region and the Source contact is separated from the Gate and OB trenches. Both Gate and OB trenches have round oxide profile at the trench corner. The oxide thickness t_{ox} is designed to vary from 0.4 to 0.6 μm. The spacing between OB trenches is in the range of 3–6 μm.

10.2.7.2. *Tunable Effects on Breakdown Voltage of TOB-Diode*

Fabrication of TOB-Diode was done to verify the TOB concept. Except for the absence of Gate structure, the device structure and dimensions are the same as in TOBUMOS. This is achieved by eliminating the masks responsible for forming MOSFETs selectively. Without external bias, the breakdown voltage of TOB-Diode based on the epi-resistivities of 0.55 Ω·cm and 0.7 Ω·cm is 83 V and 82 V, respectively. The comparisons of measured blocking characteristics of TOB-Diode under positive bias and conventional Diode fabricated on the same wafer of 0.55 Ω · cm epi are shown in Figs. 10.31 and 10.32, respectively. The V_{br} for both cases of of as-fabricated TOB-diodes are not much higher than that of the conventional Diode. However, the control voltage has significant impact.

The maximal V_{br} that TOB-Diodes can reach are 103 V at 20 V Control bias on the epi-resistivity of 0.55 Ω · cm and 112 V at 30 V Control bias on the epi-resistivity of 0.7 Ω·cm, respectively. The OB structure is sensitive to the dimensions of W, t_{ox} and n-drift doping concentration (N_d) as described in Sec. 6.6, but accompanied with the tuning effect as in TOB devices, some variations of parameters in fabrication will not bring too much influence on the device performance. Most probably due to the variations of epi-doping, the V_{br} of fabricated TOB-Diode without Control bias is about 15 V lower than the result expected by simulation. However, adjusted by the Control voltage,

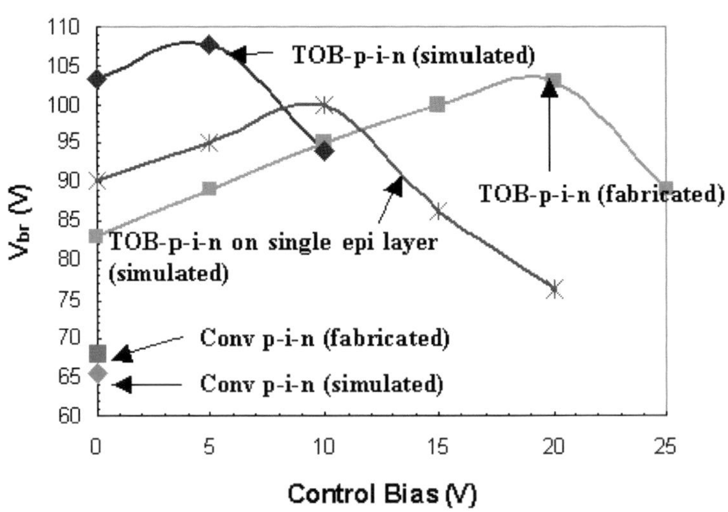

Fig. 10.31. Breakdown voltage under positive control bias for TOB structure and conventional structure on the same epi-wafer.

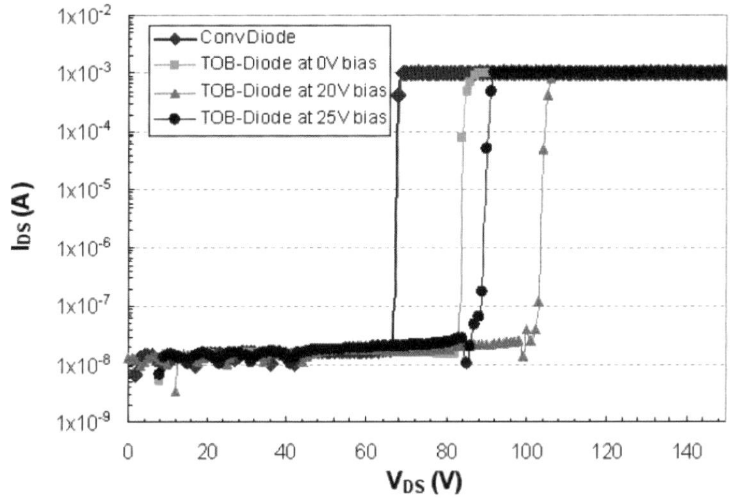

Fig. 10.32. Measured blocking characteristics of TOB-p–i–n structure under 0, 20, and 25 V control biases on epi-resistivity of 0.55 Ω · cm in comparison with the conventional structure fabricated on the same wafer.

a V_{br} higher than 100 V can be obtained. It is clear that, the introduction of TOB structure represents good tunable characteristics without bringing any additional leakage current to the device. Theoretically, TOB structure also has the ability of enhancing on-state resistance and transconductance if applied on the MOSFET. The successfully fabricated TOB-Diode forms the foundation for successful TOB-MOSFETs fabrication.

10.2.7.3. *Experimental Measurement Results on TOBUMOS First Run*

TOBUMOS devices are fabricated together with TOB-Diode on the same wafer. Unfortunately, the first separate run of TOBUMOS fabrication was not fully successful. Though the fabricated TOBUMOS shows good IV characteristics and improved specific on-resistance under certain Control voltage, it was found that the breakdown voltage is always about 30 V and the breakdown measurement is unrepeatable. By performing the electrical test and analysis on Gate region, the Gate oxide breakdown near termination region was suspected. Therefore, follow-up efforts were performed in order to solve the problem of premature breakdown.

Fabricated TOBUMOS was tested having normal IV characteristics. Figure 10.33 shows the Drain current vs Drain voltage on wafer # 06 at different Gate voltages from 0 V to 10 V at 1 V interval, without Control

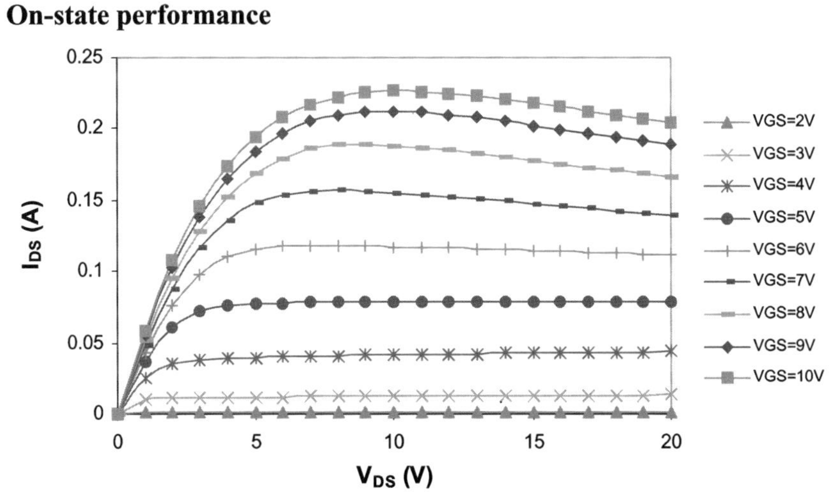

Fig. 10.33. IV characteristics of TOBUMOS with $W = 4.0\,\mu$m, Wob $= 1.5\,\mu$m, and Lg $= 0.8\,\mu$m (refer to Table 10.4) on wafer #06 at V_{GS} in the range of 0–10 V.

voltage. Measured threshold voltage as in Fig. 10.34 is about 2 V, which is higher than the simulated value of 1.2 V.

10.2.7.4. *Tuning Effects on TOBUMOS*

It was found that by applying positive voltage on Control electrode, the on-state performance of TOBUMOS is enhanced. As given in Fig. 10.35, the plots of IV characteristics of TOBUMOS under 20 V Control bias are shown in comparison to the original IV performance without external bias. The enhanced Drain current under bias can be obtained in both the linear region and the saturation region of the MOSFET. The enlarged IV curve at $V_{GS} = 10$ V at smaller V_{DS} below 0.1 V is shown in Fig. 10.36. Obviously, with the augment of I_{DS}, on-state resistance is improved when increasing the Control bias from 0 to 20 V and the decrease is larger between 0 and 10 V compared to that between 10 and 20 V Control bias.

10.2.8. Investigations for Off-State Failure

Measured breakdown voltage for most fabricated TOBUMOS was below 30 V. Only on wafer #19, one or two devices were found to have the breakdown voltage of about 80 V. There is no device showing an expected V_{br} of about 100 V as desired. Moreover, in any case, the breakdown test is unrepeatable. As shown in Fig. 10.37, the breakdown test shows that the device has a V_{br}

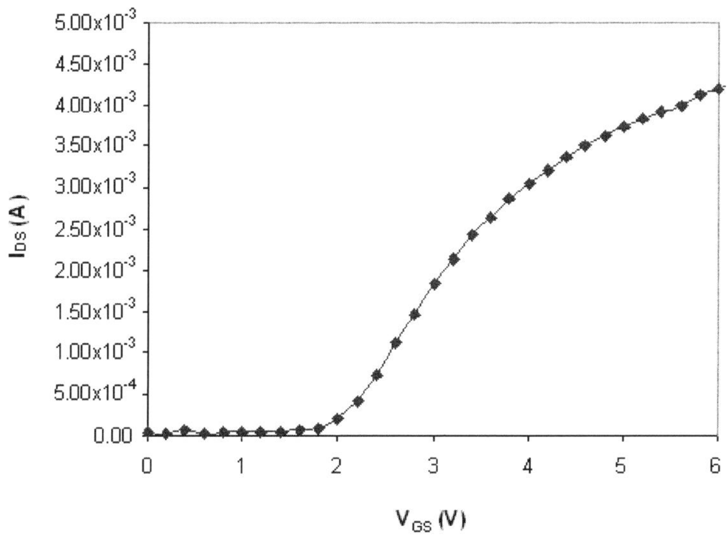

Fig. 10.34. Drain current performance of TOBUMOS on wafer #06 at V_{GS} in the range of 0–6 V and $V_{DS} = 0.1$ V.

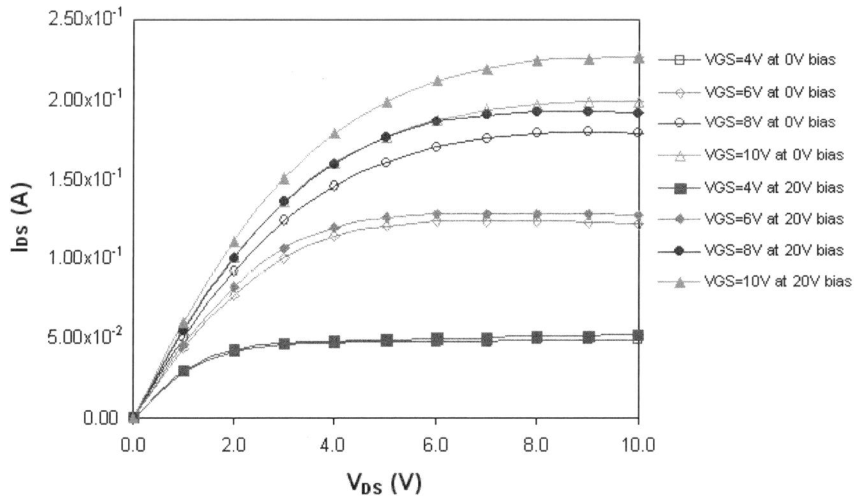

Fig. 10.35. IV performance under positive control bias of 0 and 20 V for TOBUMOS with $W = 3.5\,\mu$m, Wob $= 1.5\,\mu$m, and Lg $= 0.8\,\mu$m (refer to Table 10.4) on wafer #19, at $V_{GS} = 4, 6, 8$, and 10 V, respectively.

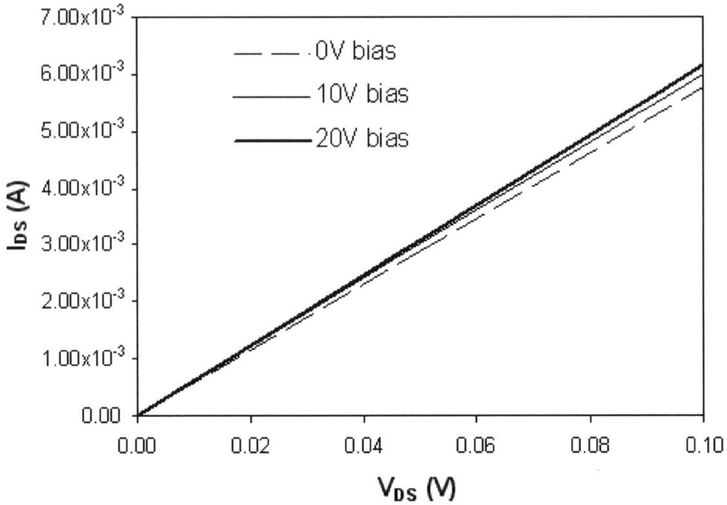

Fig. 10.36. IV performance under positive control bias of 0, 10, and 20 V for TOBUMOS with $W = 3.5\,\mu$m, Wob $= 1.5\,\mu$m, and Lg $= 0.8\,\mu$m (refer to Table 10.4) on wafer #19, at $V_{GS} = 10$ V.

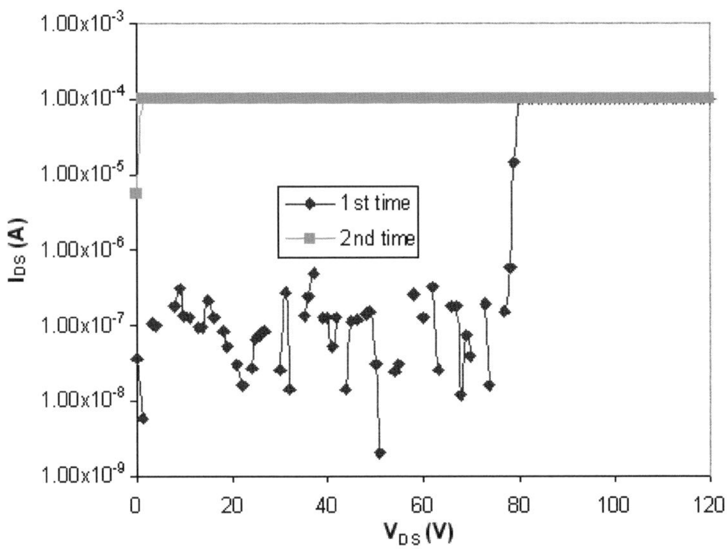

Fig. 10.37. Measured off-state performance for TOBUMOS with $W = 4.5\,\mu$m, Wob $= 1.5\,\mu$m and Lg $= 0.8\,\mu$m (refer to Table 10.4) on wafer #19.

of 78 V at first. Afterward, V_{br} reduces to below 5 V for the repeated test. It was suspected that the dielectric breakdown takes place at Gate oxide region. Some measurements were then performed in sequence to confirm this suspicion.

As shown in Fig. 10.38, a TOBUMOS on wafer #16 was tested, having a normal IV performance at first. When testing the Gate current vs Voltage performance, it shows Gate breakdown voltage of about 15 V. In theory, with 430 Å Gate oxide, the TOBUMOS should have at least 30 V Gate breakdown voltage. When the Gate current limit is set to be very small, normal IV curves are still observable afterward, as shown in Fig. 10.38(c). However, after the Drain breakdown test, Gate leakage current becomes very large at very small Gate bias (Fig. 10.38(f)), and there is no good on-state performance that can be measured. Hence, Gate oxide thinning at some locations is suspected to be a possible reason.

10.2.8.1. *Hot Spot Analysis on TOBUMOS*

To further investigate the mechanism resulting in the premature breakdown, the devices were imaged using an infrared photoemission microscope (IRPEM) operating in the wavelength range of 800–2400 nm. The hot spot images using this method clearly established that the breakdown always happens at the top termination region. Thus, it is concluded that early breakdown happens most probably because of the weakness of Gate oxide at the top termination region.

10.2.8.2. *Termination Checks and Modifications*

Device parameters for TOBUMOS at termination were checked to analyze the possible way to solve the problem of early breakdown. It was found that the minimum perpendicular OB trench gap at top termination region is 0.8 μm, which is still larger than the required maximum oxide thickness of 0.6 μm in the OB region. Therefore, after the thermal oxide growth, there is at least 0.2 μm gap left in between the thick oxide. This may result in the current leakage and a high electric field near the Gate oxide region at the termination. In addition, due to the phosphorous segregation effect during oxidation, the doping concentration nearby the termination is higher. Both the above reasons are likely to cause the electric field crowding at Gate region, which leads to the premature Gate breakdown.

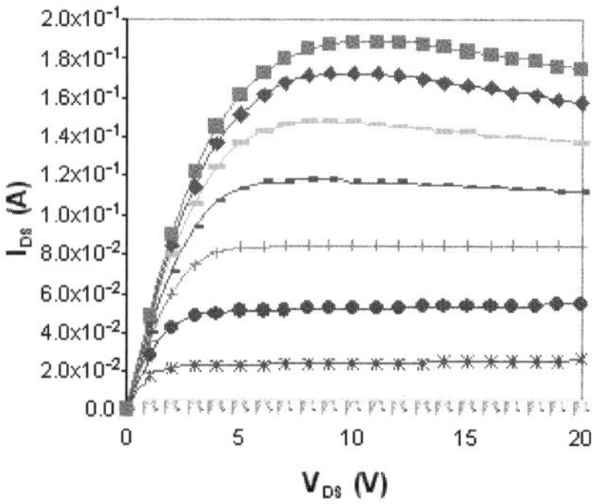

(a) Step 1: On-state characteristics

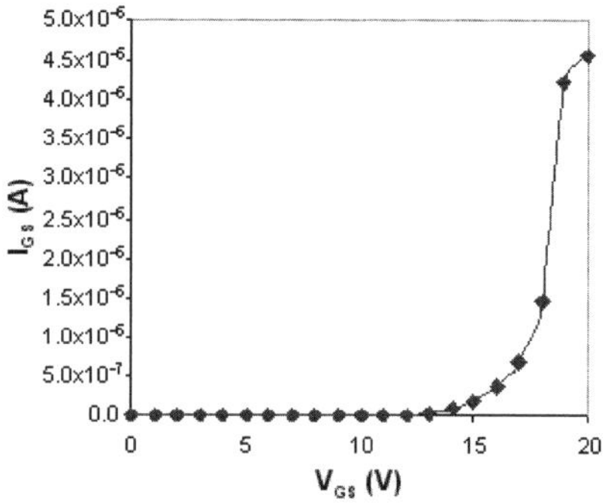

(b) Step 2: Gate breakdown

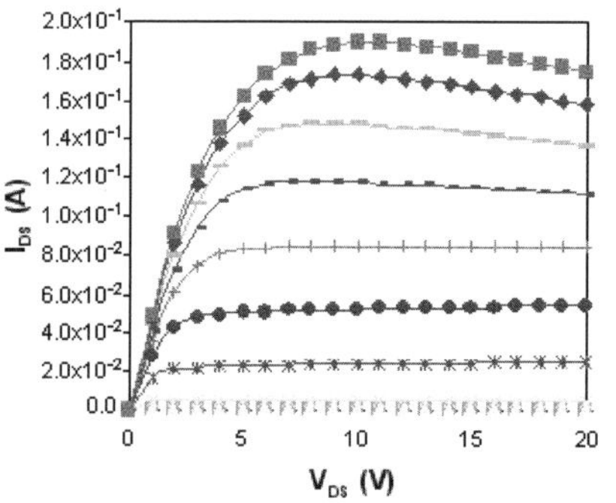

(c) Step 3: On-state characteristics

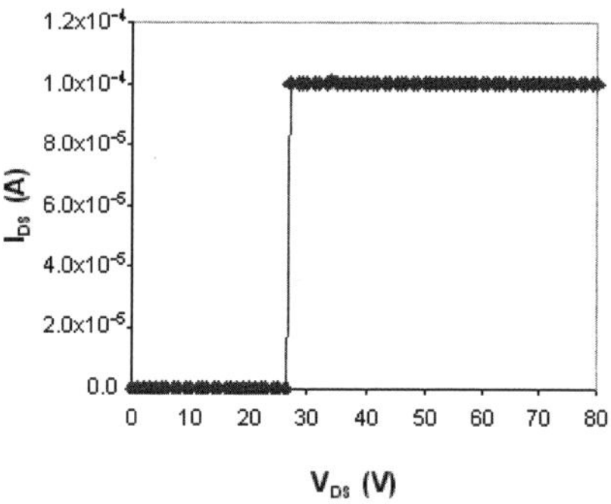

(d) Step 4: Drain breakdown

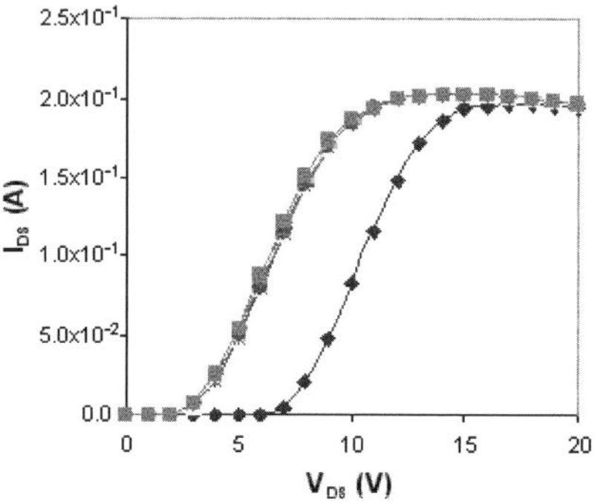

(e) Step 5: On-state characteristics

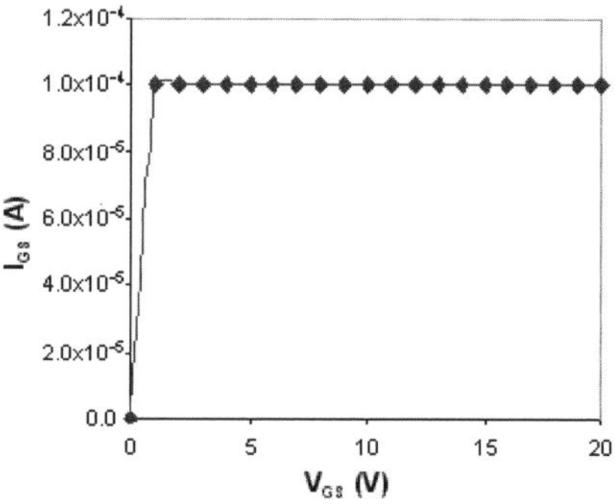

(f) Step 6: Gate breakdown

Fig. 10.38. Measurement in sequence on TOBUMOS with $W = 4.0\ \mu$m, Wob $= 1.5\ \mu$m, and Lg $= 0.8\ \mu$m (refer to Table 10.4) on wafer #16.

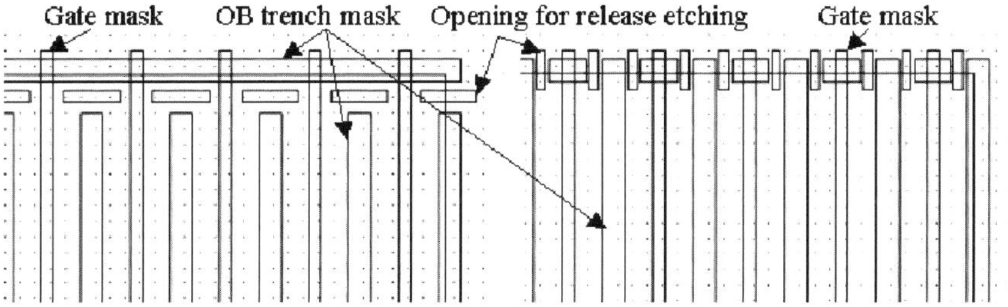

Fig. 10.39. Release etching mask added for TOBUMOS on two types of terminations.

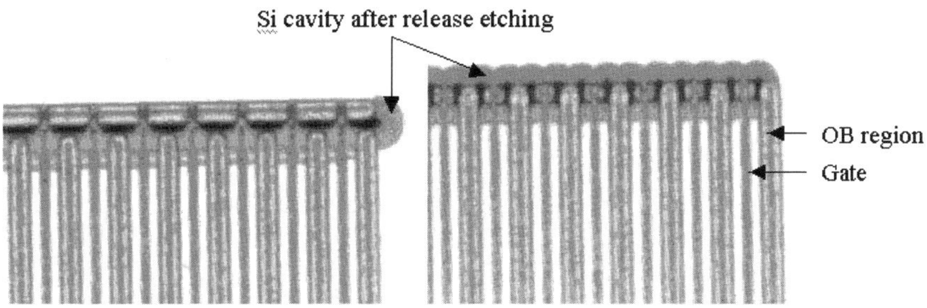

Fig. 10.40. Images under microscope after release etching for both types of terminations.

To avoid the effect of high electric field acting on the termination region, 4 μm silicon trench dry etching followed by 2–3 μm silicon release etching and oxide refill were added into the process at the top termination region.

The mask layout is shown in Fig. 10.39. In the gap of perpendicular OB trenches, there is a set of square window openings for the above-mentioned silicon etching. Depending on the space, the width of opening varies from 0.5 to 0.8 μm.

The images under microscope after release etching process are shown in Fig. 10.40. The recipe for silicon wet etching is selected so that it has high selectivity of silicon to oxide. Therefore, the etching process has no effect on Gate oxide region. It is obvious that, after release etching, Gate trenches at top termination region are surrounded by a silicon cavity underneath, which will be filled with oxide afterward. Thus, the Gate region is isolated at termination and will not be affected easily by high electric field and phosphorous segregation effect. OB-MOSFETs undergoing the process of silicon release etching have the distinct improvement on breakdown voltage.

10.2.9. Measurement Results on New Modified TOBUMOS Fabrication

TOBUMOS devices with breakdown voltages ranging from 70 to 80 V are found on Wafer #01 with the epi-resistivity of 0.55 $\Omega \cdot$ cm in the device drift region. The best result obtained on wafer #01 is the device, which has width W of 3.5 μm (Fig. 10.25). The breakdown voltage measured on this device is 76 V without Control bias and 79 V under 5 V Control voltage, while on the same wafer, breakdown voltage for conventional Diode is 66 V. The electrical on- and off-state performance tested on this device is shown in Fig. 10.41.

As shown in Fig. 10.41(a), measured threshold voltage at $V_{DS} = 0.1$ V is similar to the value obtained from the first separate run. It is 0.6 V higher than the simulation result. This TOBUMOS has normal IV curves at the on-state as in Fig. 10.41(b). The off-state performance of TOBUMOS is given in Fig. 10.41(c).

The reason why V_{br} of TOBUMOS is lower than that of TOB-Diode is probably due to the increased depth of trench Gate. Though 2 μm deep Gate trench is desired from simulation, real process results in a 0.5 μm deeper Gate trench compared to the intended depth. According to the simulation, increase of Gate trench depth leads to the premature breakdown occurring near the corner of the Gate oxide due to the high electric field nearby. As the positive Control voltage squeezes the depletion region into the upper part of the n-drift region causing the electric field crowding near the p-body–n-drift junction, TOBUMOS with the trench Gate structure in p-body region also severely restricts the increase of Control voltage. For better results, precise control on the gate trench depth is needed, which can be achieved in production by reliably tuning a recipe. Unlike the TOB-Diode structure, the tunable Control voltage is only helpful in the range of 0 to 5 V. When the external Control bias is higher than 5 V, breakdown voltage of TOBUMOS tends to drop. Since at 5 V bias, there is no improvement of breakdown voltage compared to that at 3 V bias, further increase of Control voltage is not presented in Fig. 10.41.

Experimental result shows that, maximum V_{br} measured on TOBUMOS with $W = 3.5\,\mu$m is 79 V under 5 V Control voltage. It is clear that, the introduction of TOB structure represents good tunable characteristics without bringing any additional leakage current to the device. The corresponding $R_{on,sp}$ is tested to be 0.674 m$\Omega \cdot$ cm^2 at the current density of 100A/cm^2.

With this, the description of two case studies is complete. In a nutshell, any student should never get discouraged by negative results in a first round of

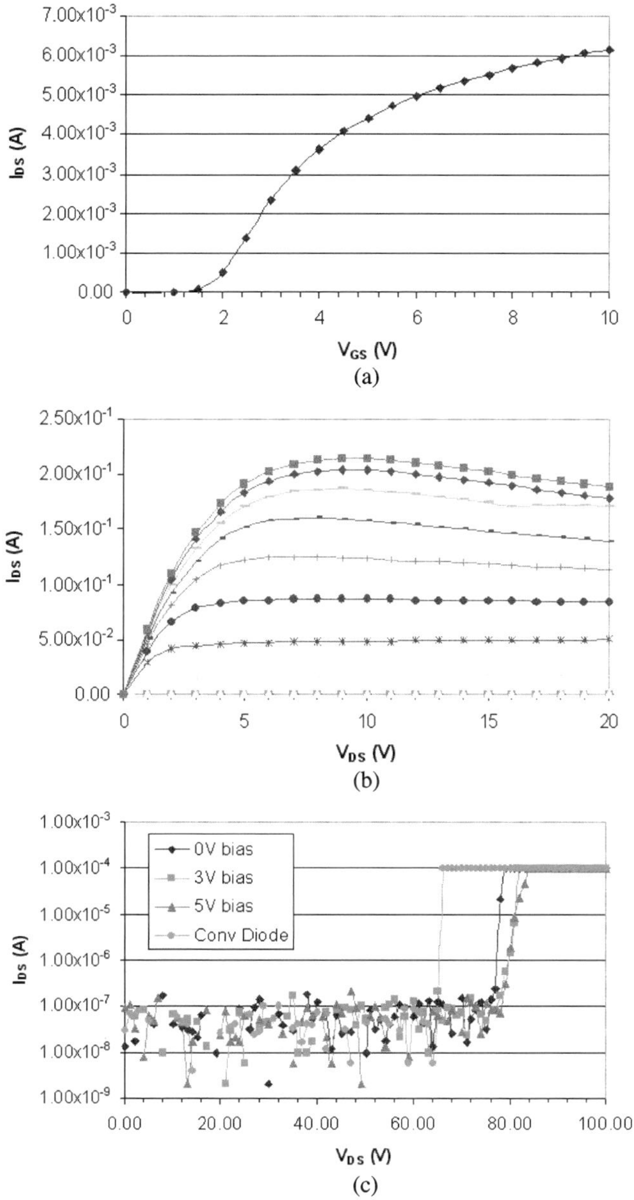

Fig. 10.41. Measurement results on TOBUMOS with $W = 3.5\,\mu$m, Wob $= 1.0\,\mu$m, and Lg $= 1.0\,\mu$m (refer to Table 10.4) on wafer #01: (a) Gate performance at $V_{DS} = 0.1$ V; (b) On-state IV performance; (c) Off-state performance at Control voltage varying from 0 to 5 V compared to that of conventional Diode on the same wafer.

fabrication even if it does not workout despite taking extra care while designing the process and masks. It is also heartening to note that after using diagnostic tools such as SEMs, the hot spot analysis and careful study of electrical characteristics as well as failure modes help researchers reach the desired goal. Clearly, simulation results set the optimal performance to achieve and are extremely useful.

References

B. J. Baliga, Power devices, in *Modern Semiconductor Device Physics*, ed. S. M. Sze (John Wiley & Sons, New York 1998), pp. 183–252.

C. Y. Chang and S. M. Sze, *ULSI Technology* (McGraw Hill, 1996).

B. E. Deal and A. S. Grove, General relationship for thermal oxidation of silicon, *J. Appl. Phys.* **36**, 3770 (1965).

G. Deboy, M. Marz, J.-P. Stengl, H. Strack, J. Tihanyi and H. Weber, A new generation of high voltage MOSFETs breaks the limit line of silicon, *IEDM Tech. Dig.*, 683–685 (1998).

K. P. Gan, Y. C. Liang, G. S. Samudra, S. M. Xu and L. Yong, Poly flanked VDMOS (PFVDMOS): A superior technology for superjunction devices, *IEEE 32nd Ann. Power Electron. Specialists Conf.* PESC **4**, 2156–2159 (2001).

K. P. Gan, X. Yang, Y. C. Liang, G. S. Samudra and L. Yong, A simple technology for superjunction device fabrication: Polyflanked VDMOSFET, *IEEE Electron Device Lett.* **23**(10), 627–629 (2002).

T. Kamins, Polycrystalline silicon for integrated circuits and displays, 2nd edn. (Kluwer Academic Publishers, 1998), pp. 199–209.

Y. C. Liang, K. P. Gan and G. S. Samudra, Oxide bypassed VDMOS (OBVDMOS): An alternative to superjunction high voltage MOS power devices, *IEEE Electron Device Lett* **22**(8), 407–409 (2001).

J. D. Plummer, M. D. Deal and P. B. Griffin, *Silicon VLSI Technology* (Prentice Hall, 2000).

P. M. Shenoy, A. Bhalla and G. M. Dolny, Analysis of the effect of charge imbalance on the static and dynamic characteristics of the super junction MOSFET, *Proc. ISPSD*, 99–102 (1999).

Silvaco process and device simulation tools — ATHENA and ATLAS.

Synopsys process and device simulation tools — TSUPREM 4 and MEDICI.

S. M. Sze, *VLSI Technology*, 2nd edn. (McGraw Hill, 1988).

X. Yang, Y. C. Liang, G. S. Samudra and L. Yong, Tunable oxide-bypassed trench gate MOSFET: Breaking the ideal superjunction MOSFET performance line at equal column width, *IEEE Electron Device Lett.* **24**(11), 704–706 (2003).

11

Practical Case Studies in Wide Bandgap Power Devices

As silicon carbide (SiC) and gallium nitride (GaN) based devices described in Chapters 7 and 8, respectively are becoming increasingly important, two detailed case studies on these devices with process and device considerations and characterization are included in this chapter.

11.1. Case Study I: Process Integration and Design of SiC DIMOSFET

In this section, the design and process flow are presented for a self-aligned SiC Double Implanted MOSFET (DIMOSFET) with a 30 μm thick 2×10^{15} cm^{-3} doped epilayer which is capable of supporting more than 3.3 kV, together with some experimental data. In Chapter 7, the problems related to the poor inversion channel mobility of SiC MOSFETs were addressed. In order to reduce the channel resistance and to fully explore the potential of SiC MOSFETs, several approaches have been proposed, one of which is to use a self-aligned process to produce a short channel length in the device. The typical cell structure for a multi-finger SiC DIMOSFET is shown in Fig. 11.1. The source electrode, which should be laid over the dielectric surrounding the gate electrode and connecting the n+ source, is not shown in the figure. Also note that the p+ contacts to short the b-base to the source electrode are also displayed. The locations of the p+ contacts are critical in order to mute the parasitic BJT and improve ruggedness of the device.

11.1.1. Process Integration to Self-Aligned SiC DIMOSFET

Before fabrication began, the process flow for a self-aligned SiC DIMOSFET was planned using some typical SiC processing equipment. The major steps for the fabrication process are described below.

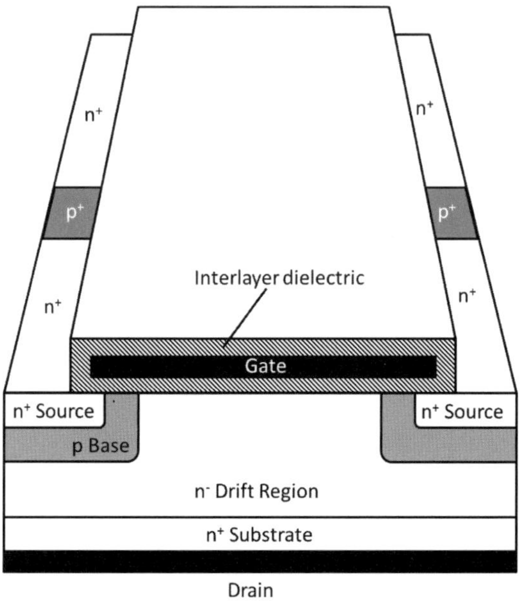

Fig. 11.1. The typical cell structure for a multi-finger SiC DIMOSFET without showing the source electrode which should be laid over the inter-layer dielectric connecting the n+ source and the p+ base contact.

(i) Start by creating a 4H-SiC Si-face epi-wafer using a suitable doping and thickness for the epilayer.
(ii) Deposit a blanket thick polysilicon, then define it using lithography followed by a dry etch.
(iii) Perform p implantation for the p-base using the remaining polysilicon as the mask.
(iv) Form a spacer with an appropriate width through the oxidation of the polysilicon.
(v) Use lithography to define the mask for the n+ source.
(vi) Perform n+ implantation for the source, which is self-aligned to the p-base, with the channel length defined by the spacer width.
(vii) Use lithography to define the mask for the p+ base contact.
(viii) Perform p+ implantation for the p+ base contact.
(ix) Remove the polysilicon mask.
(x) Deposit a blanket thick PECVD oxide, then define it using lithography followed by a dry etch.
(xi) Perform p implantation for a single-zone p-JTE using the remaining PECVD oxide as the mask.

(xii) Remove the PECVD oxide mask.
(xiii) Aneal all implantations and activate them using a graphite cap.
(xiv) Remove the graphite cap.
(xv) Form a gate dielectric using thermal oxidation followed by a nitridation anneal.
(xvi) Deposit a polysilicon gate, which is then doped and defined using lithography and a dry etch.
(xvii) Form an interlayer dielectric by oxidizing the polysilicon gate.
(xviii) Deposit the contact metal on the front surface for the source and the body.
(xix) Deposit the contact metal on the back surface for the drain.
(xx) Anneal the contact metals.
(xxi) Deposit and define the source electrode.
(xxii) Open the polysilicon gate window.

It should be noted that a typical process flow for a SiC DMOSFET is listed here, but the process has not necessarily been optimized. There are alternatives to some of the key processes since SiC DMOSFETs are still under extensive development, as discussed in Chapter 7. At present, for example, the formation of the gate dielectrics in step (xv) is most frequently accomplished through thermal oxidation, followed by a nitridation anneal in NO gas. However, several innovative gate dielectrics that are able to further enhance the MOS interface properties have been proposed in recent years, and we can be certain that others will be developed in the years to come since this issue currently remains critical and unsolved for SiC MOSFETs. In step (xiii), a graphite cap is used to preserve the surface morphology, but other thin film that is able to stand the high temperate anneal (~1650°C) could serve as an alternative for this purpose. It is also possible to rearrange the sequence of this process, or even omit some of the steps in order to reduce cost and increase throughput.

11.1.2. Simulation of a Cell Structure for a Self-Aligned SiC DIMOSFET

Using a two-dimensional (2D) process simulator ATHENA (Silvaco), process simulation was carried out based on the process flow described in the previous section, employing the targeted doping profiles and dimensions in order to identify the key process parameters and verify the cell structure. Subsequently, a device simulator ATLAS (Silvaco) was used to predict the DC characteristics of a SiC MOSFET based on this cell structure to determine whether the design requirements could be met and to identify the key design parameters.

11.1.2.1. *Process Parameters and Cell Structure*

The cell structure depicted in Fig. 11.1 was simulated based on a cell pitch of 18 μm and a channel length of 0.5 μm, created using a self-aligned process. The width of the JFET region is 5 μm, and the width of the polysilicon gate is 9 μm. This device was designed to support a breakdown voltage of more than 3.3 kV. The input file for the ATHENA process simulator is shown below, with the notation "#" indicating comments and explanations related to the process. The process parameters included in this example have been tuned so as to obtain satisfactory results.

```
# Simulation of the fabrication processes for a cell structure of a 4H-SiC
DMOSFET

# Configure the environment for the device simulator using up to 8 CPUs

go athena simflags = "−P 8"

# Define the mesh in the x-direction
# Half-cell pitch = 9 μm
line x loc = 0.00 spac = 0.20
line x loc =1.00 spac = 0.20
line x loc = 2.50 spac = 0.05
line x loc = 5.50 spac = 0.05
line x loc = 7.50 spac = 0.05
line x loc = 9.00 spac = 0.20

# Define the mesh in the y-direction
# Starting wafer thickness = 2 μm
line y loc = 0.00 spac = 0.5
line y loc = 1.00 spac = 0.5
line y loc = 2.00 spac = 0.5

# Create the starting n-type substrate
init sic_4h c.nitrogen = 1e21
structure outfile = 1_init1.str

# 30 μm thick n-type epilayer
deposit sic_4h thick = 30 c.nitrogen = 2e15 div = 50 dy = 0.01 ydy = 1
structure outfile = 2_deposit.str

# Sacrificial oxidation
deposit oxide thick = 0.05
structure outfile = 3_depositxodie.str
```

```
# Polysilicon deposition
# JFET width = 5 μm
deposit polysilicon thick = 2.5 div = 10
structure outfile = 4_depositpoly.str

# Mask 1: p-base implant mask
# Polysilicon etch
etch polysilicon left p1.x = 6.50
structure outfile = 5_poly.str

# p-base implantation
implant aluminum diverg = 0.1 n.ion = 40000 dose = 6e10 tilt = 4 rot = 0
energy = 10 bca
implant aluminum diverg = 0.1 n.ion = 40000 dose = 5e13 tilt = 4 rot = 0
energy = 360 bca
structure outfile = 6_impalumi.str

# Self-aligned spacer
# Spacer width = 0.5 μm
deposit oxide thick = 0.5
etch oxide dry thick = 0.5
structure outfile = 7_etchoxide.str

# Mask 2: n+ implant mask
# n+ opening = 2 μm
deposit photoresist thick = 1
etch photoresist right p1.x = 1.0
structure outfile = 8_etchphoto.str

# n+ source implantation
implant nitrogen diverg = 0.1 n.ion = 40000 dose = 7e14 tilt = 4 rot = 0
energy = 30 bca
implant nitrogen diverg = 0.1 n.ion = 40000 dose = 1.2e15 tilt = 4 rot = 0
energy = 80 bca
etch photoresist all
structure outfile = 9_nitrogen.str

# Mask 3: p+ implant mask
# p+ contact width = 2 μm
deposit photoresist thick = 1
etch photoresist left p1.x = 1.0
structure outfile = 10_etchphto.str
```

```
# p+ contact implantation
implant aluminum diverg = 0.1 n.ion = 40000 dose = 6e14 tilt = 4 rot = 0
energy = 20 bca
implant aluminum diverg = 0.1 n.ion = 40000 dose = 7.5e14 tilt = 4 rot = 0
energy = 60 bca
implant aluminum diverg = 0.1 n.ion = 40000 dose = 8.5e14 tilt = 4 rot =0
energy = 120 bca
implant aluminum diverg = 0.1 n.ion = 40000 dose = 9.5e14 tilt = 4 rot = 0
energy = 180 bca
etch photoresist all
structure outfile = 11_etchphoto.str

# Polysilicon and oxide etch
etch oxide dry thick = 1
etch polysilicon all
etch oxide all
structure outfile = 12_etchoxideall.str

# The process used for edge termination is not included in this simulation
# The following process steps were omitted
# 1. PECVD oxide deposition
# 2. Mask 4: JTE implant mask
# 3. PECVD oxide etch
# 4. P-JTE implantation
# 5. Oxide stripping

# The implant activation processes for SiC were not available in the Silvaco
simulator
# The following process steps were omitted, and 100% activation for all
dopants was assumed
# 1. Graphite cap formation and implant anneal
# 2. Graphite stripping

# Gate oxidation
# Gate oxide thickness = 50 nm
deposit oxide thick = 0.05
structure outfile = 13_depositoxide.str

# Polysilicon gate deposition
# Mask 5: Polysilicon gate definition
# Polysilicon etch
# Polysilicon gate width = 9 μm
```

```
deposit polysilicon thick = 1.50
etch polysilicon left p1.x = 4.5
structure outfile = 14_etchpoly.str

# Interlayer oxide formation
# Mask 6: Contact definition
# Open contact window
# Interlayer oxide thickness = 0.5 μm
deposit oxide thick = 0.5
etch oxide left p1.x = 4
structure outfile = 15_etchoxide.str

# Mirror to expand to a complete cell structure
struct mirror right
structure outfile=16_mirror.str

# Define the source contact metal
# Top metal thickness = 0.7 μm
deposit aluminum thick = 0.7

# At this point, the cell structure is complete.
# The following process steps can be omitted without causing any concerns
regarding the characteristics of the cell structure.
# 1. Contact metal deposition and lift off
# 2. Front side protection
# 3. Backside etch and metal deposition
# 4. Contact metal anneal
# 5. Mask 7: Gate window definition
# 6. Gate window etch
# 7. Mask 8: Pad metal definition
# 8. Pad metal deposition and lift off

# Define the source electrode
electrode name = Source x = 4 y = -32.2

# Define the gate electrode
electrode name = Gate x = 9.00 y = -31.00

# Define the drain electrode
electrode name = Drain backside
structure outfile = 17_athenaelectrode.str

quit
```

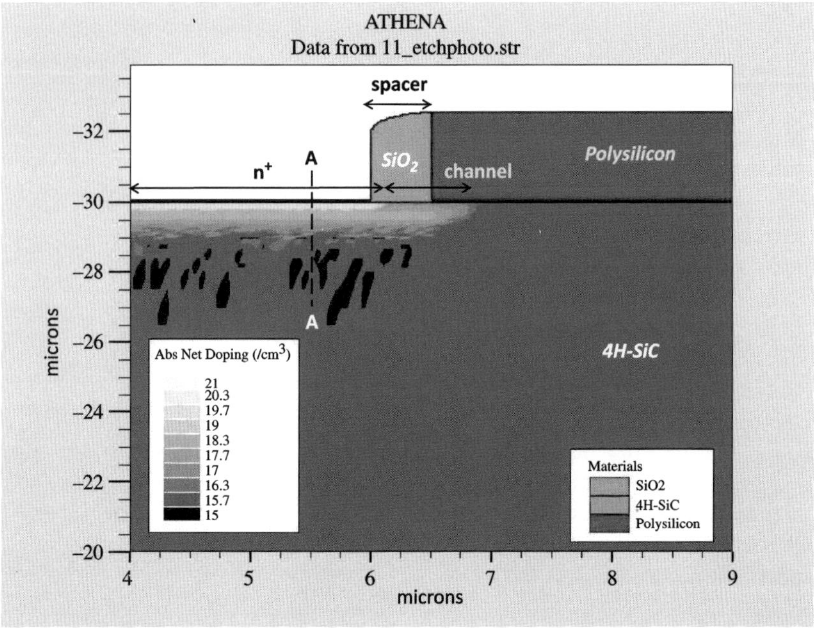

Fig. 11.2. Simulated half-cell structure for a SiC DIMOSFET from the output file "11_etchphoto.str".

The structure for half-cell following the p+ contact implantation is shown in Fig. 11.2, as generated by the Tonyplot viewing tool using one of the output files, "11_etchphoto.str". The channel length, which is defined by the SiO_2 spacer under the effects of lateral straggling from the implantation, is indicated in the figure. Some peculiar p-type islands that appeared in the n- epi-layer can be eliminated at the expense of a larger grid density and a longer computing time. In this example, however, these islands do not have any significant effect on the DC characteristics.

The doping profile in the vertical direction, along a cutline A-A, is shown in Fig. 11.3. The total implantation dose for the n+ source is $1.9 \times 10^{15}\,cm^{-2}$, with a maximum energy of 80 keV. The total implantation dose for the p-base is $5 \times 10^{13}\,cm^{-2}$, with a maximum energy of 360 keV. After accounting for the compensation from the n+ source and the n-epilayer, the effective dose in the base is about $3.9 \times 10^{13}\,cm^{-2}$, which should be more than sufficient to prevent any punchthrough in the base under the operating conditions, assuming an activation percentage of 50%.

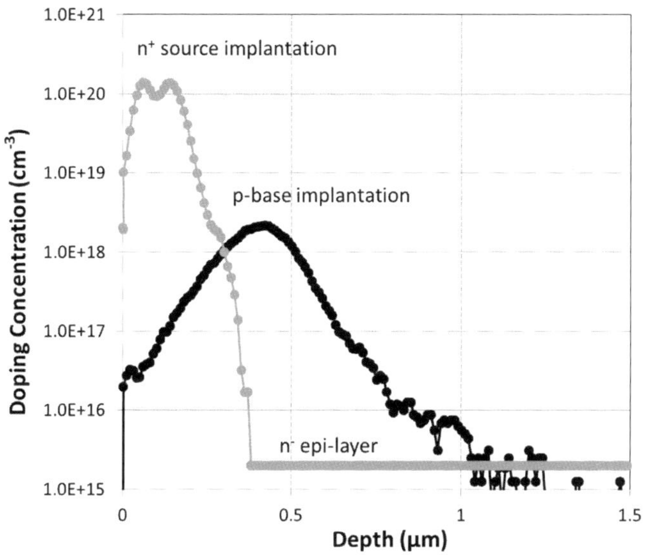

Fig. 11.3. Simulated doping profiles from the n+ source and p-base implantations into the n-epilayer.

The complete cell structure after the simulation of the entire fabrication process is shown in Fig. 11.4. A critical design parameter is the width of the parasitic JFET region, as indicated in the figure, and the effects of this will be discussed in a later section. The widths of the p+ base contact implantation and the n+ source implantation are identified. An opening in the n+ source region is created, and the p+ implantation for the p-base contact is formed through this opening in the two dimensional simulation. However, in the real design, this opening can be hidden inside the n+ stripes in a multi-finger layout, as shown in Fig. 11.1, where a polysilicon gate, an interlayer dielectric, and an overlaying source metal electrode are shown at the top.

11.1.2.2. *Simulation of the Off-state Characteristics*

Off state simulation was performed using ALTLAS based on the input file shown below. If a command line is too long, it is terminated using the notation "\" and continued on the next line. It is worth mentioning that the bandgap was adjusted to 1.12 eV for Si rather than 3.26 eV for 4H-SiC to facilitate convergence. By doing this, the intrinsic carrier density and hence the leakage current are increased. The multiplication of carriers at avalanche breakdown is easier to trigger. Otherwise extended precision with more bits to represent a number in the simulation will be needed, which takes up more memory space

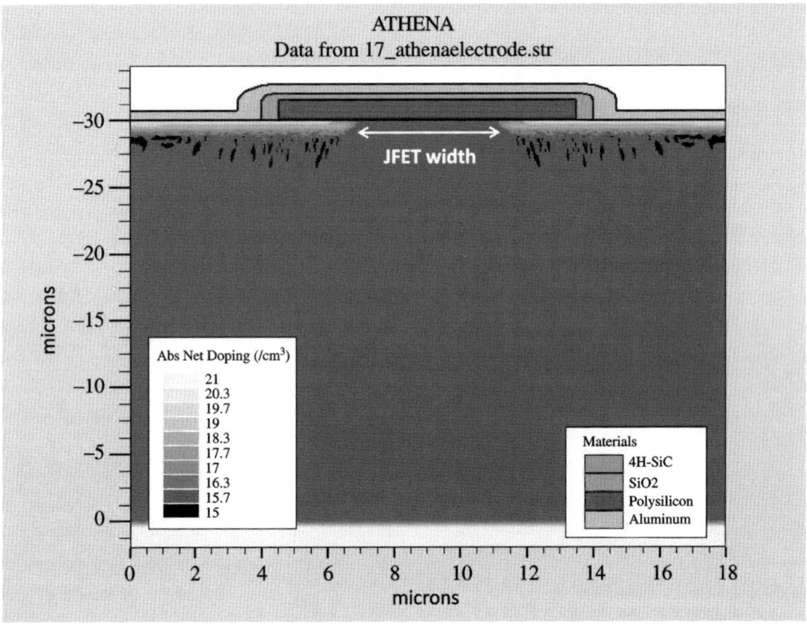

Fig. 11.4. Simulation of the complete cell structure for a SiC DIMOSFET from an output file.

and computation time. The impact ionization parameters were adjusted using Konstantinov's model, so the prediction in breakdown voltage, potential contours, and electric field distribution should still be accurate. Also note that this can only be used for breakdown voltage simulation. For accurate on-state simulations, particularly those involving build-in voltage V_{bi}, the bandgap cannot be modified.

```
# Simulation of the OFF-state characteristics for the cell structure of a
4H-SiC DMOSFET

# Load structure file generated by the process simulator
go athena
init infile = 17_athenaelectrode.str

# Set the environment for the device simulator using up to 8 CPUs
go atlas simflags = "−P 8"
log outfile = test01_for.log
```

```
# Define the 4H-SiC material properties
material material = 4H-SiC eab = 0.210 edb = 0.065 eg300 = 1.12 permitti
= 9.8 vsat = 2e7 \
  bgn.e = 9.000e-3 bgn.n = 1.000e17 \
  kaugdn = 0 kaugdp = 0 kaugcn = 5e-31 kaugcp = 9.9e-32 \
  nc300 = 1.669e19 nv300 = 2.459e19 \
  n.sch.max = 50E-9 n.sch.nref = 3E17 n.sch.gamma = 0.3 \
  p.sch.max = 50E-9 p.sch.nref = 3E17 p.sch.gamma = 0.3

mobility material = 4H-SiC mun = 947 mup = 124 tmun = 2 tmup = 2
mtn.min1 = 0 mtn.pc = 0 \
  mtn.max = 947 mtn.min2 = 0 mtn.cr = 1.94E17 mtn.alpha = 0.61 \
  mtn.mu = 0 mtn.cs = 3.43E20 mtn.beta = 2 mtp.min1 = 15.9 \
  mtp.pc = 0 mtp.max = 124 mtp.min2 = 0 mtp.cr = 1.76E19 \
  mtp.alpha = 0.34 mtp.mu1= 0 mtp.cs = 6.1E20 mtp.beta = 2 \
  n.beta0 = 1 n.betaexp = 0.66 p.beta0 = 1.213 p.betaexp = 0.17 vsatn = 2E7

# Define the simulation models
models srh conmob bgn fermi print incomplete

# Define the impact ionization parameters
impact selb an1=1.14E9 an2=1.14E9 bn1=3.8E7 \
  bn2=3.8E7 ap1=6.85E6 ap2=6.85E6 bp1=1.41E7 \
  bp2=1.41E7 betan=1 betap=1

# Define the numerical methods
method newton autonr itlimit=30 ir.tol=1e-30 ix.tol=1e-30 trap

# Solve for an initial solution
solve init
log outf=bv.log master

# Ramp the drain voltage
solve previous
solve vgate=0
solve vdrain=0 vstep=0.05 vfinal=0.5 name=drain
solve vstep=0.25 vfinal=10 name=drain
solve vstep=2.5 vfinal=100 name=drain
solve vstep=20 vfinal=5000 name=drain compliance=1e-9 cname=drain
```

```
log off

# Save and plot the output
save outfile=DMOSFET_cell_bv.str
tonyplot bv.log
tonyplot DMOSFET_cell_bv.str

quit
```

The reverse current versus voltage curve is shown in Fig. 11.5. The simulation was automatically terminated when the reverse current reached 1×10^{-9} A/μm, corresponding to a current density of 5 m A/cm^2. The extracted breakdown voltage was 4680 V, which leaves a more than sufficient safe margin for the targeted voltage of 3300 V. The electric field distribution is illustrated in Fig. 11.6. The peak electric field is about 2.6×10^6 V/cm, located at the corners of the p-base implantation beneath the polysilicon gate.

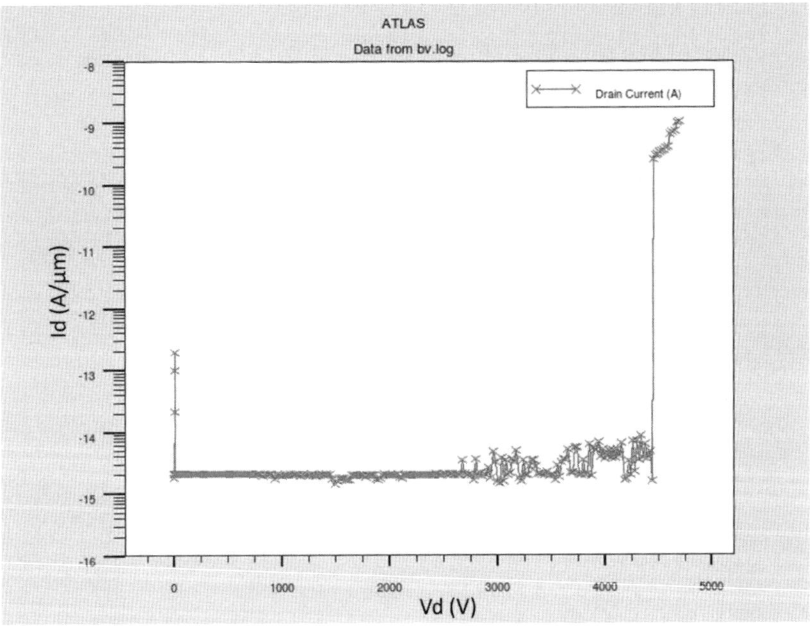

Fig. 11.5. Simulated reverse characteristics of the cell structure for a SiC DIMOSFET from the output file "bv.log".

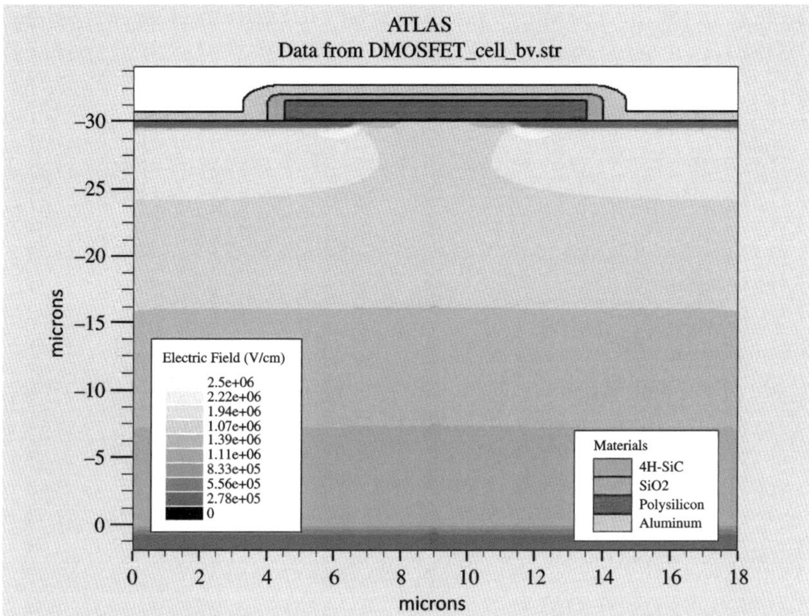

Fig. 11.6. Simulated electric field distribution at $V_d = 4680\,V$ from the output file "DMOSFET_cell_bv.str".

11.1.2.3. *Simulation of the On-state Characteristics*

The on-state characteristics of the 4H-SiC DIMOSFET were simulated using ALTLAS based on the input file shown below. Important material properties such as bulk mobility were adjusted to those of 4H-SiC. The inversion channel mobility model proposed by Lombardi *et al.* (1988) was used to account for the effects generated from bulk transportation, acoustic phonon scattering, and surface roughness. Also, the existence of single-level traps at the MOS interface with a density of $1 \times 10^{12}\,cm^{-2}$ was assumed. These models are critical in determining the properties of the MOS interface and inversion channel. However, since extensive development is still being conducted on the MOS interface in 4H-SiC, its properties are varied to a large degree by different processing conditions. The parameters provided here are for reference only and should be adjusted accordingly for different gate dielectric processes.

```
# Simulation of the on-state characteristics for the cell structure of a 4H-SiC
DMOSFET

# Load the structure file generated by the process simulator
go athena
init infile=17_athenaelectrode.str

# Configure the environment for the device simulator using up to 8 CPUs
go atlas simflags="-P 8"

# Define the 4H-SiC material properties
material material=4H-SiC eab=0.210 edb=0.065 eg300=3.26 permitti=9.8
vsat=2e7 \
  bgn.e=9.000e-3 bgn.n=1.000e17 \
  kaugdn=0 kaugdp=0 kaugcn=5e-31 kaugcp=9.9e-32 \
  nc300=1.669e19 nv300=2.459e19 \
  n.sch.max=50E-9 n.sch.nref=3E17 n.sch.gamma=0.3 \
  p.sch.max=50E-9 p.sch.nref=3E17 p.sch.gamma=0.3 \
  taun0=50e-9 taup0=50e-9

mobility material=4H-SiC mun=947 mup=124 tmun=2 tmup=2
mtn.min1=0 mtn.pc=0 \
  mtn.max=947 mtn.min2=0 mtn.cr=1.94E17 mtn.alpha=0.61 \
  mtn.mu=0 mtn.cs=3.43E20 mtn.beta=2 mtp.min1=15.9 \
  mtp.pc=0 mtp.max=124 mtp.min2=0 mtp.cr=1.76E19 \
  mtp.alpha=0.34 mtp.mu1=0 mtp.cs=6.1E20 mtp.beta=2 \
  n.beta0=1 n.betaexp=0.66 p.beta0=1.213 p.betaexp=0.17 vsatn=2E7

# Define the 4H-SiC inversion channel mobility model using Lombardi CVT
mobility material=4H-SiC mu0n.cvt=0 pcn.cvt=0 mumaxn.cvt=947
gamn.cvt=2 \
  crn.cvt=1.94E17 alphn.cvt=0.61 mu1n.cvt=0 \
  en.cvt=1 bn.cvt=1E6 cn.cvt=1.74E5 taun.cvt=0.0561 dn.cvt=0.5 \
  kn.cvt=2 deln.cvt=5.82E14 feln.cvt=1E40

# Define the interface trap density and location
inttrap e.level=0.1 acceptor density=1.e12 degen=2 \
sign=2.84e-15 sigp=2.84e-14

# Define the simulation models
models cvt fermi schsrh print incomplete
```

```
# Define the numerical methods
method gummel newton autonr itlimit=30 ir.tol=1e-30 ix.tol=1e-30

# Extract the Id–Vg curve for Vd = 1 V
solve init outf=solve_init
solve vdrain=1
log outf=idvg.log
solve name=gate vgate=0 vfinal=20 vstep=0.5
save outf=idvg.str
log off

# Extract the Id–Vd curves for different Vg values
load infile=solve_init
log off

#IdVd(Vg=0)
solve vgate=0 outf=solve_vg00
log outf=vg00.log
solve name=drain vdrain=0 vfinal=20 vstep=0.5
save outf=vg00.str
log off

#IdVd(Vg=3)
load infile=solve_vg00
solve vgate=3 outf=solve_vg03
log outf=vg03.log
solve name=drain vdrain=0 vfinal=20 vstep=0.5
save outf=vg03.str
log off

#IdVd(Vg=6)
load infile=solve_vg03
solve vgate=6 outf=solve_vg06
log outf=vg06.log
solve name=drain vdrain=0 vfinal=20 vstep=0.5
save outf=vg06.str
log off

#IdVd(Vg=9)
load infile=solve_vg06
solve vgate=9 outf=solve_vg09
log outf=vg09.log
```

```
solve name=drain vdrain=0 vfinal=20 vstep=0.5
save outf=vg09.str
log off

#IdVd(Vg=12)
load infile=solve_vg09
solve vgate=12 outf=solve_vg12
log outf=vg12.log
solve name=drain vdrain=0 vfinal=20 vstep=0.5
save outf=vg12.str
log off

#IdVd(Vg=15)
load infile=solve_vg12
solve vgate=15 outf=solve_vg15
log outf=vg15.log
solve name=drain vdrain=0 vfinal=20 vstep=0.5
save outf=vg15.str
log off

#IdVd(Vg=20)
load infile=solve_vg15
solve vgate=20 outf=solve_vg20
log outf=vg20.log
solve name=drain vdrain=0 vfinal=20 vstep=0.5
save outf=vg20.str
log off

# Plot the output
tonyplot idvg.log
tonyplot -overlay vg00.log vg03.log vg06.log vg09.log vg12.log vg15.log
vg20.log

quit
```

The simulated I_d–Vg curve for $V_d = 1\,\text{V}$ generated by the output file "idvg.log" is shown in Fig. 11.7, where the threshold voltage is about 6 V. The I_d–V_d curves for Vg values up to 20 V generated by overlaying several output files "vg00.log"–"vg20.log" are shown in Fig. 11.8. The $R_{on,sp}$ is $25\,\text{m}\Omega \cdot \text{cm}^2$, extracted through V_d/I_d at a V_d of 1 V and a Vg of 20 V. The channel resistance contributes a significant portion of the $R_{on,sp}$ because of the poor channel mobility in SiC, even though the channel length was shrunk to $0.5\,\mu\text{m}$ using a self-aligned process.

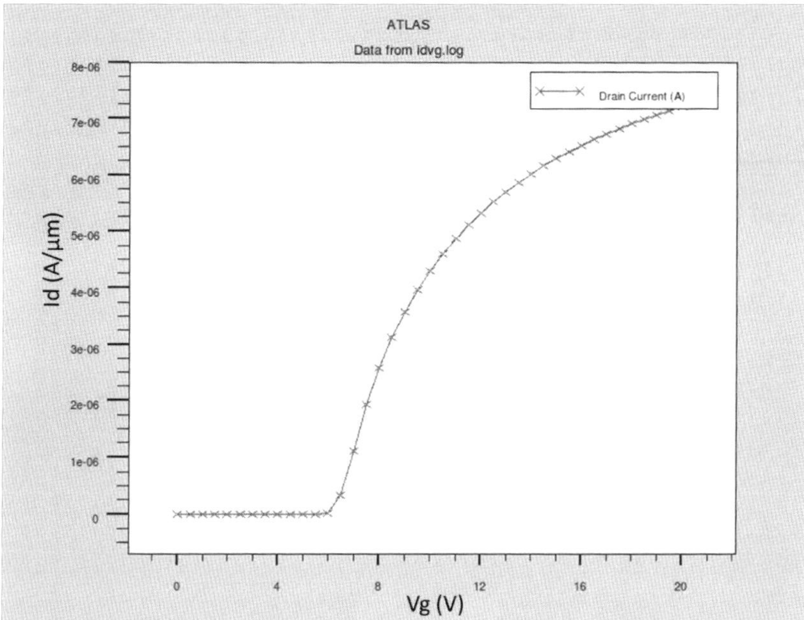

Fig. 11.7. Simulated I_D–V_G curve for $V_d = 1$ V from the output file "idvg.log".

11.1.2.4. *Design Trade-offs*

Once the device structure and characteristics have been verified via simulation, we are ready to investigate some of the design trade-offs that will be required in order to optimize the device performance, or to determine the process parameters if the models are accurate. One of the critical design issues for a DMOSFET is the width of the JFET region, as depicted in Fig. 11.4. In the following analysis, the JFET width and the polysilicon gate width are varied, while other design parameters, such as the channel length and cell pitch, are retained.

The I_d–V_d curves at a Vg of 20 V for different JFET widths are compared in Fig. 11.9. It is clear that when the JFET width is reduced, the JFET region will be pinched off by the depletion region in the lightly doped n- epilayer, which significantly reduces the drain current. It can be concluded that the JFET width should be as large as possible so as to facilitate the current conductions and reduce the $R_{on,sp}$.

However, other effects resulting from variations in the JFET width should also be considered. Figure 11.10 shows the distribution of the electric field in the gate oxide when the devices are biased at their breakdown voltage. In this

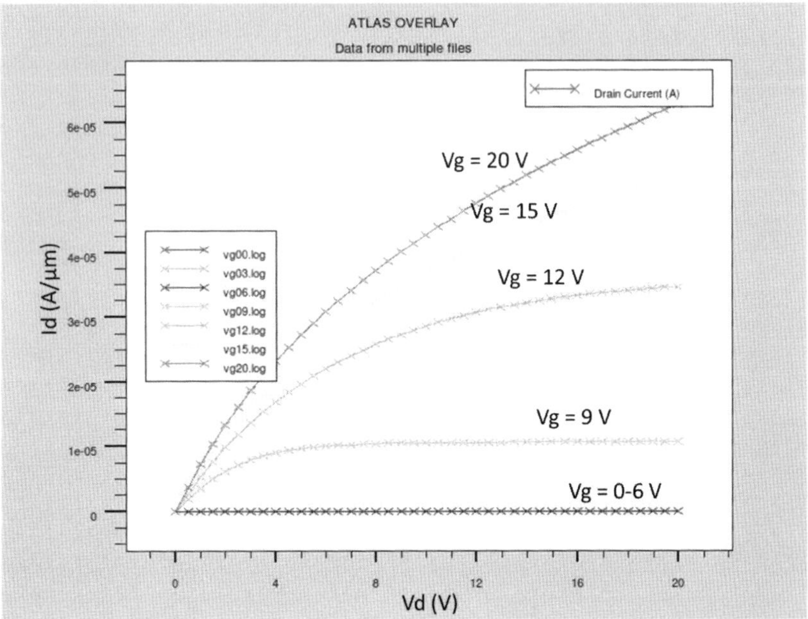

Fig. 11.8. Simulated I_d–V_d curves for Vg values up to 20 V by overlaying several output files "vg00.log"–"vg20.log".

case, the oxide field, $\hat{E}_{ox}$, increases as the JFET width increases, since the protection for the gate oxide from the p-base is weaker. The dashed line indicated in the figure represents an $\hat{E}_{ox}$ of 3 MV/cm, which is generally accepted as the maximum for reliable SiO_2 stress conditions. As can be seen in the figure, the $\hat{E}_{ox}$ exceeds 3 MV/cm for JFET widths above 3 μm, which may raise concerns about long-term reliability for these devices. Therefore, when determining an appropriate JFET width, a trade-off exists between the $R_{on,sp}$ from the on-state characteristics and the $\hat{E}_{ox}$ from the off-state characteristics, as indicated in Fig. 11.11. However, a superior trade-off solution may be established by using more innovative device structures.

11.1.3. Experimental Results

11.1.3.1. *Process Flow, Layout, and Run Sheets*

After performing simulations to identify the key design parameters and predict the device performance, fabrication was carried out to verify the process sequences and demonstrate the capability of the 4H-SiC DIMOSFET. A total

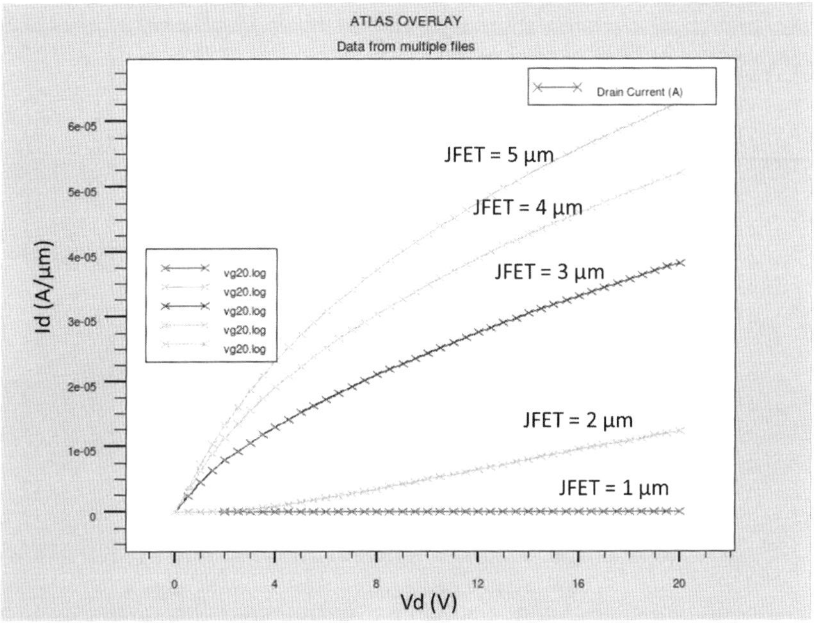

Fig. 11.9. I_d–V_d curves for devices at a Vg of 20 V based on different JFET widths.

of nine masks were used for the fabrication of a self-aligned 4H-SiC DIMOSFET, as listed below. Rather than the deposited and etched spacer approach that was implemented in the simulation, a self-aligned channel length of 0.5 μm was created during the experiment by expanding the implantation mask using wet thermal oxidation of the polysilicon defined by Mask 1, as proposed by Matin *et al.* (2004).

(1) Mask 0: Alignment keys
(2) Mask 1: p-base implant mask
(3) Mask 2: n+ source implant mask
(4) Mask 3: p+ contact implant mask
(5) Mask 4: JTE implant mask
(6) Mask 5: Polysilicon gate definition
(7) Mask 6: Contact definition
(8) Mask 7: Gate window definition
(9) Mask 8: Pad metal definition

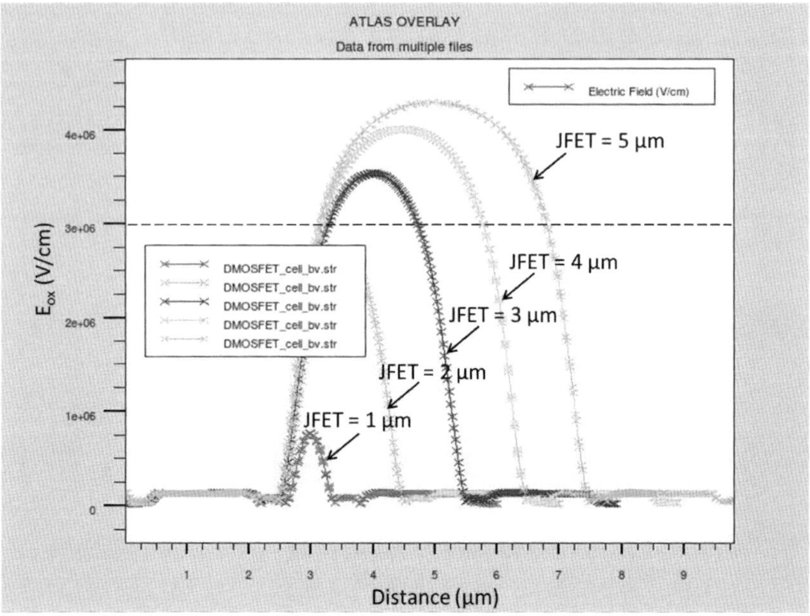

Fig. 11.10. Electric field in the gate oxide of devices based on different JFET widths when the device is OFF and biased at a V_d of 4680 V.

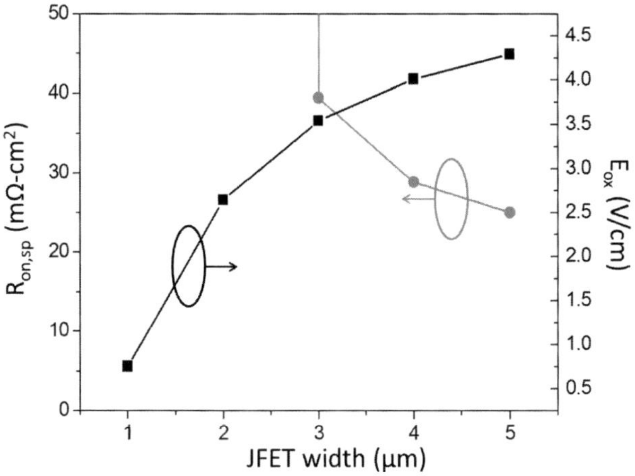

Fig. 11.11. Simulated $R_{on,sp}$ and $\hat{E}_{ox}$ for the 4H-SiC DIMOSFET structure presented in Fig. 11.4 based on different JFET widths.

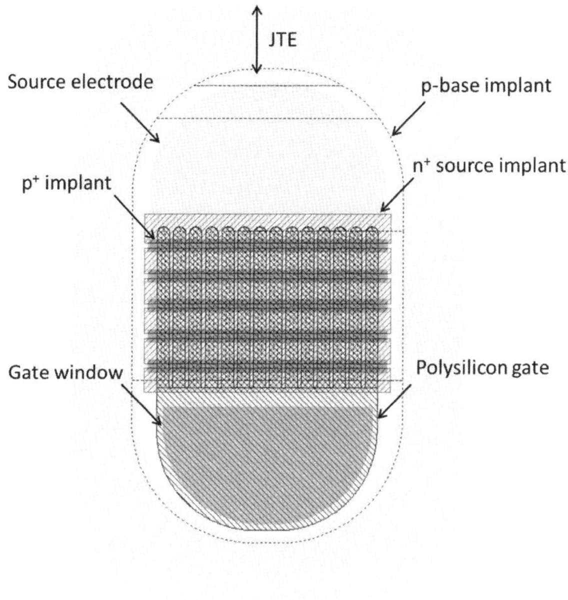

Fig. 11.12. Mask layout for a typical 4H-SiC DIMOSFET.

The layout of a typical device is displayed in Fig. 11.12. Note that only the most important layers are shown, meaning that *Mask 0 Alignment keys* and *Mask 6 Contact definition* are omitted in order to make the layout easier to understand. For the purposes of clarity, each of the layers indicated in Fig. 11.12 is displayed as a digitized area inside a pattern. In reality, however, not all the mask layers have the same polarity. All layers are "clear field", defined as chrome layers inside the patterns on a glass mask, but Mask 5 is "dark field", if a positive photoresist is used.

A run sheet was established, which is presented in Table 11.1 below, describing the complete process details. The process parameters may vary depending on the fabrication tools available. Even though simulation of the termination was not included in the previous section, it should be noted that edge termination must be included around the edge of the device. A p-type implantation at 650°C with a total dose of $8 \times 10^{12}\,cm^{-2}$ was included in the process so as to create a $50\,\mu m$ wide single-zone JTE around the device area, as shown in Fig. 11.12. This is necessary in order to enhance the breakdown voltage. A microphotograph of the fabricated device is shown in Fig. 11.13.

Table 11.1. Typical process run sheet for a Self-Aligned 4H-SiC DIMOSFET.

Step	Process Description	Sample #	Remarks
1.	**Lot start** • Wafer: n- epi on (0001) face, thick 30 μm, 2E15 /cm^3	1-4	2 cm by 2 cm
2.	**Wafer marking and pretreatment** • Mark backsides of wafers • $H_2SO_4 : H_2O_2 = 1:1$, 10 min • DI water rinse		
3.	**Polysilicon deposition** • Sacrificial oxidation, 1100°C, 90 min, wet O_2 • LPCVD polysilicon, 585°C, 10 hrs, thick 1.8 μm		
4.	**Mask 1: p-base implant mask (MJB-3)** • H_2O remove bake: 120°C, 5 min • LOR: 3000 rpm, 45 sec • LOR soft bake 170°C, 5 min • S1813: 5000 rpm, 30 sec • PR soft bake: 95°C, 2 min 30 sec • Exposure: 18 sec $\times$ 6.3 mW/cm^2 • PR Development: 24 sec • PR hard bake: 120°C, 5 min		Alignment tolerance 1 μm
5.	**Polysilicon etch (Oxford 80 plus RIE)** • 100 Watt, 30 mTorr, 6 min, SF6 40 sccm, O_2 5 sccm • PR strip		
6.	**p-base implantation** • Aluminum, 30 keV/360 keV, 6E10cm^{-2}/5E13cm^{-2}, 650°C		
7.	**Mask 0: Alignment keys** • H_2O remove bake: 120°C, 5 min • LOR: 3000 rpm, 45 sec • LOR soft bake: 170°C, 5 min • S1813: 5000 rpm, 30 sec • PR soft bake: 95°C, 2 min 30 sec • Exposure: 18 sec $\times$ 6.3 mW/cm^2 • PR Development: 24 sec • PR hard bake: 120°C, 5 min		
8.	**Alignment key etch (Oxford 80 plus RIE)** • 100 Watt, 30 mTorr, 6 min, SF6 40 sccm, O_2 5 sccm • SiC etch rate: 25 nm/min • PR strip		

(Continued)

Table 11.1. (*Continued*)

Step	Process Description	Sample #	Remarks
9.	**Self-aligned spacer/oxidation** • 1000°C, 9 hrs, wet O_2		
10.	**Mask 2: n+source implant mask** • H_2O remove bake: 120°C, 5 min • LOR: 3000 rpm, 45 sec • LOR soft bake: 170°C, 5 min • S1813: 5000 rpm, 30 sec • PR soft bake: 95°C, 2 min 30 sec • Exposure: 18 sec × 6.3 mW/cm^2 • PR Development: 24 sec • PR hard bake: 120°C, 5 min		
11.	**n+source implantation** • Nitrogen, 30 keV/80 keV, 7E14cm^{-2}/1.2E15cm^{-2} • PR strip		
12.	**Mask 3: p+contact implant mask** • H_2O remove bake: 120°C, 5 min • LOR: 3000 rpm, 45 sec • LOR soft bake: 170°C, 5 min • S1813: 5000 rpm, 30 sec • PR soft bake: 95°C, 2 min 30 sec • Exposure: 18 sec × 6.3 mW/cm^2 • PR Development: 24 sec • PR hard bake: 120°C, 5 min		
13.	**p+contact implantation** • Aluminum, 20 keV/60 keV/125 keV/180 keV, 6E14cm^{-2}/7.5E14cm^{-2}/8.5E14cm^{-2}/9.5E14cm^{-2} • PR strip		
14.	**Polysilicon strip** • $HNO_3 : HF = 1 : 1$, 10 min		
15.	**PECVD oxide deposition (Samco)** • 300°C, 10 min, 1 μm		
16.	**Mask 4: JTE implant mask** • H_2O remove bake: 120°C, 5 min • LOR: 3000 rpm, 45 sec • LOR soft bake: 170°C, 5 min • S1813: 5000 rpm, 30 sec • PR soft bake: 95°C, 2 min 30 sec		

(*Continued*)

Table 11.1. (*Continued*)

Step	Process Description	Sample #	Remarks
	• Exposure: 18 sec × 6.3 mW/cm^2 • PR Development: 24 sec • PR hard bake: 120°C, 5 min		
17.	**PECVD oxide etch (Oxford 80 plus RIE)** • 200 Watt, 40 min, CHF_3 50 sccm, Ar 50 sccm, 30 mTorr, • Oxide etch rate: ~30 nm/min		
18.	**P-JTE implantation** • Aluminum, 40 keV/120 keV/240 keV/360 keV, 0.6E12cm^{-2}/2.0E12cm^{-2}/2.1E12cm^{-2}/3.3E12cm^{-2}, 650°C		
19.	**Oxide strip** • BOE, 10 min		
20.	**Graphite cap formation and implant anneal** • AZ4620: 4000 rpm, 1min • Hard bake: 110°C, 30 min • Furnace: 900°C, 30 min, vacuum • Activation anneal 1650°C, 30 min, Ar		
21.	**Graphite strip** • Furnace: 900°C, 30 min, O_2		
22.	**Gate oxidation** • RCA clean • Thermal oxidation, 1300°C, 26 min • NO anneal, 1300°C, 30 min		
23.	**Polysilicon gate deposition** • LPCVD polysilicon: 585°C, 4 hrs, 0.7 μm thick • POCl3: 900°C, 60 min; drive-in, 900°C, 90 min, ~20 Ω/square		
24.	**Mask 5: Polysilicon gate definition** • H_2O remove bake: 120°C, 5 min • LOR: 3000 rpm, 45 sec • LOR soft bake: 170°C, 5 min • S1813: 5000 rpm, 30 sec • PR soft bake: 95°C, 2 min 30 s • Exposure: 18 sec × 6.3 mW/cm^2 • PR Development: 24 sec • PR hard bake: 120°C, 5 min		

(*Continued*)

Table 11.1. (*Continued*)

Step	Process Description	Sample #	Remarks
25.	**Polysilicon etch (Oxford 80 plus RIE)** • 100 Watt, 30 mTorr, 2.5 min, SF6 40 sccm, O_2 5 sccm • PR strip		
26.	**Interlayer oxide formation** • 1100°C, 2hrs, dry O_2, oxide 0.5 μm thick		
27.	**Mask 6: Contact definition** • H_2O remove bake: 120°C, 5 min • LOR: 3000 rpm, 45 sec • LOR soft bake: 170°C, 5 min • S1813: 5000 rpm, 30 sec • PR soft bake: 95°C, 2 min 30 sec • Exposure: 18 sec × 6.3 mW/cm^2 • PR Development: 24 sec • PR hard bake: 120°C, 5 min		
28.	**Open contact window (Oxford 80 plus RIE)** • 50 Watt, 30 mTorr, 10 min, CHF_3 50 sccm, Ar 50 sccm, • Oxide etch rate: ~6nm/min		
29.	**Contact metal deposition and loft off** • Thermal evaporation Ti/Ni: 20 nm/100 nm • Soak in PG remover: 10 min, 100°C • Soak in Acetone: 10 min		
30.	**Front side protection** • H_2O remove bake: 120°C, 5 min • LOR: 3000 rpm, 45 sec • LOR soft bake: 170°C, 5 min • S1813: 5000 rpm, 30 sec • PR hard bake: 120°C, 5 min		
31.	**Backside etch and metal deposition** • 200 Watt, 30 mTorr, 10 min, CHF_3 50 sccm, Ar 50 sccm, polysilicon etch • 100 Watt, 30 mTorr, 10 min, SF6 40 sccm, O_2 5 sccm, SiC etch • Thermal evaporation Ti/Ni: 20 nm/100 nm • PR strip		
32.	**Contact metal anneal** • RTA: 1000°C, 3 min, vacuum		

(*Continued*)

Table 11.1. (*Continued*)

Step	Process Description	Sample #	Remarks
33.	**Mask 7: Gate window definition** • H_2O remove bake: 120°C, 5 min • LOR: 3000 rpm, 45 sec • LOR soft bake: 170°C, 5 min • S1813: 5000 rpm, 30 sec • PR soft bake: 95°C, 2 min 30 sec • Exposure: 18 sec × 6.3 mW/cm^2 • PR Development: 24 sec • PR hard bake: 120°C, 5 min		
34.	**Gate window etch** • 200 Watt, 30 mTorr, 10 min, CHF_3 50 sccm, Ar 50 sccm, • Oxide etch rate: ~30 nm/min • PR strip		
35.	**Mask 8: Pad metal definition** • H_2O remove bake: 120°C, 5 min • LOR: 3000 rpm, 45 sec • LOR soft bake: 170°C, 5 min • S1813: 5000 rpm, 30 sec • PR soft bake: 95°C, 2 min 30 sec • Exposure: 18 sec × 6.3 mW/cm^2 • PR Development: 24 sec • PR hard bake: 120°C, 5 min		
36.	**Pad metal deposition and lift off** • Thermal evaporation Al: 700 nm • Soak in PG remover: 10 min, 100°C • Soak in Acetone: 10 min		

11.1.3.2. *On- and Off-state Measurement on Fabricated 4H-SiC DIMOSFETs*

After the fabrication was completed, an on-wafer probe of the devices was carried out. The results of the probe showed that the n+ source sheet resistance was 2.1 kΩ/square, the n+ contact resistivity was $5.8 \times 10^{-5} \Omega \cdot \text{cm}^2$, the p+ sheet resistance was 18.5 kΩ/square, and the p+ contact resistivity was $1 \times 10^{-4} \Omega \cdot \text{cm}^2$.

Figure 11.14 shows the measured on-state characteristics of a 4H-SiC DIMOSFET that has an active area of 333 μm × 145 μm. The total channel width is 2.2 mm and the cell pitch is 18 μm. The extracted $R_{on,sp}$ is 120 mΩ·cm^2,

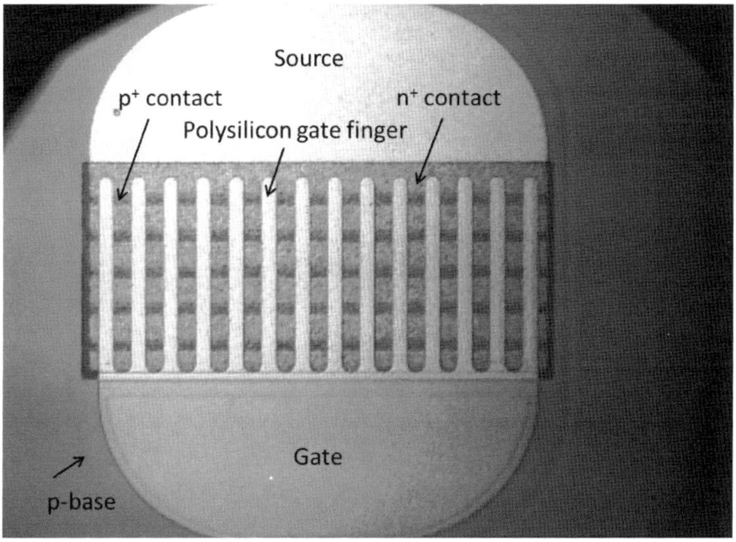

Fig. 11.13. Microphotograph of a fabricated 4H-SiC DIMOSFET.

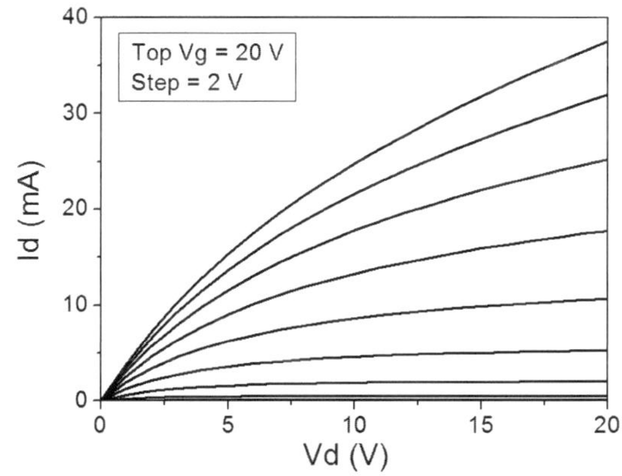

Fig. 11.14. Measured on-state characteristics of a 4H-SiC DIMOSFET that has an active area of 333 μm × 145 μm.

which is much larger than the value predicted via simulation. This discrepancy may be attributed to several factors, including:

1. The series resistance, such as the substrate resistance, not being included in the simulation.

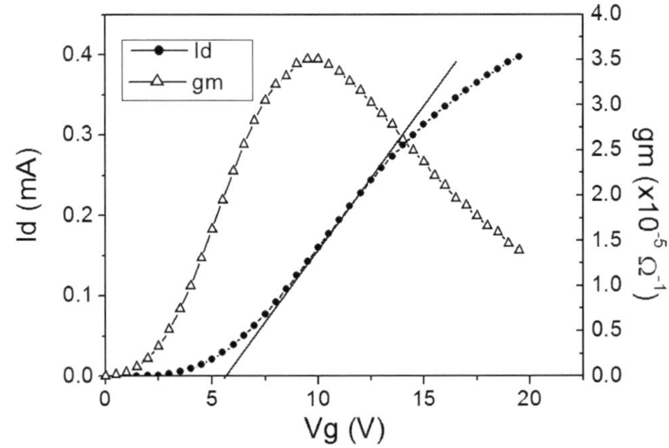

Fig. 11.15. Plot of the measured I_d and g_m versus Vg at a V_d of 0.1 V for the 4H-SiC DIMOSFET presented in Fig. 11.14.

2. Three-dimensional (3D) effects, such as the current spreading along the fingers due to a large finger resistance, and
3. A large n+ source sheet resistance, which may have occurred as a result of the nitrogen implantation, and should be able to be reduced by an order of magnitude if a phosphorus implantation is used.

The plot for I_d and g_m versus Vg at a V_d of 0.1 V is shown in Fig. 11.15. The threshold voltage is 5.34 V from the extrapolation of the I_D–V_G curve at location of the maximum g_m.

The breakdown measurement was performed at room temperature with the sample immersed in Fluorinert to prevent any arcing in air. The measured reverse characteristics are shown in Fig. 11.16. It can be seen that the leakage current is less than 100 nA, which is equivalent to 0.21 mA/cm^2. This device suffered a catastrophic breakdown at a voltage of 2240 V, which is much less than the 4480 V predicted by the simulation of the cell structure. A JTE structure was then simulated to check the sensitivity of the breakdown voltage against JTE dose and the results are shown in Fig. 11.17. The best breakdown voltage is 3510 V at a dose of 9×10^{12} cm^{-2}. From the measured breakdown voltage results, it is estimated that the active JTE dose is 5.3×10^{12} cm^{-2}, which is about 66% of the implanted dose of 5.3×10^{12} cm^{-2}. This value is consistent with those reported in the previous literature. A more advanced edge termination technique to those described in Chapter 7 is suggested as a method of further improving the breakdown voltage.

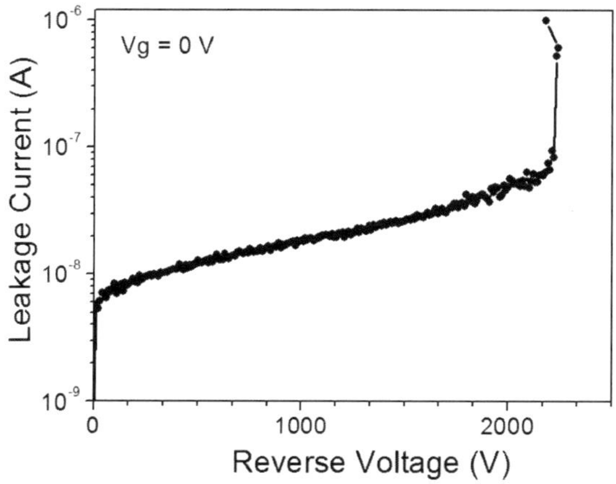

Fig. 11.16. Measured reverse characteristics of the 4H-SiC DIMOSFET presented in Fig. 11.14.

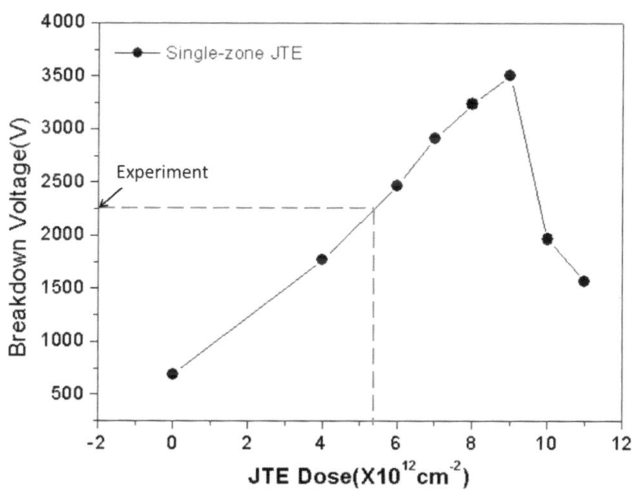

Fig. 11.17. Comparison of the simulated breakdown voltage versus the JTE dose and the measurement results.

11.2. Case Study II: Design of Normally-off GaN HEMT for Power Electronics Applications

The goal of this section on the device design and fabrication is to realize normally-off GaN High Electron Mobility Transistor (HEMT) with a high threshold voltage which sustains at higher temperature. This is important in power electronics applications as it leads to simplified gate drive circuit and improved system reliability. As GaN HEMTs are normally-on, several device structure modifications have been attempted in the past. This project intends to overcome many of the limitations found in earlier approaches to raise the threshold voltage.

One of the approaches to realize normally-off operation is of creating a selective partial or full recess in the AlGaN layer by etching say 50–100% thickness of the layer in the gate region. This leads to reduced 2D electron gas (2DEG) carrier density and normally-off operation. 2DEG density can be restored using gate field at positive gate voltage above the threshold voltage, V_{th}, if recess is partial. However, there is also a strong side effect of damage to the AlGaN-GaN interface which substantially degrades 2DEG mobility and saturated drain current in the on-state for normally-off HEMT at the same gate overdrive V_{GT} (equal to $V_{GS} - V_{TH}$) compared to that for normally-on HEMT. In this project, the damage is minimized by using low plasma energy for etching and removing only up to 70% of AlGaN layer so that damaging ions mostly confine to AlGaN and do not reach the interface in large concentration. The fabrication steps to realize such a HEMT are now described along with the schematic cross-sections of the devices after every step. Some schematic layout dimensions are also shown so that typical sizes are known to the readers. Do note that the figures are not to scale and width dimensions perpendicular to the cross-section plane not shown are much larger as the devices are typically wide. Also some of the connections are done in the width dimension.

11.2.1. Fabrication of Partial AlGaN recess Metal-Insulator-Semiconductor (MIS) HEMT

Fabrication starts with a typical GaN wafer structure as shown with 20 nm $Al_{0.25}Ga_{0.75}N$ and a thin 1 nm GaN capping layer above it. Different layers are shown using colors and fill patterns so that materials used can be easily distinguished. These are summarized in Fig. 11.18. The wafer structure cross-section is shown in Fig. 11.19.

Alignment of respective mask geometries with a reference alignment mark is one important step to be carried first. This is not a part of device structure but is essential to have successful device fabrication. This step forms 600 nm deep alignment mark. Lithography process using mask#1 is carried out and the deep mark is formed by Reactive Ion Etching (RIE) using CF_4 or CHF_3/Ar with etching depth 600 nm which maintains sidewall of the etched structure vertical. The resist is removed after etching is completed. As lithography using mask and removal of resist are common steps to all mask processes, they will not be repeated for the other masks. Also, thin 1 nm layer and structure

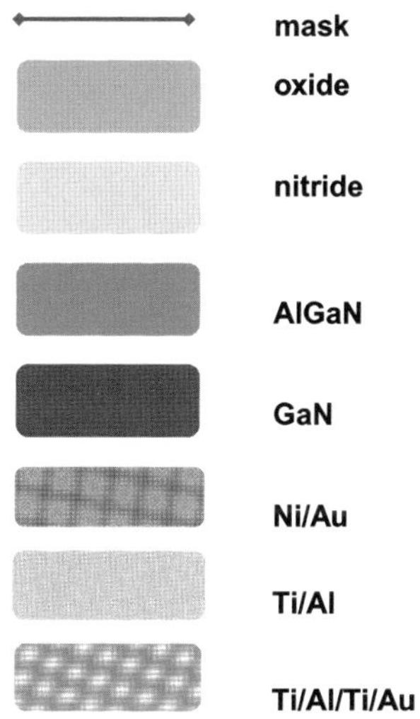

Fig. 11.18. Color and fill pattern convention used in schematic cross-section depictions.

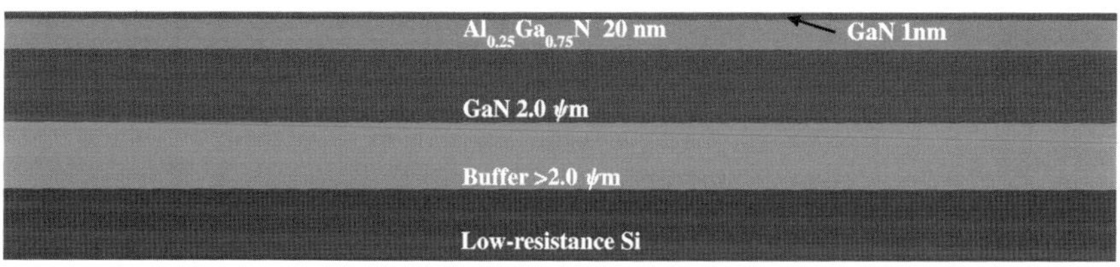

Fig. 11.19. Starting wafer structure.

below buffer layer will not be shown as the latter is not affected by fabrication. The schematic cross-section after alignment mark formation is shown in Fig. 11.20.

The next fabrication step is of mesa formation to isolate different devices from each other. Lithography process using mesa-mask#2 and etching is carried out by RIE using CF_4 or CHF_3/Ar to a depth of about 100 nm with sidewalls with slight slope to avoid any field crowding effects at the bottom corners. It is important to ascertain that desired structure is obtained after the fabrication steps. Hence, characterization is essential after important steps. In this case, measurement on etching depth was carried out using surface profiler and surface quality with low roughness was verified using Scanning Electron Microscopy (SEM) and Atomic Force Microscopy (AFM). The schematic cross-section after mesa isolation formation is shown in Fig. 11.21.

Fig. 11.20. Alignment mark formation.

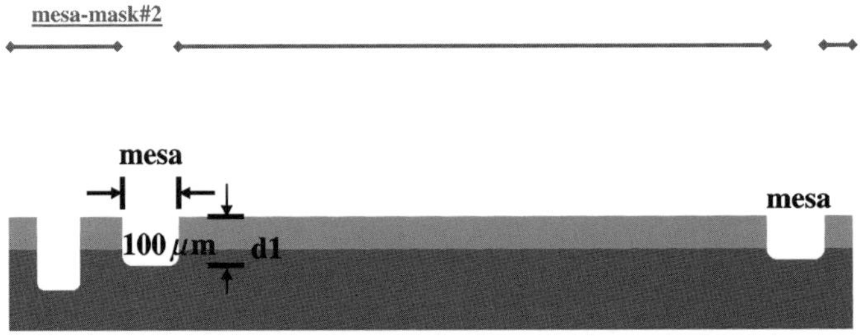

Fig. 11.21. Mesa isolation formation with depth d_1 of 100 nm.

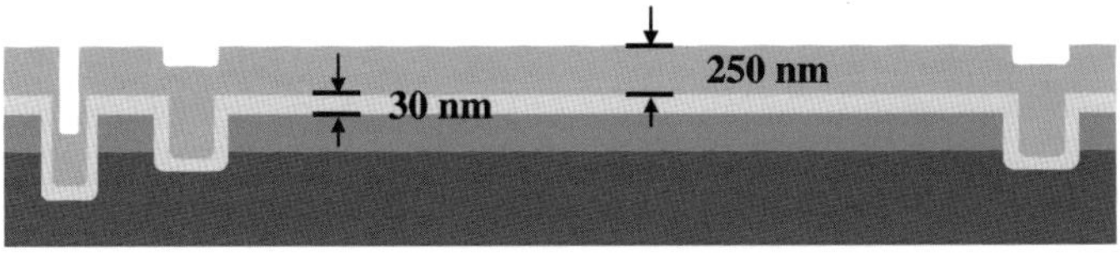

Fig. 11.22. Passivation with 30 nm nitride and 250 nm oxide.

Existence of surface traps degrades performance of HEMT devices with effects such as current collapse, reduced breakdown voltage, and current instability. Use of nitride/oxide passivation is very effective in reducing number of such states. Hence, the next step is carried out by deposition of passivation layers of 30 nm nitride and 250 nm oxide using plasma enhanced chemical vapor deposition (PECVD). The schematic cross-section after passivation is shown in Fig. 11.22.

Ohmic contact formation in the source–drain regions is now carried out using lift off process where metal will be removed with resist in the resist removal step. First lithography process using S/D-mask#3 is carried out and wet etching of nitride/oxide is done using standard chemicals HF and H_3PO_4

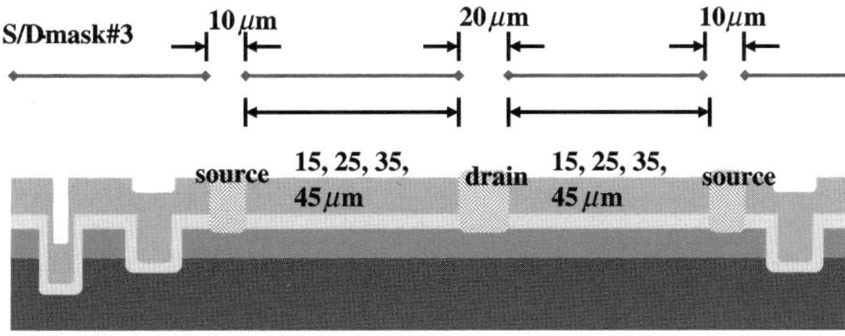

Fig. 11.23. Ohmic contact formation with 30 nm Ti and 250 nm Al.

as the mask dimensions are quite large. Electron beam (EB) deposition of Ti/Al (30/250 nm) to form ohmic contacts is done. Resist is now removed which also removes the metal in unexposed areas. The contacts are annealed in Nitrogen at 650°C for 10 min to form better contact and diffuse metal in AlGaN. It is important to include test structure and carry out measurement on ohmic contact resistance using Transmission Line Method (TLM) to ensure low contact resistivity. The schematic cross-section after ohmic contact formation is shown in Fig. 11.23.

For benchmarking, it is very important to have several device structures fabricated with different mask dimensions on parts of the wafer and different drain–gate spacing so that effectiveness of novel approach can be evaluated. In this case, the first cross-sections show two devices, one with metal-insulator-semiconductor (MIS) gate with a partial recess and normally-on device with the same MIS gate but without recess. This way we can estimate if 2DEG electron mobility is degraded because of recess by comparing on-currents at the same value of $V_{GT}(V_{GS} - V_{TH})$. As follow-up steps after this step are identical to both devices, they will not be shown. Only the cross-sections of the normally off device will be shown. The normally-on device basically has no recess in the gate region where as normally-off device has around 50% of AlGaN etched.

The gate formation does lithography process using trench-gate-mask/gate-mask and nitride/oxide removal using RIE etching so that there is no undercut like wet etching as gate length feature is around 2 μm. The gate to drain spacing is varied in the range of 10–40 μm to fabricate HEMTs with different series resistance and breakdown voltage. Only for normally-off device, gate trench is also formed using RIE etching by partial AlGaN removal. Clearly, different masks are needed for these steps, but rest of the masks are common. Then annealing at 600°C for 5 min is done to improve the surface quality and cure possible etch damage to 2DEG. Like in the mesa case, measurement on the etching surface before and after the annealing treatment is done to confirm low roughness as this step is most crucial for proper

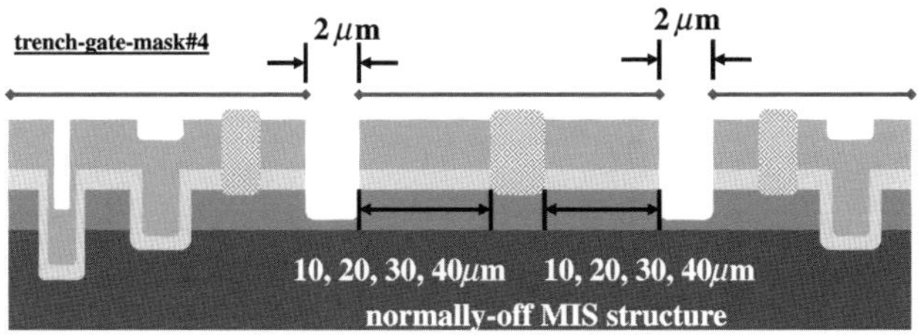

(a)

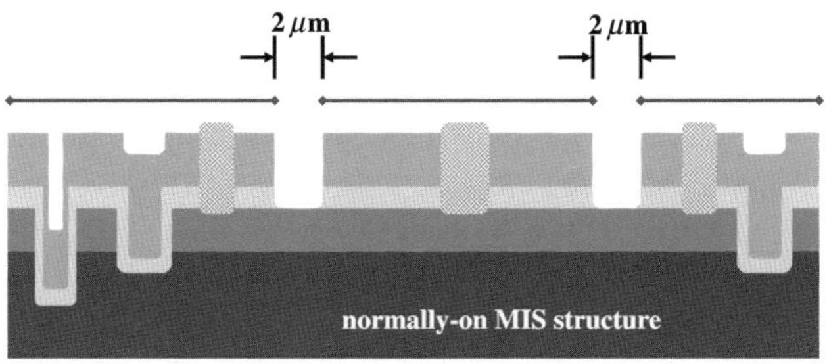

(b)

Fig. 11.24. Gate formation in (a) normally-off with gate recess (b) normally-on HEMTs.

device function. The schematic cross-sections after gate formation are shown in Fig. 11.24(a) and 11.24(b).

Now 30 nm nitride is deposited using PECVD as gate insulator in both structures. Using lithography process and nitride-removal-mask#6, the nitride is removed in the source–drain contact areas using wet etch at low etching rate so that metal below the nitride is not affected. As usual, the resist is removed after the etch. The schematic cross-section after nitride removal is shown in Fig. 11.25.

The next step forms gate metal as well as field plate and additional metal over source–drain contacts using lift-off process. The gate metal is extended over oxide/nitride layers in the drain area to form a field plate that enhances breakdown voltage through appropriate spreading of electric field. It also extends in the source area to allow for full gate trench coverage by metal even if this mask misaligns with the actual gate. As this is a lift-off process, first lithography process using gate-FP-mask is done followed by EB deposition of Ni/Au (30/1000 nm). Unwanted metal is removed with resist as in a standard lift-off process leading to final cross-section shown in Fig. 11.26.

Subsequent masks of routing and pads cannot be seen in this view and hence they are not included. It is important to note that edge termination is

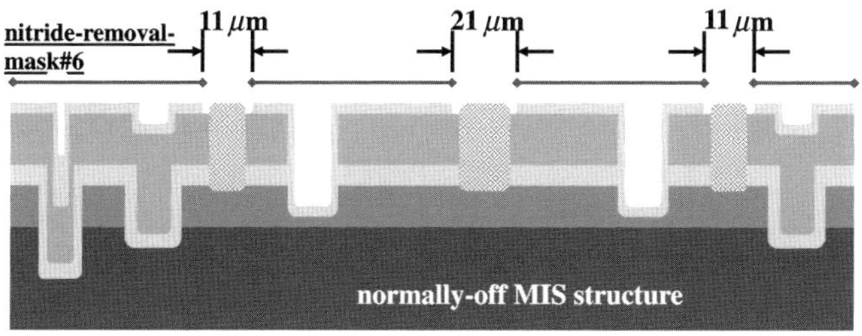

Fig. 11.25. Removal of 30 nm nitride in source–drain contact areas.

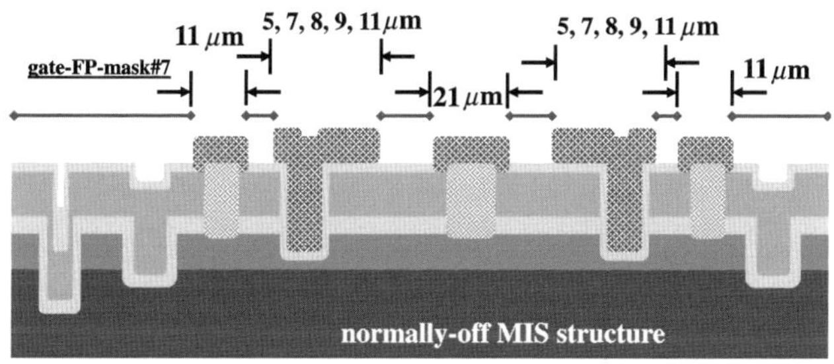

Fig. 11.26. Gate metal and field plate formation using Ni/Au (30/1000 nm).

important in power devices to achieve desired breakdown voltages which is not described here as methodology is similar in the devices.

11.2.2. Innovative Normally-off (MIS) HEMT Structure Using Multi-fluorinated Gate Stack

Many measurements with the type of structure in Sec. 11.2.1 have been reported in the literature. Here, we summarize results from (Anderson *et al.*, 2010) on the recessed gate device structure. The $Al_{0.25}Ga_{0.75}N$ barrier layer thickness was 25 nm. The thickness was 2 μm for the GaN channel layer. The gate recess was achieved using $Cl_2/BCl_3/Ar$ ICP-RIE. The largest threshold voltage V_{th} observed was about 0.2 V even when most of AlGaN was etched. Due to etch damage and other effects, the 2DEG electron mobility degraded from 1350 $cm^2V^{-1}S^{-1}$ to below 250 $cm^2V^{-1}S^{-1}$ when etch depth was increased. Hence, it is clear that only partial recess can maintain 2DEG mobility at acceptable level. However, this approach does not yield sufficiently high V_{th}. Hence, several researchers tried to raise V_{th} by introducing negative charges in the gate region by introducing fluorine ions using implant or plasma, using P+ GaN gate to deplete 2DEG and using MIS recess gate that also reduces gate

Table 11.2. Performance of reported normally-off HEMTs.

	Advantages	Disadvantages	Latest Reported V_{th}	Latest Reported I_{dmax}
Gate Recess	Easy to fabricate and control by etching time	Surface damage due to plasma degrades 2DEG mobility, introduces traps	+1.7 V (Ye *et al.*, 2013) +1.65 V (Woojin *et al.*, 2014)	500 mA/mm (Ye *et al.*, 2013) 650 mA/mm (Woojin *et al.*, 2014)
	High I_{DMAX} after surface passivation as additional plasma or implant not done	Lower V_{th} than p-GaN		
p-GaN cap	High V_{th} is possible and can be designed.	Difficult p-type Dopant Activation requires additional thermal steps of growth and/or activation	+2.44 V (Hwang *et al.*, 2013a) +3.03 V (Hwang *et al.*, 2013b)	320 mA/mm (Hwang *et al.*, 2013a) 220 mA/mm (Hwang *et al.*, 2013b)
	Gate surface protected as there is no plasma treatment needed.	Requires extra high quality epitaxial growth of p-GaN		
Fluorine plasma treatment	Easy to fabricate and control	May damage the lattice in channel region	+0.6 V (Yang *et al.*, 2013) +1.8 V (Chen *et al.*, 2011) +0.9 V (Nakajima *et al.*, 2008)	400 mA/mm (Yang *et al.*, 2013) 380 mA/mm (Chen *et al.*, 2011) 300 mA/mm (Nakajima *et al.*, 2008)
	Higher I_{DMAX} than p-GaN	Lower V_{th} than p-GaN		

effectiveness. The best reported results are summarized in Table 11.2 along with related references. The advantages and disadvantages of these approaches are also summarized in Table 11.2. Based on the reports, V_{th} is raised to maximum of 3 V at room temperature.

Therefore, this project explored novel approach to achieve high positive V_{th}. One idea was to use very low energy in RIE along with partial recess of AlGaN to minimize etch damage near AlGaN/GaN interface so that 2DEG mobility is preserved. Another modification done was to the gate stack where a possibility of multiple fluorine treatments using RIE plasma on successively deposited Al_2O_3 dielectric layers as gate insulator after each deposition using atomic layer deposition (ALD) was explored.

This novel approach of gate stack modification was first verified using device simulation. The gate stack constituted a total of four layers of gate insulator Al_2O_3. For initial exploration using simulation, first three layers of Al_2O_3 are 4 nm thick and the last layer is 3 nm thick. First three 4 nm layers of gate insulator Al_2O_3 had negative sheet charge density (SCD) distributed within top 2 nm of Al_2O_3. The last 3 nm layer had no F-treatment to ensure that RIE plasma damage will not affect metal-Al_2O_3 interface. Figures 11.27 and 11.28 show simulated I_D–V_G characteristics. With a specific SCD induced by F^- ions, simulations predicted that V_{th} above 3 V can be achieved. An equal SCD of 10^{12} cm^{-2} on top of each of the three successive Al_2O_3 layers yielded low but positive V_{th}. On the other hand, higher SCD of 3×10^{12} cm^{-2} could achieve V_{th} of about 3 V. The key advantage of this approach is that the standard process described earlier in Sec. 11.2.1 is virtually intact except for the gate stack formation.

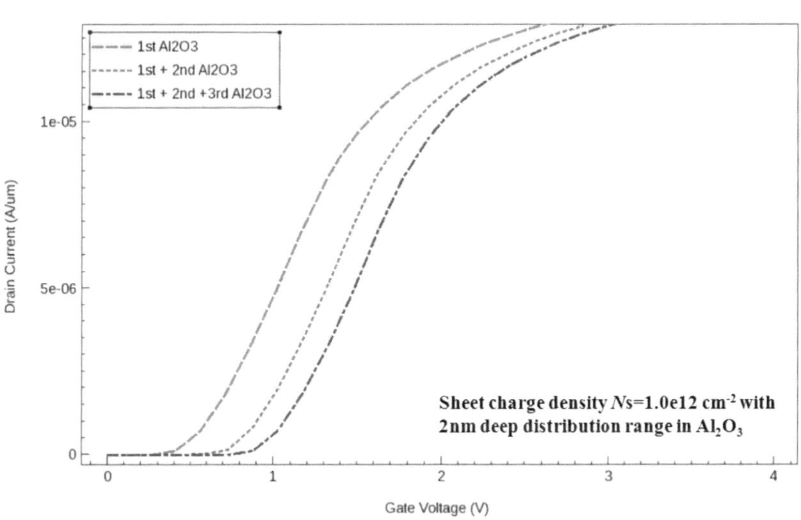

Fig. 11.27. I_D–V_G curves of the proposed devices with a fixed charge density of 10^{12} cm^{-2} at different Al_2O_3 dielectric layers.

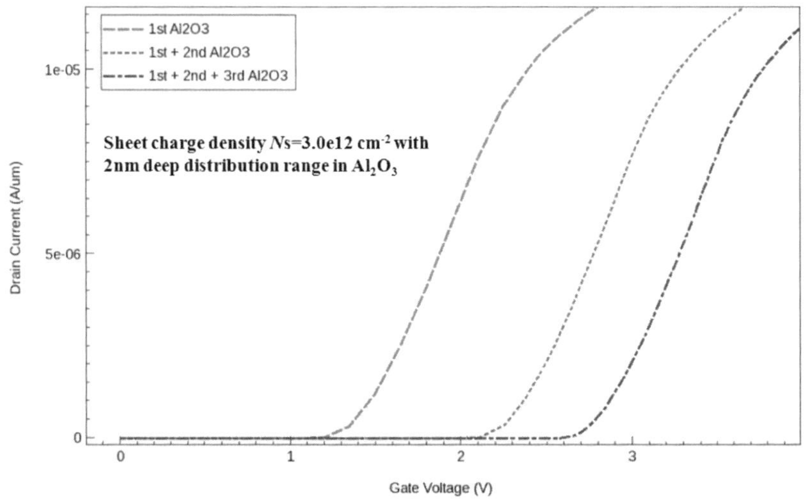

Fig. 11.28. I_D–V_G curves of the proposed devices with a fixed charge density of 3.0×10^{12} cm^{-2} at different Al_2O_3 dielectric layers.

For fabrication of such a device, it is important to know how much F^- is incorporated in the gate insulator in RIE F-based treatment which will create electron traps. Also, RIE must be carried out at sufficiently low energy to minimize etch damage. Through short loops and simulations, we were able to determine that SCD rate induced by F^- ions is fairly uniform at 4.65×10^{10} cm^{-2}s^{-1} for RIE at 30 W power using CHF_3. Process treatment time for each layer was determined based on how much SCD was to be achieved. The full details of the experimental results are described in Sec. 8.8 and reported in (Wang *et al.*, 2015a) which show achievement of large positive V_{th}.

11.2.3. Obtaining Normally-off (MIS) HEMT Structure Capable of Operating at High Temperature Using Ar and Single F-Treatment

The demonstration of high V_{th} AlGaN/GaN on Si HEMT in Sec. 11.2.2 appears to have overcome many limitations of previous research and reached the desired goal. However, one aspect still needs to be verified. One of the key advantage of the GaN based HEMTs is their ability to operate at high temperature in somewhat harsh environment such as automobile engines. Of course, on-current is expected to degrade at high temperature due to reduction in mobility. However, V_{th} is required to remain at a reasonable positive value in normally-off devices and maintain its value in normally-on devices.

This is generally true for normally-on intrinsic HEMTs and recessed gate MIS HEMTs as variation of V_{th} with temperature is quite small but V_{th} is not at a required level for normally-off operation (Fig. 11.30). However, F- treated gate stack HEMTs, that can achieve required V_{th} like ones in Sec. 11.2.2, have trapped negative charges which can de-trap at high temperature due to additional thermal energy which can significantly reduce V_{th}.

High temperature V_{th} measurements were performed on HEMTs fabricated in Sec. 11.2.2. It was found that the threshold voltage dropped a lot at high temperature and even became negative at 200°C. This clearly showed that the method used in Sec. 11.2.2 for F- incorporation mainly placed negative charge in very shallow traps which emitted electrons at higher temperature and became ineffective. Hence, a method to generate much larger density of deeper traps compared to shallow traps when performing F-treatments will be needed to overcome this problem. For a systematic solution, it is important to carry out following investigations:

(i) Characterize trap distribution from high temperature V_{th} measurements or other methods near conduction band edge of the gate insulator Al_2O_3.
(ii) Investigation on methods to create larger density of deeper traps to maintain reasonable positive V_{th} at high temperature normally expected of normally-off GaN devices.
(iii) Design a device with process simplification and Ar pre-treatment of Al_2O_3 before F-plasma treatment that will help create deeper traps for electrons to maintain high temperature operation and V_{th}.
(iv) Determine the mechanism that improves thermal stability of V_{th} to guide device designers to optimize performance of the device at high temperature.

All these steps are achieved through modifications described in this section. Full processing conditions for the device fabrication are also given at the end of this section.

11.2.3.1. *Characterization of Trap Distribution in the Bandgap Relative to Conduction Band Edge of the Gate Insulator Al_2O_3*

A method for trap characterization using gate stress voltage is used in this work to estimate the deep trap states and densities (D_{EQ}) induced by Fluorine Plasma Treatment (FPT). The basic principle is that the trapped charge will de-trap from progressively deeper states if increasing negative voltage is applied to the gate to increase attractive electric field exerted on trapped electrons. As negative gate bias voltage increases, it is possible to identify increasing $\hat{E}_{T\_MAX}$

which is the deepest energy level from which the trapped charge will be emitted. The Poole–Frenkel trap emission theory (Yeh *et al.*, 2007; Mitrofanov and Manfra, 2004; Cheong *et al.*, 2007; Specht *et al.*, 2004) governs this de-trapping process. The theory leads to Eq. (11.1) in which increased attractive electric field (ξ) leads to progressively increasing $\hat{E}_{T\_MAX}$. The negative gate bias provides the needed electric field (ξ) that enhances the emission of FPT-induced negative charges trapped within Al_2O_3. A larger reduction of V_{th} is observed with increasing gate bias voltage (V_B) due to the stronger emission field. The gate bias V_B is increased in steps from 0 V to −40 V and held for 50 s at 25°C. Then pulsed I_D–V_G measurement described below can provide vital trap distribution information.

$$\hat{E}_{T\_MAX} = \sqrt{\frac{q\xi}{\pi\varepsilon_{Al_2O_3}\varepsilon_0}} - \frac{kT}{q}\ln\left(1 - \frac{Q_{EQ,0}}{Q_{EQ,VB}}\right), \tag{11.1}$$

where

$$\xi = \left(\frac{V_G}{t_{Al_2O_3}} - \frac{\phi_b}{qt_{Al_2O_3}}\right).$$

Here $t_{Al_2O_3}$ is the gate insulator thickness, V_G is the applied negative gate voltage, ϕ_b is the barrier height at the gate insulator-AlGaN interface, q is the electronic charge, T is the temperature, $\varepsilon_{Al_2O_3}$ is relative permittivity of Al_2O_3 and ε_0 is the permittivity of the free space. Also, $Q_{EQ,0}$ and $Q_{EQ,VB}$ are the effective FPT-induced trapped negative charge before and after the gate stress is applied. These effective charge values are of total charge placed at the dielectric-AlGaN interface that produces the same effective electrostatic effect as multi-fluorinated gate stack with trapped charges at three different locations within the dielectric.

The $Q_{EQ,0}$ and $Q_{EQ,VB}$ values are found from measured V_{th} from pulsed I_D–V_G measurement after the gate stress has been applied using model Eqs. (8.11) and (8.12) in Sec. 8.8. The changes in V_{th} are due to reduction in trapped negative charge. However, to perform I_D–V_G measurement to measure V_{th}, positive V_G is applied which can inject electrons back in the traps which causes error to the characterization. In order to minimize this error, the positive gate voltage is provided in pulses as shown is Fig. 11.29. Initially, 50 s of gate stress V_B to emit the trapped negative charge is applied. Then pulsed positive gate sweep initiates and returns to the same V_B between pulses. The time for V_B between two pulses is 18 ms and the time for positive V_G pulse is 2 ms for each pulse cycle. This ensures that the charges injected by the positive V_G are emitted in each cycle. Detailed time sequence of these V_G pulses is

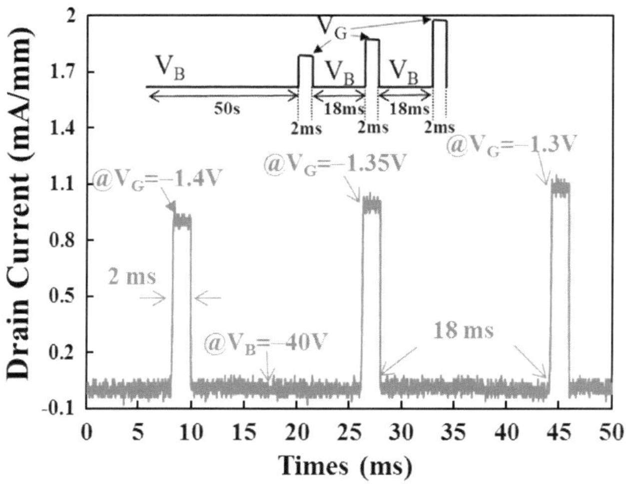

Fig. 11.29. Drain current waveform during the I_D–V_G pulse when V_G is near the V_{th} at $V_B = -40$ V and $V_D = 1$ V. It also shows the schematic V_G waveform, where the V_B is held initially for 50 s followed by the pulsed signal with a period of 20 ms (2 ms pulse V_G and 18 ms bias V_B) during each sweep. The device here has a negative V_{th}.

shown in Fig. 11.29. A sampled drain current (I_D) waveform from the characterization is also shown in Fig. 11.29. The I_D–V_G characteristic are typically measured with gate stress voltage of $V_B = 0, -10, -20, -30$, and -40 V for trap characterization.

It is possible to relate $\hat{E}_{T\_MAX}$ from the gate stress measurement to temperature effects through the deepest emission level of the FPT-induced negative charge traps corresponding to different temperature. This can be modeled in a straightforward way through Eq. (11.2) (Zafar *et al.*, 2002; Schroder, 2006). In Eq. (11.2), $\hat{E}_{T\_MAX}$ is the deepest energy level of the trapped charge that will be emitted from the trap sites, k_b is the Boltzmann's constant, and T is the device temperature.

The γ_n is found from $\gamma_n = (\nu_{th}/T^{0.5})(N_c/T^{1.5}) = 3.25 \times 10^{21}(m_n/m_0)$, where m_n/m_0 is the relative electron effective mass within Al_2O_3 which is 0.16 (Ťapajna *et al.*, 2013), ν_{th} is the thermal velocity of carriers in Al_2O_3, N_C is the effective density of states in Al_2O_3, h is the Planck's constant. Furthermore, $\sigma_{Al_2O_3} = 10^{-16}\,cm^2$ is the capture cross-section of the electrons in Al_2O_3 (Ganguly *et al.*, 2011), and $t = 50$ s is the device heating time. The 50 s time is sufficient for stable device surface temperature and reaching quasi-steady state (Yu *et al.*, 2012; Jackson *et al.*, 2013). The $\hat{E}_{T\_MAX}$ is relatively insensitive to changes in the characterized time range, which is in tens of seconds.

The thermal velocity ν_{th} is found using $\nu_{th} = (3k_bT/(m_nm_0))^{0.5}$ (Engel-Herbert *et al.*, 2010). N_C is calculated from $N_C = 2(2\pi m_n m_0 k_b T/h^2)^{1.5}$ (Engel-Herbert *et al.*, 2010).

The $\hat{E}_{T\_MAX}$ at different temperatures can be calculated and depicted in the schematic energy band diagram of Al_2O_3. It is reasonable to assume that all of the charges trapped shallower than $\hat{E}_{T\_MAX}$ can be emitted due to the thermal energy at the given temperature T. Similarly, the charges trapped deeper than $\hat{E}_{T\_MAX}$ at the same temperature T will not be emitted from the level. Thus, we now have reliable methods for trap distribution and high temperature threshold voltage behavior analysis.

$$\hat{E}_{T\_MAX} = k_b T \ln(\gamma_n \sigma_{Al_2O_3T^2t}). \tag{11.2}$$

We will use this method for characterization of SCD induced by F-treatment in this case study.

11.2.3.2. *Investigation on Methods to Create Deeper Traps to Maintain Reasonable Positive V_{TH} at High Temperature in GaN HEMTs*

We can anticipate that increasing the RF power in the RIE system to increase F flux may help replace multiple FPTs with a single FPT. Moreover, higher power RF treatment may create deeper traps as energy transferred to Al_2O_3 is much larger which will potentially break more Al_2O_3 bonds producing deeper traps. This was investigated with multi-FPT gate stack HEMTs under three different RIE conditions as shown in Table 11.3.

In Table 11.3, three different RIE recipes have been used for Dies A, B, and C respectively, with gate processing parameters described briefly. Multiple F-treatments are successively used. For Die A, 30 W of RF power with 50 sccm of CHF_3 gas is used in each of the treatments. Only higher RF power of 60 W and 90 W are used for Dies B and C respectively, to enhance the breaking of Al–O bonds. The flow of CHF_3 gas was kept the same. As the higher RF power enhanced the ion bombardment and the fluorine radical concentration ionized from CHF_3, we expect more trap creation. After correcting for the CHF_3 plasma etching of Al_2O_3, the final Al_2O_3 thickness was 18.1 nm. Scanning electron spectroscopy (SEM) characterization was used to verify the thickness of each deposited layer. As increase in surface roughness is generally a qualitative indicator of increased density of dangling bonds induced by FPT, the change in surface morphology after the FPT was assessed through the AFM. The surface roughness root mean square (RMS) was 0.648 nm for the unfluorinated sample. It is increased to 1.138 nm for the fluorinated Die A. It implies the increase in the amount of dangling bonds and possible trap sites after

Table 11.3. Processing parameters and methods for gate stack fabrication for fluorinated normally-off MIS-HEMTs Die A, Die B, and Die C.

Steps	Process	Die A	Die B	Die C
1	Al_2O_3 ALD (thickness t_1)	6.5 nm	6.5 nm	9.3 nm
2	RIE F^- treatment (140 s)	50 sccm CHF_3 @ 30 W	50 sccm CHF_3 @ 60 W	50 sccm CHF_3 @ 90 W
3	Al_2O_3 ALD (thickness t_2)	7.3 nm	10.1 nm	12.8 nm
4	RIE F^- treatment (260 s)	50 sccm CHF_3 @ 30 W	50 sccm CHF_3 @ 60 W	50 sccm CHF_3 @ 90 W
5	Al_2O_3 ALD (thickness t_3)	7.6 nm	10.6 nm	13.5 nm
6	RIE F^- treatment (280 s)	50 sccm CHF_3 @ 30 W	50 sccm CHF_3 @ 60 W	50 sccm CHF_3 @ 90 W
7	Al_2O_3 ALD (thickness t_4)	8 nm	8 nm	8 nm
Total Al_2O_3 thickness after F^- RIE treatment		**18.1 nm**	**18.1 nm**	**18.1 nm**

fluorination. For comparison, a benchmarking sample with 18 nm of Al_2O_3 deposited after gate recess without any FPT is also fabricated and labeled as the "recess-only" device.

For all these 4 samples, V_{th} was measured as a function of temperature in the range of 25–200°C and results are shown in Fig. 11.30. The recess-only device has relatively stable V_{th}. On the other hand, a decrease in V_{th} is observed for all fluorinated Dies A to C. This clearly shows that the emission of electrons from the gate stack is mainly from the FPT-induced traps within Al_2O_3. For both Dies A and B, the V_{th} becomes negative at about −1.9 V at 200°C, which is closer to the V_{th} of recess-only device. Thus, most of the FPT-induced negative charge has been emitted from the trap sites at 200°C. V_{th} is 2 V for Die C at 200°C indicating deeper trap creation when higher RIE power is used.

Using these measurements, trap characterization was done using the gate stress method described in Sec. 11.2.3.1. To obtain data in Fig. 11.31, V_B in the gate stress measurement is in the range of −40 V to 0 V with 5 V increments. The extracted trap density D_{EQ} mapping for Dies A, B, and C is shown from 0.4 eV to 1.5 eV away from the conduction band edge energy ($\hat{E}_C$) of Al_2O_3. We can only obtain electrostatically equivalent SCD as V_{th} measurement can only provide values of Q_{EQ}. The $\hat{E}_{T\_MAX}$ from Eq. (11.2) at different ambient

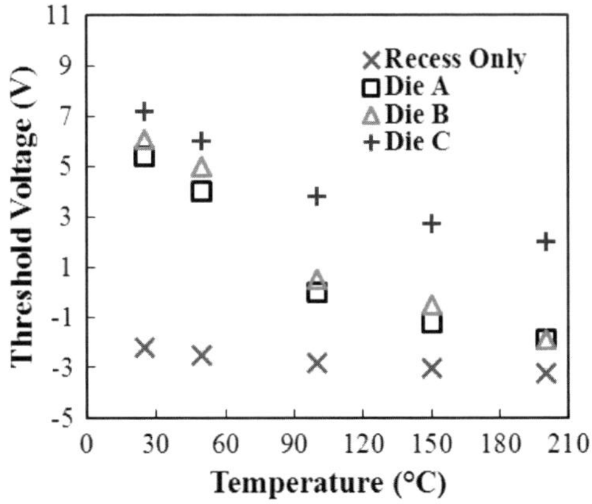

Fig. 11.30. The V_{th} obtained from the I_D–V_G transfer characteristics for FPT-processed Dies A, B, C and the unprocessed recess-only device at temperatures in the range of 25–200°C.

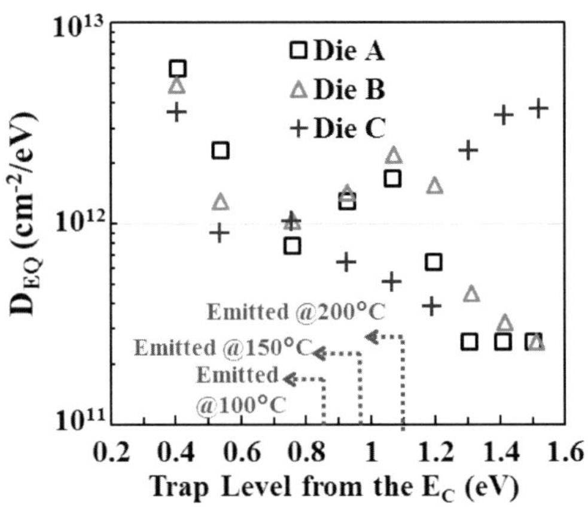

Fig. 11.31. The distribution of D_{EQ} within the Al_2O_3 energy band for Dies A, B, and C. It is quantified from the negative gate stress characterization.

temperatures are also indicated. Die A has the most D_{EQ} at shallow energy levels leading to the worse V_{th} thermal stability. On the other hand, a lot of traps are located deeper than the trap emission energy of $\hat{E}_T = 1.1$ eV at $T = 200$°C for Die C. It clearly illustrates the capability of using higher RIE power to maintain V_{th} at 2 V in Die C under high temperature.

For further verification of the trap density extracted using the gate-stressing method, the change to Q_{EQ}, ΔQ_{EQ}, can independently be extracted by integrating the D_{EQ} measured points. Due to larger steps in the trap energy, numerical integration can have large errors. To reduce numerical error, the integration was done through the best-fitted sixth-order polynomial function on the data in Fig. 11.31. The comparison of changes in Q_{EQ}, denoted as ΔQ_{EQ}, obtained from either the high-temperature measurement of V_{th} ($\Delta Q_{EQ,\,HighT}$) in Fig. 11.30 or integration of D_{EQ} mapping in Fig. 11.31 ($\Delta Q_{EQ,15}$) for Dies A–C agree within 10% validating the gate stress method. The mapping of FPT-induced trap charge distribution within Al_2O_3 energy bandgap helps in predicting the V_{th} degradation at higher temperature.

However, using higher RIE power like in Die C damages the channel quality and degrades the on-state device performance. The main reason for this is that the increase in density of RIE generated F-plasma to raise V_{th} also requires increased F-plasma ion energy. Hence, the RIE is replaced with a single Inductively Coupled Plasma ICP-RIE to carry out FPTs on Al_2O_3 in the case study presented in Sec. 11.2.3.3 inorder to overcome this problem. Unlike conventional RIE system, the ICP-RIE is able to control the amount of dissociated fluorine radicals in the plasma and the ion bombardment energy through independent control of the coil and cathode power, respectively. This method along with Argon plasma pretreatment modification to enhance deep trap formation, which yields the best results, is described in much detail in Sec. 11.2.3.3.

11.2.3.3. *Device with Process Simplification and Ar Pretreatment of Al_2O_3 Before F-Plasma Treatment to Create Deeper Electron Traps for High Temperature Operation*

The fabrication process flow and the cross-sectional schematic of the device using a single ICP-RIE step to carry out FPTs on Al_2O_3 dielectric stack are shown in Figs. 11.32 and 11.33, respectively. The device uses the same wafer and structure as in the earlier sections. Only the gate stack is simplified as there is now only one FPT. The L_G, L_{GS}, L_{FP}, and L_{GD} are 3 μm, 5 μm, 1.5 μm, and 5 μm, respectively. The wafer has a standard AlGaN/GaN-on-Si layer structure. The 2DEG carrier density and mobility are 8.5×10^{12} cm^{-2} and 1450 cm^2/Vs respectively. The first fabrication step of the device is mesa isolation by BCl_3-based ICP-RIE. It is followed by SiO_2 passivation dielectric deposition by PECVD. Ohmic contacts for source/drain were formed after etching the oxide in the contact regions using multi-metal stack of Ti/Al/Ni/Au with respective thicknesses of 25/125/45/55 nm. Annealing was done by RTA at 850°C for 30 s. After gate patterning, 10 nm (50%) of AlGaN was etched by low power BCl_3-based ICP-RIE. This step reduces the 2DEG density to 4.5×10^{12} cm^{-2} without damaging the AlGaN/GaN interface. This helps preserve carrier mobility in the 2DEG channel (Wang *et al.*, 2015a). 6.5 nm of

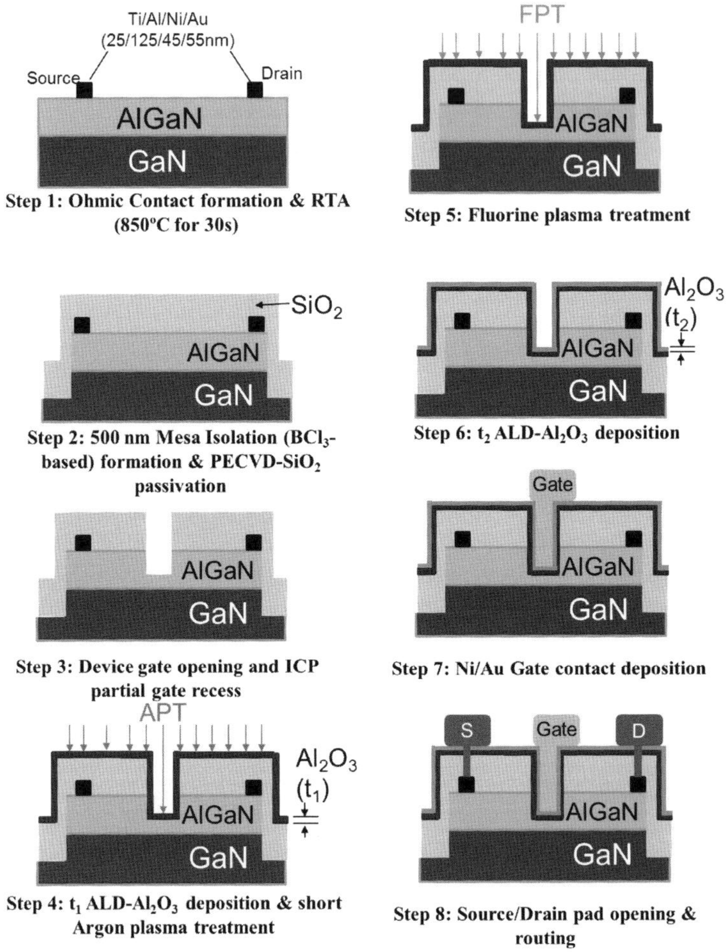

Fig. 11.32. The fabrication process flow of the normally-off Al_2O_3/AlGaN/GaN MIS-HEMT with APT-then-FPT gate process.

ALD-Al_2O_3 gate dielectric layer deposition at 250°C was then performed. As dielectric treatments were different for each die, they are separately described. Finally, Ni/Au (15/150 nm) gate-metal deposition followed by annealing at 400°C for 5 min are applied.

Table 11.4 gives summary of how each of the Dies A–E differs. Die A is recess only control device. Die B is only ICP-FPT control device. The ICP-FPT is achieved through ICP-CHF_3 plasma treatment with fixed cathode/coil power

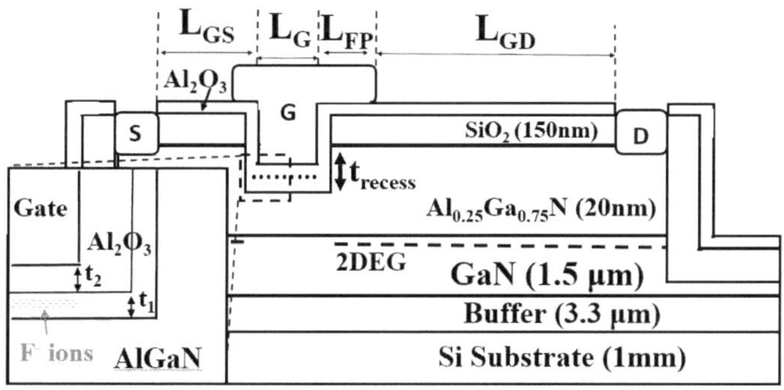

Fig. 11.33. The device cross-sectional schematics of the normally-off AlGaN/GaN MIS-HEMT with ICP-fluorinated Al_2O_3. The inset magnifies the gate region, where the fluorine-induced negatively charged ions (F^-) and the thickness of Al_2O_3 gate dielectric stack (t_1 and t_2) are shown.

Table 11.4. Processing parameters for the gate dielectric of Dies A–E.

Steps	Die A	Die B	Die C	Die D	Die E
ALD-Al_2O_3 (thickness t_1)	6.5 nm	6.5 nm	6.5 nm	6.5 nm	6.5 nm
Cathode Power (CEP) for APT. Coil Power 100 W.	N/A	N/A	50 W	75 W	100 W
ICP FPT	N/A	Coil Power 200 W. Low Cathode Power 10 W.			
ALD-Al_2O_3 (thickness t_2)	15 nm	15 nm	15 nm	15 nm	15 nm
AFM surface RMS Roughness	0.38 nm	0.53 nm	0.61 nm	0.78 nm	2.29 nm

of 10/200 W for 4 min. Dies C–E have APT before ICP-FPT identical to that of Die B. APT in Dies C, D, and E was done with 20 s ICP-Ar plasma treatment with a fixed coil power of 100 W. This novel step is argon plasma treatment (APT) before FPT.

The APT recipe implemented is based on the existing Al_2O_3 etching recipe using $BCl_3/Cl_2/Ar$ ICP-RIE. It uses the same coil power and Ar gas flow as the Al_2O_3 etching recipe to ensure the amount of dissociated Ar^+ is sufficient. The FPT recipe was obtained from various process short-loops. The resulting surface roughness RMS was measured by the AFM to obtain a qualitative signature of trap creation. It is found that by controlling the cathode power of the Argon treatment below 75 W (Die D) is able to preserve the surface roughness RMS within 1 nm. However, if the cathode power of the Ar treatment is increased to 100 W (Die E), the surface roughness RMS will significantly increase from 0.53 nm to 2.29 nm, indicating damage to the Al_2O_3 surface which will degrade device performance. Due to very large RMS roughness, Die E had much reduced drain current and hence was not analyzed further although it had the highest V_{th}. Die C had lower V_{th} than Die D but both had otherwise similar good performance. Hence, detailed studies only focused on Dies A, B, and D.

The process parameters for the gate formation and other process recipes are all included in Table 11.5 in detail. In this table, major process steps are listed first in four columns followed by sub-steps in a run sheet form in three columns to achieve the major steps. Major process steps are numbered as 1, 2(a), 2(b), etc. The sub-steps associated each major process steps are labeled with a dot as 2(b).1, 2(b).2, and so on till all sub-steps are finished. In case these sub-steps are shared between different major process steps, they are included below the very first major process step and reference is made to this in the subsequent major process steps. The structure achieved after some of the major steps is linked to the cross-sections depicted in Fig. 11.32.

As mentioned earlier, detailed studies only focused on Die A, B, and D. The main purpose for comparing three devices is to ensure that plasma treatments for Dies B and D did not degrade the two main performance parameters of on-current and breakdown voltage. The measurements showed that on-current degradation was less than 30% for devices on Die B and D compared to Die A for the same gate overdrive ($V_G - V_{th}$). The breakdown voltages for devices on Dies A, B, and D were measured to be 1110 V, 670 V, and 940 V. Clearly, FPT only device has much worse breakdown voltage. APT–FPT device breakdown is less affected compared to Die B. However, as devices on Die A have negative V_{th}, they are not of much interest in this work. Hence, devices B and D are now compared as regards to ability to create deeper traps and maintain V_{th} at high temperature of 200°C.

I_D–V_G and V_{th} measurements up to 200°C were performed and I_D–V_G characteristics are shown in Fig. 11.34. The reduction in I_D is fairly consistent with mobility degradation at high temperature. V_{th} measurements show that APT–FPT device is able to maintain it at 2.5 V at 200°C. This is the highest reported value using FPT. Moreover, the reduction in V_{th} of 1.9 V compared to that at room temperature is very low. This shows a good potential for further optimization to raise V_{th} at higher temperature by increasing its value at room temperature assuming the same 1.9 V reduction can be maintained. The guide to room temperature V_{th} is provided by the dependence of it on trapped equivalent negative charge in Fig. 11.35.

The low reduction in V_{th} at high temperature gives a clear evidence that electrons are trapped at much deeper level when APT before FPT is used.

Table 11.5. Major process flow for fluorine-treated normally-off Al_2O_3/AlGaN/GaN MIS-HEMT fabrication.

Steps	Process	Equipment Used	Process Outcome
1	Wafer Cleaning	Fume Cupboard	Cleaned wafer with particles removed
Sub-steps	**Process**	**Parameters**	
1.1	Organic Contamination Removal	Dip in H_2SO_4+H_2O_2 mix (ratio of 3:1) for 2 min; then rinse with DI water	
1.2	Native Oxide Removal	Dip in Buffered Oxide Etch (BOE) for 15 s; then rinse with DI water	
2 (a)	Lithography (Double UV Exposure)	Karl Suss MA6 Mask Aligner	Ti/Al/Ni/Au with thickness of 25/125/45/55 nm
Sub-steps	**Process**	**Parameters**	
2 (a).1	HMDS spin-coating for adhesion enhancement	3000 rpm for 20 s	
2 (a).2	AZ-5214E Photoresist spin-coating	3000 rpm for 20 s	
2 (a).3	Photoresist Pre-baking	120°C for 60 s	
2 (a).4	UV Light Exposure after alignment with the mask pattern	2 s exposure under wavelength = 320 nm	
2 (a).5	Photoresist Post-baking	120°C for 90 s	
2 (a).6	UV Light Exposure without mask	30 s exposure under wavelength = 320 nm	
2 (a).7	Pattern Development	Dip into FHD-5 positive photoresist developer for 50 s	
2 (a).8	Lift-off after metal deposition	Dip into acetone and place in ultrasonic cleaner for 20 min	
2 (a).9	Acetone Removal	Dip into Iso-Propyl Alcohol (IPA) and placed in ultrasonic cleaner for 10 min	
2 (a).10	IPA Removal	Dip into DI water and place in ultrasonic cleaner for 10 min; blow dry with N_2 gun	

(*Continued*)

Table 11.5. (*Continued*)

Steps	Process	Equipment Used	Process Outcome
2 (b)	Drain/Source Metal Deposition	ULVAC EX-400 Electron-beam Film Evaporator	

Sub-steps	Process	Comments	
2 (b).1	Sample and Crucible Loading	Load the required metal crucibles with sufficient amount of metals within	
2 (b).2	Chamber Vacuuming	Wait until the chamber pressure is under 4×10^{-6} Pa (~2 h)	
2 (b).3	E-beam location adjustment	Ensure the electron beam is bombarding the surface of the metal crucible	
2 (b).4	Metal Deposition	Carefully control the speed of metal deposition. Ensure the rate is around 0.1 nm/s.	
2 (b).5	Crucible Cooling	Allow for about 3 min of cooling time after each deposition before changing the crucible	
2 (b).6	Change of Crucible	Change the crucible to the next required metal and repeat from Step 2(b).3.	
2 (b).7	Chamber Vent and Sample Unloading	When metal deposition is completed, vent the chamber and unload the sample. Finally, vacuum the chamber after usage to avoid chamber contamination.	
2 (c)	Metal Lift-off	Fume Cupboard	
3	Rapid Thermal Annealing	BPS Nextral ADAX 60	Formation of ohmic contacts shown in Step 1 of Fig. 11.32
4 (a)	Lithography (Single Exposure)	Karl Suss MA6 Mask Aligner	~600 nm etched depth in the mesa region

Sub-steps	Process	Parameters	
4 (a).1	HMDS spin-coating for adhesion enhancement	3000 rpm for 20 s	

(*Continued*)

Table 11.5. (*Continued*)

Steps	Process	Equipment Used	Process Outcome
4 (a).2	AZ-5214E Photoresist spin-coating	3000 rpm for 20 s	
4 (a).3	Photoresist Pre-baking	120°C for 90 s	
4 (a).4	UV Light Exposure after aligned with the mask pattern	30 s exposure under wavelength = 320 nm	
4 (a).5	Pattern Development	Dip into FHD-5 positive photoresist developer for 50 s	
4 (a).6	Post-process photoresist removal	Dip into acetone and placed in ultrasonic cleaner for 10 min	
4 (a).7	Acetone Removal	Dip into IPA and placed in ultrasonic cleaner for 10 min	
4 (a).8	IPA Removal	Dip into DI water and placed in ultrasonic cleaner for 10 min; blow dry with N_2 gun	
4 (b)	Mesa Isolation	STS Multiplex ICP-RIE System	
Sub-step 4 (b).1: High Power Etch for fast etching speed			
Gas Flow	Argon		15 sccm
	BCl_3		2 sccm
	Cl_2		8 sccm
Pressure			120 mTorr
Coil Power			100 W
Cathode Power			175 W
Etch Time			190 s
Sub-step 4 (b).2: Low Power Etch for improved etched surface quality			
Gas Flow	Argon		15 sccm
	BCl_3		2 sccm
	Cl_2		8 sccm
Pressure			120 mTorr
Coil Power			30 W
Cathode Power			30 W
Etch Time			300 s
5	Surface Passivation	Oxford Plasmalab 80 Plus PECVD System	~200 nm of SiO_2 deposited shown in Step 2 of Fig. 11.32

(*Continued*)

Table 11.5. (*Continued*)

Steps	Process	Equipment Used	Process Outcome
Gas Flow	N_2O		706 sccm
	SiH_4/N_2		158 sccm
Pressure			906 mTorr
RF Power			18 W
Valve Position			26.3 deg
Temperature			300°C
Deposition Time			450 s
6 (a)	Lithography (Single UV Exposure)	Karl Suss MA6 Mask Aligner	200 nm SiO_2 etched at the gate
Follow the same sub-steps 4(a).1 through 4(a).8 of major step 4(a).			
6 (b)	Gate Region Opening	STS Multiplex ICP-RIE System	
Gas Flow	Ar		5 sccm
	SF_6		20 sccm
Pressure			50 mTorr
Coil Power			5 W
Cathode Power			100 W
Etch Time			330 s
7	AlGaN Recess at the Gate Region	STS Multiplex ICP-RIE System	~10 nm of AlGAN etched at the gate shown in Step 3 of Fig. 11.32
Gas Flow	Argon		15 sccm
	BCl_3		2 sccm
	Cl_2		8 sccm
Pressure			120 mTorr
Coil Power			30 W
Cathode Power			30 W
Etch Time			30 s
8 (a)	Gate Dielectric Deposition	Savannah 100 ALD system	Plasma-treated Al_2O_3 gate dielectric shown in Steps 4 and 5 of Fig. 11.32
Gas Flow	Trimethylaluminum (TMA)		20 sccm
	H_2O		
Temperature			250°C
Thickness per cycle			0.1 nm

(*Continued*)

Table 11.5. (*Continued*)

Steps	Process	Equipment Used	Process Outcome
8 (b)	Plasma Treatments	STS Multiplex ICP-RIE System/Oxford PlasmaPro 80 RIE System	
A: RIE Treatment			
Gas Flow	CHF_3		50 sccm
Pressure			37.5 mTorr
RF Power			This is applicable if only RIE treatment is done. Different
Treatment Time			time variations.
B: ICP-RIE Treatment			
Step 1: Argon pre-treatment			
Ar			10 sccm
Pressure			15 mTorr
Cathode Power			This is applicable only if APT treatment is done by ICP-RIE. Different time
Coil Power			variations.
Treatment Time (Variated)			20 s
Step 2: Fluorine treatment			
CHF_3			36 sccm
Pressure			37.5 mTorr
Cathode Power			10 W
Coil Power			This is applicable FPT treatment is done by ICP-RIE. Different time
Treatment Time			variations.
9 (a)	Lithography (Double UV Exposure)	Karl Suss MA6 Mask Aligner	
Follow the same sub-steps 2(a).1 through 2(a).10 of major step 2(a).			Ni/Au Contact (15/150 nm)

(*Continued*)

Table 11.5. (*Continued*)

Steps	Process	Equipment Used	Process Outcome
9 (b)	Gate Metal Deposition	ULVAC EX-400 Electron-beam Film Evaporator	shown in Step 7 of Fig. 11.32
Follow the same sub-steps 2(b).1 through 2(b).10 of major step 2(b).			
9 (c)	Metal Lift-off	Fume Cupboard	
10 (a)	Lithography (Single UV Exposure)	STS Multiplex ICP-RIE System	Exposing Pads for device characterizations shown in Step 8 of Fig. 11.32
Follow the same sub-steps 4(a).1 through 4(a).8 of major step 4(a).			
10 (b)	Pad Opening	STS Multiplex ICP-RIE System	
Sub-step 10 (b).1: Al_2O_3 Etch			
Gas Flow	Ar		10 sccm
	BCl_3		25 sccm
Pressure			15 mTorr
Coil Power			100 W
Cathode Power			175 W
Etch Time			180 s
Sub-step 10 (b).12: SiO_2 Etch			
Gas Flow	Ar		5 sccm
	SF_6		20 sccm
	Cl_2		8 sccm
Pressure			50 mTorr
Coil Power			5 W
Cathode Power			100 W
Etch Time			300 s

To verify this aspect, the trap characterization was done using the gate stress method described in Sec. 11.2.3.1. V_B in the gate stress measurement ranging from −40 V to 0 V with 5 V increments is applied to obtain the trap density D_{EQ} mapping for Dies B and D from 0.4 eV to 1.4 eV away from the conduction band energy ($\hat{E}_C$) of Al_2O_3. The extracted trap density is shown in Fig. 11.36. Figure 11.36 also shows the traps that will emit the charge at a given temperature rendering them ineffective in maintaining high V_{th}. It is clear that the trapped charge density peaks at about 1.02 eV for devices from Die B and a lot of trapped charge is emitted at 200°C. On the other hand, the peak of trapped charge SCD for devices from Die D is located beyond 1.2 eV and these

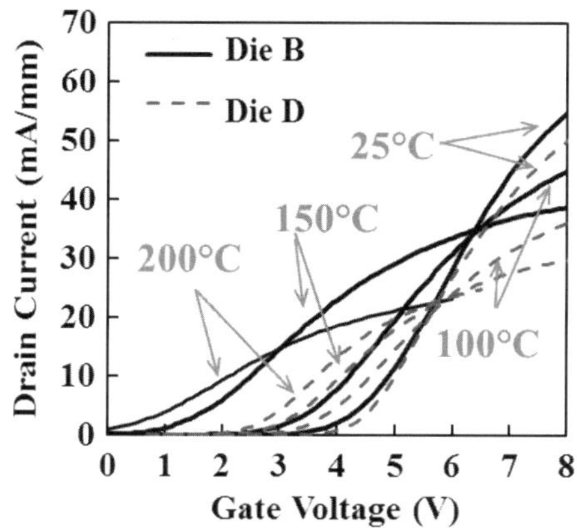

Fig. 11.34. I_D–V_G characteristics of device from Dies B and D (for $V_D = 1$ V) at ambient temperatures of 25°C, 100°C, 150°C, and 200°C.

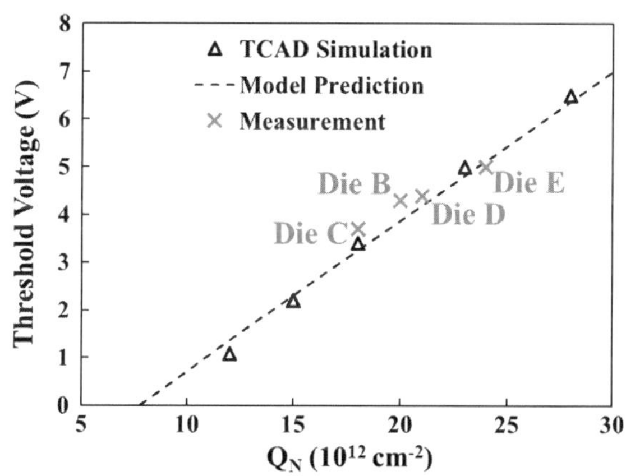

Fig. 11.35. The relationship between the Q_F and the V_{th} of Dies B–E using the model and comparison with numerical simulations using accurate TCAD tools and measurement.

high concentration trapped charges do not de-trap at 200°C which significantly improves high temperature V_{th} stability.

Although V_{th} stability at high temperature is credibly established for APT-then-FPT devices, physical mechanism that gives rise to such behavior is not established. A sound scientific reason for such behavior is needed before universality in application of APT prior to FPT can be confirmed. This particular aspect is addressed in Sec. 11.2.3.4.

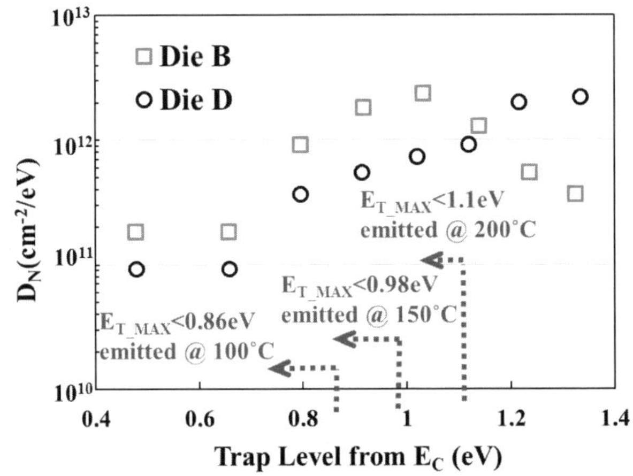

Fig. 11.36. The fluorine-induced trap state (D_F) distribution along the Al_2O_3 energy level from the $\hat{E}_C$ for Dies B and D.

11.2.3.4. *Determination of Possible Mechanism That Improves Thermal Stability of V_{TH} in APT-FPT Devices to Guide Device Designers*

As our purpose is to identify physical mechanism that leads to thermal stability in HEMTs, an experiment was carefully designed to identify the cause behind it. The characterization has to be done for Al_2O_3 stack with APT and FPT treatment conditions same as those of devices in Dies B and D. Table 11.6 shows the steps designed to identify physical mechanism which also shows purpose of each step.

The characterization methods used in this section include X-ray photo-electron spectroscopy (XPS) and secondary ion mass spectroscopy (SIMS). XPS can provide information on relative concentration of specific elements and composition of bonds in the compound to be analyzed near the surface. SIMS provides depth profile of concentration of the elements of interest.

To investigate the effect of APT to the F depth profile during the subsequent FPT process, the SIMS depth profile characterization is carried out on Samples B and D as shown in Fig. 11.37. The APT and FPT processing recipes are identical to the ones used for Dies B and C in Sec. 11.2.3.3 as mentioned earlier. As SIMS characterization sputters away materials and destroys the devices, these samples are fabricated separately from the same wafer used to fabricate Dies A, B, and D in Sec. 11.2.3.3. Here 30 nm of Al_2O_3 is deposited by ALD at 250°C before the plasma treatments and an additional 10 nm of

Table 11.6. The parameters used for the gate plasma treatments for Samples A–D.

	Sample A	Sample B	Sample C	Sample D
Purpose	This is the control sample for comparison.	This sample has the same FPT treatment as devices in Die B of Sec. 11.2.3.3 to evaluate effect of only F-treatment.	This sample has the same APT treatment as devices in Die D of Sec. 11.2.3.3 to evaluate effect of only Ar treatment.	This sample has same APT and FPT treatment as devices in Die D of Sec. 11.2.3.3 to evaluate effect of Ar treatment followed by F-treatment.
First Treatment	N/A	CHF_3-based CEP: 10 W CP: 200 W Time: 4 mins	Ar-based CEP: 75 W CP: 100 W Time: 20 s	Ar-based CEP: 75 W CP: 100 W Time: 20 s
Second Treatment	N/A	N/A	N/A	CHF_3-based CEP: 10 W CP: 200 W Time: 4 mins
Equivalence with HEMT having the same gate structure as those of Sec. 11.2.3.3	Die A	Die B	N/A as APT alone is not useful for device.	Die D

ALD-Al_2O_3 is deposited afterwards to avoid the inaccuracy of SIMS characterization on the surface when characterizing the depth profile of F atoms. In the SIMS depth profile shown in Fig. 11.37, an obvious increase in the peak of the F concentration is shown for the sample with APT-then-FPT process (Sample D). Additional F incorporation into the Al_2O_3 as compared with Sample B which underwent only FPT process is a conclusive signature of the additional dangling bonds created after Ar bombardment during APT. The F depth profile is shifted towards the surface after APT, showing the depth of the regions with additional dangling bonds created by APT and F incorporation is shallow. This helps in maintaining mobility of 2DEG as F is further away from it. The dip in O intensity where the F is present indicates the replacement

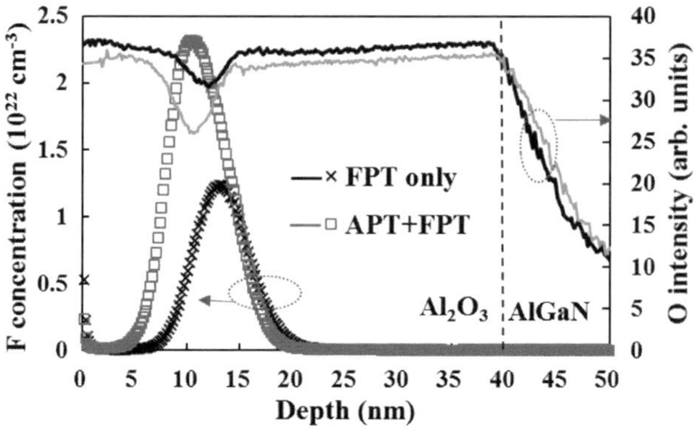

Fig. 11.37. The SIMS depth profiles for F and O atoms of the samples with (Sample D) and without APT (Sample B) prior to the FPT.

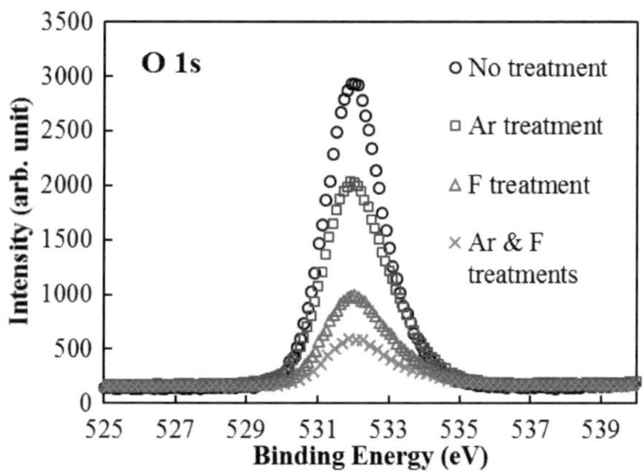

Fig. 11.38. The O 1s (binding energy = 532 eV) XPS spectra of the ALD-Al_2O_3 sample with no plasma treatments (○), FPT (△), APT (□), and combined Ar then F plasma treatments (×).

of O atoms by the F atoms within the Al_2O_3 dielectric. Hence, it also indicates a larger concentration of dangling bonds responsible for trap creation.

Although SIMS provides depth profile of F and O, it does not characterize concentration of the types of bonds that are absent or present. It gives only a good quantitative measure of F incorporation that creates dangling bonds and hence traps. Hence, XPS characterization was performed on Samples B and D to determine relative reduction in Al–O bonds through O 1s spectra in these samples as shown in Fig. 11.38. The data shows that APT actually breaks a lot of bonds and hence APT followed by FPT breaks much larger percentage of bonds. This gives an indication of the possibility that deeper traps could arise due to the potential of trap formation at sites where more than one F atom bonds with Al. It is thus important to find efficacy of Al–F bond formation when different treatments are used. Relative concentration of Al–O and Al–F bonds formed was characterized by fitting broad Al 2p spectra to their respective bond energy as shown is Fig. 11.39(a) and 11.39(b), respectively. It is observed that about 20% Al–F bonds are formed when only FPT is performed whereas this number increase to 36% when APT is performed prior to FPT.

Although data from Fig. 11.39 clearly shows formation of more Al–F bonds due to Ar pretreatment, it is not clear why it should lead to deeper traps and

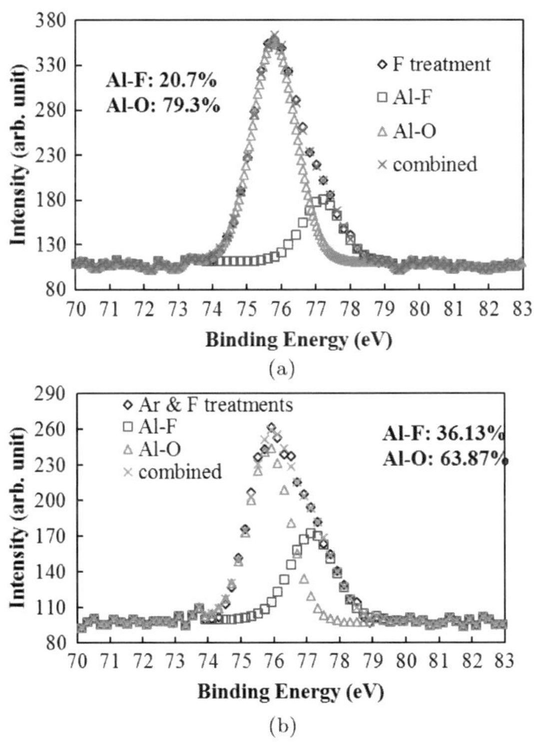

Fig. 11.39. The Al 2p XPS spectra of the ALD-Al_2O_3 sample with (a) F plasma treatment only and (b) Ar plasma treatment prior to the F plasma treatment. The measured spectrum is fit with the superposition of spectra (×) consisting of Al–O (△) and Al–F (□) chemical bonding spectrum.

not just increased trap concentration. Extracted trapped charge density in Fig. 11.36 not only shows increase trapped charge but also more trapping and possibly trap creation at much deeper level. This, however, cannot be established by the characterization methods as they do not reveal any information on traps, only on relative concentration of Al–O and Al–F bonds. Hence, this needs to be explored through fundamental first principle based simulations.

The key basis of simulation is that the Al–F bonds and lack of Al–O bonds leads to the defects or dangling bonds that form the traps to capture electrons. To evaluate the formation of trap states in Al_2O_3 bandgap, the Schrödinger equation incorporating the effect of defects needs to be numerically solved. In simulation framework, the defects basically modify potential term in the Hamiltonian of the perfect crystal through a defect potential contribution V. The Schrödinger equation incorporating effect of traps (Hjalmarson *et al.*, 1980; Yu and Cardona, 2010) is expressed as $(H_0 + V)\Phi = E\Phi$, where H_0 is the Hamiltonian of the perfect crystal, V is the potential of the defects, Φ is the wave function, and E is its energy eigenvalue. As defects are linked to more Al–F bond formation, the additional potential V arises from this phenomenon. In reality, when any impurities are introduced to the perfect lattice to substitute the host atoms, an additional potential V arises. It is related to the difference in the atomic electronegativity between original element and impurity atom. This new defect potential term leads to states in the bandgap of Al_2O_3 host with finite Density of States (DOS) forming the trap centers (Hjalmarson *et al.*, 1980; Yu *et al.*, 2010; Jaros, 1982). Hence, when more F atoms with larger electronegativity (electronegativity = 3.98) (Housecroft and Sharpe, 2008) substitute the O atoms (electronegativity = 3.44) (Housecroft and Sharpe, 2008) with lower electronegativity to form increased number of Al–F bonds by the APT-then-FPT treatments, a larger V is expected.

To study the relationship between the number of Al–F bonds to the trap state distribution within the bandgap quantitatively, the atomistic first-principle simulations were numerically carried out by the Gaussian 09 software (Frisch *et al.*, 2009) for a primitive cell with 2 Al atoms bonded to a mix of O and F when F is introduced by FPT. The accuracy of such simulation is strongly dependent on basis set. The Eigen energies of $Al_2O_xF_y$ lattice cell with different amount of Al–F bonds were found with the Hartree–Fock self-consistent method with the basis set of 6-31G for a good accuracy (Becke, 1988). Since we are exploring on a few atoms, it is reasonable to assume single crystal representation which is randomly distributed to form thicker amorphous Al_2O_3.

The numerical solution predicts the relative DOS data in pure Al_2O_3 with no impurities. It has energy bandgap ($\hat{E}_g$) of 8.6 eV, similar to the $\hat{E}_g$ of 8.7 eV for reported crystalline α-Al_2O_3 (Miyazaki, 2001). Also, there are no states

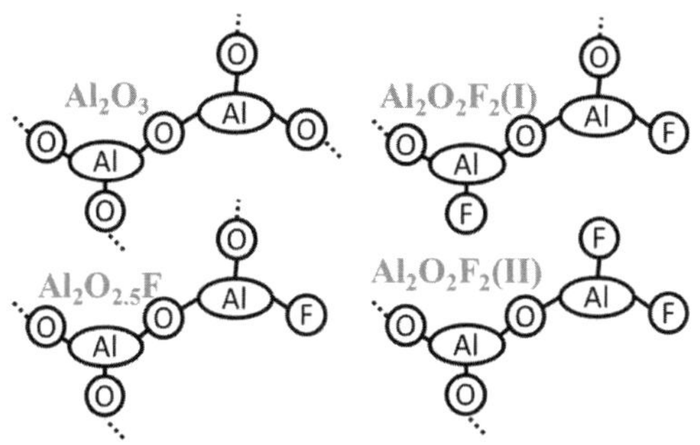

(a)

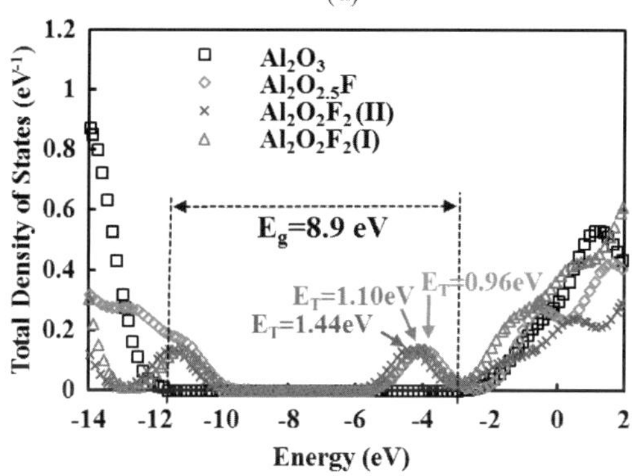

(b)

Fig. 11.40. (a) The bonding schematic of the Al_2O_3, $Al_2O_{2.5}F$, $Al_2O_2F_2$(I), and $Al_2O_2F_2$(II) primitive cells used in the Gaussian 09 simulation, where the dashed line represents the bonds connecting with adjacent primitive cells. (b) The simulated total DOS distribution of $Al_2O_XF_y$ along the energy band with the bandgap of Al_2O_3 indicated.

in the bandgap in this pure Al_2O_3 case. The bonding schematics of $Al_2O_XF_y$ primitive cells explored in simulations are shown in Fig. 11.40. Each of the O atoms at the cell edge is counted as 0.5 O in the cell formula as they are shared with adjacent primitive cells within the actual amorphous Al_2O_3 grown by ALD. The same bandgap ($\hat{E}_g$) of 8.6 eV is obtained for Al_4O_6 and Al_6O_9 cluster simulations. Therefore, the same $\hat{E}_g$ is obtained regardless of the simulated cluster size suggesting that the simulation results can be safely extended to larger Al_2O_3 thicknesses. Trap states in bandgap can be formed when the proportion of Al–F bonds against the total amount of bonds used in the simulation were similar to the Al–F bond composition of 20% and 36% of the samples that underwent FPT with (Sample D) and without (Sample B) APT shown in Fig. 11.40(a) and (b) respectively.

The simulated total relative DOS of the cells are plotted by the Multiwfn wavefunction analyzer (Lu *et al.*, 2015) in Fig. 11.40(b). For $Al_2O_{2.5}F$ simulation that has single Al–F bond percentage of 17% with three O atoms shared with adjacent primitive cells, a symmetric trap state distribution with peak DOS at trap energy level $\hat{E}_T$ of about 0.96 eV below $\hat{E}_C$ is found in this case. For $Al_2O_2F_2$ simulation with Al–F bond percentage of 33%, there are two possible bonding structures. In one case, the two F atoms can be bonded with the different Al atoms and it is denoted as $Al_2O_2F_2$ (I). When the two F atoms bond to same Al atoms, the structure is denoted as $Al_2O_2F_2$ (II). Both of the calculated $\hat{E}_T$ levels are deeper than that of $Al_2O_{2.5}F$. Deepest $\hat{E}_T$ of 1.44 eV is obtained for $Al_2O_2F_2$ (II) as the adjacency of the two F atoms provides the strongest local defect potential which hinders electron de-trapping.

The simulation now clearly identifies role played by APT in deeper trap creation. As APT leads to formation of a larger number of dangling bonds before FPT, the probability of two F atoms bonding to a single Al atom substantially increases which shifts the trap energy states in the bandgap to deeper level. Clearly, formation of any Al–F bond leads to trap states and a mix of shallow and deeper traps is always obtained but APT facilitates formation of deeper traps through more Al–F and AlF_2 bonds.

11.2.4. Concluding Remarks

The GaN case study clearly shows a highly exploratory attitude, and thorough analysis is essential to successfully achieve desired innovative research goals. The initial success of obtaining high positive threshold voltage through multiple plasma treatments on successive dielectric layers was clouded by its loss at high temperature neutralizing the key advantage of GaN-based HEMTs. However, sheer exploratory attitude and thorough investigation finally lead to realization of a good GaN/AlGaN HEMT with high temperature threshold voltage of 2.5 V using a single additional process step in fabrication.

With this, full descriptions of the wide bandgap semiconductor power device case studies is complete. In depth analysis using different characterization methods is extremely helpful in determining mechanism behind success of novel device and process design as in the case of APT before FPT in GaN HEMTs.

References

T. J. Anderson, M. J. Tadjer, M. A. Mastro, J. K. Hite, K. D. Hobart, C. R. Eddy *et al.*, Characterization of recessed-gate AlGaN/GaN HEMTs as a function of etch depth, *J. Electron. Mater.*, **39**, 478–481 (2010).

A. D. Becke, Density-functional exchange-energy approximation with correct asymptotic behavior, *Phys. Rev. A*, **38**, 3098–3100 (1988).

H. W. Chen, M. J. Wang and K. J. Chen, Enhancement-mode AlGaN/GaN HEMTs fabricated by standard fluorine ion implantation with a Si_3N_4 energy-absorbing layer, *Electrochem. Solid State Lett.*, **14**, H229–H231, (2011).

K. Y. Cheong, J. H. Moon, H. J. Kim, W. Bahng and N.-K. Kim, Analysis of current conduction mechanisms in atomic-layer-deposited Al_2O_3 gate on 4H silicon carbide, *Appl. Phys. Lett.*, **90**, 162113 (2007).

R. Engel-Herbert, Y. Hwang and S. Stemmer, Comparison of methods to quantify interface trap densities at dielectric/III–V semiconductor interfaces, *J. Appl. Phys.*, **108**, 124101, (2010).

M. J. Frisch, G. W. Trucks, H. B. Schlegel, G. E. Scuseria, M. A. Robb, J. R. Cheeseman *et al.*, *Gaussian 09*, ed. Wallingford, CT, USA: Gaussian, Inc., (2009).

S. Ganguly, J. Verma, L. Guowang, T. Zimmermann, X. Huili and D. Jena, Barrier height, interface charge & tunneling effective mass in ALD Al_2O_3/AlN/GaN HEMTs, in *Device Research Conference (DRC), 2011 69th Annual*, 121–122 (2011).

H. P. Hjalmarson, P. Vogl, D. J. Wolford and J. D. Dow, Theory of substitutional deep traps in covalent semiconductors, *Phys. Rev. Lett.*, **44**, 810–813 (1980).

C. E. Housecroft and A. G. Sharpe, *Inorganic Chemistry*, London: Pearson, (2008).

I. Hwang, J. Oh, H. S. Choi, J. Kim, H. Choi, J. Kim *et al.*, Source-connected p-GaN gate HEMTs for increased threshold voltage, *IEEE Elect. Dev. Lett.*, **34**, 605–607 (2013a).

I. Hwang, J. Kim, H. S. Choi, H. Choi, J. Lee, K. Y. Kim *et al.*, p-GaN gate HEMTs with tungsten gate metal for high threshold voltage and low gate current, *IEEE Electron Device Lett.*, **34**, 202–204 (2013b).

C. M. Jackson, A. R. Arehart, E. Cinkilic, B. McSkimming, J. S. Speck and S. A. Ringel, Interface trap characterization of atomic layer deposition Al2O3/GaN metal-insulator-semiconductor capacitors using optically and thermally based deep level spectroscopies, *J. Appl. Phys.*, **113**, 204505 (2013).

M. Jaros, *Deep Levels in Semiconductors*, Bristol: Adam Hilger, (1982).

C. Lombardi, S. Manzini, A. Saporito and M. Vanzi, A physically based mobility model for numerical simulation of non-planar devices, *IEEE Trans. on CAD* 7, 1164–1171 (1988).

T. Lu, *Multiwfn-A Multifunctional Wavefunction Analyzer*, version 3.3.8. Beijing, China: Beijing Kein Research Center for Natural Sciences (2015).

M. Matin, A. Saha and J. Cooper, A self-aligned process for high-voltage, short-channel vertical DMOSFETs in 4H-SiC,*IEEE Tran. Electron Devices*, **51**, 1721–1725 (2004).

O. Mitrofanov and M. Manfra, Poole–Frenkel electron emission from the traps in AlGaN/GaN transistors, *J. Appl. Phys.*, **95**, 6414–6419 (2004).

S. Miyazaki, Photoemission study of energy-band alignments and gap-state density distributions for high-k gate dielectrics, *J. Vacu. Sci. & Technol. B*, **19**, 2212–2216 (2001).

A. Nakajima, K. Takao, M. Shimizu, H. Okumura and H. Ohashi, Equivalent circuit model for GaN-HEMTs in a switching simulation, *Intelec 08 — 30th International Telecommunications Energy*, **1 and 2**, 488–491 (2008).

D. Schroder, Defects, in *Semiconductor Material and Device Characterization*, 3rd edn. Tempe: Wiley-Interscience Publication, 262, (2006).

M. Specht, M. Städele, S. Jakschik and U. Schröder, Transport mechanisms in atomic-layer-deposited Al_2O_3 dielectrics, *Appl. Phys. Lett.*, **84**, 3076–3078 (2004).

M. Ťapajna, M. Jurkovič, L. Válik, Š. Haščík, D. Gregušová, F. Brunner *et al.*, Bulk and interface trapping in the gate dielectric of GaN based metal-oxide-semiconductor high-electron-mobility transistors, *Appl. Phys. Lett.*, **102**, 243509 (2013).

Y.-H. Wang, Y. C. Liang, G. S. Samudra, H. Huolin, H. Bo-Jhang, H. Szu-Han *et al.*, 6.5 V high threshold voltage AlGaN/GaN power metal-insulator-semiconductor high electron mobility transistor using multilayer fluorinated gate stack, *IEEE Elect. Dev. Lett.*, **36**, 381–383 (2015a).

Y.-H.(b) Wang, Y. C. Liang, G. S. Samudra *et al.*, High output swing monolithic inverter with E-D mode MIS-HEMTs for GaN power integrated circuits, The 11th IEEE International Conference on Power Electronics and Drive Systems (PEDS), Sydney, Australia, June, 9–12 (2015b).

C. Woojin, S. Ogyun, R. Hojin, C. Ho-Young and S. Kwang-Seok, High-voltage and low-leakage-current gate recessed normally-off GaN MIS-HEMTs with dual gate insulator employing PEALD-SiN_x/RF-sputtered-HfO_2, *IEEE Elect. Dev. Lett.*, **35**, 175–177 (2014).

S. Yang, S. Huang, M. Schnee, Q.-T. Zhao, J. Schubert and K. J. Chen, Fabrication and characterization of enhancement-mode high-$LaLuO_3$-AlGaN/GaN MIS-HEMTs, *IEEE Trans. Elect. Dev.*, **60**, 3040–3046 (2013).

W. Ye, W. Maojun, X. Bing, C. P. Wen, W. Jinyan, H. Yilong *et al.*, High-performance normally-off Al_2O_3\GaN MOSFET using a wet etching-based gate recess technique, *Elect. Dev. Lett., IEEE*, **34**, 1370–1372 (2013).

C.-C. Yeh, T. P. Ma, N. Ramaswamy, N. Rocklein, D. Gealy, T. Graettinger *et al.*, Frenkel–Poole trap energy extraction of atomic layer deposited Al_2O_3 and Hf_xAl_yO thin films, *Appl. Phys. Lett.*, **91**, 113521 (2007).

P. Yu and M. Cardona, ***Fundamentals of Semiconductors.*** Berlin: Springer (2010).

Y. Yu, Y. Bo, A. Jaesoo, P. C. McIntyre, P. M. Asbeck, M. J. W. Rodwell *et al.*, A distributed bulk-oxide trap model for Al_2O_3 InGaAs MOS devices, *Elect. Dev., IEEE Trans.*, **59**, 2100–2106 (2012).

S. Zafar, A. Callegari, E. Gusev and M. V. Fischetti, Charge trapping in high k gate dielectric stacks, in *Elect. Dev. Meet., 2002. IEDM '02. International*, 517–520, (2002).

Index